W0079685

Ethics of Science and Technology Assessment
Volume 38

Book Series of the Europäische Akademie zur Erforschung
von Folgen wissenschaftlich-technischer Entwicklungen
Bad Neuenahr-Ahrweiler GmbH
edited by Carl Friedrich Gethmann

Christian Streffer · Carl Friedrich Gethmann · Georg Kamp ·
Wolfgang Kröger · Eckard Rehbinder · Ortwin Renn ·
Klaus-Jürgen Röhlig

Radioactive Waste

Technical and Normative Aspects
of its Disposal

 Springer

Series Editor
Professor Dr. Dr. h.c. Carl Friedrich Gethmann
Europäische Akademie GmbH
Wilhelmstraße 56, 53474 Bad Neuenahr-Ahrweiler
Germany

On Behalf of the Authors
Dr. Georg Kamp
Europäische Akademie GmbH
Wilhelmstraße 56, 53474 Bad Neuenahr-Ahrweiler
Germany

Desk Editor
Friederike Wütscher
Europäische Akademie GmbH
Wilhelmstraße 56, 53474 Bad Neuenahr-Ahrweiler
Germany

ISBN 978-3-642-22924-4   ISBN 978-3-642-22925-1 (eBook)

Ethics of Science and Technology Assessment   ISSN: 1860-4803
                                                    e-ISSN: 1860-4811

Library of Congress Control Number: 2011935561

© Springer-Verlag Berlin Heidelberg 2011

This work is subject to copyright. All rights are reserved, whether the whole or part of the material is concerned, specifically the rights of translation, reprinting, reuse of illustrations, recitation, broadcasting, reproduction on microfilm or in any other way, and storage in data banks. Duplication of this publication or parts thereof is permitted only under the provisions of the German Copyright Law of September 9, 1965, in its current version, and permission for use must always be obtained from Springer. Violations are liable to prosecution under the German Copyright Law.

The use of general descriptive names, registered names, trademarks, etc. in this publication does not imply, even in the absence of a specific statement, that such names are exempt from the relevant protective laws and regulations and therefore free for general use.

Cover design: eStudio Calamar S.L.

Typesetting: Scientific Publishing Services Pvt. Ltd., Chennai, India.

Printed on acid-free paper

9 8 7 6 5 4 3 2 1

springer.com

EUROPÄISCHE AKADEMIE
zur Erforschung von Folgen wissenschaftlich-technischer Entwicklungen
Bad Neuenahr-Ahrweiler GmbH

Direktor: Professor Dr. Dr. h. c. Carl Friedrich Gethmann

The Europäische Akademie

The Europäische Akademie zur Erforschung von Folgen wissenschaftlich-technischer Entwicklungen GmbH is concerned with the scientific study of consequences of scientific and technological advance for the individual and social life and for the natural environment. The Europäische Akademie intends to contribute to a rational way of society of dealing with the consequences of scientific and technological developments. This aim is mainly realised in the development of recommendations for options to act, from the point of view of long-term societal acceptance. The work of the Europäische Akademie mostly takes place in temporary interdisciplinary project groups, whose members are recognised scientists from European universities. Overarching issues, e.g. from the fields of Technology Assessment or Ethic of Science, are dealt with by the staff of the Europäische Akademie.

The Series

The series Ethics of Science and Technology Assessment (Wissenschaftsethik und Technikfolgenbeurteilung) serves to publish the results of the work of the Europäische Akademie. It is published by the academy's director. Besides the final results of the project groups the series includes volumes on general questions of ethics of science and technology assessment as well as other monographic studies.

Acknowledgement

The project "Radioactive Waste. Technical and Normative Aspects of its Disposal" was supported by the VGB PowerTech e.V. The content of the book is only the authors' responsibility.

Preface

The Europäische Akademie zur Erforschung von Folgen wissenschaftlich-technischer Entwicklungen Bad Neuenahr-Ahrweiler GmbH is concerned with the study of the consequences of scientific and technological advance both for the individual and social human life and for the natural environment. It intends to contribute to find a rational way for society to deal with the consequences of scientific and technological developments. This aim is mainly realised by proposing recommendations for options of action with long-term social acceptance. The result of the work of the Europäische Akademie is published in the series "Ethics of Science and Technology Assessment", Springer Verlag.

The issue of disposal of radioactive waste is attracting an immense public interest and has been in the focus of the Europäische Akademie GmbH since a long time. Now, though the realisation of the project started in a relatively calm phase of the debate, the project was completed and the present study was published occasionally at a time when in the sequence of the Fukushima disaster the debate, at least in Germany, altered considerably. It is an open question where the debate will lead to, but I am sure that the survey of the scientific basis the study provides, and the concise recommendations developed by the project group on this basis, will be a helpful contribution.

I would like to extend my thanks to Professor em. Dr. rer. nat. Dr. med. h. c. Christian Streffer (group chair) and my fellow group members Professor Dr.-Ing. Wolfgang Kröger, Professor em. Dr. jur. Eckard Rehbinder, Professor Dr. rer. pol. Dr. h. c. Ortwin Renn and Professor Dr. rer. nat. Klaus-Jürgen Röhlig for their good cooperation and persistent commitment as well as to Dr. phil. Georg Kamp for coordinating the project on behalf of the Europäische Akademie GmbH.

Special thanks go to the VGB PowerTech e.V. for financially supporting the Europäische Akademie GmbH and thus enabling us to initiate the project.

Bad Neuenahr-Ahrweiler, June 2011   Carl Friedrich Gethmann

Foreword

In consequence of the accident in the Japanese Fukushima Daichi nuclear power plants an intense and hot debate about the use and ethical justification of energy production by nuclear power reactors has led the German Federal Government and Parliament to decide to leave off nuclear power and to shut down the last nuclear reactor in 2022. However, independently of these decisions the responsibility remains to care for the disposal of radioactive waste from installations used in medicine, research institutions and technologies. This is especially the case for the high level radioactive waste from nuclear power stations. Although this demand is strongly accepted by a large majority of people in all countries using nuclear power including Germany the search for a site of a repository for nuclear high level waste has not been successful due to a severe resistance of the population and other organizations at the local site which is under investigation.

The present debate which has been ongoing in Germany since the seventies of last century does not give a clue for a solution. For this unsatisfactory situation an attempt is undertaken in the present study to analyze the situation and to look for possibilities in order to move the complex situation into a more favorable direction. It seems that such an effort cannot only be based on scientific and technological grounds for the construction of a repository with a long-term safety for the disposal of radioactive waste, but also ways have to be found to establish acceptance of such a repository including consideration of the legal regulation. Such an undertaking can only be pursued by a group of experts from various pertinent disciplines.

At the end of 2008 Professor Dr. phil. Dr. phil. h.c. Carl Friedrich Gethmann (University Duisburg-Essen, Germany, Europäische Akademie GmbH, Bad Neuenahr/Ahrweiler), Professor Dr.-Ing. Wolfgang Kröger (ETH Zurich, Switzerland), Professor em. Dr. jur. Eckard Rehbinder (Goethe-Universität, Frankfurt am Main, Germany), Professor Dr. rer. nat. Klaus-Jürgen Röhlig (Technical University of Clausthal, Germany) and Professor em. Dr. rer. nat. Dr. med. h.c. Christian Streffer, as elected Chairman (University Duisburg-Essen, Germany), started their cooperation in the interdisciplinary working group. From the third meeting on Professor Dr. rer. pol. Dr. h. c. Ortwin Renn (Technical University Stuttgart, Germany) joined the group and broadened the group's intellectual spectrum. The project group was very effectively co-ordinated by Dr. phil. Georg Kamp (Europäische Akademie GmbH, Bad Neuenahr-Ahrweiler) who also contributed actively to the project group's work as co-author, as valuable partner in the discussions and as collaborator in the final shaping of the project.

After a phase of general discussion and joint planning, the members of the project group prepared draft texts from the perspective of their disciplines. The drafts served as the basis for a mutual criticism in an interdisciplinary discussion of the group leading to a stepwise integration of the parts into a whole opus that reflects the scientific-technological, the sociological, legal and the normative aspects on the problem. This interdisciplinary cooperation resulted in an intense debate in order to work out a compilation as well as recommendations of general agreement across the boundaries of disciplinary preconditions and apparent biases.

The working group was formed in order to contribute to the debate on disposal of radioactive waste from the scientific point of view and to work out rationally justified criteria for the evaluation of the existing options and to give recommendations that allow for a long-term, resilient solution that is not only justifiable against the present generation but to future generations as well. In any step, the group worked in scientific independence which is a precondition not only for the reliability of the results but also for the society's trust in the recommendations. To ensure that the work is state-of-the-art and representative, plans, preliminary overviews and later on an extended outline of the study were presented to external experts and broadly discussed with them in two feedback meetings and on other occasions. On this basis the project group would like to thank Dr. Klaus-Jürgen Brammer (GNS Gesellschaft für Nuklear-Service mbH, Essen, Germany), Professor Dr. rer. nat. Michael Decker (Karlsruher Institut für Technologie/ITAS Karlsruhe, Germany), Professor Dr.-Ing. Daniel Goldmann (Institut für Aufbereitung, Deponietechnik und Geomechanik, TU Clausthal, Germany), Dr. phil. Axel Gosseries (University of Louvain, Belgium), Bengt Hedberg (Swedish Radiation Safety Authority, Stockholm, Sweden), Dr. rer. nat. Stephan Lingner (Europäische Akademie GmbH, Bad Neuenahr-Ahrweiler, Germany), Dr. Bernd Lorenz (Gesellschaft für Nuklear-Service mbH, Essen, Germany), Professor Dr. rer. nat. Rolf Michel (Leibniz University Hannover, Germany), Dr. Hans G. Riotte (OECD/NEA, Nuclear Energy Agency, Paris), Dr. rer. pol. Walter Schenkel (synergico, Zürich, Switzerland), Dr. Jochen Schulze-Rickmann (Niedersächsische Gesellschaft zur Endlagerung von Sonderabfall mbH, Hannover, Germany), Detlef Sprinz, Ph.D. (PIK Potsdam Institute for Climate Impact Research,Potsdam, Germany), Dr. Annie Sugier (IRSN, Institut de Radioprotection et de Sûreté Nucléaire, Paris, France) Dr. rer. nat. Karl A. Theis (VGB Powertech, Essen, Germany) and Dr. rer. nat. Michael Weis (VGB Powertech, Essen, Germany) for their fruitful cooperation. Their contributions have promoted the discussion of the group immensely and the study owes a lot to their valuable advice. After 13 meetings of the group and several meetings of some group members on special topics at the end of two and a half years work the project group approved the text as it appears in the present book.

Although the study focuses on the situation in Germany, it is mainly written in English, since the group is convinced that the main aspects of the presented contents are invariant to the special national preconditions and that the recommendations may be helpful for the task of a reasonable radioactive waste management in other countries. Nevertheless, the project group has permanently taken the international references into consideration, as shown in the legally comparative survey of the situation in relevant states in the Annex. The Introduction, Executive Summary, Conclusions and Recommendations have been written both in German and English.

Our acknowledgements are also very gratefully addressed to Friederike Wütscher of the Europäische Akademie GmbH who was responsible for the proof reading of the final text and did this very efficiently, as well as to Margret Pauels of the Europäische Akademie GmbH who very effectively and always helpfully took care of our meetings in Bad Neuenahr-Ahrweiler.

Bad Neuenahr-Ahrweiler, June 2011 Christian Streffer

Table of Contents

Figures

Tables

Einleitung

Die Gefährdung von Mensch und Natur durch radioaktive Stoffe wird insbesondere durch die ionisierenden Strahlen hervorgerufen, die beim radioaktiven Zerfall dieser Stoffe freigesetzt werden. Es ist daher notwendig, radioaktive Abfälle, die in der Forschung, Medizin und Technik (vor allem hoch radioaktive Abfälle bei kerntechnischen Anlagen) anfallen, sicher zu verwahren. Die Dauer des gefährdenden Zustandes hängt entscheidend von den charakteristischen physikalischen Halbwertzeiten mit einer Spanne von Bruchteilen von Sekunden bis zu vielen Millionen Jahren ab, mit der die verschiedenen Radionuklide zerfallen. Brennelemente, die den Reaktoren der Kernkraftwerke entnommen werden, enthalten Radionuklide mit sehr langen Halbwertzeiten. Seit Kraftwerke dieser Art betrieben werden, ist es daher weltweit unumstritten, dass die radioaktiven Stoffe über sehr lange Zeiträume von der Biosphäre, dem Lebensraum der Menschen und aller weiteren Organismen, abgetrennt eingeschlossen werden müssen. Die Notwendigkeit, ein Endlager zu schaffen, wird wegen des Vorhandenseins der Abfälle auch in Deutschland gesehen.

Als Konzepte der Endlagerung sind insbesondere die oberflächennahe Endlagerung, die Endlagerung in tiefen geologischen Formationen sowie die Versenkung in die Tiefsee und Einlagerung im Meeresboden diskutiert worden (SSK 1987). Weltweit wurden auch weitere Konzepte diskutiert (u. a. langfristige oder unbefristete Zwischenlagerung, Verbringung in den Weltraum, CoRWM 2006), mehrheitlich hat man sich jedoch für die Endlagerung in tiefen geologischen Formationen entschieden. Dies gilt auch für Deutschland. Für die Langzeitsicherheit dieser Endlager ist das Wirtsgestein für das Endlager von entscheidender Bedeutung, um zu verhindern, dass radioaktive Stoffe in mehr als unerheblichem Maße aus dem Endlager in die Biosphäre gelangen können. Als Wirtsgesteine stehen weltweit kristalline Formationen (z. B. Granit, Salzstöcke und Ton) zur Diskussion. In Deutschland ist bereits in den sechziger Jahren des vorigen Jahrhunderts die Endlagerung in Salzformationen vorgeschlagen worden.

Im Jahre 1977 beauftragte die Bundesregierung die Physikalisch-Technische Bundesanstalt (PTB), ein Planfeststellungsverfahren für die Endlagerung radioaktiver Abfälle im Salzstock Gorleben einzuleiten (Tiggemann 2004).

Ursprünglich sollte das Endlager Teil eines nuklearen Entsorgungszentrums mit einer Wiederaufarbeitungsanlage sein. Zur Klärung der damit verbundenen sicherheitstechnischen Fragen und Risiken des Entsorgungszentrums und damit auch eines Endlagers fand vom 28. März bis 3. April 1979 in Hannover das „Symposion der Niedersächsischen Landesregierung zur grundsätzlichen sicherheitstechnischen Realisierbarkeit eines nuklearen Entsorgungszentrums" („Gorleben-Hearing") statt. Nach sehr sachlich geprägten Diskussionen unter dem Vorsitz des Physikers Carl Friedrich Freiherr von Weizsäcker ist der damalige Ministerpräsident des Landes Niedersachsen Ernst Albrecht, der Gastgeber und Teilnehmer des Symposiums gewesen ist, u. a. zu folgendem Resümee gekommen, das er im Niedersächsischen Landtag am 16. Mai 1979 gezogen hat: „Obwohl ein Nukleares Entsorgungszentrum [...] sicherheitstechnisch grundsätzlich realisierbar ist, empfiehlt die Niedersächsische Landesregierung der Bundesregierung, das Projekt der Wiederaufarbeitung nicht weiter zu verfolgen. Stattdessen sollte unverzüglich ein neues Entsorgungskonzept beschlossen werden, dessen Grundlinien wie folgt beschrieben werden können:

(1) Sofortige Einrichtung inhärent sicherer Langzeitzwischenlager zur Entsorgung der Kernkraftwerke [...],
(2) Vorantreiben der Forschungs- und Entwicklungsarbeiten zur sicheren Endlagerung radioaktiven Abfalls,
(3) Tiefbohrungen im Salzstock und bei positivem Ergebnis bergmännische Erschließung des Salzstockes in Gorleben, und falls die Bohrungen negativ ausfallen sollten, Erkundung anderer Endlagerstätten; denn Endlagerstätten brauchen wir."

1979/1980 begann dann das Erkundungsprogramm in Gorleben – zunächst von über und von 1986 an von auch unter Tage. Das Programm wurde im Jahr 2000 durch ein maximal 10jähriges Moratorium unterbrochen. In dieser Zeit fanden immer wieder Castor-Transporte mit radioaktiven Abfällen nach Gorleben statt, die in einem dortigen Zwischenlager deponiert wurden. Die Erkundungen in Gorleben und die Transporte nuklearer Abfälle durch die Republik führten zu sehr emotionalen Debatten und vor allem umfangreichen, hitzigen, teilweise militanten Demonstrationen der Kritiker der Kernenergie im allgemeinen und des Endlagers in Gorleben insbesondere. Der Beschluss der Bundesregierung im Oktober 2010, die Erkundung des Salzstocks Gorleben wieder aufzunehmen, hat den Widerstand gegen das nukleare Endlager erneut entfacht und verstärkt. Es hat sich eine teilweise sehr aufgeheizte Atmosphäre herausgebildet, die kaum noch sachliche Dialoge zulässt.

Parallel dazu haben sich in diesem Zeitraum nationale und internationale wissenschaftliche Gremien mit Sicherheitsfragen von Endlagern für radioaktive Abfälle beschäftigt und entsprechende Kriterien festgelegt. So hat die deutsche Strahlenschutzkommission (SSK) im Juni 1985 eine Empfehlung über „Strahlenschutzaspekte bei der Endlagerung radioaktiver Abfälle in geologischen Formationen" verabschiedet. Die SSK ist von dem Prinzip ausgegangen, dass zukünftige Generationen in demselben Maße vor ionisierenden Strahlen geschützt werden müssen wie die heutigen Menschen. Für die Nachbetriebsphase und damit die Langzeitsicherheit wurde empfohlen, dass die „potentielle Strahlenexposition von Einzelpersonen der Bevölkerung nach dem Eintritt unwahrscheinlicher Ereignisse den Betrag der mittleren Schwankungsbreite der natürlichen Strahlenexposition (effektive Äquivalentdosis) in der Bundesrepublik Deutschland nicht überschreiten" sollte (SSK 1987). Dieses wird erreicht, wenn der Planungsrichtwert von 0,3 mSv pro Jahr eingehalten wird. Die Internationale Strahlenschutzkommission hat empfohlen, eine potentielle Strahlenexposition von 0,3 mSv pro Jahr (effektive Dosis) als Planungsrichtwert (dose constraint) für die Nachbetriebsphase anzusetzen (ICRP 1999).

Zur Betrachtung der Langzeitsicherheit sind zunächst Zeiträume von mehreren 10.000 Jahren in Betracht gezogen worden. Ein wesentliches Argument für eine solche Zeitperiode war, dass in diesem Zeitrahmen eine nächste Eiszeit in unserer Region erwartet werden kann. Auf der Basis prognostischer Aussagen der Geologen zu Zeitabläufen möglicher Veränderungen an den Endlagerstandorten und ggf. zur Migration von Stoffen durch die geologischen Barrieren wird die Abschätzung der Langzeitsicherheit von Endlagern für radioaktive Abfälle heute auf mehrere Hunderttausend bis zu einer Million Jahre ausgedehnt (SSK 2008; ICRP 2007; BMU 2010a). Für die potentielle Strahlenexposition in der Nachbetriebsphase ist in neuerer Zeit ein Planungsrichtwert von 0,1 mSv pro Jahr vorgeschlagen worden (SSK 2008). Die vielfältigen Modellierungen zu Endlagern für hochradioaktive Abfälle in tiefen geologischen Formationen mit verschiedenen Deckgebirgen und die Erfahrungen mit Endlagern für chemotoxische Stoffe in der Tiefe ergaben eine starke Evidenz für die Machbarkeit solcher Endlager, selbst bei den genannten strengen Kriterien.

Für chemotoxische Abfallstoffe werden schon heute in Deutschland und in anderen Ländern Endlager in tiefen geologischen Formationen betrieben. Sie finden weitgehend Akzeptanz, obwohl weniger aufwendige Sicherheitsnachweise geführt werden, diese Stoffe im Vergleich zu radioaktiven Abfällen häufig ein höheres gesundheitsgefährdendes Potential haben können und „ewige" chemische Stabilität besitzen. Dennoch ist es bisher nicht gelungen, eine gesellschaftliche und politische Akzeptanz für ein konsensuelles Konzept der Endlagerung radioaktiver Abfallstoffe zu erreichen. Dieses gilt besonders für Deutschland, aber auch weltweit für viele weitere Länder. Allerdings ist es in Finnland und in Schweden gelungen, Standorte für solche Endlager (Finnland: Olkiluoto, Schweden: Forsmark Gemeinde Östhammar) festzulegen. In Frankreich und in der Schweiz ist man offensichtlich auf einem guten Weg dahin, in den USA gab es allerdings einen erheblichen Rückschlag.

Die Gründe für die fehlende Akzeptanz von Endlagern für hochradioaktive Abfälle sind komplex. Sie liegen u. a. in der besonderen Wahrnehmung der Gefährlichkeit ionisierender Strahlen und damit von radioaktiven Stoffen in den Abfällen. Obwohl der Mensch ständig natürlicherweise durch diese Strahlen exponiert wird und täglich natürlich vorkommende radioaktive Stoffe mit der Nahrung, dem Wasser und der Atemluft aufnimmt, obwohl nahezu alle Menschen in den industrialisierten Ländern in der Medizin aus diagnostischen oder auch therapeutischen Gründen ionisierenden Strahlen ausgesetzt sind und die möglichen Gesundheits-Risiken gut abgeschätzt werden können, haben viele Menschen große Vorbehalte, wenn sie ionisierenden Strahlen aus technischen Anlagen – selbst bei niedrigen Strahlendosen – ausgesetzt werden. Bei Kernkraftwerken kommt in diesem Zusammenhang ohne Zweifel den Folgen möglicher großer Unfälle eine wesentliche Bedeutung hinsichtlich der mangelnden Akzeptanz des Gefährdungspotentials zu. Bei der Endlagerung radioaktiver Abfälle ist eine solche Gefährdung durch Explosionen, Kernschmelzen oder andere plötzliche Unfälle nicht gegeben, jedoch wirkt sich die fehlende Akzeptanz der Kernkraft auch auf die der Endlagerung aus.

Ganz offensichtlich ist die Frage nach der angemessenen Entsorgung radioaktiver Abfälle mit einem hohen Konfliktpotential behaftet. Dabei sind bei diesem Thema die emotional besetzten Positionen und das Engagement, mit dem die Argumente vorgebracht werden, nur ein äußerer Indikator. Häufig stehen hinter der Auseinandersetzung um das nukleare Endlager tieferliegende gesellschaftliche Streitfragen um den Stellenwert technischer Entwicklungen für die künftige Ausgestaltung von Wirtschaft, Energieerzeugung und gesellschaftlichem Leben. Auch die Schärfe, mit der der Konflikt gelegentlich in der Öffentlichkeit ausgetragen wird, kennzeichnet zweifelsohne die Debatte und verdient ein genaueres Augenmerk, nicht zuletzt dann, wenn man nicht lediglich an der theoretischen Entwicklung von Lösungsstrategien, sondern an der praktischen Bewältigung des Konfliktes interessiert ist. Soll eine Entscheidung durch Beantwortung von Sachfragen herbeigeführt werden, wird der Konflikt auch hinsichtlich seiner sozial-gesellschaftlichen Dimension zu analysieren sein um zu prüfen, wie die zurzeit polarisierten Positionen in einen konstruktiven Diskurs über rationale, d. h. für alle Parteien nachvollziehbare und dann auch tolerierbare Strategien überführt werden kann. Nur so wird eine faktisch legitimierte Entscheidung für die Standortfrage und die Modalitäten eines Endlagers für hochradioaktive Abfälle möglich sein.

Es ist eine Erfahrung der letzten Jahrzehnte, dass sich die Planung und der Umgang mit großtechnischen Anlagen nicht oder jedenfalls nicht allein durch die Lösung technischer Probleme erfolgreich durchführen lassen. Es ist wünschenswert, wenn nicht notwendig und in einem demokratischen System auch angemessen, die Akzeptanz oder zumindest die Toleranz bei den betroffenen Menschen zu erreichen. Dieses ist bereits beim Schlusswort zum „Gorleben-Hearing" im Jahre 1979 von dem damaligen Ministerpräsidenten Albrecht zum Ausdruck gebracht worden. Er sagte: „[...] daß die Meinungsbildung in einem demokratischen Prozeß erfolgen muß, daß es sich hier nicht um technokratische Entscheidungen handelt, sondern letztlich um demokratische Entscheidungen. Demokratie bedingt aber Argumentation, und Öffentlichkeit bedingt Transparenz." Dieses Plädoyer für Information, Partizipation, Kommunikation und Transparenz ist offenkundig in den folgenden Jahren nicht im notwendigen Maße befolgt worden. Ministerpräsident Albrecht hat aber bereits im Jahre 1979 auch gesagt, die endgültige Entscheidung „ist eine typisch politische Entscheidung. Die Verantwortung für diese Entscheidung kann niemand den politischen Instanzen abnehmen." Die Notwendigkeit dieser Handlungsfolgen wird in der vorliegenden Studie herausgearbeitet.

In dem Entscheidungsprozess über ein Endlager für hochradioaktive Abfälle kommt den Fragen der Unsicherheit der wissenschaftlichen Erkenntnisse und Modellierungen sowie einer möglichen Ambiguität der Beurteilung eines Zustandes sowie möglichen Widersprüchen der Aussagen von Experten eine erhebliche Bedeutung zu. Die Folge ist eine Verunsicherung der betroffenen Bevölkerung. In Deutschland wird dieses in erheblichem Maße durch die Vorgänge um das „Versuchs-Endlager" Asse verstärkt. Es ist der breiten Bevölkerung kaum verständlich zu machen, dass die Asse als bloßes „Versuchs-Endlager" auf einem ganz anderen Standard konzipiert worden ist, als es das geplante Endlager für hoch radioaktive Abfälle in tiefen geologischen Formationen werden wird. Es untergräbt auch die Glaubwürdigkeit, wenn festgestellt werden muss, dass bei der Asse ein fehlerhaftes Verhalten bzw. Vernachlässigung in der wissenschaftlich-technischen und behördlichen Betreuung dieses Endlagers stattgefunden hat.

Unvermeidbare Unsicherheiten der Erkenntnisse und der wissenschaftlichen Daten, ihre Darstellung und der Umgang mit ihnen sind ohne Zweifel außerordentlich schwierige Themen und erhebliche Hindernisse für das Bemühen, Akzeptanz für Endlager dieser Art zu gewinnen. Die sehr langen Zeitspannen der notwendigen Prognosen verstärken die Effekte offensichtlich noch in beträchtlichem Maße. Bereits in den kleinen und alltäglichen technischen Geräten sind nicht immer alle Parameter des Systems und alle externen Einflussgrößen bekannt. In noch so gut untersuchten singulären technischen Großinstallationen gilt dies aus prinzipiellen Gründen in verstärktem Maße. Die jüngsten Entwicklungen in Fukushima, Japan, zeigen auch, dass außerordentlich extreme Ereignisse eintreten können, die auch bei Zugrundelegung aller verfügbaren Daten nicht zu erwarten sind, gegen die aber, wegen der schwerwiegenden Folgen, Anstrengungen unternommen werden müssen, um entsprechende Vorsorge zu treffen.

Dies gilt offensichtlich insbesondere bei der Planung von Endlagern für hochradioaktive Abfälle. Es müssen Aussagen zur Langzeitsicherheit über sehr lange Zeitperioden gemacht werden, die an die Grenzen des menschlichen Vorstellungsvermögens stoßen. Prognostische Aussagen gelingen dann nur über Modellrechnungen mit entsprechend großen Unsicherheiten. Es ist keine einfache Aufgabe, das Verständnis für derartige Zusammenhänge zu erreichen.

So laufen geologische Prozesse der Evolution in wesentlich längeren Zeiträumen ab als dies für biologische oder gar soziale Prozesse der Evolution
der Fall ist. Diese Eigenheiten der möglichen geologischen Entwicklungen
und ihre Veränderungen im Zeitablauf tragen dazu bei, dass prognostische
Aussagen zur möglichen Überwindung der technischen und geologischen
Barrieren über längere Zeiträume mit hinreichend höheren Genauigkeiten
gemacht werden können im Unterschied zu Prognosen, wie die Region eines Endlagers über diese Zeitschiene besiedelt und wie die Lebensgewohn-
heiten dieser Menschen in diesen Zeiträumen sein werden. Auch noch so
große prognostische Unsicherheiten rechtfertigten jedoch nicht, die möglichen Ansprüche von Angehörigen künftiger Generationen an die Entsorgung radioaktiver Abfälle auszublenden oder als irrelevant abzutun. Eine
angemessene Entsorgungsstrategie muss vielmehr – gemäß dem ethischen
Prinzip des Universalismus – den Angehörigen künftiger Generationen in
gleichem Maße gerecht werden wie den Angehörigen der gegenwärtigen.

Mit dem vorangegangenen Hinweis auf die Überkomplexität der Kon-
fliktlage soll weder der Problematik der Entsorgung radioaktiver Abfälle
ein Sonderstatus zugesprochen noch Skepsis an einer möglichen Bewältigung dieses Konflikts angedeutet werden: Überkomplexe Kontroversen, in
denen die Debatte um die angemessene oder taugliche Lösung eines Problems nicht nur Fragen nach der Angemessenheit und der Tauglichkeit der
Lösungsstrategien aufwirft, sondern auch Fragen nach der Angemessenheit
und Tauglichkeit der Standards zu ihrer Beurteilung, sind geübte Praxis in
vielfältigen Bereichen des privaten und des öffentlichen Lebens. Sie geben
Anlass zur politischen, juristischen und sozio-ökonomischen Institutiona-
lisierung von Konfliktbewältigung. Hier sorgen Verfahrensregeln, neutrale
Beurteilung oder Wettbewerb für effiziente Entscheidungen und die Erhaltung der Handlungsfähigkeit – denn angesichts der Offenheit von Kontroversen vorerst nichts zu tun, erweist sich oft als eine eher ungünstige Option. Die Analyse und Diskussion überkomplexer Kontroversen sind aber
auch geübte Praxis in den Wissenschaften, die nachgerade als eine gesellschaftliche Hervorbringung zur Professionalisierung, d. h. zu der arbeits-
teiligen Delegation der systematischen und ausdauernden, vom Druck effi-
zienten Entscheidens befreiten Befassung mit Kontroversen dieses Typs gelten dürfen.

Es gibt also keinen Anlass, vor der Überkomplexität zu resignieren. Vielmehr sollten, solange noch kein Druck zu effizientem Entscheiden besteht,
die konfliktveranlassenden Problemkonstellationen nach den Regeln wissenschaftlicher Rationalität erfasst, die Entscheidungsfragen differenziert,
analysiert und präzisiert sowie die Optionen geordnet und geprüft werden.

In Vorbereitung auf eine Situation, in der ein effizientes Entscheiden
notwendig wird, muss sorgfältig abgewogen werden, wie die Instrumente
zur Konfliktbewältigung gewählt bzw. entwickelt werden müssen, damit
sich eine dauerhafte Bewältigung des Konflikts herbeiführen lässt. Dauer-

haft kann dabei nur sein, was auf Dauer die Zustimmung oder mindestens Tolerierung breiter Teile der Beteiligten erhält. Dies setzt voraus, dass sich die Antworten auf Sachfragen, die in die Entscheidung einfließen, nicht bald als Irrtum erweisen, und dass die Beteiligten den Prozess, über den die Entscheidung zustande gekommen ist, nicht bald zu ihren Ungunsten als „unfair" wahrnehmen und trotz ihrem früheren Einverständnis zum Prozess dem Prozessergebnis nachträglich die Legitimation absprechen. Hierbei ist an die eingangs gemachten Bemerkungen zur sozialen Dimension des Konfliktes und zur emotionalen Aufgeladenheit der Debatten zu erinnern, die zumindest in Teilen darauf zurückzuführen sein dürfte, dass hierauf in den ersten Schritten, da man die „Tiefe" des Konflikts noch nicht absehen konnte, nicht hinreichend Aufmerksamkeit verwendet worden ist. In diesem Sinne versucht die vorliegende Studie Wege aufzuzeigen und vorzuschlagen, wie man aus dem Dilemma der jetzigen Situation zu einem akzeptablen Verfahren der Standortfindung und der Errichtung eines Endlagers für hoch radioaktive Abfälle finden kann.

Im Teil A der Studie sind die Texte sowohl in englischer als auch in deutscher Sprache zu finden. Zunächst werden in umfangreicheren Zusammenfassungen die wesentlichen Aussagen der Kapitel des Teiles B dargestellt und schließlich folgen Kapitel mit Schlussfolgerungen und Empfehlungen. Es werden mögliche Wege zur Standortfindung aufgezeigt.

Im Teil B der Studie werden in getrennten Kapiteln zunächst die Strategien und technischen Konzepte der Entsorgung von hoch radioaktiven Abfällen und ihrer Endlagerung in tiefen geologischen Formationen sowie die damit zusammenhängenden Fragen des Strahlenrisikos und des Strahlenschutzes dargestellt. Der Leser soll einen Überblick über den neuesten Kenntnisstand erhalten. Es folgen Kapitel zu ethischen, normativen Aspekten der Langzeitverantwortung, zu den rechtlichen Fragen der Endlagerung radioaktiver Abfälle und zu Leitlinien für eine sozial verträgliche und gerechte Standortbestimmung. Um die Studie auch international einer wissenschaftlich geprägten Leserschaft zugänglich zu machen, sind diese Kapitel in englischer Sprache geschrieben.

Teil C bietet ergänzend strahlenbiologische Grundlagen und einen rechtsvergleichenden Überblick über die Regelungen wichtiger Kernenergie nutzender Staaten.

Eine wesentliche Schlussfolgerung der Studie ist, dass das bestmögliche Verfahren darin zu sehen ist, dass die Untersuchung von Gorleben als möglicher Standort fortgesetzt werden sollte, jedoch sollten parallel ein oder zwei weitere Standorte in Betracht gezogen und von Übertage an diesen Orten Untersuchungen durchgeführt werden. Dieses Verfahren dient vor allem dazu, bei einem möglichen Scheitern von Gorleben keine oder wenig Zeit für ein Umschalten auf einen anderen Standort zu verlieren. Es kann erwartet werden, dass ein solches Vorgehen die Akzeptanz der Standort-Suche und Findung erhöht. Zum anderen soll ein solches Verfahren auch die

Möglichkeit eröffnen, einen Alternativstandort einer vertieften, d.h. untertägigen Untersuchung zu unterziehen, wenn aufgrund der dann vorliegenden Untersuchungen zu erwarten ist, dass der alternative Standort die Auswahlkriterien deutlich besser erfüllt als Gorleben. Der Information, Kommunikation, Partizipation und Transparenz wird bei dem Prozess der Standortfindung ein hoher Stellenwert zugeschrieben. Es wird jedoch klargestellt, dass die endgültigen Entscheidungen entsprechend den gesetzlichen Regelungen der Bundesrepublik Deutschland bei den hierfür zuständigen Institutionen liegen müssen.

A Zusammenfassung, Schlussfolgerungen und Empfehlungen

1 Zusammenfassung

1.1 Technische Aspekte der dauerhaften Entsorgung radioaktiver Abfälle

Das Problem der Entsorgung radioaktiver Abfälle weist eine technische und eine gesellschaftliche Dimension auf. Lösungen müssen nicht nur technisch machbar, sondern auch gesetzlich und politisch durchsetzbar und gesellschaftlich akzeptabel sein und einen sicheren Einschluss langlebiger hochradioaktiver Abfälle für lange Zeiten gewährleisten sowie unangemessene Belastungen für zukünftige Generationen vermeiden.

Die Frage nach dem Fortbetrieb der Kernkraftwerke ist grundsätzlich von der Frage nach der angemessenen Entsorgung der radioaktiven Abfälle zu trennen, ganz gleich, welche Perspektiven einen Fortbetrieb der kerntechnischen Anlagen vorgesehen oder diskutiert werden. Die Abfälle und damit die Notwendigkeit, geeignete Entsorgungsstrategien zu entwickeln, sind unabhängig davon vorhanden. Die bei einer Entscheidung für eine Laufzeitverlängerung anfallenden zusätzlichen Mengen sind angesichts dessen nicht ausschlaggebend. Dass das Unterbreiten einer (technischen) Lösung für das Entsorgungsproblem eines der zentralen Argumente gegen die Kernkraft hinfällig machen würde, darf der Umsetzung einer solchen Lösung nicht im Wege stehen.

Bei der Nutzung spaltbaren Materials in Kernreaktoren zur Stromerzeugung entstehen verschiedene radioaktive Abfälle. Ob man die anfallenden radioaktiven Substanzen, insbesondere die bestrahlten Brennelemente, als Abfall oder als Ressource betrachtet, ist dabei häufig eine Frage der Strategie und der damit verbundenen Motivationen. Die diversen Arten von Abfällen unterscheiden sich insbesondere hinsichtlich ihres Radionuklidgehalts, ihrer chemischen Zusammensetzung und ihrer physikalischen Beschaffenheit. Diese Eigenschaften bestimmen wesentlich die Schritte, die bei einer angemessenen Entsorgung der radioaktiven Abfälle unternommen werden müssen. Die Kategorisierung der Abfälle ist in verschiedenen Ländern durchaus unterschiedlich, dabei aber immer verbunden mit den vorgesehenen Entsorgungspfaden. Im Zentrum der vorliegenden Studie stehen die wärmeentwickelnden, hochradioaktiven Abfälle, für die es in Deutschland und auch weltweit noch keine etablierte Entsorgungslösung gibt.

Hinsichtlich des Umgangs mit verbrauchten radioaktiven Brennstoffen werden gegenwärtig mit dem „once-through cycle" und dem „partially closed cycle" zwei Hauptpfade unterschieden:

– Beim „once-through cycle" werden die bestrahlten Brennelemente als Abfall angesehen. Eine Erhöhung der Abbrandrate würde dazu beitragen, die Anteile verbleibenden Spaltmaterials wie auch die Menge hochradioaktiven Hüllen- und Strukturmaterials pro Stromeinheit zu reduzieren, während die Menge an langlebigen wärmeerzeugenden minoren Aktiniden leicht anstiege.

– Der „partially closed cycle", der mit einer Wiederaufbereitung einhergeht, wurde von einigen Ländern eingeschlagen, besonders von solchen, die große und breit etablierte Nuklearprogramme haben. Von Deutschland wurde dieser Weg wieder aufgegeben. Die Castor-Transporte mit verglasten hochradioaktiven Abfällen nach Gorleben sind das Ergebnis älterer Wiederaufbereitungsverträge.

Ziel des „partially closed cycle" ist die bessere Nutzung des Energiegehalts natürlicher Ressourcen (MOX-Brennstoffe). Die Vorteile dieses Verfahrens bestehen vor allem in der Reduktion der Volumina an verbrauchtem Brennstoff und in der langfristigen Verminderung der Radiotoxizität. Auch die Perspektive eines zukünftigen Kernenergieprogramms mit möglicherweise fortgeschrittenen Kernbrennstoffkreisläufen setzt einen solchen Kreislauf voraus. Nachteile sind vor allem die erhöhte Anzahl und die größeren Volumina der Transporte radioaktiver Materialien, die potenziell gefährlich sind, daher strengen internationalen Richtlinien unterliegen und die Gegenstand öffentlicher Konflikte sind. Nachteile werden ferner in der Abtrennung von Spaltmaterial und radiotoxischen Materialien (besonders Plutonium) und deren Lagerung vor Ort gesehen.

Mit der Einführung der sogenannten „advanced fuel cycles" werden geschlossene Brennstoffkreisläufe mit fortgeschrittener Wiederaufbereitungstechnologie und verschiedene Arten von hochentwickelten Reaktoren mit einem thermischen oder schnellen Neutronenspektrum angestrebt. Es wurden zahlreiche nationale und internationale Anstrengungen unternommen, um modernere „Reaktorensysteme" zu entwickeln und zu gestalten, die ungefähr zwischen 2030 und 2040 kommerziell genutzt werden könnten.

Partitionierung und Transmutation der langlebigen minoren Aktiniden (P&T) könnte die langfristig radiotoxischen Lagerbestände erheblich, in der Größenordnung des 100-fachen, reduzieren. Der Energie-Gehalt und die Wärmeleistung der Abfälle ließen sich ebenfalls dramatisch verkleinern, sodass die Reduzierung der Lagergröße oder, im Falle größerer Programme, der Anzahl der Lagerstätten möglich wird. Hinsichtlich möglicher radiologischer Belastungen in der Biosphäre aufgrund einer eventuellen langfristigen Migration von Radionukliden durch das Wirtsgestein ist jedoch keine signifikante Verbesserung zu erwarten.

Die Unterkritikalität der bestrahlten Brennstoffe und Abfälle aller Art ist für die gesamte Entsorgungs-Kette sicherzustellen, insbesondere auch für das Endlager. Die Beurteilungen, die im Rahmen bisheriger Entsorgungs-Programme vorgenommen wurden, zeigen, dass, sofern der tatsächliche Abbrand und gegebene Unsicherheiten beachtet werden, Unterkritikalität für „once-through cycles" gewährleistet werden kann. Für nicht-offene Zyklen ist dieser Aspekt irrelevant.

Wie auch andere Arten von Abfall verlangt radioaktiver Abfall nach einer umfassenden Strategie mit dem Ziel, die menschliche Gesundheit und die Umwelt zu bewahren und Sicherheit zu gewährleisten. Die Strategie sollte die Entladung, die Behandlung, den Transport, die Lagerung und die Entsorgung als wesentliche Schritte enthalten. Die genaue Abfolge dieser Schritte unterscheidet sich von Land zu Land in Bezug auf verschiedene Abfalltypen und grundsätzliche Entscheidungen hinsichtlich des gesamten Brennstoffkreislaufes.

Die *Lagerung* geschieht (nach der Definition der IAEA) „mit der Absicht der Rückholung" und kann verschiedenen Funktionen dienen (z. B. der Verringerung des Anteils kurzlebiger Radionuklide oder der Transportlogistik, aber auch dem Abwarten, bis eine Entsorgungslösung verfügbar ist). Es sind verschiedene technische Lösungen etabliert: Über oder direkt unter der Erdoberfläche, nass oder trocken, Fremdkühlung oder Naturkonvektion. Lagerzeiträume bewegen sich im Bereich von Jahrzehnten bis hin zu einem Jahrhundert; die Lagerung über mehrere Jahrhunderte wird gelegentlich diskutiert und erwogen.

Entsorgung geschieht (nach der Definition der IAEA) „ohne Absicht der Rückholung". Das Material wird definitiv als Abfall und nicht als Ressource betrachtet.

In der Vergangenheit wurde eine Vielzahl von Entsorgungsoptionen, einschließlich sog. „exotischer" Lösungen, diskutiert. Verblieben sind danach Optionen, die sämtlich auf der grundsätzlichen Entscheidung für „Concentrate & Confine" („Konzentrieren und Einschließen") anstelle von „Dilute & Disperse" („Verdünnen und Verteilen") beruhen:

– Optionen, die technisch vergleichbar sind mit (Zwischen-)Lagerungslösungen und daher während der gesamten Lagerzeit begleitende Maßnahmen erfordern, zum Beispiel in Form von Monitoring, Kontrollen, Wartung, Erneuerungsmaßnahmen etc.;
– Optionen, die einen Zustand herstellen, bei dem keine weiteren menschlichen Eingriffe oder Folgeaktivitäten benötigt werden („passive Sicherheit"); tiefe („geologische") Endlagerung in bergwerksähnlichen Anlagen, gelegentlich auch in tiefen Bohrlöchern, wird dabei favorisiert.

Misstrauen hinsichtlich der Idee der passiven Sicherheit war einer der Beweggründe für die Betrachtung langfristiger (Zwischen-)Lagerlösungen oder auch von Ansätzen, bei denen eine Rückholung der Abfälle, wenn auch nicht beabsichtigt, so doch durch die Auslegungsmaßnahmen erleichtert wird – wenn auch nur für einen begrenzten Zeitraum nach der Einlagerung („rückholbare Endlagerung").

Internationale Lösungen (gelegentlich auch als „regional" bezeichnet) eignen sich in besonderem Maße für Länder mit geringeren Abfallmengen und/oder geografischen und geologischen Randbedingungen, die eine Implementierung nationaler Lösungen schwierig machen.

Um exemplarisch zu verdeutlichen, wie die obengenannten Elemente zu einer Strategie für den Umgang mit radioaktiven Abfällen zusammengefügt werden, soll eine eher einfache, auf dem offenen „once-through"-Brennstoffzyklus beruhende Strategie als Referenz detaillierter beschrieben werden. Abgebrannter Kernbrennstoff wird zu Abfall erklärt, es findet keine Wiederaufbereitung statt, und der Abfall wird in einem tiefgelegenen (geologischen) Endlager deponiert. Die Beschreibung des Prozesses bezieht keine anderen Abfälle mit ein:

- Entnahme und mehrjährige Abklinglagerung im Kühlbecken auf dem Reaktor-Gelände;
- Einsetzen in Transportbehälter, Kurzstreckentransport zu einer nahe gelegenen (Zwischen-)Lagerstätte oder Langstreckentransport zu einer zentralen (Zwischen-)Lagerstätte;
- Zwischen-, Puffer- und Abklinglagerung (lokal oder zentral) über mehrere Jahrzehnte – je nach Entsorgungsprogramm;
- Transport zu einer Konditionieranlage (kurze oder lange Strecke je nach Ort der Lagerung), Pufferlagerung und Konditionierung für die Endlagerung;
- Pufferlagerung, Transport zum Endlager – kurze oder lange Strecke je nach Zielort;
- Einlagerung im Endlager.

Strategien, die auf teilweise oder gänzlich geschlossenen (fortgeschrittenen) Brennstoffkreisläufen beruhen, erfordern eine Wiederaufbereitung abgebrannter Brennstoffe und die Herstellung gemischter Brennstoffe. Sie weichen daher hinsichtlich der Logistik des Material- bzw. Abfallmanagements vor der Entsorgung und hinsichtlich des Aufkommens, der Form, des radioaktiven Gehalts, der Wärmeentwicklung und anderer Eigenschaften des Materials voneinander ab.

Hinsichtlich der strategischen Entscheidung über den „Endpunkt" der Entsorgung radioaktiver Abfälle sind die wesentlichen Unterschiede zwischen einer „verlängerten (Zwischen-)Lagerung" und der Entsorgung in tiefengeologischen Schichten in dem jeweiligen Ausmaß zu sehen,

- in dem Sicherheit und Gefahrenabwehr langfristig abhängig sind von aktiven Maßnahmen (was mit Strahlenbelastungen für diejenigen, die die Maßnahmen umsetzen, einhergeht), und
- in dem die Verbringungsmaßnahmen zu unterschiedlichen künftigen Zeitpunkten reversibel (insbesondere die Abfälle rückholbar) sind.

Welcher der Optionen man dabei den Vorzug gibt, hängt letztlich auch davon ab, wie sehr man auf die Verlässlichkeit der jeweiligen Schutz- und Sicherheitsvorkehrungen vertraut und wie man sich gegenüber der Verpflichtung gegenüber künftigen Generationen stellt.

Hybrid-Lösungen wie z. B. eine rückholbare Entsorgung und/oder eine verzögerte Schließung der Endlagerstätten gehen mit Zielkonflikten einher. Ansätze hierzu werden in verschiedenen Ländern verfolgt, ein verzögerter Verschluss etwa im Vereinigten Königreich und ein stufenweiser Verschluss in Abhängigkeit von den abnehmenden Möglichkeiten bzw. dem steigenden Aufwand zur Erhaltung der Rückholbarkeit in Frankreich. Eine verlängerte Zwischenlagerung (potentiell länger als ein Jahrhundert) wird in den Niederlanden umgesetzt.

Die Vielfalt macht die inhärente Ambiguität in der Entscheidung über einen „Endpunkt" deutlich. Insbesondere wird deutlich,

- dass die Einrichtung einer Rückholbarkeit ein komplexes Unterfangen ist,
- dass es keine einfache Antwort gibt auf die Frage, ob entsorgte Abfälle rückholbar sein sollen oder nicht – eher ist von verschiedenen „Graden der Rückholbarkeit" auszugehen,
- dass der Grad der Rückholbarkeit wesentlich vom in Frage stehenden Zeitrahmen abhängt. Der ist aber (mit einigen Jahrzehnten bis zu maximal drei Jahrhunderten) in jedem Fall erheblich kürzer als derjenige, für den Sicherheit nach dem vollständigen Verschluss gewährleistet werden muss (in der Größenordnung von einigen hundert tausenden von Jahren).

Der Wunsch nach Rückholbarkeit des eingelagerten Materials einerseits und nach einer langfristig sicheren Verwahrung sowie einem langfristigen Schutz der spaltbaren Materialien vor Zugriff und Weiterverbreitung andererseits kann zu Zielkonflikten führen. Insbesondere besteht ein Gegensatz zwischen rückholbarer Einlagerung und Zugriffsschutz. Das letztendliche Ziel jeder Entsorgung und insbesondere derjenigen in tiefengeologischen Schichten besteht darin, Gefahrenabwehr und Risikovorsorge auf sehr lange Fristen zu gewährleisten und ggf. dort, wo die daraus sich ergebenden Anforderungen einander widersprechen, diese möglichst auszugleichen. Maßnahmen zur eventuellen Gewährleistung einer Rückholbarkeit sind im Falle eines Zielkonflikts demgegenüber nachrangig. Alle gegenwärtig diskutierten Konzepte streben eine Gefahrenabwehr und Risikovorsorge auf sehr lange Fristen als Hauptziele an – gleichwohl besteht eine *erhebliche Variationsbreite* aufgrund der unterschiedlichen verfügbaren Wirtsgesteine und weiterer Aspekte. Obwohl die Grundpfeiler, die die Sicherheit und Gefahrenabwehr gewährleisten sollen (z. B. Einschluss, Isolation, Migrationsbegrenzung, Beschränkung der Folgen) jeweils dieselben sind, weichen doch die Gewichtung dieser Faktoren (in Abhängigkeit von der zeitlichen Entwicklung nach dem Verschluss) und Mittel, sie im Gesamtkonzept zu gewährleisten, erheblich voneinander ab.

Grundsätzlich gilt für keines der in Frage kommenden Wirtsgesteine, dass es ausschließlich günstige Eigenschaften hinsichtlich der Vielzahl aller dieser Aspekte besitzt. Um die jeweils positiven Eigenschaften zu nutzen und die unerwünschte Einwirkung der weniger günstigen zu kompensieren, ist immer eine entsprechende – auch dem jeweiligen Zeitrahmen angemessene – Anpassung der Auslegung der Endlager und der (geo-)technischen Maßnahmen erforderlich.

Es muss mittels Langzeituntersuchungen, die das erwartbare Spektrum unerwünschter Ereignisse, Zustände und Entwicklungen vollständig in Betracht ziehen, *belegt* werden, dass die Ziele der Entsorgungsbemühungen erreicht wurden. Der Nachweis angemessener Gefahrenabwehr kann dabei anhand des Falles des (unbeabsichtigten) Eindringens in das Lager erbracht werden: Als Folge der Entscheidungen für eine konzentrierte und isolierte Lagerung der radioaktiven Abfälle muss die Möglichkeit akzeptiert werden, dass es bei einem Eindringen in die Lagerstätten zu Überschreitungen der für Strahlenexpositionen gesetzten Richt- bzw. Grenzwerte kommen kann. Die Folgen eines solchen Eindringens lassen sich dabei nicht angemessen quantifizieren, da sich die künftigen Rahmenbedingungen und andere relevante Parameter nur mit großer Unsicherheit vorhersagen lassen, auch dann nicht, wenn man die Radiotoxizität der Lagerinventare als ein Maß für ihr Gefahrenpotential heranzieht.

Die Beurteilung der langfristigen radiologischen Sicherheit, der eine herausgehobene Rolle zugeschrieben wird, erfolgt traditionell auf der Basis von Expositionsberechnungen für Freisetzungsszenarien. Da die Entwicklung einiger Systemkomponenten, insbesondere die der oberflächennahen, aber kaum zu prognostizieren ist, bestehen jedoch hohe aleatorische und epistemische Ungewissheiten. Abgeschätzte Strahlenexpositionen taugen daher kaum als Maß erwartbarer gesundheitlicher Beeinträchtigungen, sondern vielmehr als Indikatoren für das Funktionieren des Endlagers, das entscheidend von tiefer gelegenen besser prognostizierbaren Komponenten abhängt.

Das Konzept des „Safety Case" entsteht durch die Verknüpfung von sicherheitsbezogenen Elementen und Argumenten aus der Standorterkundung, Forschung, Endlagerentwicklung und -konstruktion, Sicherheitsanalyse etc. Der Safety Case entwickelt sich mit der Zeit gemeinsam mit dem Endlagerprojekt und -konzept. Der Umgang mit Unsicherheiten ist ein zentrales Element.

Die Entwicklung eines Endlagerprogramms, die damit verbundene Standorterkundung und Forschung und Entwicklung, das Endlagerkonzept und der Safety Case können als Optimierungsprozess über die Dauer mehrerer Jahrzehnte gesehen werden. Die Zielfunktion besteht aus kurz- und langfristigen Sicherheitsaspekten, Realisierbarkeit, Kosten, gesellschaftlicher Akzeptanz und rechtlicher Machbarkeit und evtl. anderen Parametern. Es muss beachtet werden, dass die Zielfunktion sich mit der Zeit verändern kann (z. B. hat sich der Zeitrahmen für die Demonstration der langfristigen Sicherheit über die vergangenen zwei Jahrzehnte um zwei Größenordnungen erhöht).

Daher ist es nicht sinnvoll, vergangene Entscheidungen nochmals aufzunehmen („retrospektive Optimierung") um danach zu fragen, ob diese Entscheidungen aufgrund heute geltender Standards erzielt wurden. Vielmehr ist die Frage essentiell, ob die Ergebnisse früherer Entscheidungen (ein Standort, eine Endlagerauslegung etc.) nach heutigen Standards noch vertretbar sind!

Der Endlagerbetrieb könnte frühestens ca. 2035 beginnen unter der Voraussetzung, dass jeder Schritt im Programm vollständig erfolgreich wäre (d. h. Entscheidung für Gorleben 2017, sofortige Planfeststellung etc.). Mögliche Verzögerungen z. B. durch die Ablehnung von Gorleben, die Suche nach einem anderen Standort (unmittelbar oder nach einer Ablehnung von Gorleben etc.) würden in Größenordnungen von mehreren Jahrzehnten liegen.

Die technische Implementierung neuer Technologien (z. B. Abtrennung und Transmutation) kann evtl. für die 2030er und 2040er erwartet werden. Solange abgebrannte Kernbrennstoffe nicht endgelagert werden (d. h. mindestens bis 2035) wird es keine Einschränkung der Flexibilität selbst bei Unterstellung eines „schnellen" Endlagerprogramms geben.

Dies gilt allerdings nur unter der Voraussetzung, dass ein Plan mit klar definierten Entscheidungspunkten besteht. Andernfalls wird das Hin und Her der Vergangenheit und Gegenwart, mit seiner starken Abhängigkeit von der politischen Ausrichtung der jeweiligen Regierung, die mit jeder Wahl wechseln könnte, weitergehen, und der Prozess könnte endlos weiter gehen, ohne zu einer Lösung zu führen.

1.2 Strahlenrisiko und Strahlenschutz

Für die Evaluation radio-toxikologischer Effekte ist die Kenntnis der Dosiswirkungsbeziehung notwendig. Im Strahlenschutz werden zwei grundsätzlich verschiedene Klassen dieser Beziehungen verwendet. Der prinzipielle Unterschied besteht darin, ob eine Dosis-Wirkungs-Kurve mit oder ohne Schwellendosis vorliegt. Die erste Möglichkeit trifft für *deterministische* Effekte (akute Effekte, fibrotische und ähnliche Gewebe-Effekte, sowie entwicklungsbiologische Effekte) zu. Die Schwellendosen für diese Effekte liegen in einem Bereich von mehr als 100 mSv. Keine Schwellendosis wird dagegen für die Verursachung von *stochastischen* Effekten (genetischen Effekten und Krebs) angenommen, so dass unter dieser Annahme auch bei sehr kleinen Strahlendosen Effekte auftreten können. Es wird für diese Effekte eine lineare Dosiswirkungsbeziehung ohne Schwellendosis (LNT-Modell) vorgesehen.

Die fundamentale Größe im Strahlenschutz ist die absorbierte Dosis (D) in Gray (Gy), sie gibt die absorbierte Energie in einem Massenelement (m) an. Da verschiedene Strahlenqualitäten (Strahlenart und Strahlenenergie) bei gleicher absorbierter Dosis zu unterschiedlichen Effekten führen, wird eine Gewichtung der absorbierten Dosis vorgenommen und man erhält dann die Äquivalentdosis bzw. Organdosis (H) in Sievert (Sv). Das für niedrige Strahlendosen bedeutsame Krebsrisiko ist in den verschiedenen Organen bei gleicher Organdosis unterschiedlich. Daher werden die Organdosen mit entsprechenden Gewebe-Wichtungsfaktoren multipliziert. Die Summe dieser Produkte ergibt die effektive Dosis (E) in Sv, die mit dem stochastischen Strahlenrisiko in Beziehung steht. Die effektive Dosis ist ein brauchbares Instrument, um regulatorische Prozesse z. B. der Optimierung abzuwägen. Es können durch sie alle Expositionspfade beim Menschen erfasst und in einem Gesamtwert angegeben werden.

Die Dosisgrenz- und Richt-Werte werden als effektive Dosen angegeben. Die Grenzwerte betragen während der Betriebsphase des Endlagers (Einlagerung der radioaktiven Stoffe) für Beschäftigte am Arbeitsplatz 20 mSv pro Jahr und für Personen der Bevölkerung in der unmittelbaren Umgebung kerntechnischer Anlagen während der Betriebsphase 1 mSv pro Jahr. Bei Endlagern für hochradioaktive Abfälle wird ein „Dosis-Constraint" (Richtwert) von 0,1 bis 0,3 mSv pro Jahr in der Nachbetriebsphase (Tausende bis Millionen Jahre) nach Verschluss des Lagers von verschiedenen internationalen Gremien vorgesehen. Das BMU hat neuerdings für diese Dosis 10 μSv pro Jahr vorgeschlagen. Eine solche Dosis bezeichnet die ICRP als „triviale" Dosis, der keine Bedeutung zukommt.

Zur Optimierung von Strahlenquellen im Strahlenschutz wird häufig die effektive Kollektivdosis herangezogen. Es wird der Mittelwert der effektiven Individualdosen mit der Zahl der exponierten Personen multipliziert und die Dosis dann in „man Sv" angegeben. Zur besseren Abschätzung der Situation sollten die Bereiche der Personendosen, die zeitliche Periode, in welcher die betreffenden Dosen berücksichtigt werden, und andere Faktoren bei der Ermittlung der Kollektivdosen benannt werden. Für die Abschätzung von Strahlenrisiken insbesondere im niedrigen Dosisbereich sind die Kollektivdosen nicht geeignet.

Die Radiotoxizität der Radionuklide wird bestimmt durch die Strahlung, die beim radioaktiven Zerfall auftritt. Hinsichtlich des Langzeitrisikos sind Radionuklide mit einer langen physikalischen Halbwertzeit von entscheidender Bedeutung, da diese Radionuklide in einigen Fällen über mehr als eine Million Jahre in beträchtlicher Menge weiter bestehen und über lange Zeiträume radioaktiv zerfallen. Nach Aufnahme dieser Stoffe in den menschlichen Körper kann vor allem dann eine Strahlenexposition über längere Zeitperioden auftreten, wenn neben einer langen physikalischen Halbwertzeit auch eine lange biologische Halbwertzeit besteht. Es muss dann die „Folgedosis" über einen längeren Zeitraum (bei Erwachsenen 50 und bei Kindern bis zum Alter von 70 Jahren) abgeschätzt werden. Die Strahlendosis wird in diesen Fällen mit Hilfe des Folgedosiskoeffizienten (committed effective dose coefficient, CED, angegeben in Sv/Bq) ermittelt. Dieser Koeffizient ist ein gutes Maß für die Radiotoxizität eines Radionuklids (International Commission for Radiological Protection ICRP).

Zwei Faktoren sind für die Exposition des Menschen und der Umwelt durch Endlager von entscheidender Bedeutung:

(1) Die Dauer der physikalischen Halbwertzeit des betreffenden Radionuklids
(2) Die Mobilität des radioaktiven Materials nach Freisetzung aus dem Behälter im Endlager und die folgende Migration durch die technische sowie anschließend durch die geologische Barriere in die Biosphäre.

Modellrechnungen mit Granit, Salz und Ton als Wirtsgestein und Deckgebirge haben ergeben, dass im Allgemeinen wegen ihrer geringen Mobilität nicht die schwer löslichen Oxide der α-strahlenden Transurane entscheidend zur Exposition des Menschen in der späten Phase eines Endlagers beitragen. Es sind vielmehr die besonders mobilen, häufig als Ionen vorliegenden Radionuklide wie Jod-129, Chlor-36, Kohlenstoff-14, Selen-79 und Cäsium-135. In den vorliegenden Modellrechnungen werden im Allgemeinen die höchsten Strahlenexpositionen durch Jod-129 in dem Zeitraum von etwa 20.000 bis etwa 100.000 Jahren ermittelt. Die Strahlendosen werden wiederum als effektive Dosen angegeben. Das bedeutet für Jod-129, dass aus Gründen der Biokinetik und des Stoffwechsels durch dieses Radionuklid nahezu ausschließlich die Schilddrüse exponiert wird. So liegt die Strahlendosis bei radioaktivem Jod in der Schilddrüse um etwa den Faktor 1.000 höher als in den anderen Organen und Geweben. Die effektive Dosis liegt etwa um den Faktor 20 niedriger als die Dosis in der Schilddrüse. Die Verursachung von Krebs in der Schilddrüse durch radioaktives Jod ist vor allem nach der Reaktorkatastrophe von Tschernobyl und bei nuklearmedizinischen Studien gut untersucht. Die Freisetzung und Migration der mobilen Radionuklide aus dem Endlager in die Biosphäre und ihre Modifizierung bedarf weiterer intensiver Untersuchungen.

Mikrodosimetrische Überlegungen führen dazu, dass bei Strahlendosen unterhalb von 1 mSv sehr heterogene Dosisverteilungen auftreten. Mit weiter abnehmender Dosis werden immer weniger Zellen eines Gewebes exponiert. Bei einer Dosis von 1 µSv wird unter 1.000 Zellen weniger als 1 Zelle noch exponiert, während die Dosis pro exponierter Zelle gleich bleibt. Welche Bedeutung dieses für die Verursachung von Gesundheitseffekten hat, ist bisher nicht geklärt.

Bei Endlagern hochradioaktiver Stoffe sind in der Spätphase nach Abschluss des Lagers keine Strahlenexpositionen in Höhe der Schwellendosen für deterministische Effekte zu erwarten. Daher können diese Effekte nicht auftreten. Es können nur stochastische Effekte (insbesondere die Verursachung von Krebs) durch Extrapolationen rechnerisch abgeschätzt werden. Epidemiologische Untersuchungen haben ergeben, dass für eine allgemeine Bevölkerung eine signifikante Erhöhung der Krebserkrankungen nach einer Strahlendosis von etwa 100 mSv beobachtet werden kann. Da es im individuellen Falle einer Krebserkrankung keine spezifischen Merkmale gibt um zu erkennen, ob der individuelle Krebs durch ionisierende Strahlen verursacht worden ist oder nicht, ist nur die Angabe von Verursachungs-Wahrscheinlichkeiten für die Erkrankung möglich.

Diese Untersuchungen haben zur Bestimmung von Risikofaktoren geführt. Es ergibt sich ein Wert von 5 x 10^{-2} pro Sv. Das bedeutet, dass bei einer Dosis von 1 mSv das Krebsrisiko 5 x 10^{-5} und bei 0,1 mSv 5 x 10^{-6} beträgt. Dieses Risiko sollte man mit dem allgemeinen Lebenszeit-Risiko von etwa 0,4 vergleichen, mit dem die Menschen in Deutschland an Krebs erkranken. Die potentiellen Krebsrisiken, die sich durch ein Endlager ergeben können, liegen weit unterhalb eines „Messbereiches", in dem mögliche, zusätzliche Krebserkrankungen erkannt werden können.

Alle Lebewesen werden seit jeher durch ionisierende Strahlen aus natürlichen Quellen exponiert. In Deutschland beträgt diese Exposition für jeden Bürger im Mittel etwa 2,3 mSv pro Jahr (effektive Dosis). Etwa 50 % dieser Dosis wird durch die Inhalation von Radon mit seinen radioaktiven Folgeprodukten hervorgerufen. Die anderen Komponenten der Expositionen sind externe kosmische Strahlung durch die Sonne, externe terrestrische Strahlung durch radioaktive Stoffe im Boden und interne Strahlenexpositionen durch die Aufnahme radioaktiver Stoffe mit der Nahrung und dem Trinkwasser. Jeder Mensch in Deutschland trägt in seinen Geweben mit sich im Mittel etwa 9.000 Bq an natürlichen radioaktiven Stoffen (vorwiegend Kalium-40 und Kohlenstoff-14). Die Strahlenexpositionen aus natürlichen Quellen sind regional sehr unterschiedlich. Die externen Expositionen können in Deutschland um den Faktor 5 bis 8 und beim Radon noch größer über dem Durchschnittswert liegen.

Neben den Expositionen aus natürlichen Quellen erhalten die Menschen vor allem in den Industrieländern Strahlenexpositionen aus weiteren zivilisatorischen Situationen. Den größten Anteil hierzu liefert die Medizin, vor allem die Röntgendiagnostik. Der Durchschnittswert pro Kopf der Bevölkerung in Deutschland liegt jetzt etwa im Bereich der mittleren Exposition aus natürlichen Quellen mit weiter steigender Tendenz.

Es gibt keinen wissenschaftlichen Zweifel, dass die Strahlenexpositionen aus natürlichen Quellen dieselben biologischen bzw. gesundheitlichen Effekte verursachen können wie die zivilisatorischen Expositionen. Ein Vergleich der Strahlendosen aus den verschiedenen Quellen mit den daraus folgenden Wirkungen ist daher in vollem Maße gerechtfertigt.

Die Unsicherheiten der Dosis- und Risiko-Abschätzungen sind in dem niedrigen Dosisbereich von 1 mSv und unterhalb dieser Strahlendosis sehr groß. Da die möglichen Strahlendosen bei Endlagern vor allem durch Inkorporation radioaktiver Stoffe hervorgerufen werden, nehmen die Unsicherheiten weiterhin zu, da die Dosis-Abschätzung nur über Modellrechnungen erfolgen kann. Andererseits gibt es für die radioökologischen Expositionspfade von Jod und Caesium relativ gute Daten aus den Erfahrungen nach der Reaktorkatastrophe in Tschernobyl und dem Fallout der Atombombentests.

Die Dosis- und Risiko-Abschätzungen werden mit Hilfe von Referenz-Modellen und -Personen durchgeführt. Daher wird die individuelle Variabilität innerhalb der menschlichen Population nicht berücksichtigt. Andererseits kann davon ausgegangen werden, dass der Stoffwechsel und damit auch die Biokinetik radioaktiver Stoffe im menschlichen Organismus selbst nach Tausenden bis Millionen Jahren sich nicht so dramatisch ändern werden, wie das für den Lebensstil und allgemeine menschliche Verhaltensregeln der Fall sein wird. Ebenso ist es wohl kaum möglich, Aussagen über die Größe menschlicher Populationen und weitere Bedingungen menschlichen Lebens am Ort des Endlagers über einen langen Zeitraum zu machen, der die Perioden mit soliden Kenntnissen zur Kultur der zurückliegenden Menschheitsgeschichte übersteigt.

1.3 Entsorgung hochradioaktiver Abfälle unter dem Aspekt der Langzeitverantwortung

Dem Appell an „die Ethik" in Streitlagen und dem Hinweis auf die Notwendigkeit, sich in einer Entscheidungslage von gesellschaftlichem Belang an „ethischen Maßstäben" orientieren zu müssen, liegt häufig ein unangemessenes Verständnis der angesprochenen Disziplin zugrunde. Ethik kann keine höheren Prinzipien, keine zeitlos gültigen Imperative oder Werte formulieren, aus denen sich für die in Frage stehende Situation top-down angemessene Handlungsanweisungen deduzieren ließen. Die Domäne der Ethik ist vielmehr die Entwicklung aussichtsreicher Strategien zur Konfliktbewältigung. Konflikterzeugend ist nicht selten gerade die Orientierung an den Regeln, die das Handeln innerhalb einer Kultur bestimmen und dem Einzelnen als Legitimationsgrund für sein Handeln dienen („Moralen"), die in verschiedenen Kulturen sowie deren Binnen- und Subkulturen aber auf unterschiedliche Weise etabliert sind. Die kritische Analyse von Moralen und ihrer Reichweite ist daher ein prominenter Gegenstand der Ethik.

Im Konflikt um die Entsorgung hochradioaktiver Abfälle ist – nebst den divergierenden Meinungen über die Wirkungen und Nebenwirkungen der einzusetzenden Mittel – auch mit divergierenden Zwecksetzungen sowie mit divergierenden moralischen Vorstellungen, letztlich mit unterschiedlichen Verständnissen dessen, was legitimationsbedürftig ist und was als Legitimationsgrund taugen kann, zu rechnen. Eine rationale Bewältigung dieses Konflikts setzt damit nicht nur eine Verständigung über die Ursache-Wirkungs-Zusammenhänge voraus, sondern auch eine kritische Reflexion der Zweck-Konstellationen und der vorgebrachten Legitimationsgründe für die Mittelwahl („Entsorgungsstrategien").

Es wäre ein Kategorienfehler, in generationenübergreifenden Problemstellungen vom Typ der Entsorgung hochradioaktiver Abfälle ungeprüft

Maßstäbe vorauszusetzen, die den etablierten lebensweltlichen Moralpraxen entnommen sind – jedenfalls solange die Handlungserfordernisse noch (zeitliche) Ressourcen freilassen, um unter neutraler Beleuchtung möglichst vieler Aspekte möglichst rationale Vorschläge zu entwickeln und ggf. Handlungs- und Entscheidungsbedingungen so zu verändern, dass eine optimierte Konfliktbewältigung möglich wird.

Drei solche Kategorienfehler durchziehen die Debatte um die Entsorgung hochradioaktiver Abfälle:

(1) Die Forderung nach einer unverzüglichen Problemlösung geht mit erheblichen Rechtfertigungslasten einher, die nur eingelöst werden können, wenn man die situativen Handlungsbedingungen nach dem Muster einer Notfallsituation deutet. Notfallsituationen rechtfertigen oft Zumutungen gegenüber Dritten, die sonst in längerfristiger Planung nicht hinzunehmen wären.

(2) Angesichts der handlungsdruck-enthobenen Entscheidungslage für Fragen der gerechten Verteilung von Lasten (Kosten, Risiken) sind nicht lediglich die etablierten Prinzipien der Gerechtigkeit auf eine feste Verteilsituation mit einer festen „Verteilmasse" anzuwenden. Vielmehr gilt es einen Plan zu entwerfen, der – unter der Maßgabe der begründeten Ausnahme von der Gleichheit – die gezielte Optimierung der Verteilsituation und der „Verteilmasse" herbeizuführen geeignet ist. Die unter Appell an moralische Prinzipien vorgetragenen Anspruchskonflikte lassen sich so in Erwägungen (oder auch: Verhandlungen) über konfliktbefriedende Allokationen aller möglichen in Frage stehenden Lasten und Güter überführen.

(3) Angesichts der vielen künftigen von den Risiken Betroffenen kann es – entgegen der im Verursacherprinzip ausgedrückten moralischen Intuition – gerade ein Gebot der Fairness sein, nicht dem Verursacher der Abfälle auch deren Entsorgung aufzutragen. Sind Andere aufgrund höherer Kompetenz oder günstigerer Voraussetzungen besser geeignet, eine den Ansprüchen künftiger Generationen gerecht werdende Entsorgung zu gewährleisten, dann könnte es aus ethischer Sicht sogar geboten sein, dass diesen gegen eine zwangsfrei akzeptierte Kompensation die Entsorgung übertragen wird.

Angesichts der weit in die Zukunft hineinragenden Folgen einer Entscheidung über Strategien zur Entsorgung hochradioaktiver Abfälle muss auch und gerade eine Reflexion über die möglichen Konflikte mit Angehörigen künftiger Generationen wesentlicher Teil der Entscheidungsvorbereitung sein. Eine Reichweitenbeschränkung auf diejenigen, die mit uns gleichzeitig interagieren oder auf eine festgelegte zeitliche Reichweite (z. B. n Generationen) ist wegen der Irrelevanz aller Gründe einer Grenzziehung für ethisch-rationale Konfliktbewältigungsinteressen nicht tragfähig.

Mit Blick auf konkrete Handlungen besteht eine Verpflichtung daher immer genau so lang, wie ihre Folgen Konfliktpotential erzeugen. Die Verpflichtungen, die mit der Entsorgung hochradioaktiver Substanzen einhergehen, bestehen zeitlich also so lange wie das Gefährdungspotential der radio- und chemotoxischen Substanzen.

Es ist *aus ethischen* Gründen zwischen der universalistisch, d. h. unbegrenzt bestehenden *Verpflichtung* einerseits und einem nach räumlicher, sozialer und temporaler Ferne abnehmenden Grade ihrer *Verbindlichkeit* zu unterscheiden. Indem wir im Rahmen unserer moralischen Praxis die Verpflichtung gegenüber den nahestehenden Generationen mit einem höheren Grade an Verbindlichkeit versehen (und diesen eine Verpflichtung gegenüber den ihnen nahestehenden zuschreiben), organisieren wir die Langzeitverpflichtung und machen sie praktisch bewältigbar. Dabei ist allerdings das Maß, in dem Verbindlichkeit abnimmt, nicht fest mit der zeitlichen Entfernung, sondern mit unseren Potentialen, Verpflichtung planvoll und kontrolliert wahrzunehmen, korreliert. Wie weitreichend dieses Potential ist, darauf haben die moralischen Akteure Einfluss – und es ist Teil ihrer Verpflichtung, im Rahmen der Verhältnismäßigkeit der Mittel diese Einflussmöglichkeiten wahrzunehmen. So besteht, wenn wir technisch handeln wollen, auch eine Verpflichtung, uns Wissen über die Folgen und deren Konfliktpotential zu beschaffen. Ebenso geht mit der Vergrößerung der Reichweite unseres Handelns durch Kollektivierung und Technisierung die Verpflichtung einher, Institutionen auszubilden und mit den erforderlichen Ressourcen auszustatten, die auf der Grundlage des gewonnenen Wissens oder rational begründeter Vermutungen die gesellschaftliche Langzeitverpflichtung organisieren können, ohne von den Ressourcenbeschränkungen der einzelnen Akteure abhängig zu sein. Insbesondere erfordert auch die advokatorische Vertretung künftiger Ansprüche eines durch prozedurale Organisation legitimierten, institutionell verankerten und gesellschaftlich kontrollierten Mandats – Vertretungsrechte können nicht einfach durch Erklärung der eigenen Kompetenz, Zuständigkeit oder (vermeintlichen) Betroffenheit reklamiert werden. Die Akteure schaffen damit „Verantwortung", treten aber nicht ihre Verpflichtung ab – es verbleibt vielmehr die weitere Aufgabe, über die Dauer sicherzustellen, dass die Verpflichtung auch verantwortlich wahrgenommen wird.

Bei der Optimierung komplexer Planungen mit hohem Konfliktpotential ist die Partizipation aller Orts- und Sachkundigen unabhängig von zertifizierter Qualifikation und Profession im Sinne der Planoptimierung und Verfahrenskontrolle prinzipiell wünschenswert. Die Einbeziehung von direkt Betroffenen in komplexe Planungen kann dazu beitragen, durch Abbau von Misstrauen und Ängsten die sozialen Voraussetzungen für einen rationalen Diskurs zum Zwecke der Konfliktbewältigung zu schaffen. Partizipative Verfahren sind dabei so zu gestalten, dass das Entscheidungsverhalten der Entscheidungsverantwortlichen nicht durch implizite oder explizite Interessenvertretung der Präsentischen verzerrt („gebiased") wird, sondern an den Ansprüchen aller, auch der nicht partizipierenden Angehörigen künftiger Generationen ausgerichtet bleibt.

Der von Teilen der Öffentlichkeit und Teilen der Fachcommunity vorgebrachten Erwartungen, Partizipation könne auch zu Legitimation von Entscheidungen beitragen, ist mit Skepsis zu begegnen. Insbesondere sollte die Einbeziehung partizipativer Verfahren in die Entscheidungsvorbereitung nicht dazu führen, dass die Verantwortung, die von den Bürgern als Träger der Langzeitverpflichtung delegativ auf Institutionen übertragen wurde, von den Institutionen auf die Bürger zurückübertragen wird („Kompetenzzuweisungs-Zirkel"). Die als Begründung vorgebrachte Auffassung, der Einzelne sei doch letztlich der beste Kenner seiner Bedürfnisse und Interessen und könne sie daher auch am besten vertreten („Eigenkompetenzthese") verkennt dabei, dass es gerade um die Gestaltung eines Ausgleichs zwischen einer disparaten Menge von Interessen geht, für die keine Teilklasse repräsentativ sein kann und die daher auch nicht durch einzelne Interessenträger repräsentativ vertreten werden kann. Modelle, die die Entscheidungskompetenz auf den Bürger zurückübertragen wollen, gehen dabei nicht selten von einem Bild aus, das die gesamtgesellschaftliche Entscheidungsbildung als eine antagonistische Verhandlung zwischen gesellschaftlichen Teilsystemen beschreibt, unter der „die Politik", „die Wissenschaft" und „die Öffentlichkeit" quasi nur als gleichberechtigte Stämme unter anderen ihre jeweiligen Interessen vertreten („Tribalisierungsthese"). Dabei werden der methodisch organisierte Erkenntnisdiskurs der Wissenschaften und der institutionell organisierte Entscheidungsdiskurs der Politik, die gerade den zentrifugalen Prozessen subjektiver Meinungs- und Interessenbildung entgegenwirken sollen, mit bloßer Meinungs- und Interessenvertretung seitens der beteiligten Individuen verwechselt. Gerade aber in Entscheidungsfragen, in denen die Ansprüche einiger Konfliktparteien (wie etwa die der Angehörigen künftiger Generationen) nur begründet zugeschrieben und advokatorisch vertreten werden können, ist die legitimatorische Kraft von Prozeduren von großer Bedeutung. Eine Veränderung oder Erweiterung der Prozeduren, die durch Partizipation präziser und differenzierter auch auf die konfliktrelevanten Ansprüche und die Orts- und Sachkunde der Präsentischen reagiert, ist wünschenswert, muss aber mit der Verpflichtung gegenüber den Angehörigen zukünftiger Generationen abgewogen werden.

1.4   Rechtsfragen

Die Endlagerung hochradioaktiver Abfälle wirft komplexe Rechtsfragen auf. Von prinzipieller Bedeutung sind dabei die Ausgestaltung der Verantwortung des Staates, die Vermeidung institutioneller Konflikte, die Prinzipien der nuklearen Entsorgung und der – verfassungsrechtlich gebotene oder rechtspolitisch sinnvolle – Grad an Verrechtlichung. In diesem groben Rahmen stellt sich eine ganze Reihe komplexer Teilfragen.

Auf *internationaler Ebene* ist das Gemeinsame Übereinkommen über die Sicherheit des Umgangs mit abgebrannten Brennstäben und die Sicherheit der Entsorgung radioaktiver Abfälle von 1997 (International Atomic Energy Agency – IAEA) für die Entwicklung der Grundsätze der nuklearen Entsorgung und des institutionellen Rahmens der Regulierung von besonderer Bedeutung. Außerdem gibt es eine Reihe relevanter Empfehlungen der IAEA und anderer Organisationen wie insbesondere der International Commission for Radiological Protection (ICRP).

Die *Europäische Union* hat bisher nur in begrenztem Umfang Regelungen für die Entsorgung radioaktiver Abfälle erlassen. Zu nennen sind die Gesundheitsnormen nach der Euratom-Richtlinie 96/29, die für den Strahlenschutz gelten, und die Richtlinie über die Sicherheit der Zwischenlagerung und Behandlung abgebrannter Brennstäbe. Ein Kommissionsvorschlag von 2010 sieht nunmehr auch die Einführung von Regelungen über die nukleare Endlagerung vor, die zu gewissen Anpassungen des deutschen Rechts nötigen werden.

Die Untersuchung ist *rechtsvergleichend* angelegt. Sie vergleicht die Regulierung der Entsorgung hochradioaktiver Abfälle in wichtigen Nuklearstaaten wie USA, Frankreich, Großbritannien, Schweiz, Schweden, Finnland, Spanien und Japan (erfolgt im Haupttext nur zusammenfassend, während die Einzelheiten im Anhang wiedergegeben werden). Die Vergleichung zeigt eine Reihe von Gemeinsamkeiten, aber doch auch erhebliche Unterschiede, die auf vielfältigen Faktoren beruhen. Diese Vielfalt schließt es aus, allgemein gültige Folgerungen aus den ausländischen Erfahrungen zu ziehen. Jedoch vermitteln diese Erfahrungen eine Reihe wichtiger Denkanstöße. Hervorzuheben ist, dass als Reaktion auf die überall, wenngleich in unterschiedlicher Stärke, spürbaren Akzeptanzprobleme bei der Standortwahl es eine deutliche Tendenz zur Einführung neuer Formen der Partizipation bei der Strategieentwicklung und insbesondere der Standortauswahl gibt. Die guten Erfahrungen in Schweden und Finnland mit diesen Verfahren lassen sich allerdings aus mancherlei Gründen nicht verallgemeinern. In den anderen betreffenden Staaten steht das neue Entscheidungsmodell noch auf dem Prüfstand.

Die *Verantwortung des Staates* für die Sicherheit der nuklearen Entsorgung in Deutschland ist in erheblichem Maße verfassungsrechtlich determiniert. Art. 20a GG macht den Schutz künftiger Generationen ausdrücklich zur Staatsaufgabe. Das Atomgesetz enthält anspruchsvolle Anforderungen an die Zulassung nuklearer Endlager. Nach der Rechtsprechung ist die bestmögliche Gefahrenabwehr und Risikovorsorge nach dem Maßstab der praktischen Vernunft Standard der staatlichen Pflichten. Allerdings räumt die Rechtsprechung der Legislative und im Rahmen des Atomgesetzes auch der Exekutive die Letztverantwortung für die Ermittlung und Bewertung der betreffenden Risiken und die Entscheidung über ihre Tolerabilität ein.

Das Atomgesetz schreibt vor, dass der Bund Endlager für radioaktive Abfälle selbst oder durch obere Bundesbehörden wie das Bundesamt für Strahlenschutz betreibt. Auch Dritte können mit der Erfüllung dieser Pflichten beauftragt oder sie können hiermit beliehen werden. Im Hinblick auf die Langfristigkeit der Aufgaben erscheint das Modell der Beleihung trotz unbestreitbarer Vorzüge allerdings nicht unproblematisch.

Das Atomgesetz konzentriert Managementaufgaben und Überwachungspflichten beim Betrieb von Endlagern für hochradioaktive Abfälle beim Bundesamt für Strahlenschutz, das grundsätzlich weisungsgebunden ist. Dies ist mit dem in dem Gemeinsamen Übereinkommen und den IAEA-Empfehlungen verankerten Grundsatz der effektiven Unabhängigkeit von Management und Regulierung noch vereinbar, weil die Überwachung im Bundesamt innerorganisatorisch selbständig ist. Eine Reform der Zuständigkeiten erscheint jedoch sinnvoll.

Die *grundlegenden strategischen Optionen* für die Entsorgung hochradioaktiver Abfälle sind im Atomgesetz niedergelegt (teilweise, wie hinsichtlich des Grundsatzes der geologischen Endlagerung, allerdings nicht explizit). Dagegen unterliegen viele wichtige Strategieelemente der Entscheidung der Exekutive, die sich hierbei am Grundsatz der bestmöglichen Gefahrenabwehr und Vorsorge orientieren muss. Für den Grundwasserschutz gilt der Besorgnisgrundsatz nach dem Wasserhaushaltsgesetz. Das Bundesumweltministerium hat im Herbst 2010 Sicherheitsanforderungen an die Endlagerung hochradioaktiver Abfälle veröffentlicht, die im Zuge der weiteren Erkundung von Gorleben fortentwickelt werden sollen.

Problematisch in rechtlicher Sicht sind insbesondere der Zeitrahmen der Vorsorge in der Nachbetriebsphase und damit verbunden die Anforderungen an den Sicherheitsnachweis. Ein praktischer Ausschluss jeglichen Risikos muss nicht bis zum (fast) völligen Abklingen der Radioaktivität der eingelagerten Abfälle, sondern nur für einen solchen Zeitraum sichergestellt sein, wie dies nach dem gegenwärtigen Stand von Wissenschaft und Technik, einschließlich der Prognosefähigkeit, möglich ist. Darüber hinaus müssen die Risiken so weit wie vernünftiger Weise möglich reduziert werden. Dies stellt bestmögliche Gefahrenabwehr und Risikovorsorge im Sinne des Atomgesetzes dar.

Die *Standortwahl für atomare Endlager* in Deutschland zeichnet sich bisher durch einen geringen Grad an Verrechtlichung aus. In Reaktion auf die als technokratisch empfundene Auswahl des Salzstocks in Gorleben als Standort für das Endlager für hochradioaktive Abfälle in den Achtziger Jahren sind verschiedene Vorschläge gemacht worden, die Standortsuche neu zu beginnen. Obwohl die Bundesregierung nunmehr entschieden hat, die untertägige Erkundung in Gorleben ohne Berücksichtigung von Alternativen wieder aufzunehmen, erscheint es sinnvoll, sich erneut mit der Standortwahl zu befassen, da weder die Eignung von Gorleben feststeht noch bekannt ist, ob es nicht eine besser geeignete Alternative gibt.

Aus verschiedenen Gründen kommt weder das atomrechtliche Planfeststellungsverfahren noch die Raumordnung des Bundes für ein Standortauswahlverfahren in Betracht. Sinnvoll und rechtlich möglich wäre es dagegen, ein der Planfeststellung vorgeschaltetes Fachplanungsverfahren entsprechend dem abfallrechtlichen Bewirtschaftungsplan einzuführen. Verfassungsrechtlich geboten ist ein solches Auswahlverfahren im Hinblick auf die Letztverantwortung des Gesetzgebers und der Exekutive für die Tolerabilität nuklearer Risiken nicht. Bei der Ausgestaltung des Verfahrens sollte sich der Gesetzgeber grundsätzlich am Optimierungsgebot orientieren, das aus dem gesetzlichen Grundsatz der bestmöglichen Gefahrenabwehr und Risikovorsorge abzuleiten ist. Allerdings ist die Sicherheit eines Endlagers das Produkt einer Kombination von geologischen und technischen Barrieren an einem bestimmten Standort. Das Optimierungsgebot kann daher nicht isoliert in Bezug auf das Wirtsgestein an einem bestimmten Standort verfolgt werden.

Obwohl Gründe der Legitimität und Akzeptanz für die völlige Wiederaufnahme des Standortauswahlverfahrens mit einer neuen Verfahrensstruktur sprechen mögen, hält sich die Entscheidung, die untertätige Untersuchung von Gorleben fortzusetzen, im Rahmen der Letztverantwortung der Exekutive und kann auch pragmatisch gerechtfertigt werden. Sinnvoll ist es aber, parallel zu der weiteren untertägigen Untersuchung von Gorleben Alternativstandorte zu prüfen und hierfür ein *modernes Auswahlverfahren* zu schaffen, das mit dem Fortgang der Untersuchungen in Gorleben verknüpft ist. Diese Vorgehensweise wirkt übermäßigen Verzögerungen der Standortsuche für den Fall des Scheiterns von Gorleben entgegen. Darüber hinaus ermöglicht sie aus Gründen guter Sachpolitik und Legitimität, dass auch bei Eignung von Gorleben eine intensive Untersuchung und gegebenen Falls die Wahl einer Standortalternative durchgeführt wird, die sich im parallelen Auswahlverfahren als vermutlich eindeutig überlegen erwiesen hat.

Das Auswahlverfahren könnte auf der Grundlage von Auswahlkriterien, die die Exekutive unter Beteiligung einer unabhängigen Expertenkommission und der Öffentlichkeit vorzugeben hat, in acht bis neun Verfahrensstufen ablaufen, sollte transparent sein und eine umfassende Beteiligung aller Betroffenen und Interessenträger ermöglichen. Die herkömmliche Partizipation sollte zur Verbesserung der Legitimität und Akzeptanz der betreffenden Entscheidungen angereichert werden. Zu den Elementen einer Neugestaltung gehören insbesondere die Einschaltung einer unabhängigen Expertenkommission, die Möglichkeit einer frühzeitigen Einflussnahme der Betroffenen und Interessenträger auf die Gestaltung und Planung des Suchprozesses, auf lokaler/regionaler Ebene die Errichtung eines permanenten Dialogforums und die Zugriffsmöglichkeit der Betroffenen und Interessenträger auf einen Pool unabhängiger Sachverständiger.

Abgesehen von der Feststellung der Eignung des Standorts soll das gesetzlich vorgesehene Planfeststellungsverfahren sicherstellen, dass *Bau und*

Betrieb der Anlage den gesetzlichen Anforderungen entsprechen. Hierzu gehört die Einhaltung der Anforderungen der Strahlenschutzverordnung ebenso wie der Nachweis der Anlagensicherheit und der Grundwasserschutz, für den der Besorgnisgrundsatz maßgeblich ist. Die gesetzlichen Anforderungen gelten im Grundsatz, wenngleich mit Modifikationen, die durch die Langfristigkeit der Regulierungsaufgabe bedingt sind, auch für die Nachbetriebsphase der Anlage. Die Strahlenschutzverordnung, die auf nukleare Anlagen mit laufendem Betrieb zugeschnitten ist, kann zumindest als Richtschnur für den Strahlenschutz in der Nachbetriebsphase herangezogen werden.

Das Atomgesetz weist die *finanzielle Verantwortung* für die Endlagerung radioaktiver Abfälle den Betreibern der Kernkraftwerke zu. Sie sind verpflichtet, finanzielle Beiträge für die Deckung der erforderlichen Kosten, unter anderem auch für die Planung der Anlagen, zu erbringen. Da die Standortauswahl darauf abzielt, die Voraussetzungen für die Errichtung eines Endlagers zu schaffen und den Betreibern daraus Vorteile erwachsen, handelt es sich bei den Kosten der Standortauswahl um notwendige Kosten, soweit das Verfahren gesetzlich geboten und nicht nur rein politisch etabliert ist.

Eine Schwäche des geltenden Rechts liegt darin, dass sich die Pflichten der Betreiber in Bezug auf künftige Finanzierungslasten auf die Bildung von Rückstellungen beschränken. Im Hinblick auf die Langfristigkeit der Endlagerung hochradioaktiver Abfälle wirft diese Reglung Probleme für den Fall der Insolvenz oder Liquidation eines Abfallerzeugers auf, soweit es sich um Tochtergesellschaften der Energieerzeuger handelt. Bei Beendigung des bestehenden Unternehmensvertrags aufgrund Insolvenz oder Liquidation beschränkt sich die Haftung der Muttergesellschaft auf bereits begründete Verbindlichkeiten. Es ist zweifelhaft, ob die abstrakte gesetzliche Verpflichtung zur Tragung der künftigen, noch nicht absehbaren Kosten der Entsorgung als eine bereits „begründete" Verbindlichkeit angesehen werden kann. Diese Zweifel könnten durch eine Mithaftung der Muttergesellschaft des Betreibers unabhängig vom Bestand des Unternehmensvertrags, insbesondere auch wenn der Betrieb nicht fortgesetzt wird, gelöst werden. Die gesetzliche Haftung der Muttergesellschaft für Verbindlichkeiten der Tochtergesellschaft im Fall eines Gewinnabführungsvertrages bleibt natürlich unberührt. Von Fondlösungen ist abzuraten.

1.5   Leitlinien für eine sozial verträgliche und gerechte Standortbestimmung

Die Endlagerfrage mobilisiert Menschen, und zwar nicht nur in Deutschland, sondern weltweit. Die Endlagerfrage ist symbolisch überhöht: Es geht nicht mehr allein und auch nicht mehr vordringlich um die Frage der technischen Machbarkeit, nicht einmal mehr um die langfristige Sicherheit, sondern um grundlegende Perspektiven gesellschaftlicher Entwicklung:

Wollen wir weiterhin zentrale, hoch effiziente, mit hoher Energiedichte versehene, aber gleichzeitig riskante Technologien in der Energieerzeugung? Oder wollen wir lieber auf dezentrale, oft wenig effiziente, auf geringe Energiedichte basierende und in ihren Auswirkungen nicht unbedingt risikoarme, aber lokal begrenzte Technologien setzen?

Diese Ausgangslage bestimmt die Bedingungen für eine künftige Lösung der Frage nach der Entsorgung hochradioaktiver Abfälle. Das Thema „Endlagerung" ist emotional hoch besetzt; es löst bei vielen Menschen Ängste aus. Dazu einige empirische Ergebnisse:

- Bei allen Befragungen belegt die nukleare Endlagerlösung Spitzenplätze in der öffentlichen Wahrnehmung von Bedrohlichkeit. Das ist weltweit so, interessanterweise auch in Finnland, wo das Problem der Endlagerung trotz dieser öffentlichen Besorgnis politisch weitgehend gelöst werden konnte.
- Die Komplexität dieses Sachverhalts wird deutlich, wenn man die Ergebnisse einer repräsentativen Umfrage aus den Jahren 2001 und 2002 betrachtet: Während zum Zeitpunkt des Surveys circa 65 % der Befragten davon ausgingen, dass innerhalb der nächsten zehn Jahre ein Endlager für hochradioaktive Abfälle zur Verfügung stehen wird, lehnten gleichzeitig 81 % der Befragten ein Endlager in ihrer unmittelbaren Wohnumgebung ab. Dieses klassische *NIMBY-Syndrom* („Not in my backyard!") ist ein Kennzeichen von Standortfindungsprozessen für großtechnologische und risikobezogene Anlagen. Die Notwendigkeit der Technologie wird im Prinzip bejaht, jedoch möglichst weit weg vom eigenen Wohnort.
- Bei Untersuchungen zur Stakeholder-Mobilisierung gibt es weltweite Unterschiede, aus denen man viel lernen kann. Einige Länder wie Finnland, Schweden und die Schweiz haben Fortschritte bei der Lösung der Endlagerung gemacht. Eine institutionell befriedigende und für die meisten Menschen tolerierbare Lösung ist bei richtiger Vorgehensweise nicht unmöglich. Es ist aber nicht einfach, ein Verfahren zu finden, das auf Akzeptanz stoßen wird. Es gibt auch niemals eine Garantie für ein Gelingen. Nur: Wenn man es falsch macht, gibt es die Garantie, dass man scheitert.

Warum ist die Risikowahrnehmung der nuklearen Endlagerung so emotional hoch geladen? Aus psychologischer Sicht sind die Risiken der nuklearen Stromerzeugung insgesamt aber auch der Endlagerung in der Wahrnehmung der Bevölkerung dem semantischen Muster „Schwert des Damokles" zuzuordnen. Semantische Muster haben ähnliche Funktionen wie Schubladen in einem Aktenschrank. Wenn man mit einem neuen Risiko konfrontiert wird oder wenn man eine neue Information zum Risiko aufgenommen hat, versuchen Menschen in der Regel, diese neuen Informationen in eine der bestehenden Schubladen einzuordnen.

Darunter fällt auch das Muster des Damokles-Schwertes. Dabei geht es um technische Risiken, bei denen, unabhängig davon, ob diese Zuschreibung gerechtfertigt ist oder nicht, ein hohes Schadenspotential mit einer sehr geringen Eintrittswahrscheinlichkeit verbunden wird. Die stochastische Natur eines solchen Ereignisses macht eine Voraussage über den Zeitpunkt des Eintritts unmöglich. Folglich kann das Ereignis in der Theorie zu jedem Zeitpunkt eintreten, wenn auch mit jeweils extrem geringer Wahrscheinlichkeit. Wenn wir uns jedoch im Bereich der Wahrnehmung von seltenen Zufallsereignissen befinden, spielt die Wahrscheinlichkeit eine geringe Rolle: Die Zufälligkeit des Ereignisses ist der eigentliche Risikofaktor. Die Vorstellung, das Ereignis könne zu jedem beliebigen Zeitpunkt die betroffene Bevölkerung treffen, erzeugt das Gefühl von Bedrohtheit und Machtlosigkeit. Instinktiv können die meisten Menschen mental (ob real mag hier dahin gestellt bleiben) besser mit Gefahren fertig werden, wenn sie darauf vorbereitet und darauf eingestellt sind.

Dazu kommt noch die schleichende Gefahr der Radioaktivität, die man nicht sinnlich wahrnehmen kann. Damit sind wir bei einem zweiten semantischen Muster, das häufig angstauslösend wirkt. Im Rahmen dieses Risikomusters nehmen Menschen zu Recht an, dass wissenschaftliche Studien schleichende Gefahren frühzeitig entdecken und Kausalbeziehungen zwischen Aktivitäten oder Ereignissen und deren latente Wirkungen aufdecken können.

Im Falle des semantischen Musters „Schleichende Gefahr" sind die betroffenen Menschen auf Informationen durch Dritte angewiesen. Sie können die Gefahren in der Regel nicht sinnlich wahrnehmen, noch die Behauptungen der sich häufig widersprechenden Experten nachprüfen. Bewerten Laien diese Risiken, dann stoßen sie auf eine Schlüsselfrage: Vertraue ich den Institutionen, die mir dazu die notwendigen Informationen geben, ja oder nein? Ist die subjektive Einschätzung negativ, dann wird kompromisslos ein Nullrisiko gefordert. Denn wer bei der Bewertung solcher Risiken auf Informationen durch Dritte angewiesen ist, diesem Dritten aber nicht vertraut, der lässt sich auf keine Kosten-Nutzen-Bilanz ein, sondern fordert die Nullbelastung. Ist er dagegen unentschieden, ob er vertrauen kann oder nicht, dann werden periphere Merkmale besonders wichtig, Merkmale, die mit der Entscheidungslage sachlich nicht verknüpft sind. Der Laie hat aber keine andere Möglichkeit, als Vertrauen nach peripheren Merkmalen zu verteilen, denn er kann das Risiko, durch radioaktive Strahlung zu Schaden zu kommen, nicht selbst untersuchen. Er muss irgendeiner Seite trauen oder gar nicht trauen.

Diese Muster sind tief in unbewusste Bewertungsprozesse der Wahrnehmung eingebunden. Sie lassen sich nur dann überwinden, wenn die Menschen diese Wahrnehmungsmuster selbst begreifen lernen und deren Wirkung als unbewusste Bewertungsmaßstäbe der eigenen Urteilsfindung erkennen können. Risikokommunikation kann sich daher nicht auf die Vermittlung der wissenschaftlichen Einsichten über Risiken beschränken, sondern muss auch die Mechanismen der Risikowahrnehmung plastisch vermitteln.

Selbst wenn man, wie dies in jüngster Zeit zunehmend geschieht, in einen intensiven Risikodialog eintritt, ist damit die Konfliktsituation keineswegs aufgelöst. Die Schaffung einer Kommunikationsbasis ist vielmehr die Voraussetzung, aber keineswegs die hinreichende Bedingung dafür, dass es zu einer allgemein akzeptierten Lösung kommen kann. Im Prinzip lassen sich aufgrund der gegebenen Verhältnisse drei prinzipielle Vorgehensweisen zur Endlagerung hochradioaktiver Abfälle skizzieren:

Im *Top-Down-Ansatz* haben die durch die demokratische Gesellschaftsordnung gewählten Staatsvertreter die alleinige Entscheidungsbefugnis inne. Kraft ihres Amtes entscheiden sie zum Wohle des Volkes. Eine aktive Beteiligung der Bürger ist hierbei, wenn überhaupt, nur sehr restriktiv vorgesehen. Allerdings beruht auch diese Lösung auf einer transparenten und den Bürger einbeziehenden Risikokommunikation. Die Bürger dürfen ihre Meinung in Anhörungen oder Erörterungsterminen einbringen, jedoch gibt es keine Garantie auf Einflussnahme in der finalen Entscheidungsfindung. Die Entscheidungsträger müssen auch nachweisen, dass alle Einwände ordnungsgemäß behandelt wurden. Dann aber liegt es in den Händen der Entscheidungsträger, eine Entscheidung unter Offenlegung der Argumente für und wider zu treffen.

In einer *Muddling-Through-Strategie*, einer pragmatischen Mischung aus Top-Down- und Bottom-up-Ansatz, kann man sich auf die im politischen Meinungsprozess gewachsenen Minimalkonsense (Muddling Through) verlassen. Als legitim werden nur solche Entscheidungsoptionen angesehen, die den geringsten Widerstand in der Gesellschaft hervorrufen. Gesellschaftliche Gruppen nehmen in dieser Steuerungsvariante insoweit auf den Prozess der Willens- und Entscheidungsbildung Einfluss, wie sie anschlussfähige, d. h. dem Sprachcode und dem Verarbeitungsstil des politischen Steuerungssystems angepasste, Vorschläge liefern und öffentlichen Druck mobilisieren. In der Politik setzt sich dann der Vorschlag durch, der sich im Wettstreit der Vorschläge am besten behauptet, d. h. der für die politischen Entscheidungsträger die geringsten Einbußen an Unterstützung durch Interessengruppen mit sich bringt. Die bisherige Auseinandersetzung um die Endlagerung scheint weitgehend einem solchen Muddling Through zu entsprechen.

Die dritte Variante – der *Bottom-up-Ansatz* zur diskursiven Standortbestimmung – setzt auf eine diskursive Lösung und auf den Versuch einer fairen Aushandlung der Standortfindung zwischen den beteiligten Gruppen. Diskursive Verfahren erheben den Anspruch, rationalere (im Sinne eines diskursiven Vernunftverständnisses), gerechtere (im Sinne eines verhandlungsbasierten Gerechtigkeitsverständnisses) und kompetentere Lösungen von Problemen zu ermöglichen. Gleichgültig welche

Ansprüche man im Einzelnen mit diskursiven Prozessen verbindet: Sie müssen nach bestimmten Regeln strukturiert sein, um ihre Leistungsfähigkeit zu gewährleisten, um etwa konstruktive Problemlösungen sachgerecht und fair bereitzustellen und mehrere Entscheidungsoptionen offen zu halten, und um strategische Verhaltensweisen der Teilnehmer so weit wie möglich zu verhindern. Im Prinzip ist die Legitimation von kollektiv verbindlichen Normen an drei Bedingungen geknüpft: Zustimmung aller Beteiligten, substanzielle Begründung der im Diskurs gemachten Aussagen sowie angemessener Ausgleich von negativ betroffenen Interessen und Werten.

Wie ließe sich eine sinnvolle Kombination aus bottom up und top down realisieren? Das gesamte Auswahlverfahren muss transparent und nachvollziehbar sein (Kriterium der effektiven Risikokommunikation). Das Auswahlverfahren muss gegenüber Nichtbeteiligten als fair (alle beteiligten Interessen- und Wertgruppen kommen zu Wort), kompetent (dem Problem angemessen und mit der notwendigen Sachkenntnis versehen) und effizient (die Mittel bez. Entscheidungskosten sind den Zielen angemessen) erscheinen. Die Auswahl selbst muss in ihrem normativen wie kognitiven Gehalt nachvollziehbar und intersubjektiv begründbar sein und sollte die pluralen Wertvorstellungen der betroffenen Bürger im Sinne eines fairen Konsenses oder Kompromisses widerspiegeln.

Wollte man alle diese Forderungen zur Legitimation des Standortauswahlprozesses erfüllen, so dürfte ein einziges politisches Steuerungsinstrument mit Sicherheit nicht ausreichen. Vielmehr verlangen Entscheidungen von so großer Reichweite eine Aneinanderreihung verschiedener Steuerungsinstrumente, die jeweils unterschiedliche Teilforderungen abdecken.

Um die genannten Grundsätze einzuhalten, müssen mehrere Schritte und Komponenten kombiniert werden. Zunächst bedarf es einer wissenschaftlich-technischen Übereinkunft über die Eignung von Standortkonzepten und über die Kriterien in Form von Schwellenwerten, welche erreicht werden müssen, damit ein Standort unter dem Aspekt der Langzeitverantwortung als geeignet gelten kann. Diese Kriterien müssen festgelegt werden, bevor die Ergebnisse der Eignungsprüfung vorliegen. Das sollte auf der Grundlage von konsensorientierten Methoden der wissenschaftlichen Prüfung erfolgen, die unabhängig, sachbezogen und transparent angewandt werden müssen. Um das zu institutionalisieren, braucht man eine neutrale Plattform unter professioneller Führung, bei der Wissenschaftler auf nationaler Ebene unter Einbeziehung internationaler Experten (das erhöht die unverzichtbare Glaubwürdigkeit) zusammenkommen, mit dem Ziel, den hier geforderten Wissenskonsens herbeizuführen. Um auch gegenüber der Öffentlichkeit zu dokumentieren, dass hier keine einseitige Auswahl der Experten stattfindet, kann man ein Nominierungsrecht von Stakeholdern vorsehen.

Zum zweiten benötigt man einen fairen Ausgleich für die Übernahme von Unsicherheiten. Ziel ist es hier, eine robuste, allseits akzeptierte Lösung zu finden, um vorausschauend mit Unsicherheiten umzugehen. Verfahren, wie die Mediation oder die Einrichtung eines Runden Tisches mit Stakeholdern können zu einem als gerecht empfundenen Ausgleich von Unsicherheitsfolgen führen. Hier ist es besonders wichtig, die Unsicherheiten, vor allem die über die Langzeiteffekte, nicht zu verschweigen, sondern offen anzusprechen und Ausgleichsmöglichkeiten zu schaffen, indem man gleichzeitig etwas für die Wirtschaftsförderung oder für die Standortentwicklung tut. Dabei geht es nicht um „Ablasshandel" oder eine korrumpierende Bezahlung von Risikoübernahme, sondern um auch in anderen Lebensbereichen übliche Formen des sozialen Ausgleichs: Diejenigen, die unsichere Folgen und Belastungen in Zukunft für die Allgemeinheit tragen sollen, sollen dafür auch von der Gemeinschaft anerkannt bzw. unterstützt werden. Dieses Vorgehen zeigt, dass die Übernahme von Unsicherheit respektiert und honoriert wird. Man kann nicht verlangen, dass die Unsicherheit schweigend „geschluckt" wird. Dafür sind neutrale Dialogforen mit den von den Folgen betroffenen Gruppen am besten geeignet. Am Beispiel Schwedens lässt sich lernen, dass diese Foren möglichst lokal besetzt sein sollten. Experten werden dabei als Wissensquellen und Auskunftspersonen fallweise hinzugezogen.

Dann folgt der dritte und letzte Bestandteil einer diskursiven Lösung: ein Forum zur gesellschaftlichen Orientierung über künftige Energieversorgung und postindustrielle Lebensstile. Die Debatte um Endlagerung ist mehr als eine Debatte um Abfallbehandlung, es geht vielmehr um die Frage: Wie wollen wir in Zukunft leben? Wie kann das Thema Endlagerung in einen konstruktiven Entwurf künftiger Lebensstile und Lebensbedingungen eingeordnet werden? Hier könnten diskursorientierte Methoden wie Bürgerforen, Runde Tische oder Konsensuskonferenzen, die sich in anderen Ländern teilweise gut bewährt haben, zum Einsatz kommen.

Die Reduzierung der Komplexität mithilfe eines Konsenses in der Wissenschaft über das beste Auswahlverfahren, die Bewältigung der Unsicherheit durch faire Angebote an diejenigen, die unter den Folgen der Unsicherheit werden leben müssen, und die Behandlung der Ambiguität durch einen offenen und ehrlich geführten Zieldiskurs über die Zukunft der Energieversorgung sind die Stichworte, die am Anfang einer neuen Initiative zur Lösung der Frage nach der verantwortlichen Entsorgung hochradioaktiver Abfälle stehen müssen. Nicht zuletzt ist ein gesellschaftlicher Diskurs über die Frage erforderlich, wie wir als rohstoffarmes Land in Zukunft bestehen können.

2 Schlussfolgerungen und Empfehlungen

2.1 Ethische Grundlagen

(1) **Verpflichtungen gegenüber künftigen Generationen gelten prinzipiell unbefristet. Entsorgungsstrategien für radioaktive Abfälle sind gleichwohl für befristete Zeiträume zu entwickeln.**

Verpflichtungen, die den Akteuren eine umsichtige Entsorgung radioaktiver Abfälle auferlegen, gelten prinzipiell unbefristet und bestehen – wenn auch in ihrer Verbindlichkeit graduell abnehmend – auch gegenüber den Angehörigen ferner Generationen. Die bei der Entwicklung von Entsorgungsstrategien einzubeziehenden komplexen Verläufe (die „Folgenräume") sind gleichwohl aus rationalen Erfordernissen der Planung und aus Gründen der Effizienz zeitlich zu befristen. Eine solche Befristung sollte sich am voraussehbaren künftigen Wirkungspotential der Folgen orientieren, und damit am relativen, mit den Phasen des Zerfalls-Prozesses und der gewählten Entsorgungsstrategie variierenden Gefährdungspotential der Lagerinventare und möglicher Expositionen in der Biosphäre.

(2) **Die jetzige Generation als primäre Nutznießerin der Kernenergie hat die Verpflichtung, die Lösung des Entsorgungsproblems einzuleiten. Die Forderung nach einer *unverzüglichen* Entsorgung hochradioaktiver Abfälle bürdet der jetzigen Generation jedoch nicht zu rechtfertigende Lasten auf.**

Die Inanspruchnahme moralischer Prinzipien, die den Angehörigen der gegenwärtigen Generation als Verursacher- und Nutznießergemeinschaft die vollständige Entsorgung auferlegen, ist aus ethischer Sicht keineswegs selbstverständlich. Sofern und solange nach belastbaren Prognosen über Generationen- oder Gemeinschaftsgrenzen verlässliche Tauschverhältnisse organisiert werden könnten und dieses nicht zu Lasten Dritter ginge, wäre etwa eine Übertragung der „Entsorgungsverantwortung" gegen einen zwangsfrei akzeptierten Ausgleich ethisch unbedenklich. Die Forderung nach der Unverzüglichkeit einer Problemlösung ist keinesfalls selbstverständlich und bedarf der Rechtfertigung. Ist es wahrscheinlich, dass eine künftige Generation, mit der wir in kontrollierbarer Interaktion stehen, über „bessere" Entsorgungsstrategien verfügt, kann es gar geboten sein, diese Option zu ergreifen. Gerechtigkeitserwägungen erlegen dann allerdings dem Verursacher eine Pflicht zur angemessenen Kompensation auf. Dasselbe gilt auch, mutatis mutandis, für internationale Austauschbeziehungen: Für eine angemessene Wahrnehmung der Langzeitverpflichtung sind nicht die – angesichts der in Frage stehenden Zeiträume ohnehin eher historisch kontingent erscheinenden – Nationengrenzen, sondern die Verfügbarkeit von Kompetenzen und Ressourcen die relevanten Größen.

(3) Legitimierende Verfahren zur Lösung der Endlagerfrage sind so zu
gestalten, dass sie allen in gleichem Maße gerecht werden und insbesondere auch die Ansprüche künftiger Generationen angemessen
miteinbeziehen.

Eine Ausgestaltung der legitimierenden Prozeduren, die auch auf die konfliktrelevanten Ansprüche und die Orts- und Sachkunde der gegenwärtig Lebenden reagiert, ist wünschenswert, muss aber mit den Verpflichtungen gegenüber den nicht an den Beratungen Beteiligten, insbesondere
den Angehörigen zukünftiger Generationen, abgewogen werden. Maßstab
für die Entscheidung darf daher nicht allein die prozedural hergestellte
faktische Zustimmung von Angehörigen der gegenwärtigen Generationen
sein. Die Entscheidung muss vielmehr auch im Sinne einer rational
dargelegten, universalistisch geführten Begründung akzeptabel sein. In
der öffentlichen Debatte sind diejenigen, für die eine Beteiligung nicht
möglich ist oder für die es keine Anreize gibt, sich für ihre Ansprüche zu
engagieren, advokatorisch zu vertreten. Die Verantwortung dafür ist legiti-
miert zu übertragen und darf nicht durch einzelne Interessenvertreter oder
Interessengruppen einfach für sich reklamiert werden.

(4) Eine grundsätzliche Zurückweisung aller Lösungsvorschläge für die
Entsorgung hochradioaktiver Abfälle ist nicht mit der Verpflichtung
gegenüber zukünftigen Generationen verträglich.

Die Entsorgung radioaktiver Abfälle ist eine kollektive, gesamtgesellschaftliche Aufgabe. Daraus ergibt sich nicht nur die Sorgfaltspflicht, unterbreitete
Vorschläge daraufhin zu prüfen, ob sie im Sinne der Verpflichtung aussichtsreich sind, sondern auch die Verpflichtung, sich konstruktiv an der Entwicklung geeigneter Vorschläge zu beteiligen bzw. Strukturen zu bilden oder zu
fördern, die eine solche Beteiligung ermöglichen. Das Vetorecht derer, die
mit der mangelnden Eignung eines vorgeschlagenen Standortes oder auch
mit der Untauglichkeit des Entsorgungskonzepts an sich argumentieren, ist
an die Erwartung geknüpft, sich an der Entwicklung alternativer Vorschläge
konstruktiv zu beteiligen. Aussichtsreiche Projekte und Prozesse, die auf die
Entwicklung alternativer Vorschläge zielen, sind mit den erforderlichen Ressourcen zu unterstützen. Eine grundsätzliche Ablehnung aller Vorschläge ig-
noriert die Verpflichtung gegenüber künftigen Generationen.

(5) Die Instrumentalisierung künftiger Generationen für Argumente gegen längere Laufzeiten von Kernkraftwerken ist unzulässig.

Es besteht gegenüber den Angehörigen künftiger Generationen eine Verpflichtung zur Entsorgung der bereits vorhandenen radioaktiven Abfälle. Die
im Falle einer Laufzeitverlängerung von Kernkraftwerken zusätzlich anfallenden Mengen wären zu gering, als dass sie auf die Wahl der Entsorgungsstrategien wesentlichen Einfluss nehmen könnten. Die in Frage stehenden
Verlängerungszeiträume sind angesichts der für die Endlagerung ohnehin zu

erwartenden langen Fristen ohne Bedeutung. Es wäre daher nicht zulässig, die Interessen künftiger Generationen zu instrumentalisieren, um auf Entscheidungen über die Laufzeiten von Kernkraftwerken steuernd einzuwirken.

2.2   Sicherheitsanforderungen und –ziele

(6)   Die umsichtige Entsorgung radioaktiver Abfälle setzt die Entwicklung einer angemessenen Gesamtstrategie hinsichtlich der Sicherheit und des Gesundheits- und Umweltschutzes voraus. Das Grundgesetz und das Atomgesetz geben einen klaren gesetzlichen Rahmen für eine bestmögliche Gefahrenabwehr und Risikovorsorge.

Die umsichtige Entsorgung radioaktiver Abfälle setzt die Entwicklung einer angemessenen Gesamtstrategie voraus, die alle einschlägigen technischen Bedingungen und Optionen einbezieht und auf der Grundlage schon getroffener Entscheidungen den Erfordernissen der technischen, sozialen und politischen Umsetzbarkeit genügt. Sowohl das deutsche Grundgesetz als auch das Atomgesetz geben hierfür einen eindeutigen normativen Rahmen, indem sie jede Lösung an die gesetzliche Forderung der bestmöglichen Gefahrenabwehr und Risikovorsorge binden. Allerdings wird dabei der Legislative und der Exekutive ein breiter Ermessensspielraum zugestanden.

Die Sicherheit eines Endlagers ist das Ergebnis einer Kombination geologischer und technisch hergestellter Barrieren an einem bestimmten Standort. Aufbauend auf dem internationalen Konsens hinsichtlich der herausgehobenen Bedeutung der geologischen Barrieren lässt sich sicher behaupten, dass die besten geologischen Barrieren ausgewählt und mit den besten technischen Barrieren verbunden werden sollen. Diese Kombination muss darum allerdings noch nicht notwendigerweise auch das höchste Maß an Sicherheit gewährleisten. So könnten etwa die Beschaffenheit des Wirtsgesteins und die lokalen geophysikalischen Bedingungen als limitierender Faktor für den Zugewinn an Sicherheit durch die technischen Barrieren wirken. Letztendlich ausschlaggebend muss die Sicherheit des Gesamtsystems sein.

(7)   Für die Beurteilung der langfristigen Sicherheit ist von Szenarien auszugehen, die auf der Beurteilung natürlicher Entwicklungen hinsichtlich des Potentials für einen Schadstofftransport durch Wirtsgesteine und Deckgebirge sowie potentielle radiologische Expositionen in die Biosphäre beruhen.

Die Bewertung der Strategien muss auf der Grundlage von Kriterien erfolgen, die die kurzfristigen wie die langfristigen Folgen für die Umwelt berücksichtigen und den Sicherheits- und Schutz-Bedürfnissen der Gegenwärtigen ebenso gerecht werden wie denen der Angehörigen künftiger Generationen. Für die Beurteilung der langfristigen Sicherheit sind Szenarien über natürlich induzierte Verläufe zu Grunde zu legen. Es ist zu untersuchen, in wie weit diese einen allmählichen Transport radioaktiver Substanzen durch geologische Wirtsgesteine und Deckgebirge herbeiführen können und daher mit dem Risiko einer Freisetzung in die Biosphäre und den entsprechenden Risiken für den Menschen bei Strahlenexposition an der Oberfläche oder bei unbeabsichtigtem Eindringen verbunden sind. Ferner muss Unterkritikalität des ins Endlager eingebrachten bestrahlten Brennstoffs und radioaktiven Abfalls gewährleistet werden; die entsprechenden Nachweise scheinen unproblematisch zu sein.

(8) **Die menschlichen Fähigkeiten, Prognosen über zukünftige Entwicklungen geologischer oder anthropogener Systeme über extrem lange Zeiträume zu entwickeln, sind begrenzt.**

Zu den ethischen Grundlagen gehört auch die Einsicht, dass die Einbeziehung ferner Gefahrenpotentiale, insbesondere der mit einer möglichen Freisetzung radioaktiver Substanzen verbundenen Risiken für den Menschen, aus planerischen Erfordernissen zeitlich zu befristen ist. In den bisherigen Debatten finden sich solche Fristsetzungen oft begründet mit dem Rückgang der Radiotoxizität über die Zeit hinweg. Die in den bisherigen Debatten angeführten Größenordnungen von 100.000 bis zu einer Million Jahren orientieren sich dabei meist an der Radiotoxizität natürlicher Uranerz-Vorkommen. Dem Vergleich liegt zu Grunde, dass bei diesem Zeitrahmen dieses Maß von den eingelagerten radioaktiven Abfällen unterschritten werden wird. Die menschliche Fähigkeit, Prognosen über die Entwicklung geogener oder anthropogener Systeme über sehr lange Zeiträume zu entwickeln, sind indes begrenzt. Modellrechnungen, die z. B. den Einschluss von Abfällen oder die Migration von Radionukliden in die Biosphäre beschreiben, werden zunehmend bedeutungslos, da die zugrunde liegenden Annahmen mit länger werdenden Prognosezeiträumen zunehmend ihre Rechtfertigung verlieren. Zwar lassen sich prinzipiell Anhaltspunkte für die weitere Entwicklung einer Verlaufskurve gewinnen, indem man den möglichen Verlauf über das Ende des vorhersehbaren Zeitrahmens hinaus extrapoliert. Demgegenüber steht aber die abnehmende Verlässlichkeit der Informationen, die aus solchen Berechnungen gewonnen werden können.

(9) **Der Zeitrahmen von einer Million Jahren für die Beurteilung der Langzeitsicherheit hochradioaktiver Abfälle erscheint als ein angemessener Kompromiss zwischen den ethisch begründeten Forderungen nach Langzeitverantwortung und den Grenzen der praktischen Vernunft.**

Der vielfach angeführte, insbesondere auch vom Bundesministerium für Umwelt (BMU) propagierte Zeitrahmen von einer Million Jahren ist mit der Fähigkeit begründet, Lagerstätten in einer geologischen Umgebung ausmachen zu können, von der nach aktuellem geowissenschaftlichen Kenntnisstand der Fortbestand ihrer günstigen Eigenschaften über einen Zeitraum dieser Größenordnung angenommen werden darf. Dieser Ansatz erscheint als ein vernünftiger Kompromiss zwischen den ethisch begründeten Forderungen auf der einen, und der Einsicht in die Beschränktheit der praktischen Fähigkeiten auf der anderen Seite.

(10)   Der Vorschlag, die Referenzdosis in Deutschland auf 10 µSv (0,01 mSv) festzusetzen, ist aus strahlenbiologischer Sicht abzulehnen; er weicht zudem stark vom internationalen Konsens von 100 µSv (0,1 mSv) pro Jahr ab.

Der vorgeschlagene Richtwert einer effektiven Dosis von 10 µSv (0,01 mSv) pro Jahr für Lagerstätten wärmeerzeugender radioaktiver Abfälle während der Nachbetriebs-Phase sollte neu überdacht werden. Eine solche Strahlendosis stellt weniger als ein Prozent der durchschnittlichen Strahlendosis dar, die aus natürlichen Quellen (in Deutschland und vielen anderen Regionen weltweit) stammt und liegt damit weit im Schwankungsbereich der Expositionen aus natürlichen Quellen. Eine solche Strahlendosis ist um den Faktor 10 geringer als der international übliche Wert von 100 µSv (0,1 mSv) für wenig wahrscheinliche Situationen der Nachbetriebs-Phase. Die ICRP nennt die Dosis von 10 µSv (0,01 mSv) eine „triviale" Dosis.

(11)   Untersuchungen über die Langzeitsicherheit sollten sich vorwiegend auf Radionuklide konzentrieren, die für die potentielle Strahlenexposition von Menschen von besonderer Relevanz sind.

Anhand von Modellen lässt sich abschätzen, dass einige Radionuklide, die als leicht wasserlösliche Ionen vorliegen, für die Strahlenexposition der Menschen von besonderer Relevanz sind. Weitere Untersuchungen sollten sich besonders den Fragen widmen, wie ^{129}I, ^{14}C und ^{36}Cl im Lager festgehalten werden können, und erforschen, wie die Migration dieser Radionuklide besser verstanden und modelliert werden kann.

2.3   *Entsorgungsprogramm und zeitlicher Ablauf*

(12)   Es besteht eine hohe Evidenz für die technische Umsetzbarkeit einiger bereits entwickelter Konzepte für die Endlagerung.

Die entwickelten Konzepte für eine Entsorgung hochradioaktiver Abfälle, die den Sicherheits- und Schutz-Bedürfnissen der gegenwärtig Lebenden ebenso gerecht wird wie denen der Angehörigen künftiger Generationen, erscheinen prinzipiell als technisch umsetzbar. Umsetzungsstrategien müssen folgende Aspekte einbeziehen:

(1) Unterscheidungen zwischen solchen radioaktiven Materialien, die als Abfall, und solchen, die als Ressource behandelt werden sollen;

(2) eine Entscheidung, radioaktive Abfälle zu konzentrieren, in geeigneter Weise einzuschließen und sie in tiefengeologischen Formationen zu deponieren;

(3) Unabhängigkeit von aktiven Maßnahmen nach Verschluss der Lagerstätten;

(4) einen expliziten, gut abgestimmten Zeitplan für die Durchführung eines Entsorgungsprogramms.

(13) Die Entwicklung eines Rahmenprogramms und eines Zeitplans für Deutschland wird empfohlen. Das Programm sollte so gestaltet werden, dass eine realistische Chance für sein Fortbestehen auch über Regierungswechsel hinweg besteht.

Da es in Deutschland insbesondere an einem abgestimmten Rahmenprogramm und Zeitplan mangelt, ist die Entwicklung eines Programms zu empfehlen, das explizit die Prinzipien aufführt, die dem Konzept zugrunde liegen, die Optionen zur Umsetzung dieser Prinzipien vollständig erfasst, einen Ablauf festlegt, der bestimmt, welche Zeiträume der Forschung zur Verfügung gestellt werden müssen, wann zwischen alternativen Optionen eine verbindliche Auswahl getroffen werden muss, und der das Verfahren zur Standortbestimmung im Rahmen der bestehenden rechtlichen Regelungen strukturiert (vgl. hierzu die Entscheidungspfade in Abschnitt A 2.8). Ein solches zeitlich gegliedertes Programm entspräche der vorgeschlagenen EU-Richtlinie zum Umgang mit abgebrannten Brennstoffen und radioaktiven Abfällen.

Ein umfassendes Programm würde einen Anreiz schaffen für die Ausbildung eines systematischen und koordinierten Entsorgungs-Ansatzes. Er könnte zur erhöhten Transparenz des politischen Entscheidungshandelns beitragen, eine Beteiligung der Öffentlichkeit bei der Entscheidungsvorbereitung unterstützen und damit insgesamt zur Verbesserung der faktischen Akzeptanz wie der ethischen Akzeptabilität beitragen.

(14) Es muss klargestellt werden, auf welcher wissenschaftlich-technischen Grundlage, auf welcher Faktenbasis, mit welchem Verfahren und durch welche Institutionen die Entscheidungen getroffen werden sollen.

Für jede im Programm vorzusehende Entscheidung ist offenzulegen, auf welcher wissenschaftlich-technischen Grundlage und Faktenbasis (z. B. Ergebnisse der Standorterkundung, der konzeptionell-technischen Endlagerentwicklung und der Sicherheitsbewertung, Inhalt des Genehmigungsverfahrens), mit welchem Verfahren und durch welchen Entscheidungsträger Entscheidungen getroffen werden sollen.

(15) Das Rahmenprogramm sollte die Entsorgungsstrategie, das Verfahren und den Zeitplan für die Standortauswahl enthalten. Es sollte auch klarstellen, welche Entscheidungen schon getroffen und welche Maßnahmen schon ergriffen worden sind und welche davon ggf. als reversibel zu betrachten sind.

Das Rahmenprogramm sollte explizit die bereits ergriffenen Maßnahmen und getroffenen Entscheidungen aufführen und kennzeichnen, welche davon als potentiell revidierbar zu betrachten sind und unter welchen Bedingungen eine Revision möglich wäre. Nach Einschätzung der Autoren führen die bisher in Deutschland getroffenen Entscheidungen zu wohlbestimmten Mengen radioaktiver Abfälle, die zur Entsorgung anstehen. Die sich ändernde Politik zur Wiederaufbereitung abgebrannter Brennstoffe hat dazu geführt, dass sowohl verglaste hochradioaktive Abfälle aus der Wiederaufarbeitung als auch abgebrannte Brennstoffe auf ihre Entsorgung warten. Die abgebrannten Brennstoffe sind teils auf dem Reaktorgelände, teils in zentralen Anlagen zwischengelagert, in letzteren werden auch die verglasten Abfälle gesammelt. Abgebrannte Brennstoffe werden als Abfall eingestuft und sind damit zur Entsorgung in tiefengeologischen Formationen ohne aktive Kontrollmaßnahmen in der Nachbetriebsphase vorgesehen. Weitere Entsorgungsschritte unterliegen den folgenden politischen Vorgaben:

- den Prinzipien und Forderungen, die in den jüngsten Sicherheitsanforderungen des BMU aufgestellt werden, insbesondere denjenigen, die den einschlusswirksamen Gebirgsbereich, die Maßnahmen zur Bergung sowie die Entwicklungs- und Optimierungsprozesse bei der Inbetriebnahme des Lagers betreffen;
- die Strategie, als Lagerstätten bevorzugt steile Steinsalzformationen in Aussicht zu nehmen.

(16)   Eine Entsorgungsstrategie sollte für mögliche positive Anpassungen offen sein. Dies gilt auch für mögliche Fortschritte in Forschung und Entwicklung.

Es gibt verschiedene Möglichkeiten, bei gegebenen geologischen Bedingungen die Aufnahmekapazität einer Lagerstätte zu beeinflussen. Zu diesen Möglichkeiten gehört die Veränderung der Abbrandrate, der Abklingzeit im Zwischenlager oder des Lagerdesigns zur besseren Ausnutzung des verfügbaren Wirtsgesteins. Eine Entsorgungsstrategie sollte entwickelt werden, ohne dabei die Verfügbarkeit dieser Optionen zu unterstellen, aber für mögliche positive Adaptionen offen gehalten werden. Dies gilt in gleicher Weise für mögliche Fortschritte in Forschung und Entwicklung wie etwa die Abtrennung und Transmutation, die die an ein Endlager zu stellenden Anforderung in Bezug auf Sicherheit und Gefahrenabwehr erheblich reduzieren könnten.

(17)   Das Rahmenprogramm sollte hinsichtlich der Standortauswahl und Errichtung eines Endlagers darauf ausgerichtet sein, die Lösung so schnell wie vernünftigerweise erreichbar herbeizuführen.

Das empfohlene Programm für die Standortauswahl und die Errichtung einer Endlagerstätte sollte darauf ausgerichtet sein, die Lösung so schnell herbeizuführen, wie dies unter Einbeziehung der relevanten technischen, rechtlichen und planerischen Gesichtspunkte und der gesellschaftlichen und politischen Belange vernünftigerweise vertretbar ist.

Dabei sollten auch neue wissenschaftliche Entwicklungen einbezogen werden, die die Anforderungen an ein Endlager ändern könnten. Das Programm sollte für unvorhergesehene Ereignisse offen sein und die Möglichkeit einschließen, dass sich Gorleben als ungeeignet erweisen könnte.

Es sollten genügend Zeitreserven für gerichtliche Entscheidungen, für Partizipation und Kommunikation eingebaut werden. Zeiträume für jeden Abschnitt des Partizipationsverfahrens sollten vorab mit den Beteiligten vereinbart werden, um eine Strategie der endlosen Verzögerung von Entscheidungen durch einzelne Akteure zu erschweren.

Theoretisch besteht immer die Möglichkeit, dass sich ein anderer Standort als (noch) besser erweist. Aus praktischen Gründen jedoch ist eine unbegrenzte Fortsetzung der Standort-Erkundung nicht möglich. Daher kann eine Optimierung der Standortauswahl kaum nach absoluten Maßstäben angestrebt werden, sondern nur im Rahmen einer Planungsweisung – wie auch in seinem klassischen Anwendungsgebiet (dem Strahlenschutz) das Minimierungsgebot nicht absolut gilt.

2.4 Auswahlverfahren, Kriterien

(18) **Die in der Diskussion in Deutschland zu Tage tretenden Konflikte erfordern ein auf Legitimation und Konfliktminderung ausgerichtetes Entscheidungsverfahren, das auch die Berücksichtigung tieferliegender Konflikte ermöglicht.**

Die Wahl von Gorleben als Endlagerstandort ist in der öffentlichen Diskussion höchst umstritten. Zum einen wird vorgetragen, dass diese Vorauswahl überwiegend nach politischen und nicht nach wissenschaftlich-technischen Kriterien vorgenommen sei, zum anderen wird kritisiert, dass eine Eignungsprüfung bei nur einem Standort keine verlässliche Methode darstelle, um einen geeigneten Standort zu finden. Eine Entscheidungsfindung ist höchst konfliktbeladen: eine Eingrenzung der Eignungsprüfung auf Gorleben nährt den Verdacht, dass nicht objektiv nach einem relativ besten Standort gesucht wird, eine Ausdehnung auf andere Standorte wird aller Voraussicht nach dort ebenso weitreichende Proteste auslösen. Dieses Dilemma erscheint nicht kurzfristig auflösbar und erfordert umso dringender ein auf Legitimation und Konfliktminderung ausgerichtetes Entscheidungsverfahren. Entscheidungsprozesse müssen darüber hinaus die tieferliegenden Konflikte um Modernisierung und Lebensstile, energiepolitische Weichenstellung und Fragen der Legitimation für die kollektive Entscheidungsfindung mit berücksichtigen.

(19) **Bei den Entscheidungen müssen höchstmögliche Transparenz und Nachvollziehbarkeit sowie Chancen der direkten Beteiligung der Öffentlichkeit ohne Einschränkung der Verantwortlichkeit des Staates gewährleistet werden.**

Im Rahmen der Entscheidungsfindung und des Auswahlverfahrens ist vor allem auf eine höchstmögliche Transparenz und Nachvollziehbarkeit der einzelnen Verfahrensschritte und die konstruktive Beteiligung der Zivilgesellschaft zu achten, ohne dabei die klare Verantwortlichkeit der durch das Grundgesetz und Atomgesetz legitimierten Institutionen im Rahmen des repräsentativen und föderalen Systems der Bundesrepublik Deutschland einzuschränken. Die Einbindung strukturierter Formen der Beteiligung gesellschaftlicher Gruppen und der lokalen Bevölkerung an der Willensbildung und Entscheidungsvorbereitung steht nicht im Widerspruch zu der repräsentativen Demokratie, sondern kann sie ergänzen und bei geeigneten Vorgehensweisen bereichern.

(20)   Es sollten verschiedene Optionen für das Verfahren der Standortsuche diskutiert und gegeneinander abgewogen werden.

Prinzipiell kommen folgende Möglichkeiten der Problemlösung, insbesondere der Standortbestimmung in Frage:

(a) Auslotung einer gemeinsamen europäischen Lösung,
(b) Fortsetzung der untertägigen Untersuchungen und Überprüfung des Standortes Gorleben nach anerkannten Zielvorgaben für die Eignungsfähigkeit,
(c) wie (b), aber parallel dazu sofortige Ermittlung und Erkundung weiterer Standorte und
(d) Aufnahme eines völlig neuen Suchverfahrens, um mehrere möglicherweise geeignete Standorte (unter Einschluss oder Ausschluss von Gorleben) parallel zu erkunden und relativ zueinander zu bewerten (vgl. ergänzend die Darstellungen im Abschnitt A 2.8).

(21)   Eine gemeinsame europäische Lösung ist nur dann glaubhaft, wenn Deutschland auch einen möglichen Standort einbringt.

Eine gemeinsame europäische Lösung ist nur dann glaubhaft und fair zu vertreten, wenn Deutschland auch eine Standortoption im eigenen Land einbringt. Insofern würde auch diese Lösung eine Standortwahl in Deutschland voraussetzen. Damit reduziert sich die Zahl der zu erwägenden Möglichkeiten auf drei.

(22)   Es wäre nicht sinnvoll, auf der Standorterkundung in Gorleben ohne Prüfung von Alternativen zu bestehen.

Nachdem die deutsche Bundesregierung bereits entschieden hat, die unterirdische Erkundung von Gorleben ohne die Erwägung alternativer Standorte wieder aufzunehmen, scheinen Vorschläge zur Initiierung einer neuen Suche politisch fragwürdig. Gleichwohl erscheint es mit Blick auf die verbreiteten Akzeptanzmängel sowie dem Anspruch, einen für Deutschland besonders geeigneten Standort zu finden, sinnvoll, weitere Standortoptionen mit in die Standortsuche zu integrieren. Zudem haben die Überlegungen zur räumlichen und regionalen Entwicklung, die die Auswahl des Standortes Gorleben Ende der 1970er Jahre bestimmt haben, inzwischen viel von ihrer Überzeugungskraft verloren. Damit stellt sich die Frage einer raumplanerisch angemessenen und umweltgerechten Lösung erneut. Daher sollte die Wahl zwischen den drei genannten Möglichkeiten der Problemlösung (b), (c) und (d) neu eröffnet und unter breiter Beteiligung der Öffentlichkeit erörtert werden. Erst im Anschluss an eine organisierte nationale Debatte, die auch um Formen der Internet-Partizipation ergänzt werden könnte, sollte die Frage, welche der Optionen zu ergreifen ist, durch die Bundesregierung neu entschieden werden.

(23) Die Projektgruppe empfiehlt einen Hybridansatz, der die Weitererkundung des Standorts Gorleben bei gleichzeitiger Untersuchung alternativer Standorte von Übertage vorsieht.

Die Arbeitsgruppe empfiehlt einen Hybridansatz („Gorleben plus"), der folgende Komponenten einschließt:

(a) Weitererkundung des Standorts Gorleben, Entwicklung und Bewertung von Entsorgungsstrategien für diesen Standort mit dem Ziel abzuklären, ob er sich als geeignet erweist und den Sicherheitsanforderungen und planerischen Erfordernissen genügt.

(b) gleichzeitig: Untersuchung alternativer Standorte als zusätzliche Entsorgungs-Optionen und Beginn von Erkundungen dieser Standorte von Übertage, insbesondere unter Aufnahme der Vorschläge des „Arbeitskreis Auswahlverfahren Endlagerstandorte" (AKEnd) und der durch die Bundesanstalt für Geowissenschaften und Rohstoffe als aussichtsreich eingestuften Regionen und Standorte.

(c) ein präziser bestimmter Zeitpunkt für die prinzipielle Entscheidung, ob und ggf. wie die Erkundung am Standort Gorleben und/oder an den alternativen Standorten fortgesetzt werden soll; die Entscheidung sollte auf eine Weise vorbereitet und getroffen werden, die sie möglichst robust macht gegen politische Wechsel.

(d) ein Konzept, nach dem alle zuvor ermittelten Ergebnisse und Argumente als auch die wissenschaftlich-technischen Rahmenbedingungen an alle Akteure kommuniziert werden können, damit sie sich angemessen an der Debatte und – soweit wie es die rechtlichen Bestimmungen vorsehen – am Entscheidungsprozess beteiligen können.

(24) Die Empfehlung, den Standort Gorleben zu erkunden und zusätzliche Standort-Optionen zu bestimmen, zielt auf die Maximierung der Chancen, in den nächsten Jahrzehnten über ein Endlager verfügen zu können.

Die Empfehlung, sowohl den Standort Gorleben weiter zu erkunden als auch zusätzliche Entsorgungs-Optionen zu bestimmen, zielt vor allem auf die Maximierung der Chancen, in den nächsten Jahrzehnten über ein Endlager für radioaktive Abfälle verfügen zu können. Gorleben nicht zu berücksichtigen würde zum Verlust einer möglichen Standortoption führen, ohne dass gewährleistet ist, dass innerhalb einer angemessenen Zeit ein alternativer Standort gefunden werden kann. Auf alternative Standortoptionen bei der laufenden Standortsuche zu verzichten, würde zu einem erheblichen Zeitverlust (in der Größenordnung einiger Jahrzehnte) führen, falls Gorleben in einem späten Stadium des Erkundungsprojekts scheitert. Die zeitliche Staffelung der Erkundung ist sicherheitstechnisch durchaus vertretbar, wenn die Zwischenlager entsprechend nachgerüstet werden (vgl. ergänzend die Darstellungen im Abschnitt A 2.8).

(25)   **Die parallele Suche spart im Falle des Scheiterns von Gorleben nicht nur Zeit ein, sondern erhöht auch die Glaubwürdigkeit des Auswahlverfahrens.**

Die parallele Suche erhöht auch die Glaubwürdigkeit des Prüfverfahrens insgesamt. Gleichwohl ist einzuräumen, dass – insbesondere auf der lokalen und regionalen Ebene – auch dann die öffentliche Akzeptanz noch zurücksteht hinter derjenigen, die für ein vollständig offenes Suchverfahren zu erwarten ist. Dennoch stellt „Gorleben plus" eine Lösung in Aussicht, die zeitnah und ressourcenschonend ist, die für die Meinung der Öffentlichkeit und das Engagement der Stakeholder sensibel ist.

(26)   **Die Festlegung der Kriterien für die Standorte sollte in einem transparenten Verfahren unter Einbeziehung eines internationalen „Peer Reviews" vorgenommen werden.**

Die Festlegung der Kriterien, um die Eignung oder Nicht-Eignung der Standorte festzustellen, wird in einem transparenten Verfahren unter Einbeziehung eines internationalen „Peer Reviews" vorgenommen. Dabei werden in Analogie zum Sachplan in dem Schweizer Verfahren zur Bestimmung eines Standortes sicherheitstechnische, geologische und raumplanerische Kriterien parallel entwickelt. Kriterien und Schwellenwerte müssen so definiert werden, dass sie nicht nur auf Gorleben, sondern auf alle Standorte anwendbar sind.

(27)   **Es sollte festgelegt werden, dass Erkundungen untertage an einem anderen Standort unternommen werden, wenn die oberirdischen Erkundungen erwarten lassen, dass dieser Standort die Auswahlkriterien eindeutig besser als Gorleben erfüllen könnte.**

Aus pragmatischen Gründen sollte das Schwergewicht darauf gelegt werden, einen Standort zu finden, der nach entwickelten Kriterien geeignet ist, hochradioaktiven Abfall sicher einzulagern. Optimierungsgesichtspunkte sind insbesondere bei der Prüfung der Reservestandorte zur Geltung zu bringen. Es sollten Vorkehrungen getroffen werden, dass unterirdische Erkundungen dann an einem der anderen Standorte unternommen werden, wenn und falls die Erkundung von Übertage zu dem Ergebnis kommt, dass dieser alternative Standort (mit Salz oder einem anderen Wirtsgestein) die Auswahlkriterien nachweislich besser erfüllt als Gorleben.

(28) **Programme für Forschung und Entwicklung sollten bedarfsgerecht gestaltet und regelmäßig überprüft werden, aber hinreichend breit und flexibel angelegt sein, damit etwaige Revisionen von Entscheidungen nicht ausgeschlossen werden.**

Programme für Forschung und Entwicklung sollten in Abstimmung mit den aus dem Ablaufplan des Rahmenprogramms ableitbaren Erfordernissen gestaltet und regelmäßig überprüft werden. Sie sollten darauf ausgerichtet sein, die durch Beschluss bestimmten Programmoptionen zu unterstützen, zugleich aber hinreichend breit angelegt und flexibel, um einer Revision getroffener Entscheidungen nicht im Wege zu stehen. Insbesondere sollte, um alle in Frage kommenden zusätzlichen Optionen offen zu halten, auch die laufende Erforschung der Einlagerung in verfestigtem Ton fortgesetzt werden.

2.5 Transparenz, Risikokommunikation, Partizipation

(29) **Eine umfassende Information der Öffentlichkeit ist zu gewährleisten. Zur Unterstützung der erforderlichen Transparenz sollte ein Informationszentrum eingerichtet werden.**

Für einen erfolgreichen Abschluss des Verfahrens ist Transparenz von besonderer Bedeutung. Eine umfassende Information und Kommunikation mit der Öffentlichkeit und allen Betroffenen über die bei der Ausgestaltung des Endlagers zugrunde gelegten Standards für Sicherheit und Gefahrenabwehr sind unerlässlich. Alle Aktivitäten der Entscheidungsfindung sollen durch ein proaktives und auf Dialog basierendes Kommunikationsprogramm begleitet werden. Noch nicht festgelegte Personen erhalten Informationsangebote, die notwendig sind, um sich selbst ein fundiertes Urteil für die Entscheidungsfindung bilden zu können. Insbesondere sollten Informationen über die nachfolgenden Aspekte in einer für Laien verständlichen Darstellungsform – auch unterstützt durch ein einzurichtendes Informationszentrum (Kommunikationszentrum) – verfügbar gemacht werden:

- die geologischen Voraussetzungen des Standortes und ihr Zusammenwirken mit der technischen Auslegung des Lagers sowie der verschiedenen Barrieren;
- die anstehenden Verfahrensschritte und ihre zeitliche Abfolge;
- die Referenzdosiswerte (Richt- und Grenzwerte) für die verschiedenen Phasen der Lagerstätte sowie ihre Stellung und Begründung im Gesamtsystem des Strahlenschutzes;
- Modelle und Methoden für die Abschätzung von Strahlendosen für die ausgewählten Freisetzungsszenarien und deren Ergebnisse;
- Forschung und Entwicklung, die in Deutschland und im Ausland unternommen werden, Entsorgungsoptionen, die international diskutiert sowie Strategien, die in anderen Ländern ergriffen wurden und werden.

**(30)   Bei der Standortauswahl sollte eine wirksame Partizipation stattfin-
den, ohne die letztgültige Verantwortung des Parlaments und der Exekutive in Frage zu stellen.**

Das Standortauswahl-Verfahren sollte für eine wirksame Partizipation offen sein, ohne die letztgültige Verantwortung der Parlamente und der Exekutive in Frage zu stellen. Um den Prozess nicht durch Überkomplexität
zu lähmen, sollten übermäßig schwerfällige Verfahrensordnungen vermie-
den werden, etwa ein auf verschiedene Ebenen gestaffelter Diskurs und eine
Vielzahl formeller Institutionen.

**(31)   Über Fragen von nationaler Bedeutung sollte eine organisierte öf-
fentliche Debatte vorgesehen werden.**

Auf nationaler Ebene sollte, etwa durch die Entsorgungs- und die Strahlenschutzkommission oder das Bundesamt für Strahlenschutz, eine öffentliche
Debatte über die Fragen von nationaler Bedeutung initiiert und organisiert
werden, ergänzt um neue Formen der Partizipation. Gegenstände der Debatte sollten v. a. die Entscheidung über Optionen des Auswahlverfahrens,
die Kriterien für die Standortauswahl, die Auswahl für weitere (untertägige)
Charakterisierung des Standortes sowie Fragen der Rückholbarkeit sein.

**(32)   Eine weitergehende Mitwirkung von lokalen Vertretern der Zivilge-
sellschaft ist erst dann sinnvoll, wenn Fragen von lokaler Bedeutung
zur Diskussion stehen.**

Eine weitergehende Mitwirkung von lokalen Gruppen der Zivilgesellschaft
ist erst dann sinnvoll, wenn Fragen von lokaler Bedeutung in Frage stehen.
Dies gilt für die fortlaufende Bewertung der Untersuchungsergebnisse für
Gorleben, für die Bewertung der Ergebnisse der Erkundung alternativer
Standorte von Übertage und für die Entscheidung, ob Gorleben als Standort geeignet ist oder nicht. Im Falle einer Entscheidung für Gorleben sollte
auch bei der lokalen und regionalen Umsetzung (hinsichtlich des „wie")
eine Bürgerbeteiligung stattfinden. Die Vorschläge, die der AKEnd dazu
gemacht hat, könnten hier aufgegriffen werden.

**(33)   Bei der parallel zur Erkundung des Standortes Gorleben erfolgenden
Standortauswahl sollten freiwillige alternative Kandidaten, die geeignet erscheinen, vorrangig betrachtet werden.**

In Analogie zu dem in Finnland und Schweden gewählten Verfahren wäre
die Einbeziehung potentieller freiwillig optierenden Standortgemeinden
oder regionen wünschenswert. Daher sollten im Standortauswahlverfahren, das parallel zur weiteren Erkundung des Standorts Gorleben durchgeführt werden sollte, freiwillig optierende alternative Kandidaten, sofern sie
die für den jeweiligen Stand des Verfahrens anzulegenden Kriterien erfüllen, vorrangig betrachtet werden.

2.6   Institutionen im Verfahren, Expertengruppen

(34)   Es sollte ein Gremium von Experten mit ausgewiesener wissenschaftlicher bzw. technischer Kompetenz gebildet werden.

Die Arbeitsgruppe empfiehlt, eine Expertengruppe einzuberufen, deren Aufgabe es v. a. ist, die Kriterien für die Eignungsprüfung aufzustellen und am Ende der Erkundung eine Empfehlung auszusprechen, ob die Kriterien erfüllt sind oder nicht. Da auf diesem Gremium ein hohes Maß an Verantwortung ruht und es in einer so konfliktreichen Situation auf hohe Legitimation angewiesen sein wird, ist auf ausgewiesene wissenschaftliche bzw. technische Kompetenz, Unabhängigkeit und Ausgewogenheit zu achten. Um dies zu gewährleisten, empfiehlt die Arbeitsgruppe:

- Das Gremium soll bei einer unabhängigen Institution, z. B. bei der Nationalen Akademie der Wissenschaften, angesiedelt sein.
- Mitglieder können von den einschlägigen Wissenschaftsorganisationen, von zivilgesellschaftlichen Gruppen und den Betreibern vorgeschlagen werden.
- Die Sitzungen des Gremiums sind öffentlich.
- Das Gremium erhält das Recht und die notwendigen Ressourcen, Gutachten einzuholen sowie Hearings abzuhalten, wenn es dies für notwendig ansieht.
- Die Empfehlungen des Gremiums sind nicht bindend. Allerdings müsste der zuständige Entscheidungsträger schon stichhaltige und öffentlich nachvollziehbare Argumente vorweisen können, um von dem Votum des Gremiums abzuweichen.

(35)   Auf lokaler Ebene sollte für jeden zur Wahl stehenden Standortkandidaten ein Dialogforum als institutionelle Anlaufstelle für die regionale Bevölkerung eingerichtet werden.

Auf lokaler Ebene sollte als eine ergänzende Institution der öffentlichen Partizipation für jeden noch zur Wahl stehenden Standortkandidaten (also auch für Gorleben) ein Dialogforum eingerichtet werden. Die lokalen Dialogforen sollen als institutionelle Anlaufstelle die wirksame Beteiligung der regionalen Bevölkerung erleichtern und in den verschiedenen Phasen des Auswahlverfahrens den Zugang zu unabhängigem Sachstandswissen ermöglichen. Hierzu könnten die Foren öffentliche Debatten anregen, öffentliche Diskussionsrunden organisieren und den Zugang zu Expertenpools ermöglichen.

2.7   Behördenorganisation

(36)   Die Verteilung der Zuständigkeiten der mit der Entsorgung hochradioaktiver Abfälle befassten Behörden und sonstigen Organisationen sollte verbessert werden.

Eine verbesserte Verteilung der Zuständigkeiten zwischen allen damit befassten Organisationen würde die Chancen zur Durchführung des vorgeschlagenen Programms zur Entsorgung radioaktiver Abfälle erheblich erhöhen. Dies würde sowohl dem Gemeinsamen Abkommen als auch der vorgeschlagenen EU-Richtlinie zur Entsorgung radioaktiver Abfälle entsprechen. Eine klare und deutlich wahrnehmbare Trennung und die Einhaltung des erforderlichen organisatorischen Abstandes zwischen Antragsteller und Regulierungsbehörden dürften zu einer Verbesserung der Akzeptanz sowie zu einem höheren Vertrauen in das Verfahren und die Verfahrensbeteiligten beitragen. Die unterstützende Forschung sollte in einer Weise organisiert sein, die die Wirksamkeit dieser veränderten Strukturen maximal fördert. In dieser Hinsicht dürfte sich die gegenwärtig etablierte Trennung zwischen standortspezifischer Forschung und Grundlagenforschung als unangemessen erweisen.

(37)   **Als Teil der Reform des institutionellen Rahmens sollte das Bundesamt für Strahlenschutz in eine unabhängige Regulierungsbehörde umgewandelt werden.**

Ein zentrales Anliegen der angeregten Reformen des institutionellen Rahmens für die Regulierung und die Durchführung der Entsorgung hochradioaktiver Abfälle ist die Überführung des Bundesamtes für Strahlenschutz in eine tatsächlich unabhängige Regulierungsbehörde, die zentrale Aufgaben im Bereich der Entsorgung radioaktiver Abfälle, insbesondere für die Regulierung, Standortwahl und Aufsicht übernehmen kann. Zu diesem Zweck sollte die gegenwärtige Vermischung ihrer verschiedenen Aufgaben in der Regulierung und Durchführung zugunsten einer klaren Trennung ihrer Funktionen aufgegeben werden.

(38)   **Der Betrieb des Endlagers könnte einer neuen Bundesoberbehörde, einer Körperschaft des öffentlichen Rechts oder einer privatrechtlichen Körperschaft, die mit entsprechenden hoheitlichen Befugnissen ausgestattet ist, übertragen werden.**

Die Verwaltung des Endlagers könnte einer neuen Agentur auf Bundesebene, oder, in dem vom Atomgesetzes vorgesehenen Rahmen, einer Körperschaft des öffentlichen Rechts oder einer privatrechtlichen Körperschaft, die mit entsprechenden hoheitlichen Befugnissen ausgestattet ist, übertragen werden. In einer solchen Körperschaft könnten die Unternehmen, die direkt oder indirekt (über Tochtergesellschaften oder Joint Ventures) die Kernkraftwerke in Deutschland betreiben, und – im Interesse des öffentlichen Vertrauens – die Bundesregierung gemeinsam Anteilseigner werden. Die Anteilsmehrheit sollte dabei in jedem Falle von der öffentlichen Hand gehalten werden.

2.8 Entscheidungs-Diagramme

2.8.1 Vorbemerkungen

Die nachfolgenden Entscheidungsdiagramme sollen die Rahmenbedingungen für die Entwicklung eines Entsorgungsprogramms (vgl. die Schlussfolgerungen und Empfehlungen in Abschnitt A.2.3) graphisch illustrieren. Die Diagramme sollen eine Übersicht geben (i) über Entscheidungen, die in der Vergangenheit schon getroffen wurden und über deren Beibehaltung oder Revision heute nachgedacht werden darf (kann), (ii) über die Optionen, die für künftige Entscheidungen offen stehen. Die in die Diagramme eingetragenen zeitlichen Angaben beruhen auf den in Abschnitt B.1.5 ausführlich dargelegten und begründeten Abschätzungen.

Die Entscheidungsdiagramme sind abgeleitet worden von Überlegungen, die im Rahmen des EU COMPAS-Projekts (Dutton et al. 2004) entwickelt wurden. Sie beziehen jedoch Aspekte mit ein, die für die Situation in Deutschland spezifisch sind. Obwohl die Logik der COMPAS-Studie grundsätzlich übernommen wurde, wurde eine wesentliche Abweichung hinsichtlich der Einkapselung eingeführt. Die Einkapselung der Abfälle kann nur dann auf angemessene Weise erfolgen, wenn die Entsorgungs-Lösung bereits bekannt ist.

Die Entscheidungs-Diagramme sind auf die Behandlung abgebrannter Kernbrennstoffe zugeschnitten. Dabei wurden folgende Annahmen bezüglich des Deutschen Entsorgungsprogramms vorausgesetzt:

- Es werden keine weiteren Kernkraftwerke in Deutschland gebaut. Wie in Teil B dieses Buches dargelegt, werden die Laufzeiten der bestehenden Anlagen die Voraussetzungen für die anstehenden Entscheidungen nicht wesentlich verändern.
- Die sichere Entsorgung radioaktiver Abfälle in tiefengeologischen Schichten ist prinzipiell technisch machbar.

Entscheidungen, die bereits in der Vergangenheit getroffen wurden und/ oder durch technische Erfordernisse determiniert sind, sind durch dunkelgraue Flächen gekennzeichnet. Fette Linien zeigen die in der vorliegenden Studie gegebenen Empfehlungen an – welche früher schon getroffenen Entscheidungen besser nicht revidiert und welche künftigen Entscheidungen getroffen werden sollten, wenn das Ziel darin besteht, die Chancen für ein Endlager in den nächsten Jahrzehnten zu maximieren.

Einige der im Diagramm I (Fig. A.1) dargestellten Entscheidungen werden nachfolgend eingehender erläutert. Die Nummern beziehen sich dabei auf diejenigen in den Entscheidungsknoten des Diagramms.

⟨1⟩ Gemäß dem Atomgesetz finden seit 2005 keine weiteren Transporte zu Wiederaufarbeitungsanlagen in Frankreich und in Großbritannien statt. Entsprechend ist die Entscheidung zugunsten von „keine Wiederaufarbeitung" als bereits in der Vergangenheit getroffen gekennzeichnet. Allerdings sind Abfälle aus früherer Wiederaufarbeitung vorhanden und ebenfalls zu entsorgen.

⟨2⟩ Die Wiederaufarbeitungs-Anlage in Wackersdorf hat den Betrieb nie aufgenommen. Vor 2005 wurden abgebrannte Brennstoffe (SNF) in französische und britische Wiederaufarbeitungs-Anlagen geliefert. Vor 1989 wurden alle SNF aus Kernkraftwerken der DDR in die Sowjet-Union verbracht.

⟨3⟩ Während alle Abfälle, die aus der Wiederaufarbeitung in französischen oder britischen Anlagen nach Deutschland zurückkommen, werden keine Abfälle aus der Wiederaufarbeitung der SNF, die in der DDR angefallen sind, nach Deutschland zurückkehren. Verbleibende Brennstoffe aus ostdeutschen Beständen werden im Zwischenlager Nord nahe Lubmin/Greifswald gelagert. Internationale Verträge sehen vor, dass Brennstäbe aus Forschungsreaktoren der DDR an Russland übergeben werden.

⟨4⟩ Gegenwärtig werden SNF in Deutschland lokal und nahe den Reaktoranlagen zwischengelagert. In der Vergangenheit wurden allerdings schon einige SNF in die zentralen Zwischenlager von Ahaus und Gorleben verbracht und dort eingelagert. In Gorleben werden auch die verglasten hochradioaktiven Abfälle aus der Wiederaufarbeitung gelagert und es ist vorgesehen, die verbleibenden Abfälle aus der Wiederaufarbeitung ebenfalls dort zu sammeln. Alle Zwischenlager haben eine Genehmigung für vierzig Jahre, beginnend mit dem Aufstellen des ersten Behälters. Diese Genehmigungen sind an die Laufzeiten für die Castor-Behälter geknüpft. Eine Verlängerung der Betriebszeiten für Zwischenlager würde daher einer neuen Genehmigung bedürfen.

⟨5⟩ Hier wird lediglich unterschieden zwischen Entsorgungs-Optionen, die ihrer Art nach prinzipiell keine Rückholung der Abfälle erlauben und solchen, die einer späteren Rückholung der Abfälle prinzipiell nicht entgegenstehen (unabhängig davon, ob in der konkreten Umsetzung dann die Rückholbarkeit vorgesehen wird oder nicht). Das Entscheidungsdiagramm sieht lediglich die *Option* der Rückholbarkeit vor; Entscheidungen, die tatsächlich eine Rückholbarkeit vorsehen, werden dagegen nicht betrachtet.

⟨6⟩ De facto ist in Deutschland eine Entscheidung zugunsten einer Entsorgung in tiefen geologischen Anlagen getroffen. Dies steht im Einklang mit den hier gegebenen Empfehlungen.

⟨7⟩ Zu den Notwendigkeiten, die mit einer internationalen Lösung einhergehen, vgl. Schlussfolgerungen/Empfehlungen § 21: Eine gemeinsame europäische Lösung ist nur dann glaubhaft, wenn Deutschland auch einen möglichen Standort einbringt.

⟨8⟩ Die Entscheidung für ein Wirtsgestein ist zu treffen mit Blick auf eine Reihe von Rahmenbedingungen und Kriterien, einschließlich der Verfügbarkeit in geologisch stabilen Regionen und Sicherheits-Analysen, die auf dem gesamten Entsorgungskonzept (einschließlich technischer Komponenten) basieren. Besonders hinzuweisen ist auf eine wechselseitige Abhängigkeit von und mit Entscheidung á9ñ über die Rückholbarkeit. Die fett gezeichnete Pfeillinie deutet die Präferenz für Salz als Wirtsgestein in Deutschland an, die sich nicht nur in den Erkundungen in Gorleben, sondern auch in zahleichen anderen Forschungs- und Entwicklungs-Aktivitäten manifestiert.

Der Vorschlag des AKEnd zielt auf ein Standortauswahl-Verfahren ohne Vorentscheidung für ein Wirtsgestein. Gleichwohl ist der Begriff eines „einschlusswirksamen Gebirgsbereichs", wie er zunächst durch das AKEnd geprägt wurde und der nun Eingang gefunden hat in die „Sicherheitsanforderungen an die Endlagerung wärmeentwickelnder radioaktiver Abfälle" des BMU (2010a), geht mit einer Vorauswahl für Salz und Ton als Wirtsgestein einher, da es als sehr unwahrscheinlich gelten muss, dass kristalline Formationen in Deutschland den Anforderungen an einen solchen „einschlusswirksamen Gebirgsbereichs" entsprechen können.

(9) Die „Sicherheitsanforderungen an die Endlagerung wärmeentwickelnder radioaktiver Abfälle" des BMU (2010a) fordern, dass während der gesamten Betriebsphase die Rückholbarkeit gewährleistet und die „Handhabbarkeit der Abfallbehälter bei einer eventuellen Bergung aus dem stillgelegten und verschlossenen Endlager für einen Zeitraum von 500 Jahren" gegeben bleibt (§ 8.6). Es ist damit zu rechnen, dass zukünftige Diskussionen über die Entsorgung radioaktiver Abfälle diese Frage erneut eröffnen werden. Hinzuweisen ist in diesem Zusammenhang auf die Interdependenz dieser Aspekte mit der Wahl des Wirtsgesteins.

⟨10⟩ Entscheidungen, die in näherer Zukunft anstehen. Die fetten Pfeillinien deuten die durch mehrere Gründe bestimmte Präferenz der Arbeitsgruppe für einen Hybridansatz an („Gorleben plus", vgl. die Schlussfolgerungen und Empfehlungen, §§ 23, 24, 25). Die nachfolgende Übersicht des Diagramms soll deutlich machen, wann und in Abhängigkeit von welchen Entwicklungen wesentliche Prozessziele erreicht werden können. Die erwartbare zeitliche Entwicklung ist auch durch die vertikalen Abstände der Figuren angedeutet.

Diagramm II (Fig. A.2) gibt einen ausführlicheren Überblick über die Phasen und die möglichen Resultate des Standortauswahl-Verfahrens im Rahmen des von der Arbeitsgruppe empfohlenen Hybridansatzes „Gorleben plus".

2.8.2　Entscheidungsdiagramm I

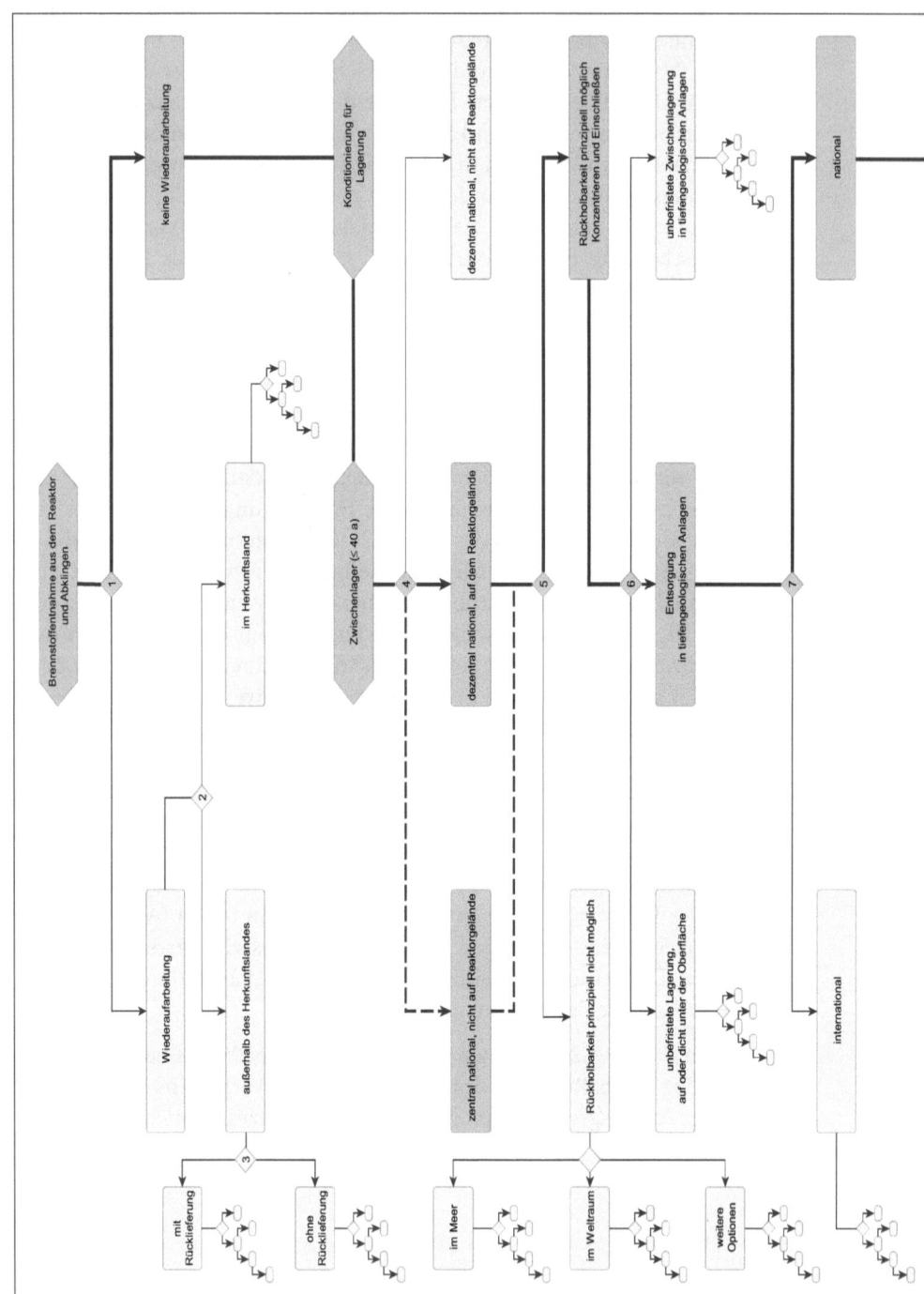

Abb. A.1:　Entscheidungsdiagramm I: Rahmenbedingungen für die Entwicklung eines Entsorgungsprogramms (vgl. die Schlussfolgerungen und Empfehlungen in Abschnitt A.2.3)

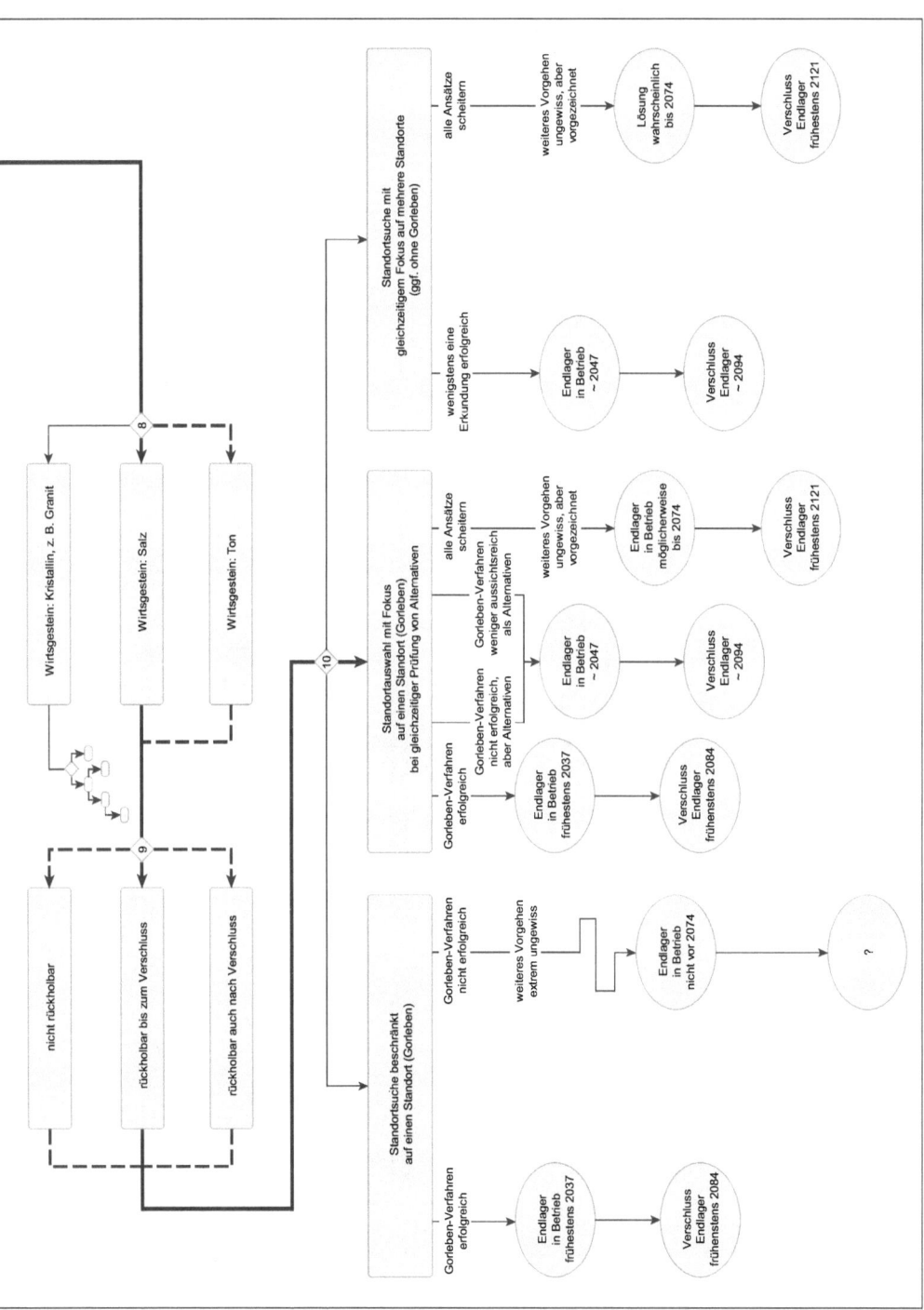

2.8.3   Entscheidungsdiagramm II

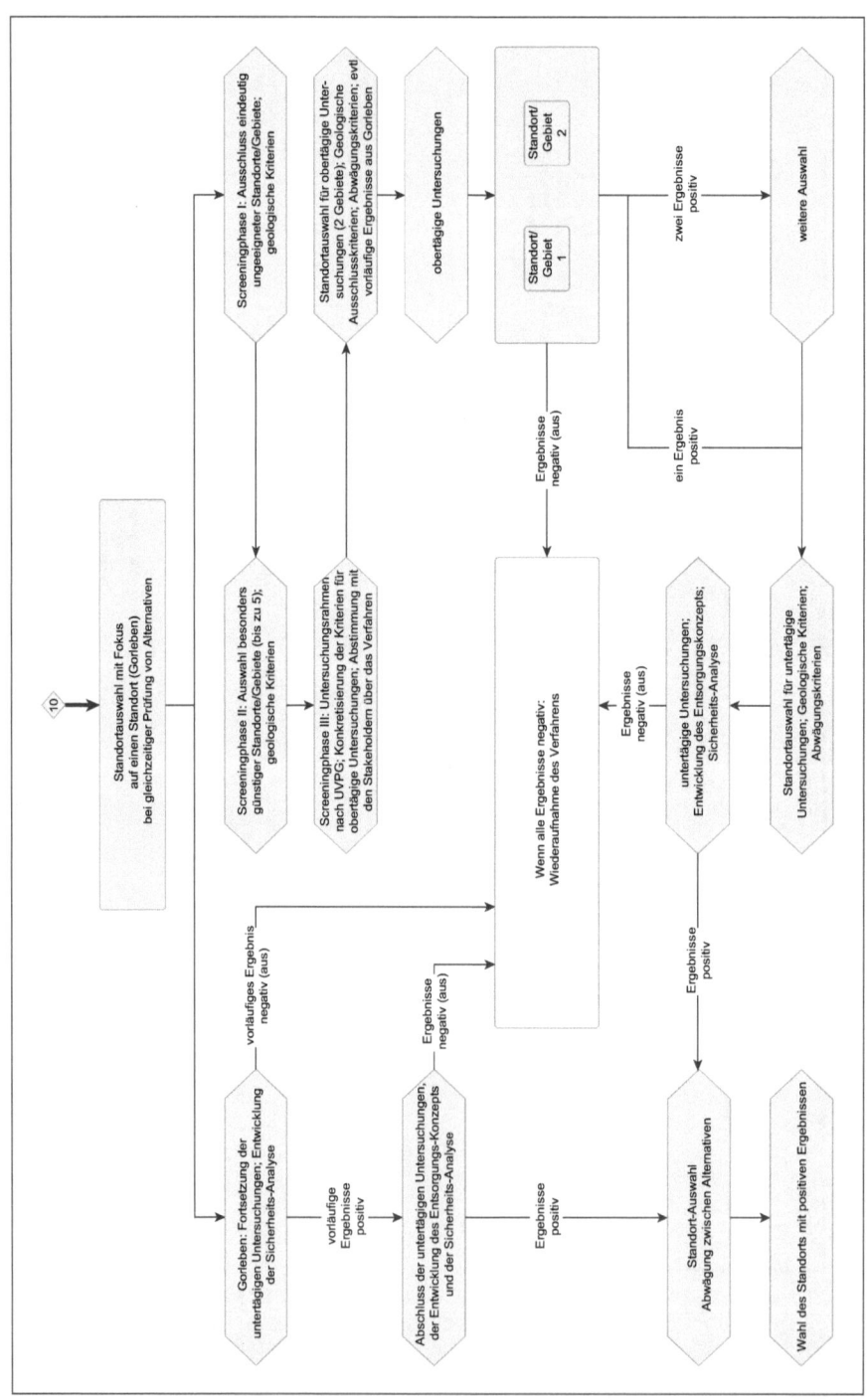

Abb. A.2:   Diagramm II: Phasen und mögliche Resultate des Standortauswahl-Verfahrens im Rahmen des Ansatzes „Gorleben plus" (vgl. hierzu section B 4.5.5)

Introduction

The threat posed to humans and nature by radioactive material is particularly a result of the ionizing radiation released during the radioactive decay of this material. It is therefore necessary to safely store radioactive waste accumulated in research, medicine and technology (particularly high level waste in nuclear facilities). The decisive factor determining the duration of the hazardous state is the characteristic physical half-life time in which the various radionuclides decay, ranging from fractions of seconds to many millions of years. Fuel elements taken from the reactors of nuclear power plants contain radionuclides with very long half-lives. It has therefore been an accepted fact worldwide, since power plants of this type have been in operation, that the radioactive material needs to be confined in isolation from the biosphere, the habitat of humans and all other organisms, for very long periods of time. As waste is also present in Germany, it is seen as necessary to create a final disposal site here too.

The concepts for final disposal which have been discussed most are near-surface disposal, disposal in deep geological formations, and deep sea or sub-seabed disposal (SSK 1987). Internationally, other concepts have also been discussed (e.g. long-term or indefinite storage, disposal in outer space, CoRWM 2006), the majority have, however, chosen disposal in deep geological formations. This is also the case in Germany. Of vital importance for the long-term safety of these disposal facilities is the host rock, which must prevent radioactive materials from escaping from the disposal facility into the biosphere in any significant quantity. Internationally, the host rocks under discussion are crystalline formations (e.g. granite), salt domes and clay. In Germany disposal in salt formations was proposed as early as the 1960s.

In 1977 the German federal government commissioned the national metrology institute, the Physikalisch-Technische Bundesanstalt (PTB) to begin the planning approval procedure for the disposal of radioactive waste in the Gorleben salt dome (Tiggemann 2004).

Originally the disposal facility was intended to be part of a nuclear waste management centre with a reprocessing facility. To clarify the connected safety issues and risks of the waste management centre, and thus of a disposal facility, a "Symposium of the Government of Lower Saxony on the fundamental safety-related feasibility of a nuclear waste management centre" (the "Gorleben Hearing" was held in Hanover from 28 March to 3 April

1979. After matter-of-fact discussions chaired by the physicist Carl Friedrich Freiherr von Weizsäcker, the following conclusion (amongst others)
was reached by Ernst Albrecht, the Minister President of Lower Saxony at
the time, who hosted and took part in the symposium. He presented his
findings to the Parliament of Lower Saxony on 16 May 1979: "Although a
nuclear waste management centre [...] is in principle feasible in terms of
safety, the Government of Lower Saxony recommends that the Federal
Government not pursue the reprocessing project. Instead a new waste management concept should be resolved upon immediately, along the following lines:

(1) Immediate establishment of inherently safe long-term storage sites to
    manage waste from the nuclear power plants [...],
(2) Promotion of research and development on the safe disposal of radioac-
    tive waste,
(3) Deep drilling in the salt dome and, if the results are positive, mining ex-
    ploration of the salt dome in Gorleben; if the drilling should have a neg-
    ative outcome, exploration of other disposal sites; because we do need
    disposal sites."

The exploration programme in Gorleben began in 1979/1980 – first from
the surface, and then, from 1986 on, underground. The programme was
interrupted by a moratorium of a maximum of 10 years in 2000. In this
period numerous Castor shipments of radioactive waste were made to Gorleben, and deposited in an interim storage site there. The explorations in
Gorleben and the shipments of nuclear waste through Germany led to emotionally charged debates, and to extensive, heated, sometimes militant
demonstrations by critics of nuclear energy in general and of the Gorleben disposal site in particular. The Federal Government's decision, in October 2010, to resume the exploration of the Gorleben salt dome reignited
and strengthened resistance to the nuclear disposal facility. The result is
a highly charged atmosphere which makes matter-of-fact dialogue nearly
impossible.

    In a parallel development in this period, national and international sci-
entific committees have examined safety issues relating to disposal facilities
for radioactive waste, and established corresponding criteria. Thus in June
1985 the German Commission on Radiological Protection (Strahlenschutzkommission, SSK) approved a recommendation about "aspects of radiological protection when disposing of radioactive waste in geological formations". The SSK worked on the principle that future generations have to be
protected from ionizing radiation to the same extent as people today. For
the post-closure phase and thus for long-term safety, it was recommended
that the "potential radiation exposure for individual members of the population after the occurrence of improbable events should not exceed the
amount of the average range of variation of natural exposure to radiation

(effective dose equivalent) in the Federal Republic of Germany" (SSK 1987). This is achieved if the dose constraint of 0.3 mSv per year is observed. The International Commission on Radiological Protection recommended setting a potential radiation exposure of 0.3 mSv per year (effective dose) as the dose constraint for the post-closure phase (ICRP 1999).

In order to look at long-term safety, periods of several tens of thousands of years were initially taken into consideration. A key argument for such a period was that a new ice age can be expected in our region within this time frame. On the basis of geologists' prognostic statements on the time-scales of possible changes in the sites of disposal facilities and on the possible migration of material through the geological barriers, assessments of the long-term safety of disposal facilities for radioactive waste have now been expanded to a range between hundreds of thousands and one million years (SSK 2008; ICRP 2011; BMU 2010). More recently a dose constraint of 0.1 mSv per year has been proposed for potential radiation exposure in the post-closure phase (SSK 2008). The various models of disposal facilities for high level waste in deep geological formations with various overlaying rocks, and the experiences with deep geological disposal facilities for chemotoxic materials offer strong evidence for the feasibility of such disposal sites, even with the strict criteria stipulated.

In Germany and other countries deep geological disposal facilities for chemotoxic waste material are already in operation today. They are largely accepted, despite the absence of comparably elaborate safety cases, and the fact that this material possesses "perpetual" chemical stability and can pose a greater potential threat to health than radioactive waste. Nonetheless, it has not yet been possible to achieve social and political acceptance for a consensual concept for the disposal of radioactive waste material. This is particularly the case for Germany, but also for many other countries worldwide. It has been possible in Finland and Sweden, however, to locate sites for such disposal facilities (Finland: Olkiluoto, Sweden: Forsmark, Östhammar municipality). France and Switzerland are evidently progressing well towards this goal, but there has been a serious drawback in the USA.

The reasons for the lack of acceptance for disposal facilities for high level waste are complex. They lie partly in the special perception of the danger of ionizing radiation and thus of radioactive material in the waste. Although humans are permanently exposed to this radiation from natural sources, and incorporate naturally occurring radioactive materials every day with their food, their water, and the air they breathe, and although nearly all humans in industrialized countries are exposed to ionizing radiation in medicine, for diagnostic or therapeutic reasons, and the possible health risks can be reasonably well estimated, many people have serious reservations when it comes to being exposed to ionizing radiation from technical facilities – even if the radiation doses are low. In the case of nuclear power plants there is no doubt that in this context the consequences of possible major accidents play a significant role in the non-acceptance of their hazard potential. The disposal of radioactive waste does not involve any danger of explosions, meltdowns or other sudden accidents, and yet the lack of acceptance of nuclear power impacts on the acceptance of nuclear waste disposal.

It is very obvious that the issue of the appropriate management of radioactive waste is attended by considerable potential for conflict. Yet when it comes to this topic the emotionally charged positions and the passion with which arguments are put forward are only an external indicator. Often the discussion about the nuclear disposal facility masks deeper social debates about the significance of technological developments for the future shape of the economy, energy production and life in society. The acrimony with which this conflict is carried on in public is also doubtlessly characteristic of the debate and is worthy of more careful attention, especially if one is interested not just in the theoretical development of problem-solving strategies, but also in the practical resolution of the conflict. If a decision is to be brought about through the answering of factual questions, the social dimensions of the conflict will also have to be analysed in order to discover how the current polarized positions can be converted into a constructive discourse about rational strategies, i.e. ones which can be understood and tolerated by all parties. This is the only possible way to achieve an effectively legitimized decision on the site issue and modalities of a final disposal facility for high level waste.

Recent decades have shown that planning and dealing with large-scale technological facilities cannot be successfully carried out purely by solving technical problems. It is desirable, if not necessary, and also appropriate in a democratic system to attain the acceptance or at least the tolerance of the people affected. This idea was already expressed in Minister President Albrecht's concluding remarks on the "Gorleben Hearing" in 1979. He said: "[...] that the formation of opinions must occur in a democratic process, that this is not about technocratic decisions, but ultimately about democratic decisions. Democracy, however, requires argumentation, and publicity requires transparency." This plea for information, participation, communication and transparency was clearly not followed closely enough in the years that followed. Minister President Albrecht also said, however, back in 1979, that the final decision "is a typically political decision. The responsibility for this decision lies with the political authorities, and no one can relieve them of it." The need for these sequences of action is elaborated in the present study.

In the decision-making process regarding a disposal facility for high level waste, considerable importance is accorded to the uncertainty of scientific knowledge and models, to possible ambiguity in the evaluation of a condition, and to possible contradictions in the statements of experts. As a consequence, the affected population feels insecure. In Germany this has been reinforced considerably by the events surrounding the "experimental disposal facility" at Asse. It is scarcely possible to explain to the wider public that Asse, being only an "experimental disposal facility", was conceived with quite different standards than the planned geological disposal facility for high level waste will be. Credibility is also undermined when it emerges that there has been misconduct or neglect in the scientific/technical and administrative management of the Asse disposal facility.

Unavoidable uncertainties in knowledge and scientific data, and the way these are presented and dealt with, are without doubt extremely difficult topics and substantial obstacles for the attempt to gain acceptance for disposal facilities of this kind. The very long time spans of the necessary prognoses clearly reinforce these effects considerably. Even in small and everyday technical instruments, the parameters of the system and the external influencing variables are not always fully known. For reasons of principle, this applies even more to singular large-scale technical installations, however thoroughly they may have been examined. The most recent developments in Fukushima, Japan, also show that exceptionally extreme events can occur which cannot be expected even if all available data are used as a basis for planning, but against which precautionary efforts must nonetheless be undertaken because their consequences are so severe.

Clearly this applies particularly to the planning of disposal facilities for high level waste. Statements must be made about long-term safety over periods of time so long that they reach the limits of the human capacity of imagination. Prognostic statements are then only possible by means of model calculations with correspondingly high levels of uncertainty. Reaching an understanding of these matters is no easy task.

Geological processes of evolution occur over considerably longer periods of time than biological or social processes of evolution. These characteristics of possible geological developments and their changes over time mean that prognostic statements about the possible breaching of technical and geological barriers over longer periods of time can be made with sufficiently higher degrees of accuracy than prognoses about how the region of a disposal site will be inhabited over this period, and what the lifestyles of these people will be like. But even such great prognostic uncertainties do not justify ignoring the possible claims of future generations to adequate management of radioactive waste, or dismissing these claims as irrelevant. On the contrary, an adequate waste management strategy must – in keeping with the ethical principle of universalism – give the same consideration to members of future generations as it does to those of the present generation.

The above discussion of the extreme complexity of the conflict is not intended to either accord a special status to the problem of radioactive waste management, or indicate scepticism about the likelihood of resolving this conflict: highly complex controversies, in which the debate over the appropriate or suitable solution to a problem not only throws up questions about the appropriateness and suitability of the problem-solving strategies, but also questions about the appropriateness and suitability of the standards for evaluating these, are established practice in various areas of private and public life. They give rise to the political, judicial and socio-economic institutionalization of conflict resolution. Here procedural rules, neutral evaluation or competition ensure that efficient decisions are made and the capacity to act is maintained, – since responding to ongoing controversy by doing nothing often turns out to be an unfavourable option. The analysis and discussion of highly complex controversies are also established practice in science, however, which can virtually be seen as something brought forth by society for the professionalization, i. e. the delegation, of the systematic and long-lasting engagement with controversies of this type, freed from the pressure to make efficient decisions.

There is no reason, then, to give up in the face of extreme complexity. On the contrary, as long as there is no pressure to reach efficient decisions, the conflict-producing constellations of problems should be registered following the rules of scientific rationality; the questions requiring decisions should be differentiated, analysed and stated more precisely, and the options put in order and examined.

In preparation for a situation in which efficient decision-making will become necessary, care must be given to how the instruments of conflict resolution are chosen or developed so that a lasting resolution of the conflict can be brought about. But the only lasting solution can be one that is accepted or at least tolerated in the long term by a large proportion of those involved. This presupposes that the answers to the technical questions which have influenced the decision do not quickly turn out to be erroneous, and that those involved do not suddenly begin to perceive the process through which the decision came about as "unfair" and disadvantageous to them, and retrospectively deny the legitimacy of the result despite their earlier agreement with the process. Here we should recall the remarks made at the outset about the social dimension of the conflict and the emotionally charged nature of the debates, which may be at least partially due to the fact that not enough attention was paid to this in the early stages, when the "depth" of the conflict could not be foreseen. With this in mind, the present study attempts to point out and propose ways to find a way out of the dilemma of the current situation and devise an acceptable procedure for locating a site and constructing a disposal facility for high level waste.

In part A of the study the texts are available in both English and German. First the key information from the chapters of part B is presented in comprehensive summaries, followed by chapters with conclusions and recommendations. Possible ways to find a site are pointed out.

In part B of the study the strategies and technical concepts for the management of high level waste and its disposal in deep geological foundations are presented in separate chapters, along with the related issues of radiation risk and radiation protection. The aim is to give the reader an overview of the latest state of knowledge. Chapters follow on ethical, normative aspects of long-term responsibility, on legal issues of the disposal of radioactive waste, and on guidelines for a socially sustainable and fair means of selecting a site. In order to make the study accessible to an international readership from a scientific background, these chapters are written in English.

Part C offers supplementary information on the fundamentals of radiobiology and a comparative law overview of the regulations of important states using nuclear energy.

A key conclusion of the study is that the best possible procedure would be for the investigation of Gorleben as a possible site to continue, but with one or two other sites also being considered and surface-based exploration being carried out at these locations. The advantage of this procedure would be that if Gorleben were to fail, little or no time would be lost switching to another site. It can be expected that such an approach would increase acceptance of the site selection process. On the other hand such a procedure would also offer the possibility of submitting an alternative site to a more in-depth, i.e. sub-surface exploration, if the available studies show that the alternative site fulfils the selection criteria much better than Gorleben. In the site-finding process a high level of significance is ascribed to information, communication, participation and transparency. It is made clear, however, that the final decisions must, in accordance with the legal regulations of the Federal Republic of Germany, lie with the responsible institutions.

A Executive summary, conclusions and recommendations

1 Executive summary

1.1 Technical issues of long-term radioactive waste management

A solution to the problem of long-term radioactive waste management (RWM) comprises a technical and social dimension, i.e. it must not only be technically achievable, but also legally and politically feasible and publicly acceptable. The technical solutions have to ensure beyond reasonable doubt safe and secure containment of long-lived highly radioactive waste for the indefinite/distant future and avoidance of undue burdens on future generations. Despite the perceived link between RWM and the controversial debate on nuclear power production the problem of RWM is considered as one which has to be solved no matter which perspectives are foreseen or debated concerning nuclear power production: the additional amount of radioactive waste due operating time extension is small compared to the existing stocks of radioactive waste. The fact that a solution of the RWM problem would disprove a key argument against nuclear power is not a justified reason to hinder such a solution. There is a wide variety of radioactive wastes arising from several activities, the most important one (in Germany) being nuclear power production. For some radioactive materials it is a matter of definition and strategy whether they are considered as resource or waste. In particular, this applies to irradiated spent fuel (SNF).

The types of waste mentioned above show a great variety in terms of radionuclide content, chemical composition, and physical condition. These features have implications w.r.t. all steps of RWM. Waste categorization varies from country to country often linked to the disposal route foreseen. The focus of the report is heat generating waste (called high-level waste, HLW) for which a disposal solution is still lacking in Germany.

Presently we distinguish between "once-through cycle" and "partially closed cycle":

- For the former, any irradiated fuel is considered as HLW and an increase of fuel burn-up would help to reduce the fractions of remain-ing fissile material as well as the amount of highly radioactive cladding and structural material per unit electricity while the amount of long lived, heat-generating minor actinides would slightly increase.
- The latter, involving reprocessing, has been selected by a number of countries, particularly those with large and well-established nuclear power programmes but abandoned by Germany. The infamous CASTOR shippings to Gorleben go back to earlier reprocessing con-tracts, resulting in vitrified HLW to be taken back by the country of origin.

The partially closed cycle concept aims to better utilize the energy content of natural resources by MOX fuel and to reduce the volume and toxicity of the spent fuel and waste to be stored. It leads to an increased number and volume of transports of radioactive materials which are potentially hazardous and subject to strict international regulation but also to public intervention. Separation of fissile and radiotoxic material (in particular plutonium) and its storage on site are considered further disadvantages. Advanced fuel cycles strive for fully closed fuel cycles connected with further developed proliferation-resistant reprocessing technology and various types of advance reactors either with thermal or fast neutron spectrum. Numerous national and international efforts are under way to develop advanced "reactor systems" being commercially deployable between 2030 and 2040.

Partitioning and transmutation of long-lived minor actinides (P&T) could significantly reduce, i.e. a hundred-fold, long-term radiotoxicity inventories of the disposal; the energy content and heat capacity of the final waste could also be reduced dramatically, resulting in potential decrease of the repository size or, in the case of large programmes, number of repositories needed. Regarding potential radiation exposures to the biosphere in the very long run due to a possible migration of radioactive nuclides through the host geology no further signification improvements are to be expected, though.

Subcriticality of irradiated fuel and wastes of all kinds has to be ensured for all steps of the RWM chain, especially for the final repository. Assessments as an element of national RWM programmes show that, if burn-up credits and given uncertainties are accepted, subcriticality can be ensured for "once-through cycles"; for non-open cycles this issue becomes irrelevant. As other wastes, proper radioactive waste management (RWM) requires a strategy ultimately targeted on protecting human health and the environment and ensuring security. It should comprise unloading, processing, transportation, storage and disposal as important steps. The exact sequence of these steps varies for different countries, different waste types and basic decisions and choices made concerning the whole fuel cycle.

Storage is (by IAEA definition) done "with the intent of retrieval" and may be of different purpose (e.g. in order to decrease the content of short-lived radionuclides or transport logistics, but also awaiting a disposal solution). Established technical solutions vary from above or shallow below ground, wet or dray, forced cooling or natural convection. Storage times are in the range of decades up to one century; even several centuries are sometimes discussed or envisaged.

Disposal is (by IAEA definition) done "without the intent of retrieval". This implies that the material is indeed declared as waste and not as resource. A variety of disposal options was discussed in the past, including exotic solutions. Remaining options for HLW/SNF follow the basic decision "concentrate & confine" instead of "dilute & disperse". They are

- those technically comparable to storage and this requiring human activities such as monitoring, control, maintenance, refurbishment etc. over the whole timeframe considered;

– those leading to a state at which no further human intervention or follow-up activity is needed ("passive safety"); deep ("geologic") disposal in
mine-type facilities, sometimes also deep borehole disposal are favored.

Lack of confidence in "passive safety" was one of the motivations behind
promoting long-term storage solutions or of going for disposal approaches
for which waste retrieval, although not intended, is – although only for
some time after emplacement – eased by design measures ("retrievable disposal"). International solutions (sometimes also called regional) make in
particular sense for countries with smaller waste amounts and/or geographical and/or geological boundary conditions which make national solutions
difficult to implement.

To explain how the above mentioned elements of a RWM strategy match
and to illustrate associated safety and security challenges, a rather simple
strategy is described as reference case in greater detail, based on the "once
through" cycle. SNF is declared as waste, no reprocessing takes place, and the
waste is to be finally disposed of in a deep (geologic) repository. Not addressing other waste, the SNF-related process comprises the following steps:

– Decay storage for several years in cooling pond at reactor site after discharge from reactor;
– Emplacement in transport casks, short-distance transport to local or
long-distance transport to central storage site;
– Interim, buffer, and decay storage (local or central) for several decades
(dependent on disposal programme);
– Transport to conditioning facility (short or long distance dependent on
location), buffer storage, conditioning for disposal;
– Buffer storage, transport to disposal facility (long or short distance dependent on location);
– Emplacement in disposal facility.

Strategies based on partially or fully closed (advanced) fuel cycles call for
reprocessing of SNF and fabrication of mixed fuels and therefore result in
differences concerning logistics of pre-disposal material/waste management and waste amount, form, radioactivity content, heat generation, etc.

As to strategic decision on the "end point" of RWM the two most remarkable differences between the options "prolonged storage" and the "disposal
in deep geologic formations" are the degree to which

– safety and security in the long run depend on active measures (which involve radiological exposures of workers), and
– emplacement measures are reversible (and, in particular, the waste is
retrievable) at different future points in time.

Preferences concerning these issues depend on the confidence one has into
the safety and security of the different options and on the consequences

one derives from the awareness of the duties to future generations. Potential conflicts of targets become also visible when considering hybrid solutions such as retrievable disposal and/or delayed closure of repositories. Approaches pursued in different countries are, e.g., delayed closure in the UK and staged closure of repository going along with decreasing possibility/increasing effort of reversibility in France. Prolonged storage (potentially for more than one century) is implemented in the Netherlands. The variety clearly demonstrates the inherent ambiguity in the decision on the "end point". In particular, it becomes evident

- that implementing reversibility or retrievability is a complex under-taking,
- that there is no simple answer to the question whether disposed waste is retrievable but rather a varying "degree of retrievability" can be considered, and
- that this degree of retrievability strongly depends on the timeframes considered which, however, are considerably shorter (in the order of decades up to three centuries) compared to the timeframes post-closure safety has to be ensured (up to thundered thousands of years).

The desire for retrievability of the emplaced materials on one hand and for a disposal which is safe and secure in the long run might lead to conflicts of goals or trade-offs. In particular there is an antagonism between retrievable disposal and protection against access (security). The ultimate goal of any disposal and deep geological disposal in particular, is to ensure safety and security in the long run and to balance conflicting aspects, where appropriate. This is the same for all concepts, but there is still room for considerable variations due to different available host rocks and other reasons. But in case of a target conflict measures that allow for retrievability should be subordinated. All concepts presently discussed strive for safety and security in the long run – although the pillars of safety and security (containment, isolation, migration delay, limitation of impacts) are the same, the weight to be put on each of them (depending on the point in time after closure) as well as the ways they are achieved vary. Generally spoken, no host rock will provide only favorable conditions with regard to all of these aspects, designs and (geo-)technical measures have to be adjusted in order to take advantage of the favorable features of the host rock and to compensate for the less favorable ones, taking different time frames into account.

Evidence must be provided that the ultimate goals of RWM have been achieved by long term assessments taking the whole spectrum of undesired events, states and processes into account. The proof of adequate security can focus on human intrusion: Following the decision for the concept of concentrating and isolating the radioactive waste in a repository, the possibility inevitably has to be accepted that radiation exposure limits may be exceeded in the event of intrusion into the repository. In addition, it is not possible to quantify appropriately the consequences associated with human intrusion due to the lack of predictability of the boundary conditions and other parameters to be assumed. However, the radiotoxicity of the disposed material may serve as a measure for the hazard involved.

Emphasis on assessing radiological safety in the long run traditionally concentrates on calculating potential exposure for scenarios, but lack of predictability for some components of the systems, especially those close to the surface, lead to high aleatory and epistemic uncertainties. Thus, calculated doses are not measures of real health detriment, but rather indicators of repository performance which mainly depends on the behaviour of better predictable components situated at greater depth. The concept of Safety Case is created by bringing together safety-related elements and arguments from site exploration, research, disposal site development and construction, safety analysis, etc. The Safety Case is evolving over time together with the repository programme and concept. Handling uncertainties is a pivotal element of the safety case.

The evolution of a repository programme, associated site investigation and R&D case can be seen as optimization process over some decades. The target function is composed of short term and long term safety and security issues, feasibility, costs, societal acceptance, legal feasibility and perhaps others. The target function itself can change over time (e.g. the timeframe of concern for long-term safety proof increased by two orders of magnitude over the last two decades); the knowledge base and ability to bound uncertainties also evolve over time. Thus, it does not make sense to revisit past decisions ("retrospective optimization") in order to ask whether these decisions were achieved according to today's standards. Rather the question whether the result of past decisions (site, layout, etc.) is still defensible under today's standards is essential.

In Germany repository operation could start in 2035 at the earliest, provided that every programme step would be fully successful (i.e. acceptance of Gorleben in 2017, planning permit without delay, etc.). Delays caused, e.g., by the rejection of Gorleben, the search for a new site (immediately or after rejection of Gorleben), etc. would lie in the order of magnitude of one or two decades.

Technical implementation of new technologies (e.g. P&T) can be expected in the 2030ies or 2040ies. As long as SNF is not emplaced (i.e. at least until 2030) there is no decrease of flexibility caused by a "fast" disposal programme. This is, however, only valid if a schedule with well-defined decision points exists. Otherwise, the forth-and-back of the past and the present with its strong dependence on the political orientation of governments which might change with each election would continue, the process could become indefinite without obtaining a solution.

1.2 *Radiation risk and radiological protection*

To evaluate radio-toxicological effects it is necessary to know the dose-effect relationship. In radiological protection two fundamentally different classes of these relationships are used. The fundamental difference consists in whether a dose-effect curve with or without threshold dose is present. The first option applies to *deterministic* effects (acute effects, fibrotic and similar tissue effects, and effects in the area of developmental biology). The threshold doses for these effects lie in a range of over 100 mSv. There is, on the other hand, no threshold dose assumed for *stochastic* effects (genetic effects and cancer); according to this premise, effects may occur even at very low radiation doses. A linear dose-effect relationship without a threshold dose (LNT model) is envisaged for these effects.

The fundamental quantity in radiological protection is the absorbed dose (D) in gray (Gy), it indicates the energy absorbed in an element of mass (m). Since various qualities of radiation (radiation type and radiation energy) lead to different effects from the same absorbed dose, the absorbed dose is weighted, giving the dose equivalent dose (H) in sievert (Sv). The risk of cancer significant for low radiation doses differs from organ to organ at the same equivalent dose. Hence the equivalent doses are multiplied by corresponding tissue weighting factors. The sum of these products gives the effective dose (E) in Sv, which relates to the stochastic radiation risk. The effective dose is a useful instrument for estimating regulatory processes such as optimization. It allows us to measure all pathways of human exposure and show them as a total value.

The dose limits and dose constraints are given as effective doses. The limits for on-site employees during the operational phase of the disposal facility (i.e. the placing of the radioactive material in the repository) are 20 mSv per year, and for members of the public in the immediate vicinity of nuclear facilities during the operational phase 1 mSv per year. For disposal facilities for high level waste various international committees envisage a "dose constraint" of 0.1 to 0.3 mSv per year in the post-operational phase (thousands to millions of years) after the closure of the disposal facility. The BMU (German Ministry for the Environment) recently suggested 10 µSv per year for this dose. The ICRP deems this a "trivial" dose of no significance.

To optimize sources of radiation in radiation protection, the effective collective dose is often cited. The average of the effective individual doses is multiplied by the number of persons exposed, and the dose is then given in "man Sv". To better judge the situation, the areas of the personal doses, the time period in which the relevant doses were considered, and other factors should be mentioned when establishing the collective doses. Collective doses are not suitable for assessing radiation risks, especially in the lower dose range.

The radiotoxicity of the radionuclides is determined by the radiation which occurs during radioactive decay. With regard to the long-term risk, radionuclides with long physical half-lives are of crucial importance, since these radionuclides continue to exist in considerable quantities for more than a million years, in some cases, and undergo radioactive decay over long periods of time. After this material has been incorporated into the human body, radiation exposure can occur over fairly long periods of time, particularly if a long physical half-life is accompanied by a long biological half-life. It then becomes necessary to estimate the "committed dose" over a long period of time (50 years for adults and for children up to the age of 70 years). In these cases the radiation dose is estimated with the help of the committed effective dose coefficient (CED, given in Sv/Bq). This coefficient is a good measure of the radiotoxicity of a radionuclide.

Two factors are of decisive importance for the exposure of humans and the environment through disposal facilities:

(1) The duration of the physical half-life of the relevant radionuclide
(2) The mobility of the radioactive material after it is released from the container in the disposal facility, and the subsequent migration through the technical and eventually the geological barrier into the biosphere.

Model calculations with granite, salt and clay as host rock and overlaying rock have found that, because of their low mobility, it is generally not the poorly soluble oxides of the α-emitting transuranium elements which contribute decisively to human exposure in the late phase of a disposal facility. On the contrary, it is the very mobile radionuclides, often present as ions, such as iodine-129, chlorine-36, carbon-14, selenium-79 and caesium-135. In the available model calculations the highest radiation exposure comes from iodine-129 generally found to occur in the 20,000 to 100,000 year time span. The radiation doses are again given as effective doses. For iodine-129, this means that, for biokinetic and metabolic reasons, it is almost only the thyroid gland which is exposed due to this radionuclide. Thus the radiation dose for radioactive iodine is higher by a factor of around 1,000 in the thyroid gland than in the other organs and tissues. The effective dose is lower than the dose in the thyroid gland by a factor of about 20. The causation of thyroid cancer by radioactive iodine has been thoroughly investigated, especially after the Chernobyl reactor disaster and in studies in the field of nuclear medicine. The release and migration of the mobile radionuclides from the disposal facility into the biosphere and their modification requires further intensive investigations.

Microdosimetric considerations mean that very heterogeneous dose distributions occur with radiation doses less than 1 mSv. As the dose decreases further, fewer and fewer cells of a tissue are exposed. With a dose of 1 μSv, less than 1 cell per 1,000 is exposed, while the dose for each exposed cell remains the same. The significance of this fact for the causation of health effects has not yet been clarified.

For disposal facilities of highly radioactive material, no radiation exposures at the level of the threshold doses for deterministic effects are to be expected in the late phase after the closure of the facility. Hence these effects cannot occur. Only stochastic effects (especially the causation of cancer) can be estimated using extrapolative calculations. Epidemiological studies have found that, for a general population, a significant rise in cancer cases can be observed after a radiation dose of around 100 mSv. Since there are, in an individual cancer case, no specific features to show whether the individual cancer was caused by ionizing radiation or not, it is only possible to give the probability of causation for the disease.

These studies have led to the definition of risk factors, resulting in a value of 5×10^{-2} per Sv. This means that, given a dose of 1 mSv, the cancer risk is 5×10^{-5}, and given a dose of 0.1 mSv, it is 5×10^{-6}. This risk should be compared with the general lifetime risk of about 0.4, the rate at which people in Germany develop cancer. The potential cancer risks which may arise through a disposal facility lie far below a "measurement range" in which possible, additional cancer cases might be recognized.

All living beings have always been exposed to ionizing radiation from natural sources. In Germany this exposure is about 2.3 mSv per year (effective dose) for every citizen. About 50 % of this dose is caused by the inhalation of radon with its radioactive progeny. The other components of the exposure are external cosmic radiation through the sun, external terrestrial radiation from radioactive material in the earth, and internal radiation exposure through the incorporation of radioactive material with food and drinking water. Every human being in Germany is carrying around, on average, about 9000 Bq of natural radioactive material in his or her tissues (mainly potassium-40 and carbon-14). Radiation exposure from natural sources differs greatly from region to region. In Germany the external exposure can be 5 to 8 times greater than the average value, even more with radon.

As well as exposure from natural sources, humans, particularly in the industrialized countries, are also exposed to other kinds of man-made radiation. Medicine, particularly diagnostic radiology, delivers the greatest proportion. The average value per head of population in Germany now lies roughly in the range of the median exposure from natural sources, and continues to rise.

There is no scientific doubt that radiation exposure from natural sources can cause the same biological/health effects as exposure to man-made radiation. A comparison of radiation doses from the various sources with the ensuing effects is thus completely justified.

In the low-dose range of 1 mSv and below, estimations of doses and risks are attended by great uncertainty. Since the possible radiation doses related to disposal facilities are mainly produced by the incorporation of radioactive material, this uncertainty becomes even greater, since the dose can only be estimated using model calculations. On the other hand there are relatively good data for the radioecological pathways of iodine and caesium, from experiences after the reactor disaster in Chernobyl and the fallout from tests of nuclear weapons.

Estimations of doses and risks are carried out with the aid of reference models and reference persons. Hence individual variability within the human population is not taken into consideration. On the other hand it can be assumed that metabolism, and thus the biokinetics of radioactive material in the human organism, will not change as dramatically, even after thousands to millions of years, as lifestyles and general human rules of conduct. It is also scarcely possible to make statements about the size of human populations and other conditions of human life at the location of the disposal facility over a time span reaching much further into the future than our knowledge of human history reaches into the past.

1.3 Management of high level waste with reference to long-term responsibility

The appeal to "ethics" in conflict situations is often based on an inappropriate understanding of this discipline, as is the idea that it is necessary to take one's bearings from "ethical standards" when making decisions of importance to society. Ethics cannot formulate higher principles, timelessly valid imperatives or values from which appropriate directives for action for the situation in question might be deduced from the top down. On the contrary, the domain of ethics is the development of promising strategies for conflict resolution. After all, the conflict is often caused by following the rules ("morals") which determine action within a culture and serve to legitimate the actions of the individual, but which are established differently in different cultures and in their subcultures. The critical analysis of morals and their scope is thus a prominent object of ethics.

In the conflict over the management of high level waste we can expect not just divergent opinions about the effects and side effects of the means to be employed, but also divergent aims, divergent ideas of morality, and ultimately different understandings of what requires legitimation and what constitutes acceptable grounds for legitimation. A rational solution to this conflict thus presupposes not only agreement on the cause-effect relations, but also critical reflection on the desired ends and on the grounds for legitimation put forward for the choice of means ("waste management strategies").

In cross-generational problems such as the management of high level waste it would be a category mistake to take for granted, without further testing, criteria taken from established "life-world" moral practices – at least as long as the requirements for action still leave (temporal) resources available to develop suggestions which are as rational as possible, while neutrally considering as many aspects as possible, and if need be to change the conditions of action and decision-making in such a way that optimal conflict resolution becomes possible.

Three category mistakes of this kind pervade the debate on the management of high level waste:

(1) The demand for an immediate solution to the problem bears a considerable burden of justification which can only be lifted if the situative conditions of action are interpreted according to the model of an emergency situation. Emergency situations often justify impositions on third parties which would otherwise be unacceptable in longer-term planning.

(2) In view of the decision-making situation for issues of the just distribution of burdens (costs, risks), where there is no urgent need to act, it is not simply a matter of applying the established principles of justice to a fixed situation of distribution with a fixed "volume to be distributed". On the contrary it is necessary to devise a plan which – working on the principle that exceptions to the rule of equality must be justified – is suited to bringing about the strategic optimization of the situation of distribution and of the "volume to be distributed". The conflicts over claims, put forward with an appeal to moral principles, can thus be transformed into deliberations (or negotiations) about conflict-pacifying allocations of all possible burdens and commodities.

(3) In the light of the many people who will be affected by the risks in the future, fairness may dictate – contrary to the moral intuition expressed in the "polluter pays" principle – that the producer of waste not be given the task of disposing of it. If others are, because of greater expertise or more favourable conditions, in a better position to provide a form of waste management which will do justice to the claims of future generations, then it could even be advisable, from an ethical point of view, to transfer the task of waste management to them, in return for freely accepted compensation.

Considering the far-reaching consequences of a decision about strategies for high level waste management, reflection about the possible conflicts with members of future generations must be an essential part of the preparation for decision-making. Given the irrelevance of all reasons for setting limits, a restriction of scope to those people who interact with us, or to a fixed time span (e.g. n generations) is not sustainable for ethical, i.e. rational, conflict resolution.

With regard to concrete actions, an obligation thus exists for exactly as long as the consequences of these actions produce potential for conflict. The obligations attendant on the management of high level waste therefore exist for as long as the hazard potential of the radiotoxic and chemotoxic substances.

For *ethical* reasons it is necessary to distinguish between on the one hand the universalistic, i.e. indefinitely existing *obligation* and on the other hand the degree of its *binding force*, which diminishes with spatial, social and temporal distance. When, in the framework of our moral praxis, we ascribe a higher degree of binding force to the obligation towards the generations closely following ours (and ascribe to these generations an obligation towards those which closely follow them), we are organizing our long-term obligations and making them practically manageable. The extent to which binding force diminishes, however, is not firmly correlated with temporal distance, but with our potential to fulfil obligation in a tactical and controlled manner. Moral agents have an influence on how far-reaching this potential is – and it is part of their obligation to make use of these opportunities for influence, within the bounds of proportionality of means. Thus, if we want to act in a technological way, there is also an obligation to procure knowledge about the consequences and the conflict these may cause. In the same way the increase in the scope of our actions through technologisation and collectivization comes with the obligation to form organizations and equip them with the requisite resources so that they can, on the basis of the knowledge gained, or of rationally founded suppositions, organize society's long-term obligations, without being dependent on the resource constraints of the individual actors. In particular, advocacy for future claims requires a mandate which is legitimized by procedural organization, anchored in institutions, and controlled by society – rights of representation cannot simply be claimed by declaring oneself to be competent, responsible, or personally affected. The actors thus create "responsibility", but do not relinquish their obligation – on the contrary, they still have the task of ensuring in the long term that their obligations are fulfilled responsibly.

In optimizing complex plans with high potential for conflict, the participation of all those with local or technical expertise, regardless of actual certification or profession, is in principle desirable, for purposes of plan optimization and process control. The inclusion of those directly affected in complex planning can, by dispelling mistrust and fears, help to create the social prerequisites for a rational discourse aimed at conflict resolution. Participatory processes must be designed in such a way that the decision-making behaviour of those responsible for making decisions is not distorted (biased) by implicit or explicit lobbying by the present generation, but that it remains focused on the claims of all, including the non-participating members of future generations.

The expectation that participation might also help to legitimize decisions, as expressed by parts of the public and of the specialist community, should be met with scepticism. In particular, the incorporation of participatory processes into the preparations for decision-making should not lead to a situation where the responsibility delegated to institutions by the citizens as bearers of long-term obligations is then transferred back to the citizens by the institutions ("circle of allocation of powers"). The view put forward in justification, that the individual is ultimately the one who knows most about his own needs and interests and is thus best able to represent them ("thesis of self-competence") overlooks the fact that this is all about creating a balance between a disparate mass of interests for which no sub-group can be representative, and which can therefore not be representatively replaced by individual stakeholders. Models which seek to transfer decision-making powers back to the citizen are often based on an image of social decision-making as an antagonistic negotiation between social sub-systems, with "politics", "science" and "the public", as tribes of equal standing amongst others, each representing its own interests ("tribalisation thesis"). This, however, confuses the methodically organized discourse of knowledge in the sciences, and the institutionally organized discourse of decision-making in politics, which are actually intended to counteract the centrifugal processes involved in the subjective formation of opinions and interests, with the mere representation of opinions and interests on the part of the individuals involved. And yet it is precisely in decision-making questions in which the claims of some parties to the conflict (such as those of the members of future generations) can only be reasonably imputed and represented by means of advocacy, that the legitimating power of procedures takes on great significance. It is desirable that procedures be changed or expanded to increase participation and respond more precisely and in a more differentiated way to the claims (insofar as they are relevant to the conflict) and to the local and technical expertise of the present generation. This must, however, be weighed up against the obligation towards the members of future generations.

1.4   Legal questions of managing high level radioactive waste

The final disposal of high level radioactive waste raises difficult legal questions. Of primary importance are the shaping of the responsibility of the state, the avoidance of institutional conflicts, the principles of radioactive waste management and the – constitutionally required or politically desirable – degree of legalisation. Within this framework a number of complex legal questions of detail is addressed. At international level the Joint Convention on the Safety of Spent Nuclear Fuel Management and on the Safety of Radioactive Waste Management of 1997 (International Atomic Energy Agency – IAEA) is most important as regards the development of principles of nuclear waste management and the institutional framework of regulation. Besides, there are a number of relevant recommendations adopted by IAEA and other organisations such a, in particular, the International Commission for Radiological Protection (ICRP).

The European Union has regulated in the field of radioactive waste disposal only to a limited extent. To be noted are the basic health standards laid down in the Directive 96/29/Euratom and the Directive relating to the safety of interim storage and treatment of spent fuel. A Commission proposal sub-mitted in 2010 also provides for the introduction of regulation in the field radioactive waste disposal which will require certain adjustments of German law.

The present study compares the regulation of high level waste disposal in major nuclear countries such as the United States, France, the United Kingdom, Switzerland, Sweden, Finland, Spain, and Japan (a summary is presented in the main text, the details are contained in the Annex). The comparison reveals a number of common features but also important differences that can be attributed to a number of factors. This variety rules it out that one could draw generally valid conclusions from the foreign experience. However, the experience of foreign countries can give important impulses to the discussion. It is to be noted that as a response to widespread, albeit differently strong, lack of public acceptance of previous site selection attempts, there has been a clear tendency to introduce new forms of participation relating to the development of strategies and in particular site selection. For various reasons, the good experience made in Finland and Sweden in this respect cannot be generalized. In the other countries under review, the new decision-making model still is on probation.

Under German law, the responsibility of the state for radioactive waste disposal has strong roots in the Constitution. Article 20a of the Federal Constitution explicitly obliges the state to protect the natural bases of life of future generations. The Atomic Energy Act contains demanding requirements for the authorization of radioactive waste repositories. According to case law best possible prevention of significant risks and precaution against low risks according to practical reason are the standards that govern the state duty to protect. However, the German courts recognise that the ultimate responsibility for determining and evaluating the relevant (potential) risks and deciding on their tolerability is vested in the legislature and – in the framework of the Atomic Energy Act – to the executive.

The Atomic Energy Act provides that the Federal State is responsible for establishing and operating a high level radioactive waste repository, either on its own or through a federal agency such as the Federal Radiation Protection Agency. The State can charge third persons with performing these obligations or even confer the sovereign powers necessary for performing them to third persons. In view of the very long-term nature of the relevant tasks the model of assignment is not without problems. The Atomic Energy Act concentrates management and supervisory functions regarding the operation of high level radioactive waste repositories in the Federal Radiation Protection Agency which in principle is subject to directions from the Government. This institutional set-up is still consistent with the principle of effective independence of waste management and regulation as laid down in the Joint Convention and IAEA recommendations. The reason for this assessment is that the supervisory function within the Agency is vested in a separate department that is independent. However, a reform of the relevant competences appears advisable.

The basic strategic options of high level radioactive waste management in Germany are set out in the Atomic Energy Act (partly, as regarding the principle of geological disposal, not explicitly). By contrast many important elements are left to decision by the executive which has to observe the principle of best possible prevention and precaution. As regards groundwater protection, the principle of "no concern" under the Water Resources Management Act is applicable. In autumn 2010 the Federal Ministry for the Environment published safety requirements for the final disposal of high level radioactive wastes. These requirements are to be further developed during the ongoing underground investigations at Gorleben. Problematic from a legal perspective are in particular the timeframe of precaution in the post-operation phase and associated to it the requirements for the demonstration of safety. A practical exclusion of any risk of future harm does not need to be ensured until the (almost) complete decay of radio-activity of the wastes disposed of, but only for such a period of time as possible according to the present state of science and technology, including prognostic feasibility. Beyond, the relevant risks must be reduced to a reasonable extent. This constitutes best possible prevention of risk and precaution against significant risk in the meaning of the Atomic Energy Act.

The selection process for the site of a repository for radioactive waste in Germany has been characterised by a low degree of legalisation. In response to the selection of Gorleben as the site for a final repository in the 1980s that was considered as technocratic, various proposals have been made to reopen the site selection process. Although the German Government in the meantime decided to resume the underground investigations at Gorleben without considering alternatives, it appears reasonable to discuss site selection anew because the suitability of Gorleben is not yet established and it is not known either whether there is no alternative that is better suited than Gorleben.

For various reasons a site selection procedure cannot be based on the planning permission procedure under the Atomic Energy Act or on general spatial planning at federal level. However, following the pattern of the waste management plan under the Life Cycle Economy and Waste Act, it would make sense and be legally feasible to introduce a sectoral planning procedure that would precede the planning permission: It is to be noted that such a site selection procedure is not mandated by the Constitution since the legislature and executive have the ultimate responsibility for deciding on tolerability of risk. In shaping the proposed procedure the legislature should orient itself at the requirement of optimisation which is derived from the principle of best possible prevention of risk and precaution against potential risk However, the safety of a repository is a product of a combination of geological and engineered barriers at a particular site. Therefore, optimisation cannot be pursued in an isolated manner in relation to a particular host rock and site.

Although considerations of legitimacy and acceptance may militate for the complete reopening of the site selection process, the decision of the Federal Government to continue the underground investigation of Gorleben is in keeping with the ultimate responsibility of the executive and can also justified by pragmatic reasons. However, it appears reasonable to investigate, in parallel to Gorleben, alternative sites and introduce, for that end, a modern site selection procedure that is linked to the progress of investigations carried out at Gorleben. This mode of action counteracts further postponing of site selection should Gorleben fail. Moreover, on the grounds of good governance and legitimacy, it makes it possible that even in the case of suitability of Gorleben an intensive investigation and eventually selection of an alternative site can be carried out where it can be demonstrated that this site may fulfil the site selection criteria evidently better than Gorleben. The site selection could proceed in eight or nine steps and should be based on selection criteria that the executive, with the participation of an independent expert commission and the public, would prescribe. The procedure should be transparent and allow for a comprehensive participation of all affected persons and stakeholders.

For improving the legitimacy and acceptance of the relevant decisions, the traditional rules on public participation should be enriched. The major elements of such enrichment should in particular be the establishment of an independent expert commission, an opportunity of affected persons and stakeholders to exert an early influence on the shaping and planning of the finding process, the establishment of a permanent dialogue forum at local-regional level and access of affected persons and stakeholders to a pool of independent experts.

Apart from reviewing the suitability of the site of the repository, the planning permission procedure is designed to ensure that the construction and operation of the repository is consistent with the legal requirements. In particular it must be ensured that during normal operations the requirements of the Radiation Protection Regulation are complied with. Furthermore a demonstration of safety and groundwater protection, which is governed by the principle of "no concern", is required. The relevant statutory requirements are in principle, that is, subject to modifications mandated by the long-term nature of the regulatory task, also applicable to the post-operation phase of the facility. The Radiation Protection Regulation that is geared to current operations of nuclear facilities can at least serve as a guideline for radiation protection in the post-closure phase.

The Atomic Energy Act places the financial responsibility for final disposal of radioactive waste on the operators of nuclear power plants. They are obliged to make financial contributions for covering the necessary expenses for waste disposal, among others also for the planning of the disposal facilities. Since site selection aims at creating the prerequisites for the construction of a repository and the repository confers advantages on the operators, it is submitted that the expenses of site selection are necessary insofar as mandated in accordance with the relevant statutory site selection procedure – as opposed to a purely political procedure.

A weakness of existing law is that the obligations of operators relating to future financial burdens are limited to establishing reserves in their balance sheets. In view of the very long-term nature of final disposal of high level radioactive waste, this financing model raises some problems in case of insolvency or dissolution of waste generators insofar as these are subsidiaries of the public utilities. In case of termination of the existing enterprise contract between the parent company and its subsidiary due to insolvency or dissolution of the latter the liability of the parent company is limited to obligations that have already arisen. It is doubtful whether the abstract statutory obligation to cover the future, not yet foreseeable costs of waste management can be deemed to be an obligation that already has "arisen". These doubts could be solved by a joint and several liability of the parent company independent from the validity of the enterprise contract, in particular when operations are not continued. Of course, the existing statutory liability of the parent company for the debts of the subsidiary in case of a contract to transfer the profits would remain unaffected. Fund solutions are not recommended.

1.5   Guidelines for a socially acceptable and fair site selection

Nuclear waste disposal mobilises people, not only in Germany, but all around the globe. The question of final disposal is symbolically charged: it is no longer primarily about technological viability, or even about long-term safety, but rather about fundamental perspectives of societal development. As a society, do we wish to continue pursuing technologies in power generation that are central, highly efficient and energy-dense, but risky? Or should we opt for decentralized technologies, which are often not very efficient, low in energy density, and, while not necessarily low-risk, are locally restricted with respect to disaster impact?

This starting position defines the terms for a future solution to the question of how to dispose of high-level radioactive waste. According to some empirical results:

– In all existing surveys, solutions to final nuclear waste disposal rank very highly in the public perception of what poses a threat. This is the case throughout the world, even, interestingly enough, in Finland, where the problem of final waste disposal has been largely resolved on a political level despite these public concerns.
– The complexity of this situation becomes apparent when considering the results of a German representative survey from 2001 and 2002: at the time of the survey, roughly 65 % of those interviewed assumed that over the next decade a final nuclear waste repository would be established,

while 81 % objected to such a repository being created in their vicinity. This classic *NIMBY* ("Not in my backyard!") syndrome is typical in site selection processes for large-scale technological and risk-related facilities. In principle, the public endorses the need for such technology, but on condition that it is as far removed as possible from their domicile.

- Surveys on stakeholder mobilisation show differences worldwide, from which much can be learned. Some countries, such as Finland, Sweden and Switzerland, have made progress with regard to finding solutions for final nuclear waste disposal. It is not impossible to reach an institutionally satisfactory solution that is tolerable for the majority of the population if the right approach is taken. However, it is not easy to find an approach that will gain acceptance. And success is never guaranteed. If the wrong approach is chosen, however, failure is certain.

Why is the risk perception of final nuclear waste disposal so emotionally charged? From a psychological perspective, the risks of nuclear power generation in general, as well as those of final nuclear waste disposal, are associated in the eyes of the public with the semantic pattern of "the sword of Damocles". In the way they function, semantic patterns are not unlike drawers in a filing cabinet. If confronted with a new risk, or taking in new information regarding risk, people, in general, will try to file away this information into one of the existing drawers.

The "sword of Damocles" is such a pattern. This is to do with technological risks where, irrespective of whether or not such attributions are justified, a high potential for damage is combined with a very low probability of occurrence. The stochastic nature of such an event makes it impossible to predict the time of its occurrence. Subsequently, the event may theoretically occur at any point in time, even though the probability for each of these points is extremely low. However, when in the realm of the perception of rare random events, probability plays but a minor role: the randomness of the event is the actual risk factor. The idea that an event could hit the affected population at any random point in time generates a feeling of threat and loss of control. Instinctively, most people can mentally (in real life this may be questionable) cope much better with danger if they are prepared for and attuned to it.

Another factor is the nature of radioactivity, which defies sensory perception. This takes us to a second semantic pattern that can often trigger fear and is called "insidious or creeping danger". In the context of this risk pattern, people rightly assume that scientific studies can detect creeping dangers in good time and discover causal relationships between activities or events and their latent effects.

In the case of the semantic pattern of "creeping danger", the people affected must rely on information provided by third parties. In general, they can neither perceive the hazards sensorially nor verify the claims of various experts, which tend to be contradictory. When laypeople evaluate such risks they are faced with a key question: can I, or can I not, trust the institutions that provide me with the necessary information? If the subjective evaluation is negative, then there will be an uncompromised call for zero risk. This is because a person who relies on third-party information for risk assessments, but does not trust the third party, will not accept any cost-benefit balance, but rather call for zero risk. In contrast, if that person is undecided as to whether or not he can trust the third party, then peripheral aspects assume particular importance, aspects that factually have no connection to the circumstances of the decision. A layperson has no option but to distribute trust according to peripheral aspects, because she/he is not able to assess the risk of being harmed by radiation. She/he must trust either one side, or none at all. These patterns are deeply embedded in unconscious processes for evaluating reality. They can only be overcome when people learn to understand these perception patterns and recognize the impact they have as unconscious evaluation benchmarks when forming judgements. Risk communication cannot, therefore, merely be confined to relaying scientific insights about risks, but must also vividly communicate the mechanisms of risk perception.

Even if the parties involved engage in an intensive risk dialogue, as has occurred to an increasing extent in recent times, this certainly does not mean a resolution of the conflictual situation. Rather, the establishment of a basis of communication is the prerequisite, but by no means sufficient condition for reaching a universally acceptable solution. In principle, three main approaches to the final disposal of high-level radioactive waste can be outlined based on current conditions:

In the *Top-Down Approach,* the representatives elected within the democratic system have the sole right of decision-making. By virtue of their office they act in the best interests of the people. Any active participation by the population, if envisaged at all, would be within a very narrow scope. However, this solution is also based on transparent risk communication involving the population. Citizens may voice their opinions during hearings or discussions, but without having any guaranteed say in the final decision-making process. The decision-makers must also prove that all objections have been duly dealt with. Then, however, it is in the hands of the decision-makers to come to a decision, while being obliged to disclose all arguments in favour and against.

The *Muddling-Through Strategy,* a pragmatic mixture of the top-down and bottom-up approaches, relies on minimum consensus (muddling through), which emerges from the political opinion process. The only options to be considered legitimate are those that induce the least amount of opposition within society. In this form of management, societal groups may influence the process of political decision-making in the extent to which they provide proposals that offer connectivity, i.e., that are adapted to the language code and processing style of the political control systems, and mobilise public pressure. In politics, the proposal that will then be accepted will be the one that best holds its ground in the competition of proposals, i.e., the proposal that entails the least loss of support among interest groups for the political decision-makers. Previous debates on final waste disposal seem to conform most closely to this form of muddling through.

The third variety – *the bottom-up approach* – is based on a discursive solution and an attempt at fair negotiation of the site selection between the different groups involved. Discursive methods claim to enable more rational (in the sense of a discursive understanding of reason), just (in the sense of an understanding of justice based on negotiation) and competent solutions to existing problems. No matter which specific claims we associate with discursive processes: They must be structured according to certain rules in order to ensure their effectiveness, for instance to provide constructive solutions to problems in an appropriate and fair manner keeping open more than one possible decision, and to prevent, as far as possible, strategic behaviour amongst the participants. In principle, the legitimisation of collectively binding norms depends upon three conditions: the agreement of all parties involved, a substantial justification of the statements delivered within the discourse as well as suitable compensation for negatively affected interests and values.

How could a meaningful combination of bottom up and top down be implemented? The entire selection procedure must be transparent and comprehensible (effective risk communication criterion). The selection procedure must appear fair (all shared value and interest groups involved have a say), competent (the problem is treated appropriately and with the necessary expertise) and efficient (the means or costs of the decision must be proportionate to the objective) to non-participants. The selection itself must be comprehensible and inter-subjectively justifiable whilst reflecting the plurality of moral concepts of the affected population in the sense of a fair consensus or compromise. If we seriously aimed at fulfilling all these demands for legitimising the site selection process, a single political regulatory instrument would certainly not suffice. Rather, such wide-reaching decisions call for a sequence of different regulatory instruments, each of which would cover a different aspect of these criteria.

Adhering to the principles mentioned requires a combination of several steps and elements. First, it requires scientific-technological agreement over the suitability of site location plans and criteria in the form of threshold values that would have to be obtained for a site to count as suitable in terms of long-term responsibility. These criteria would have to be established before the results of the suitability test were presented. This should be based on consensus-oriented methods of scientific tests, which must be conducted in an independent, factual and transparent fashion. Institutionalizing it would require a professionally led, neutral platform, which would bring together scientists on a national level, and include international experts (thus increasing essential credibility), with the goal of arriving at the knowledge consensus demanded here. Also, in order to demonstrate to the general public that the experts are not making a one-sided selection, provision can be made for giving the stakeholders the right of nomination.

Second, fair compensation for accepting uncertainty is needed. The goal here is to find a robust solution that is acceptable to all concerned, in order to anticipate and deal with uncertainty. Methods such as mediation or establishing a round table with stakeholders can lead to compensation for the consequences of uncertainties in a way perceived of as fair. Here it is particularly important to not conceal insecurities, particularly regarding long-term effects, but to address them openly and to ensure compensation by promoting economic or location development. We are not referring here to the "sale of indulgences" or the corrupting practice of paying people to carry the risk, but rather about types of social compensation common to other areas of life: Those who will in future bear the uncertain consequences and burdens for the general public should in return receive recognition and support from society. This procedure would show that accepting uncertainty is respected and honoured. We cannot expect people to silently "swallow" the uncertainty. To this end, unbiased dialogue forums with the groups affected by the consequences would be preferable. We can learn from Sweden in this respect, where such forums are staffed locally. In some cases, experts are involved as sources of knowledge and information.

Here follows the third and final component of a discursive solution: a forum for societal orientation about future energy supply and post-industrial lifestyles. The debate about final disposal is about more than waste managment, rather it involves the question of how we want our lives to be in the future. How can the subject of final disposal be integrated into a constructive blueprint for future lifestyles and living conditions? This could involve discourse-oriented methods, such as citizens' forums, round tables or consensus conferences, which have shown a degree of success in other countries.

At the beginning of any new initiative for resolving the question of how to dispose of high-level radioactive waste we must concentrate on such keywords as: reducing complexity by means of a scientific consensus as to the best selection procedure, coping with uncertainty through fair offers to those who will have to live with the consequences of this uncertainty, and treatment of ambiguity through an open and sincere discourse on objectives regarding the future of our energy supply. Not least, it requires discourse in society about how Germany as a country with few natural resources can remain successful in the future.

2 Conclusions and recommendations

2.1 Ethical framework

(1) Obligations to future generations are generally valid indefinitely. Management strategies for radioactive waste must nonetheless be developed for limited time spans.

Obligations which commit the stakeholders to prudent management of radioactive waste are in principle valid indefinitely and continue to exist towards the members of distant generations – though their binding force gradually decreases over time. The complex sequences (the "sequence spaces") which must be included in the development of waste management strategies must nonetheless have time limits set, to meet the rational requirements of planning and for reasons of efficiency. Such a time limit should be based on the foreseeable future effect of the consequences, and thus on the relative hazard potential of the contents of the disposal facility and of possible exposure in the biosphere, which vary with the phases of the decay process and the chosen waste management strategy.

(2) The present generation as the primary beneficiary of nuclear energy has the obligation to initiate the solution of the disposal problem. The demand for *immediate* disposal of high level waste, however, imposes an unjustifiable burden on the present generation.

From an ethical point of view, the utilization of moral principles which impose the entire burden of disposal on the present generation as a community of originators and beneficiaries is by no means self-evident. As long as – based on authoritative forecasts – reliable relations of exchange could be organized across generational or communal boundaries, without detriment to third parties, there could be no ethical objection to, for example, transferring the "responsibility for disposal" in exchange for freely accepted compensation. The demand for an immediate solution to the problem is not self-evident and requires justification. If it is probable that a future generation with which we are in verifiable interaction has access to "better" waste management strategies, then it may even be advisable to take this option. Considerations of justice do however impose a duty on the originator to offer adequate compensation. This also applies, mutatis mutandis, to international relations of exchange: for an adequate perception of long-term obligations, the relevant factors are not national borders – which in any case seem rather historically contingent in view of the time spans in question – but the availability of skills and resources.

(3) Processes of legitimisation to solve the question of disposal facilities must be designed in such a way that they do justice to everyone equally, and in particular give adequate consideration to the claims of future generations.

It is desirable that the legitimising procedures should be designed to respond to the claims and the local and technical knowledge of those alive today, where these are relevant to the conflict, but this must be weighed up against obligations towards those who are not involved in the consultations, in particular members of future generations. Thus the criterion for the decision must not be solely the agreement of members of the present generation, achieved through procedures. On the contrary, the decision must also be acceptable in the sense of a rationally presented justification based on universalist arguments. In the public debate, there must be advocacy for those who cannot participate, or who have no incentives to stand up for their requirements. This responsibility cannot simply be claimed by individual stakeholders or groups of stakeholders; the transfer of this responsibility must be legitimated.

(4)   A categorical rejection of all proposed solutions for the disposal of high level waste is incompatible with our obligations towards future generations.

The management of radioactive waste is a collective duty shared by society as a whole. This gives rise not only to the duty of care, which dictates that proposals should be tested to see whether they fulfil this obligation, but also to the obligation to participate constructively in the development of suitable proposals, and/or to create or support structures which make this kind of participation possible. The right of veto held by those who base their arguments on the unsuitability of a proposed site or of the disposal concept per se is linked with the expectation that they will participate constructively in the development of alternative proposals. Promising projects and processes which aim at the development of alternative proposals must be supported with the necessary resources. A categorical rejection of all proposals ignores our obligations towards future generations.

(5)   The instrumentalisation of future generations for arguments against longer operating lives of nuclear power plants is unacceptable.

There is an obligation towards the members of future generations to dispose of already existing radioactive waste. The additional quantities which would accumulate if the operating life of nuclear power plants were extended would be too small to have any substantial impact on the choice of disposal strategies. In the light of the long delays to be expected for disposal, the extension periods in question are insignificant. It would therefore be unacceptable to instrumentalise the interests of future generations to exert a controlling influence on decisions about the operating life of nuclear power plants.

2.2   Safety requirements and goals

(6)   The prudent disposal/management of radioactive waste presupposes the development of an appropriate overall strategy with regard to safety, health protection, and environmental protection. The German Constitution and the Atomic Energy Act give a clear legal framework for the best possible prevention of significant risks and precaution against low risks.

The prudent disposal/management of radioactive waste presupposes the development of an appropriate overall strategy which considers all relevant technical factors and options and, taking into account previously made decisions, meets the requirements of technical, social and political feasibility. Both the German Constitution and the Atomic Energy Act establish a clear normative framework for this by linking each solution to the legal requirement of best possible prevention of significant risks and precaution against low risks. The legislature and executive are, however, accorded a broad margin of discretion in this matter.

The safety of a repository is a product of a combination of geological and engineered barriers at a particular site. Based on the international consensus as to the particular importance of geological barriers, one could indeed assert that one should select the best possible geological barrier and couple it with the best engineered barriers. However, this combination does not necessarily provide the best possible safety since the nature of the host rock and the local geophysical conditions may place limits on the additional safety that can be provided by the engineered barriers. What ultimately counts is the safety of the whole system.

(7)   For evaluating long-term safety, it is necessary to work from scenarios based on the evaluation of natural developments with regard to the potential for the transport of contaminants through host rocks and overlaying formations, and for the potential release of radiation into the biosphere.

The evaluation of strategies must be carried out on the basis of criteria which take into account short and long term environmental impact and which meet the safety and protection needs of the present generation just as much as those of members of future generations. The assessment of long-term safety must be based on scenarios about naturally induced processes. It is necessary to investigate to what extent these can lead to a gradual transport of radioactive substances through geological host rocks and overlaying formations, and to what extent they are therefore connected to the risk of a release of radiation into the biosphere and to the corresponding risks for humans when exposed to radiation on the surface or in the case of inadvertent intrusion. Furthermore, subcriticality needs to be ensured for the irradiated fuel and radioactive material brought into the disposal facility; the corresponding proofs seem to be unproblematic.

(8)   Humans have a limited ability to develop prognoses about future de-
       velopments of geological or anthropogenic systems over extremely
       long time spans.

One of the ethical principles is the insight that, due to planning requirements,
a time limit must be set on the consideration of distant hazard potentials, especially of the risks for humans connected with a possible release of radioactive
substances. In discussions held so far, such time limits are often justified by
the decrease in radiotoxicity over time. The orders of magnitude cited in these
discussions, ranging from 100,000 to one million years, are usually based on
the radiotoxicity of natural uranium ore bodies. The basis for the comparison
is the fact that, for this time frame, the buried radioactive waste will fall short
of this measurement. The human ability to predict the evolution of geogenic
or anthropogenic systems over extremely long time periods is limited, however. Models describing e. g. the confinement of waste or radionuclide migration to the biosphere become increasingly meaningless since the underlying
assumptions lose their justification as the forecast periods increase. It is possible in principle to derive clues for the subsequent development of charted data
by extrapolating the possible course beyond the end of the foreseeable time
frame. This must be weighed, however, against the increasing unreliability of
the information which can be gained by such calculations.

(9)   The time frame of one million years for the evaluation of the long-
       term safety of high level waste seems an appropriate compromise
       between the ethical requirements for long-term responsibility and
       the limits of practical reason.

The rationale behind the often quoted time frame of one million years proposed by the German Ministry for the Environment (Bundesministerium
für Umwelt, Naturschutz und Reaktorsicherheit, BMU) is the ability to find
sites in a geological environment which is, according to current geo-scientific
knowledge, believed to conserve its favourable features for a period of that
order of magnitude. This approach appears to be a reasonable compromise
between the requirements coming from ethical considerations on the one
hand and the realization of the limits of practical capabilities on the other.

(10)   The proposal to set the reference dose in Germany at 10 μSv (0.01 mSv)
       should be rejected from a radiobiological point of view; it also diverges
       considerably from the international consensus of 100 μSv (0.1 mSv) per
       year.

The annual reference effective dose of 10 μSv (0.01 mSv) proposed for disposal
facilities with heat-generating radioactive waste in the post-closure phase should
be reconsidered. Such a radiation dose is less than one per cent of the average
radiation dose which originates from natural sources (in Germany and many
other regions worldwide) and thus lies within the range of variation for exposure from natural sources. This radiation dose is lower by a factor of 10 than the

customary international value of 100 µSv (0.1 mSv) for unlikely situations in the post-closure phase. The ICRP calls a dose of 10 µSv (0.01 mSv) a trivial dose.

(11)   Investigations of long-term safety should mainly concentrate on radionuclides which are of particular relevance for potential radiation exposure in humans.

It is possible to estimate by means of models that there are certain radionuclides, present as ions readily soluble in water, which are particularly relevant for human exposure to radiation. Subsequent investigations should focus on how ^{129}I, ^{14}C and ^{36}Cl can be kept immobile in the disposal facility, and examine how the migration of these radionuclides can be better understood and modelled.

2.3   *Waste management programme and timescale*

(12)   There is a high degree of evidence for the technical feasibility of some concepts for disposal which have already been developed.

The concepts developed for the disposal of HLW, meeting the safety and security needs of both present and future generations, seem in principle to be technically feasible. The strategies for implementation have to include the following aspects:

(1) distinctions between radioactive material which is to be treated as waste and that which is to be treated as a resource;
(2) a decision to concentrate the radioactive waste, confine it properly and deposit it in deep geological formations;
(3) independence from active measures after closure of the disposal facilities;
(4) an explicit, well-coordinated schedule for the execution of a waste management programme.

(13)   The development of a general programme and a schedule for Germany is recommended. The programme should be designed in such way that there is a realistic chance of it surviving changes in government.

As Germany particularly lacks a coordinated general programme and schedule, it is recommended that a schedule be developed which explicitly lists the principles upon which the concept is built, covers all the options for implementing these principles, and establishes a sequence defining what time periods must be allowed for research and when a binding choice must be made between alternative options, and structuring the site selection process in the framework of the existing legal regulations (cf. the decision-making pathways in section A 2.8). Such a schedule would be in line with the proposed EU Directive on the management of spent fuel and radioactive waste.

A comprehensive national programme would provide an incentive for developing a systematic and coordinated approach to radioactive waste management. It could also help to make political decision-mak-

ing more transparent, facilitate public participation in the preparation
for decisions, and thereby help to improve actual acceptance and ethical
acceptability.

**(14)   It must be made clear on which scientific/technical and factual ba-
           sis, by which procedure and by which institutions the decisions are
           to be made.**

For each decision to be taken according to the programme it is necessary to
clarify on which scientific/technical and factual basis (e. g. site investigation
results, outcomes of conceptual/technical developments, safety assessment
results, content of license application), by which procedure and by whom
the decision is to be taken.

**(15)   The general programme should contain the waste management strat-
           egy, the procedure and the schedule for the site selection. It should
           also make clear which decisions have already been made and which
           measures have already been taken, and which of these can be con-
           sidered reversible if need be.**

The general programme should explicitly list the measures already taken
and decisions already made, and indicate which of these can be considered
potentially open to revision, and in which circumstances a revision might be
possible. In the view of the authors, decisions previously taken in Germany
result in well-defined amounts of radioactive waste awaiting disposal. The
changing policy on the reprocessing of spent nuclear fuel (SNF) in the past
has led to a situation in which both vitrified high level waste from reprocessing and SNF are awaiting disposal. SNF is stored partly at reactor sites,
and partly in central facilities, while the vitrified waste is stored in the latter facilities. SNF is considered as waste and it is thus intended that it will be
disposed of in deep geological formations without relying on active controls
after the closure of the facility. Further disposal measures are subject to the
following political parameters:

– the principles and requirements stated in the most recent BMU Safety
   Requirements, in particular those concerning the confinement of the
   waste in a "rock zone", the measures for recovery, as well as the development and optimization processes to be followed when commissioning the facility;
– the strategy of favouring steep rock salt formations for disposal.

**(16)   A waste management strategy should be open to possible positive
           adaptations. This also applies to possible advances in research and
           development.**

There are various options for influencing the capacity of a disposal facility with a given set of geological conditions. Such possibilities include
changing the burn-up rate, the decay time in the storage facility, or the

disposal facility design, allowing for better utilization of the available host rock. A waste management strategy should be developed which does not assume the availability of these options, but remains open for potential positive adaptations. This applies equally to possible advances in research and development, such as partitioning and transmutation, which may significantly reduce the safety and security requirements to be met by a disposal facility.

(17)   **With regard to site selection and the construction of a disposal facility, the general programme should be focused on bringing about a solution as fast as is reasonably achievable.**

The recommended programme for site selection and construction of a disposal facility should be focused on bringing about a solution as fast as is reasonably viable given the relevant technical, legal and planning-related aspects and the social and political concerns.

New scientific developments which might change the requirements for a disposal facility should also be incorporated. The programme should remain open for unforeseen events and should include the possibility that Gorleben might turn out to be unsuitable.

Sufficient reserves of time should be built in for legal decisions, participation and communication. Time spans for each section of the participation process should be agreed on in advance with those involved, to make it more difficult for individual stakeholders to pursue a strategy of perpetually delaying decisions.

There always is the theoretical possibility of finding a better site. However, for practical reasons, an infinite continuation of site investigations is not possible. Therefore, optimisation through siting arguably cannot be pursued absolutely but only in the form of a planning directive – as in its classical area of application (radiation protection) the minimisation principle is not absolute.

2.4   Selection process, criteria

(18)   **The conflicts emerging in the discussion in Germany demand a decision-making process focused on legitimation and conflict mitigation, one which also makes it possible to take into consideration more deep-seated conflicts.**

The choice of Gorleben as site of a disposal facility is highly controversial in the public discussion. One argument is that this preliminary selection was made largely on the basis of political, not scientific/technical criteria, another criticism is that carrying out a test of suitability on only one site is not a reliable means of finding a suitable site. Reaching a decision is fraught with conflict: restricting the test of suitability to Gorleben fuels the suspicion that no objective search for the (relatively) best site is taking place, but extending the search to other sites will in all probability provoke equally great protests there. This dilemma does not seem soluble in the short term, which makes it all the more urgent that a decision-making process focused on legitimation and conflict mitigation be devised. Decision-making processes must, moreover, take into consideration the deeper-seated conflicts over modernization and lifestyles, the general direction of energy policy, and questions about the legitimation of collective decision-making.

(19) Decisions must be as transparent and comprehensible as possible, offering opportunities for direct public participation, but without restricting the responsibility of the state.

In the framework of the decision-making and the selection process it is important that the individual steps in the process be made as transparent and comprehensible as possible, and that civil society be constructively involved, without limiting the clear responsibility of the institutions legitimated by the German Constitution and the Atomic Energy Act in the framework of the representative and federal system of the Federal Republic of Germany. The integration of structured forms in which groups in society and the local population can participate in building a consensus and preparing for decision-making is not incompatible with representative democracy, which it can supplement and even enhance if a suitable approach is taken.

(20) Various options for the process of seeking a site should be discussed and weighed up against one another.

In principle the problem-solving options, particularly for the site selection, are as follows:

(a) exploration of a joint European solution,
(b) continuation of sub-surface exploration and testing of the Gorleben site following recognized target values for suitability,
(c) as in (b), but with further sites to be found and explored immediately in parallel, and
(d) beginning a completely new search process to explore several possibly suitable sites at the same time (including or excluding Gorleben) and evaluate them in relation to each other (cf. for further details the remarks in section A 2.8).

(21) A joint European solution is only plausible if Germany also contributes a possible site.

A joint European solution can only be put forward convincingly and fairly if Germany also contributes a site in its own country. Thus this solution would also require the selection of a site in Germany. This reduces the number of options to be considered to three.

(22) **It would not be sensible to insist on the site investigation in Gorleben without testing alternatives.**

Since the Federal Government of Germany has already decided to resume underground investigations at Gorleben without considering alternatives, proposals to initiate a new search seem politically questionable. However, in view of wide-spread acceptance problems, and the aspiration to find a site particularly suitable for Germany, it seems reasonable to integrate further site options into the search. Furthermore, the considerations about spatial and regional development that determined the selection of the Gorleben site at the end of the 1970s have lost much of their persuasiveness in the meantime. This means that it is once again necessary to seek a solution which is appropriate in terms of land-use planning and environmentally acceptable. For these reasons, the selection among the three remaining options (b), (c) and (d) should be reopened and become subject to broad public participation. Only after an organized national debate, which could also be supplemented by forms of Internet participation, should the Federal Government decide anew which option is to be pursued.

(23) **The project group recommends a hybrid approach allowing for the continued exploration of the Gorleben site while at the same time carrying out surface investigations of alternative sites.**

The working group recommends a hybrid approach ("Gorleben plus") which comprises the following components:

(a) further exploration of the Gorleben site, development and assessment of waste management strategies for this site in order to clarify whether it is suitable and fulfils the safety and planning requirements.
(b) at the same time: investigation of alternative sites as additional disposal options, and beginning of above-ground investigations of these sites, building especially upon the German "Arbeitskreis Auswahlverfahren Endlagerstandorte" (AKEnd) proposals and the regions and sites identified as "promising" by the Federal Institute for Geosciences and Natural Resources (Bundesanstalt für Geowissenschaften und Rohstoffe).
(c) a well-defined point in time for taking a decision in principle about whether and how to pursue the exploration of the Gorleben site and/ or the alternative sites; the decision should be prepared and taken in a manner that makes it as robust as possible in the face of political changes.
(d) a concept for communicating all previously obtained results and arguments as well as the scientific and technical constraints to all stakeholders so that they can participate appropriately in the debate and – as far as the legal provisions allow for this – in the decision-making process.

(24)   The recommendation to explore the Gorleben site and determine additional site options is aimed at maximizing the chances of having a disposal facility available in the next few decades.

Recommending both further investigation of the Gorleben site and the identification of additional waste management options primarily aims to maximize chances of having a disposal facility for radioactive waste available within the next few decades. Excluding Gorleben would result in losing a potentially suitable site without any guarantee that an alternative would be found within a reasonable time period. Excluding alternative site options while the site search is still underway would result in a considerable loss of time (in the order of several decades) if the Gorleben project should fail at a late stage in the investigation. From a safety point of view it is quite reasonable to stagger the exploration over time, if the storage facilities are upgraded accordingly (for more on this cf. the remarks in section A 2.8).

(25)   The parallel search not only saves time if the Gorleben project fails, but also increases the credibility of the selection process.

The parallel search increases the overall credibility of the testing procedure. However, it must be admitted that there are, even then, certain shortcomings in terms of public acceptance, especially at local and regional level, compared to an entirely open site selection process. Nevertheless, "Gorleben plus" offers the prospect of a solution that is timely, conserves resources, and is sensitive towards public opinion and stakeholder engagement.

(26)   The establishment of the criteria for the sites should be carried out in a transparent process incorporating an international "peer review".

The establishment of the criteria for determining the suitability or non-suitability of the sites will be carried out in a transparent procedure incorporating an international "peer review". As in the Swiss procedure for selecting a site, safety-related, geological and land-use planning criteria will be developed in parallel. Criteria and threshold values must be defined in such a way that they are not just applicable to Gorleben but to all sites.

(27)   It should be stipulated that sub-surface explorations will be carried out at another site if the surface explorations give rise to the expectation that this site might fulfil the selection criteria considerably better than Gorleben.

For pragmatic reasons the focus should be on finding a site which is, according to the criteria developed, suitable for the safe burial of high level waste. Aspects of optimization should be brought to the fore, particularly when examining the reserve sites. Arrangements should be made for sub-surface exploration to be carried out at one of the alternative sites if and when the surface exploration finds that this alternative site (with salt or another host rock) fulfils the selection criteria demonstrably better than Gorleben.

(28) Programmes for research and development should be designed to meet requirements and monitored regularly, but should be set up with sufficient breadth and flexibility that any revisions of decisions are not excluded.

Programmes for research and development should be designed in coordination with the requirements derivable from the schedule in the general programme, and should be monitored regularly. They should be focused on supporting the programme options determined by resolution, but at the same time be sufficiently broad and flexible that they do not impede a revision of earlier decisions. In particular, the ongoing investigation of burial in reinforced clay should be continued, in order to keep all eligible additional options open.

2.5 Transparency, communication of risks, participation

(29) A comprehensive supply of information to the public must be ensured. To support the required transparency, an information centre should be set up.

Transparency is of particular importance for the successful completion of the process. Comprehensive information and communication with the public and all involved parties about the safety and security standards which have been laid down for the disposal facility are essential. All decision-making activities should be accompanied by a proactive and dialogue-based communication programme. Persons (yet to be specified) will be offered the necessary information to be able to form a well-founded judgement for decision-making. In particular, information about the following aspects should be made available in a way which is easily understandable for non-experts. This should be supported by an information centre (communication centre), which has yet to be set up. Public information should include:

- The geological setting of the site and its interplay with the technical design of the disposal facility and the different barriers;
- The relevant processes and their time frames;
- The reference doses (dose constraints and dose limits) for the various phases of the disposal facility and their position and justification within the overall system of radiological protection;
- Models and methods for estimating doses for the selected release scenarios, and their results;
- Research and development undertaken in Germany and abroad, waste management options discussed internationally, and solutions implemented in other countries in the past and present.

(30)   Effective participation should take place in the selection of the site, without calling into question the ultimate responsibility of parliament and executive.

The site selection process should allow for effective participation, while not calling into question the ultimate responsibility of parliament and the executive. In order to avoid paralysing the process by excessive complexity, overly cumbersome procedural structures, e. g. ones entailing several levels of discourse and a multitude of formal institutions, should be avoided.

(31)   An organized public debate should be provided for all questions of national significance.

At the national level, a public debate on questions of national importance, supplemented by new forms of participation, should be initiated and organized, for example by the German Nuclear Waste Management Commission (Entsorgungskommission), the German Commission for Radiological Protection (Strahlenschutzkommission), and the Federal Agency for Radiation Protection (Bundesamt für Strahlenschutz). Items for debate should include the decision about options for the site selection process, the site selection criteria, and the selection of sites for further (sub-surface) characterization, as well as issues of retrievability.

(32)   Closer involvement of local representatives of civil society only makes sense when issues of local importance are at stake.

Closer involvement of local civil society groups only makes sense when issues of local importance are at stake. This applies to the ongoing evaluation of the investigation results for Gorleben, the evaluation of the results of the exploration of alternative sites, and the decision about whether Gorleben is suitable as a site or not. If a decision is made in favour of Gorleben, public participation should also take place around the local and regional implementation (regarding the "how"). The relevant proposals made by AK-End could be taken up here.

(33)   In the site selection process taking place in parallel to the exploration of the Gorleben site, voluntary alternative candidates who appear suitable should be considered first.

Voluntarism on the part of a potential host community or region would be desirable, in analogy to the procedure in Finland and Sweden. Thus during the selection process for alternative sites which is to be conducted in parallel to the investigation of the Gorleben site voluntary alternative candidates should be considered first, provided they fulfil the selection criteria in the relevant stage of the procedure.

2.6   Institutions in the procedure, expert groups

(34)   A committee of experts with proven scientific or technical expertise should be formed.

The working group recommends convening a group of experts whose main task is to establish the criteria for the test of suitability, and to make a recommendation at the end of the exploration as to whether the criteria are fulfilled or not. Since this committee will have a high level of responsibility, and will be reliant on a high degree of legitimation in such a conflict-laden situation, the focus should be on proven scientific or technical expertise, independence and balance. To ensure this, the working group recommends:

- The committee should be attached to an independent institution, e.g. the National Academy of Science.
- Members can be nominated by the relevant scientific organizations, by civil society groups, and by the companies operating nuclear power plants.
- The meetings of the committee are public.
- The committee will be given the right and the necessary resources to request expert advice, and to hold hearings if it sees these as necessary.
- The recommendations of the committee are not binding. However, the responsible decision-maker would have to be able to demonstrate cogent and publicly comprehensible arguments in order to diverge from the committee's vote.

(35)   At the local level a dialogue forum should be set up for every site under consideration for selection as an institutional point of contact for the regional population.

A further institution of public participation at the local level should be a dialogue forum for each site still under consideration for selection (including Gorleben). The local dialogue forums should function as an institutional point of contact to facilitate effective participation of the regional population and provide access to independent knowledge on the state of the proceedings in the various stages of the selection process. To this end, the forums could instigate public debates, organize public panel discussions and allow access to pools of experts.

2.7   Administrative structure

(36)   The division of responsibilities of the government agencies and other organizations dealing with the management of HLW should be improved.

A streamlined distribution of responsibilities between organisations involved in waste management would considerably increase the chances of implementing the proposed radioactive waste management programme.

This would also comply with the Joint Convention as well as with the proposed EU Directive on radioactive waste management. A clear and visible distinction and the observance of the necessary organizational distance between the applicant and the regulatory agencies should contribute to a greater acceptance of, and trust in, the process and those involved in it. The supporting research should be organised in a manner that boosts the effectiveness of these altered structures as much as possible. In this respect, the presently established distinction between site-specific and basic research may turn out to be inappropriate.

(37) **As part of the reform of the institutional framework, the Federal Agency for Radiation Protection (Bundesamt für Strahlenschutz) should be turned into an independent regulatory agency.**

One major objective of reforms of the institutional framework for the regulation and execution of high level radioactive waste management should be the transformation of the Federal Agency for Radiation Protection into a genuinely independent regulatory agency that can assume core responsibilities in the field of radioactive waste management, especially for regulation, siting and supervision. To this end, the present commingling of its various duties in regulation and execution should be abandoned in favour of a clear separation of functions.

(38) **The operation of the disposal facility could be transferred to a new higher federal authority, a public body, or a private-law corporation vested with public powers.**

The management of the disposal facility could be transferred to a new federal agency or, in the framework envisaged by the Atomic Energy Act, to a public corporation or a private law corporation that is vested with corresponding public powers. Shares could be held in this corporation by both the companies that directly or indirectly (through subsidiaries or joint ventures) operate the nuclear power plants in Germany and – in the interests of creating public trust – the Federal Government. In any case, however, the majority of shares should be held by the public authorities.

2.8   Decision-diagrammes

2.8.1   Preliminary Remarks

In order to illustrate the boundary conditions concerning the development of the proposed radioactive waste management programme (cf. conclusions and recommendations, section A 2.3), a decision tree was developed which aims at visualising (i) decisions already made in the past which might or might not be subject to revision and (ii) options for decisions to be made in the future. The time estimates shown in the tree are based upon the considerations of section B 1.5 "Timescales and potential roadmap".

The decision tree has been derived using ideas developed in the EU COMPAS project (Dutton et al. 2004) but taking issues specific for the situation in Germany into account. Although the logic of the COMPAS project was in principle adhered to, a major deviation was introduced with regard to encapsulation. This has been done because encapsulation for disposal only can take place adequately if the disposal solution is already known.

The decision tree focuses on the management of Spent Nuclear Fuel (SNF). The following assumptions concerning the German programme were anticipated:

- No new nuclear power plants will be built in Germany. As explained in the Part B of this book, the lifetimes of existing nuclear power plants will have no principal impact on the decisions to be made.
- The safe disposal of radioactive waste in deep geologic formations is technically feasible.

Decisions which were already taken in the past and/or are technically determined are indicated by darker grey panels. Bold arrow lines indicate the recommendations given in the present study – which decisions taken in the past better should not be revised and which decisions should be taken in the future, if maximizing the chances of having a disposal facility available in the next few decades is the aim.

In the following, some of the decisions shown in diagramme I (Fig. A.1) are further explained. The numbers correspond to the ones in the decision nodes of the diagramme I.

⟨1⟩ According to the German Atomic Energy Act, since 2005 no further transports to reprocessing plants in France and in the UK are taking place. Thus, the decision "no reprocessing" has been marked as "taken in the past". Nevertheless, waste from earlier reprocessing exists and has to be managed as well.

⟨2⟩ The reprocessing plant at Wackersdorf went never into operation. Before 2005, SNF was sent to reprocessing plants in France and the UK. Before 1989, all SNF arising form nuclear power operation in Eastern Germany was sent back to the Soviet Union.

⟨3⟩ While the waste arising from reprocessing in France and the UK was or will be taken back to Germany, no waste resulting from the reprocessing of SNF which arose in Eastern Germany was or will be returned. Remaining SNF from Eastern German nuclear power plants is being stored in the "Zwischenlager Nord" storage facility near Lubmin/Greifswald. According to international treaties, SNF from Eastern German research reactors has to be returned to Russia.

(4) Presently, the policy in Germany is to store SNF locally close to reactor sites. However, in the past some SNF was already transported to central storage facilities at Ahaus and Gorleben and is being stored at these facilities. At Gorleben, also vitrified high-level reprocessing waste is being stored and it is planned to also ship the remaining waste from reprocessing to Gorleben. All storage facilities are licensed for 40 years, starting with the emplacement of the first cask. The licenses are linked to the lifetime for which the CASTOR casks are licensed. Therefore, a lifetime extension of storage facilities would require new licenses.

(5) Here, it will only be distinguished between those options for waste management endpoints for which waste retrieval is impossible due to the nature of these options on one hand and those for which retrieval is in principle possible (regardless whether retrievability will actually be implemented or not). The decision tree does only consider the *option* of retrievability; in contrast, decisions connected to *actual retrieval* are not considered.

(6) Practically, Germany has taken the decision in favour of deep geological disposal. This is in accordance with the recommendations made here.

(7) For the implications of considering international solutions cf. Conclusions and Recommendations, § 21: A joint European solution is only plausible if Germany also contributes a possible site.

(8) The decision for a host rock has to be made accounting for a number of boundary conditions and criteria, including availability in geologically stable regions and safety evaluations based on the disposal concept as a whole (including technical components). Note that there is an interplay with the decision <9> on retrievability. The bold arrow indicates the preference for rock salt as a host rock in Germany which is manifested not only by the Gorleben investigations but also by numerous R&D activities. The AKEnd proposal attempted at a site selection process without determining the host rock *a priori*. Nevertheless, the concept of a "confining rock zone" which was first mentioned by the AKEnd and now found its way into the BMU Safety Requirements (2010a: "Sicherheitsanforderungen an die Endlagerung wärmeentwickelnder radioaktiver Abfälle") results in a focus on salt and clay as host formations while it is unlikely that crystalline formations in Germany can offer such a confining rock zone.

(9) The BMU Safety Requirements (BMU 2010a) require retrievability during the operational phase and containers remaining for 500 years in a condition allowing recovery (§ 8.6). It is likely that future discussions on the waste management programme will re-open the issue. Note the interdependence of the issue with the choice of the host rock.

⟨10⟩ Decision to be taken in the near future. The bold arrow indicates the preference of the working group for a hybride approach ("Gorleben plus") for a number of reasons (cf. Conclusions and Recommendations, §§ 23, 24, 25). The subsequent part of the diagram will indicate the time when and in dependence of which outcomes central targets of waste management presumably will be reached. The vertical distance between the tree elements give a rough idea of the timeframes to be anticipated.

Diagramme II (Fig. A.2) describes in greater detail the phases and various possible outcomes of the site selection process under the hybrid model "Gorleben plus" advocated by the group.

2.8.2   Decision diagramme I

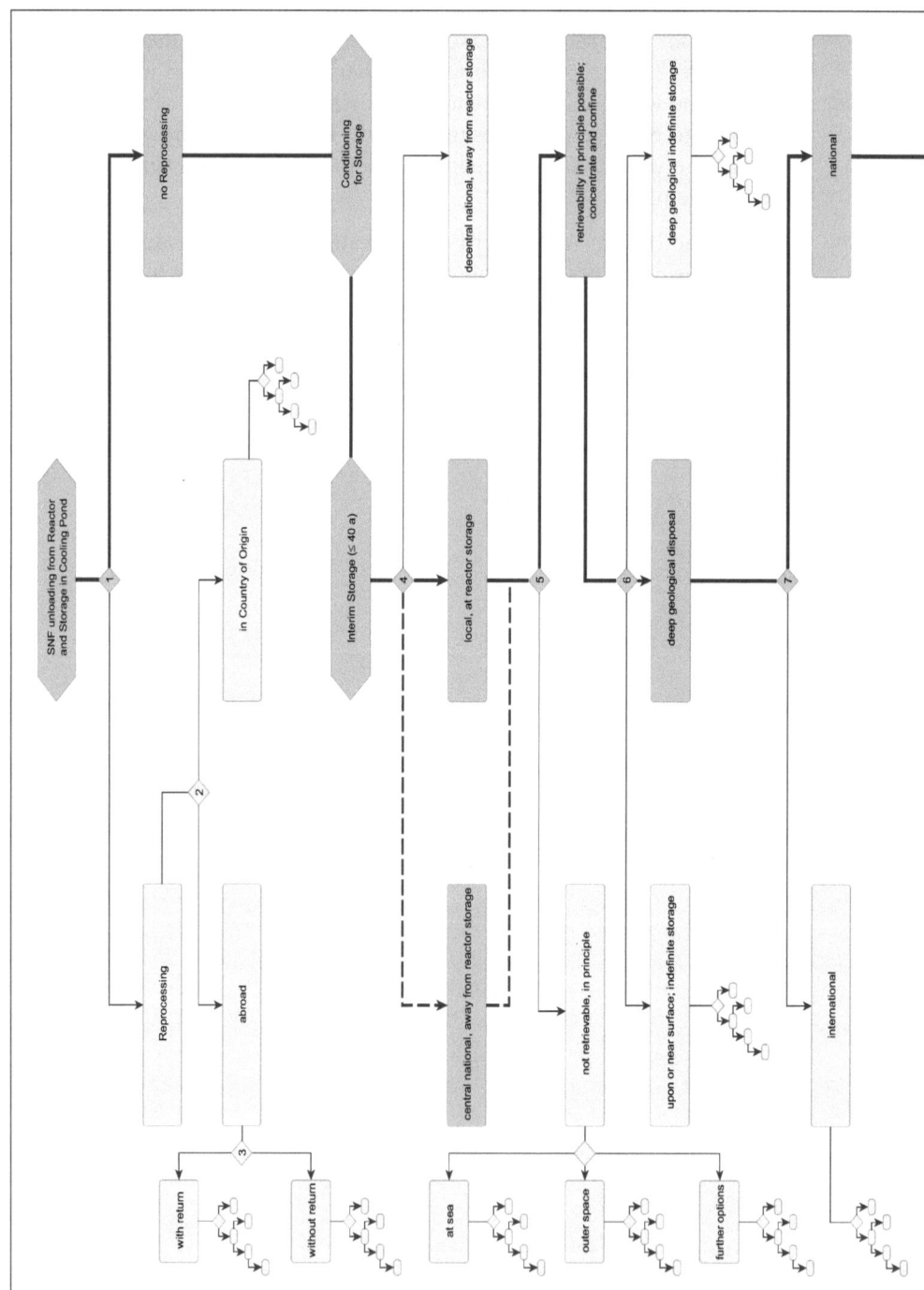

Fig. A.1:   Decision diagramme I: Boundary conditions concerning the development of the proposed radioactive waste management programme (cf. conclusions and recommendations, section A. 2.3)

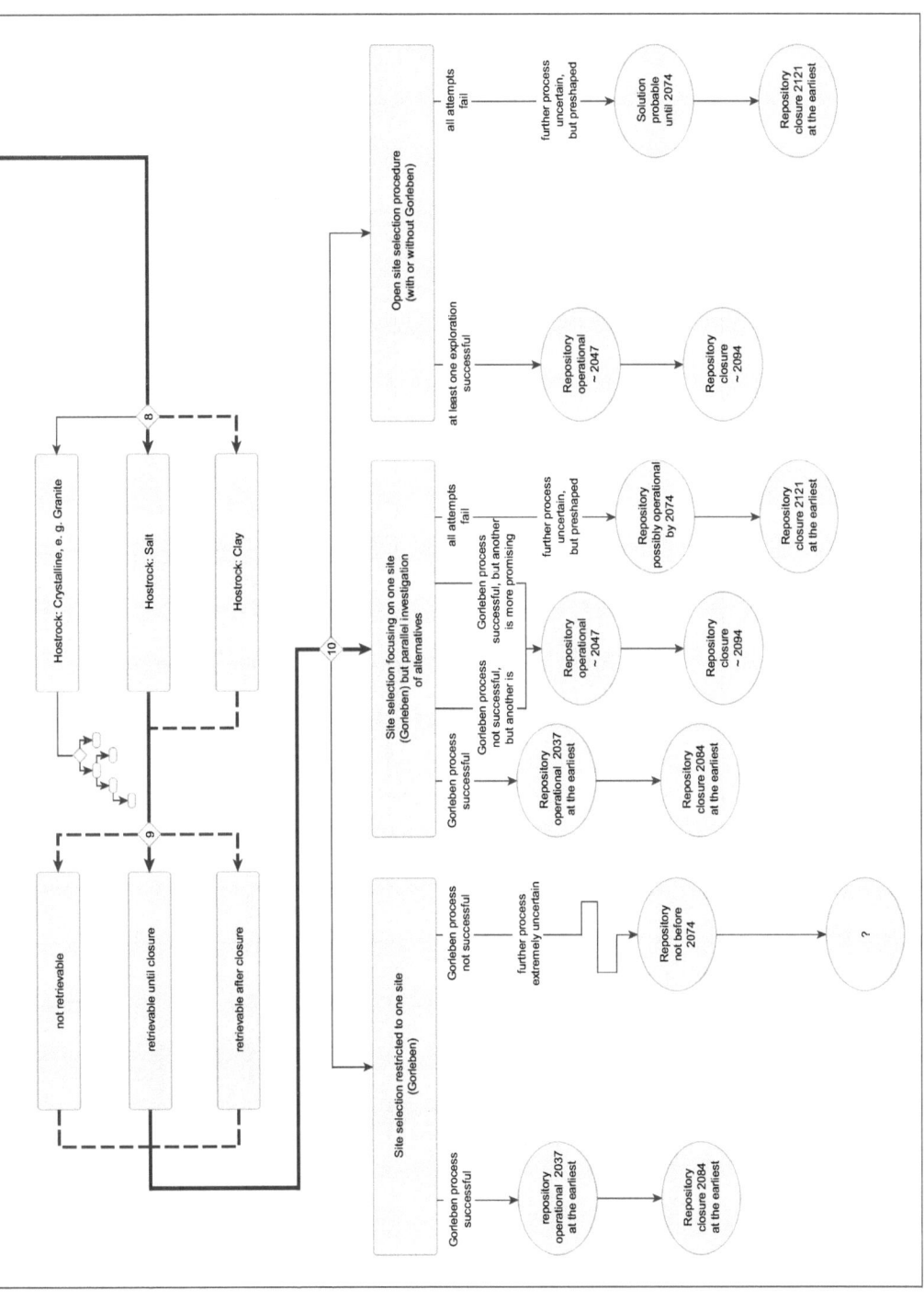

2.8.3 Decision diagramme II

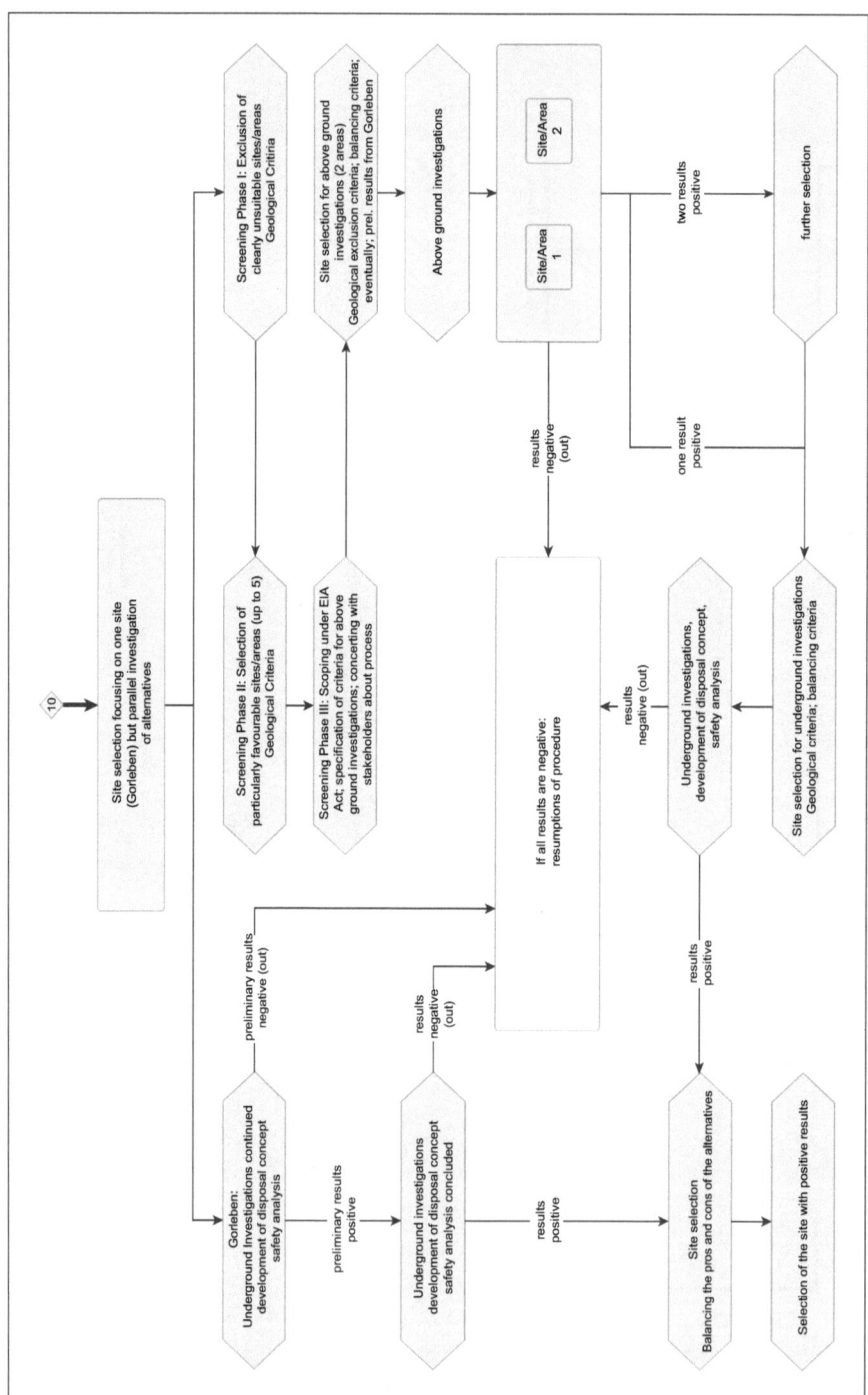

Fig. A.2:   Decision diagramme II: Phases and various possible outcomes of the
site selection process under the model "Gorleben plus" (cf. section
B 4.5.5)

B Technical and normative foundations

1 Waste management strategies and disposal design

1.1 Background, basic approach

A solution to the problem of long-term radioactive waste management (RWM) comprises a technical and social dimension, i. e. it must not only be technically achievable, but also publicly acceptable. The technical solutions have to demonstrate beyond reasonable doubt that a method/concept exists to ensure safe and secure containment of long-lived highly radioactive waste for the indefinite/distant future and that undue burdens on future generations are avoided (taken from Flowers Report (Flowers 1976) but modified).

The social problem's association with the (continuous) use of nuclear energy for power production and weapons (including proliferation) and the risks to health from radioactivity (and low doses, respectively) make RWM basically an issue of controversy and conflict. Besides protecting people and the environment robust public confidence needs to be gained which includes confidence in the societal process and trust in the acting organizations and persons.

The following chapter will concentrate on technical issues. It will

- focus on waste streams following civilian/commercial use of nuclear power and associated fuel cycles;
- categorize waste, focus on high-level radioactive waste (HLW) and spent nuclear fuel (SNF) and provide information about amounts and compositions;
- describe RWM-strategies including storage and transport with a view about how to integrate final disposal options;
- explore final disposal options discussed in the past and pursued in the present;
- focus on key safety and security concerns;
- base expositions and assessments on best scientific knowledge available to the authors;
- screen, shortlist and evaluate strategies and options focusing on technical considerations but allow also linkage of technical aspects (facts) to social issues (ethical reasons such as cross-generational justice and sustainability for which indicators such as burden placement, flexibility and depletion of resources may serve as examples).

Addressing uncertainties appropriately is key for confidence in both, the scientific/technical results and societal attitude (preferences, trust); therefore, geo-scientific/technical and cultural (institutional) uncertainties will be distinguished and addressed when assessing options. For instance, inherent safety features (based on physical laws) and passive safety and security features (barriers) might be less uncertain to be assessed and more preferred than active safety and security means (monitoring and control system), although some stakeholders rather tend to the opposite position.

Uncertainties issues entail the inventory for disposal, the safety case based on detailed knowledge of the local geology and the technical features coupled with understanding, predictability and modeling of mechanical, hydraulic, and chemical processes relevant for safety (e. g., the transport by groundwater), the robustness of the repository concept as a whole, the results of research and whether or not basic scientific premises will change. Institutional and regulatory issues and uncertainties related to low dose impact are addressed elsewhere. Further uncertainties are related to the question whether irradiated fuel will still be considered solely as waste and treated as such instead of being regarded as future resource or both.

1.2   Fuel cycle options and influence on basic aspects of radioactive waste management

1.2.1   Classification of radioactive waste

According to the German Atomic Energy Act[1], material is considered *radioactive* if it contains "one or more radionuclides and whose activity or specific activity in conjunction with nuclear energy or radiation protection cannot be disregarded". The Act refers to the possibilities of clearance or exemption in the case that certain levels of activity or specific activity are not exceeded and of disregarding activities if substances of natural origin were *used* for reasons other than their activity content.

Mobile matters become *waste* if their owner disposes them of, wants to dispose them of, or has to dispose them of because they were either generated in a course of action which was not directed at their generation or the purpose for which they were generated is no longer valid and has not been replaced by another purpose. Although following the lines of an act which is concerned with waste other than radioactive[2], this description seems to be applicable to radioactive waste as well.

In many cases, the decision about whether or not a material is considered waste will be made by its owner. This is, however, not always the case for radioactive materials: In Germany, since 2005 the Atomic Energy Act does not allow to ship SNF to reprocessing facilities. As a consequence, the owners of this fuel are forced to consider this material as waste and not as a resource. The status of the uranium and plutonium extracted from earlier reprocessing of irradiated nuclear fuel is not yet defined.

---

[1] "Gesetz über die friedliche Verwendung der Kernenergie und den Schutz gegen ihre Gefahren", adopted by German Parliament 23 December 1959, latest amendement 8 December 2010.

[2] "Gesetz zur Förderung der Kreislaufwirtschaft und Sicherung der umweltverträglichen Beseitigung von Abfällen" (Kreislaufwirtschafts- und Abfallgesetz), adopted by German Parliament 11 October 2010.

Radioactive waste in Germany arises from a variety of activities including nuclear power production, research, industrial applications of radioactivity, and medical applications.

Waste from nuclear power production includes, in general,

- Spent Nuclear Fuel (SNF),
- operational wastes such as evaporator bottoms, installation components, filters, cleaning substances, protective clothing,
- waste arising during the production of nuclear fuel such as mine and mill tailings, enriched and depleted uranium,
- waste from the reprocessing of SNF such as vitrified High-Level Waste (HLW), hulls and end-cups from fuel assemblies, extracted plutonium and uranium, and operational wastes, and
- waste from the decommissioning of Nuclear Power Plants (NPPs).

Obviously, the types of waste mentioned above (and others) show a great variety in terms of radionuclide content, chemical composition, and physical condition. Different waste types require different ways of conditioning, packaging, shipping, storage, and disposal. As examples might serve:

- The radiation (especially the γ radiation) emitted from the waste defines shielding requirements.
- The amount of volatile radionuclides contained in the waste defines requirements concerning the tightness of containers used for transportation, storage, and disposal.
- The amount of long-lived radionuclides contained in the waste defines the timeframes over which the waste needs to be isolated and/or monitored.
- The heat production of the waste defines cooling requirements and the layout of transport means, storage and disposal facilities as well as requirements concerning the handling of the materials in all steps of its management.
- The chemical composition and physical condition of the waste needs to be considered and often altered by conditioning measures in order to achieve favourable physical and chemical conditions for packaging, transportation, storage, and disposal.
- The potential radiological impact caused by the different radionuclides determines the radiological hazard the waste represents and thus influences its management. The same applies of course to the chemotoxical potential.
- The concentration of fissile nuclides in the waste and the geometric configuration of matters containing such nuclides determine the potential for criticality. Configurations with such a potential have to be avoided.

Therefore, proper radioactive waste management asks for an adequate waste characterization and categorization in the first place. Many distinguishing features of the waste, however, are similar to those of non-radioactive waste. As a consequence, waste categorization schemes recommended internationally focus on radioactivity, activity content and accompanying heat production as well as on lifetime of the radionuclides contained as the determining factors. The IAEA Classification of Radioactive Waste Safety Guide No. 111-G-1.1 (classification scheme shown in Tab. B.1) as well as the very similar scheme proposed by the European Union are very much orientated towards an adequate and safe disposal solution.

Tab. B.1:   Typical waste characteristics (based upon IAEA 1994)

Waste classes	Typical characteristics	Disposal options
1. Exempt waste (EW)	Activity levels at or below clearance levels based on an annual dose to members of the public of less than 0.01 mSv	No radiological restrictions
2. Low and intermediate level waste (LILW)	Activity levels above clearance levels and thermal power below about 2 kW/m^3	
2.1. Short lived waste (LILW-SL)	Restricted long lived radionuclide concentrations (limitation of long lived alpha emitting radionuclides to 4,000 Bq/g in individual waste packages and to an overall average of 400 Bq/g per waste package)	Near surface or geological disposal facility
2.2. Long lived waste (LILW-LL)	Long lived radionuclide concentrations exceeding limitations for short lived waste	Geological disposal facility
3. High level waste (HLW)	Thermal power above about 2 kW/m^3 and long lived radionuclide concentrations exceeding limitations for short lived waste	Geological disposal facility

This is even more the case for the new IAEA classification scheme (IAEA 2009a) (Fig. B.1) which names for each waste category a corresponding disposal option. A country in which this philosophy is to a large extent mirrored in national practice is France (Fig. B.2).

In contrast, radioactive waste management in Germany allows only deep (geological) disposal (at several 100 meters depth) as option for waste exceeding exemption or clearance levels. A first coarse distinction is being made between "heat-generating waste" and "waste with negligible heat generation". In this context, the term "negligible" is defined with reference to the disposal route foreseen for this waste: For the deep repository Konrad (the operation of which shall commence in 2013) the temperature increase at the sidewall of the emplacement chambers in the facility is limited to 3 K for the next 100,000 years. Thus, the conditions *in the disposal facility* (in particular heat conductivity of the rock) are a determining factor for waste categorization. It should also be noted that this way of categorizing waste does not only refer to the matter itself but also to the way it is packaged because the degree to which radionuclides responsible for heat production are dispersed over the waste packages determine the packages' heat output and thus the waste category they belong to. There are, however, limitations for such a dispersion of heat-producing radionuclides in space since the total volume available for disposal at Konrad and the total activity to be disposed of are also limited.

Besides heat production, other restricting factors for accepting waste at Konrad refer to local dose rate, contamination with certain α emitters at the package surface, physical and chemical conditions as well as nuclide-specific limits for the activity amount allowed per package. These latter limits were derived considering heat production but also other issues such as transport safety and potential releases from the repository in the long-term.[3]

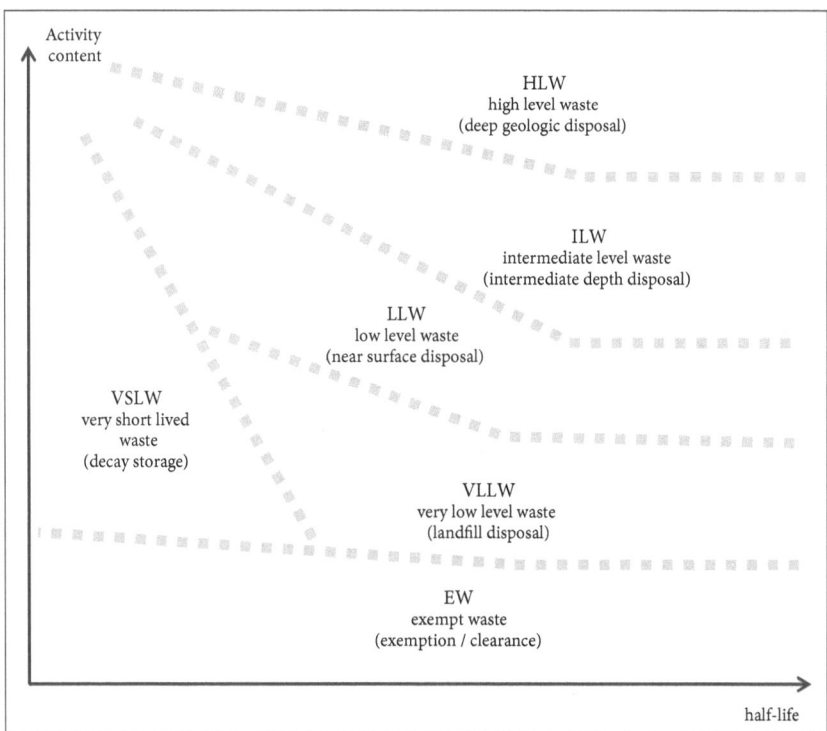

Fig. B.1:   IAEA waste classification scheme as currently established (based upon IAEA 2009a)

---

[3]   For overview see the informations for experts on the conditions of final storage ("Experteninformationen: Endlagerungsbedingungen") on http://www.endlager-konrad.de (in German, last visited February 21, 2011).

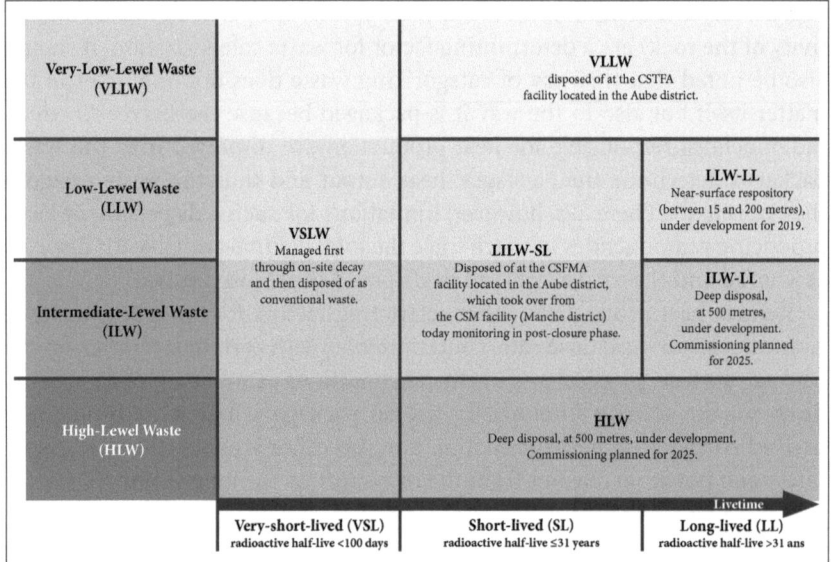

Fig. B.2:   Waste classification used in France (online-documentation, www. andra.fr, last visited April 2011). Certain waste when it contains a too large amount of tritium (radioactive hydrogen) must be stored before disposal in order to allow for the decay of this tritium (approximately 12-year physical half-life).

In a nutshell, it can be concluded that Germany knows two broadly defined waste categories: Waste which is suitable for Konrad (mostly "waste with negligible heat generation"), and waste which is not (mostly "heat-generating waste").

It can be roughly estimated that the former category of "Konrad waste" represents about 90 % of the volume to be managed (around 300,000 m³) but less than 1 % of the radioactivity (around $5 \cdot 10^{18}$ Bq, including around $1.5 \cdot 10^{18}$ Bq α emitters).[4] The latter category (waste not suitable for Konrad) is the one covering SNF as well as the vitrified HLW from reprocessing, both emitting considerable amounts of heat[5], but also other materials such as certain Medium-Level Waste (e. g. vitrified waste from water treatment in reprocessing plants or compacted structural parts and sleeves from SNF reprocessing). For a phase-out according to the Atomic Energy Act from 2002, a total amount for heat-producing waste of about 30,000 m³ or about 17,000 tonnes heavy metal is estimated.

---

[4]   Online-documentation, URL=www.bfs.de, last visited on November 30, 2010.

[5]   One SNF element produces heat in the order of some 100 to some 1,000 W. The actual amount depends on a number of factors such as the fuel element type (and, thus, the reactor type), the burn-up rate, and the point in time after discharge from the reactor.

These figures are, however, subject to change if the German policy concerning nuclear power production would change. According to the German Radiation Protection Agency (Bundesamt für Strahlenschutz, BfS) the 17 presently operating German reactors produce about 370 tonnes heavy metal SNF per year (www.bfs.de, last visited November 30, 2010). The exact amount of heavy metal SNF per year depends on the burn-up rate – e. g. an average burn-up rate of 52 MWd/kg would result in 360 tonnes heavy metal SNF per year (VGB, personal communication).

Thus, a lifetime increase of one year would result in slightly more than 20 tonnes heavy metal per reactor or an increase of less than 1.5 ‰. In other words, a lifetime increase *of 10 years for all 17 reactors* would increase the amount of heat-producing waste by slightly more than 20 %. The increase of the amount of waste with negligible heat generation would only be in the order of magnitude of some per cent (according to the online-documentation of the German Radiation Protection Agency (www.bfs.de) about 45 m^3 per reactor and year), since the major part of the 300,000 m^3 already to be managed is either existing waste or waste from the decommissioning of nuclear facilities which will arise in any case.

It can be concluded that lifetime increases in the order of a few decades do not imply a substantial change with regard to the challenge of radioactive waste disposal. In particular, this applies to waste with negligible heat generation but even the increase of the amount of heat-producing waste by slightly more than 20 % per decade lifetime increase does not result in a situation significantly different from the one Germany is in right now. It should also be noted that there are several possibilities of influencing the capacity of disposal facilities: This capacity is determined by the volume of available suitable host rock of a particular site on one hand and, as the most important factor, heat production and the need to disperse the heat-producing waste in space on the other. Means to adjust the capacity of a repository at a given site include changes of burn-up rate, of decay storage time, and of repository designs allowing for better utilisation of the available host rock volume.

Heat-generating waste not suitable for Konrad including SNF as well as the vitrified HLW from reprocessing is the category of waste in focus of this publication because it is the one for which a solution is still being looked for in Germany. It causes frequent and heated debates – very often with relationship to the question of nuclear power production as a whole and sometimes leading to spectacular protest actions, in particular at the occasions of the yearly "CASTOR shippings" of vitrified HLW back from reprocessing to the interim storage facility at Gorleben.

1.2.2   *Current options for irradiated fuel management*

Before dealing with current options for irradiated spent fuel (SNF) and radioactive waste management it makes sense to refer to the fundamentals of nuclear fission. Uranium-235 is the only natural material being fissionable by neutrons of low energy:

$$_{92}U^{235} + {}_0n^1 \rightarrow 2 \text{ fission products} + 2...3 \, {}_0n^1 + 207 \text{ MeV} \tag{2.1}$$

Natural uranium contains 0.71 % U-235; its fraction needs to be enriched to 3–5 % before insertion as uranium-oxide (UOX) into thermal reactors, i.e. thermal light-water reactors (LWR). Absorption of neutrons by U-238 (99.28 % of natural uranium) leads to the production of a "man made" fissionable material, that is, Plutonium-239

$$_{92}U^{238} + {}_0n^1 \rightarrow {}_{92}U^{239} \xrightarrow[(23\,\text{min})]{\beta^-} {}_{93}Np^{239} \xrightarrow[(2.3\,\text{days})]{\beta^-} {}_{94}Pu^{239} \tag{2.2}$$

which contributes to energy production by in-site conversion. The composition of the discharge fuel depends on the initial enrichment and fuel burn-up. With an initial enrichment of about 3.5 % and average burn-up of about 33GWd/tHM[6] the fraction of U-235 will be down to almost 1 % (see Fig. B.3).

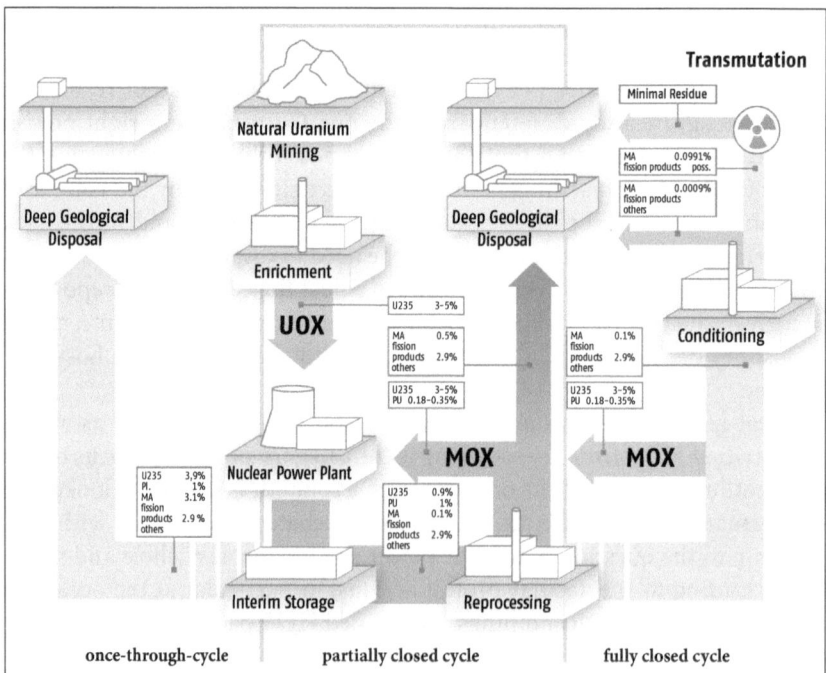

Fig. B.3: Fuel cycle options and fractions of fissile material, minor actinides (MA) and fission products (from: Basler/Hofmann 2008)

---

6   Gigawatt-days per ton of heavy metal.

In a *once-through cycle*, the irradiated fuel is considered as waste (cf. previous section) and treated as such despite the relatively large fraction of fissionable material (e.g. almost 30 % of the initial U-235); increase of fuel burn-up would help to reduce the fractions of remaining fissile material as well as the amount of highly radioactive cladding and structural material per unit electricity while the amount of long lived, heat-generating minor actinides (see section B 1.2.3) would slightly increase (see Tab. B.2) as well as pre-cooling times (see below). The change in irradiated fuel properties must be recognized and accounted for in the design and operation of subsequent back-end facilities.

Tab. B.2: Exemplary impact of increasing burn-up on the back-end of the fuel cycle (GWd: gigawatt-days; tHM: tonnes of heavy metal; TWhe: terawatt-hours of electricity). Source: Debes 2006

Impact Areas			
Burn-up (GWd/tHM)	33	45	60
Fission products (kg/TWhe)	140	140	140
Cladding and structural material (kg/TWhe)	1 210	890	660
Recyclable uranium (kg/TWhe)	3 830	2 810	2 100
Minor actinides (kg/TWhe)	4.3	4.5	4.7
Recyclable plutonium (kg/TWhe)	37	32	27

After unloading from the reactor the irradiated fuel is stored in the reactor pool where it is cooled for several years to allow a large part of the fission products to decay. It can then be sent to a facility for extended storage, to pending conditioning and emplacement in long-term storage facilities or, mostly favoured, to a deep geological disposal facility. With this policy, the fissile material remaining in the spent fuel is not separated, which is sometimes considered favourable in terms of non-proliferation. As no removal of the more radioactive components takes place, this results in a relatively large volume of high-level waste (HLW). In addition, the radiotoxicity of the waste is particularly high as a result of the presence of plutonium and other actinides. The amount of fuel in storage will become significant in countries with a large nuclear programme.

The *once-through fuel cycle* is usually selected by countries with a small nuclear power programme, but there are some notable exceptions. For example, Belgium, the Netherlands and Switzerland have opted for reprocessing contracts with other countries facilities and recycling of the fissile material as mixed oxide fuel.

The country with the largest number of reactors, the United States, is currently favouring a once-through cycle. The USA policy since 1977 has been to ban reprocessing of irradiated fuel in an effort to discourage the spread of technologies which were considered to increase the risk of nuclear proliferation and to treat all irradiated fuel as high-level waste (Gen–IV FCCG 2001). However, following a comprehensive discussion initiated in recent years, this policy was reviewed in the context of a new initiative, the Global Nuclear Energy Partnership, launched in 2006. GNEP is aiming, inter alia, at recycling the used nuclear fuel to minimize waste while developing new technologies to reduce the proliferation concern.[7]

Tab. B.3:   Current fuel cycle options in selected countries (NEA 2008)

Country	Number of reactors	Material	Destination
United States	104 in operation	spent fuel	deep geological disposal
France	59 operation 1 under construction	vitrified HLW	deep geological disposal
		separated U and Pu	recycling
Japan	55 in operation 3 under construction	vitrified HLW	deep geological disposal
		separated U and Pu	recycling
Russian Federation	31 in operation 7 under construction	vitrified HLW	deep geological disposal
		separated U and Pu	recycling
Korea	20 in operation 6 under construction	spent fuel	deep geological disposal
United Kingdom	19 in operation	vitrified HLW	deep geological disposal
		separated U and Pu	to be determined
Canada	18 in operation	spent fuel	deep geological disposal
Germany	17 in operation	Vitrified HLW (past)	deep geological disposal
		separated U and Pu (past)	recycling
		Spent fuel (future)	deep geological disposal

---

[7]   Recently the US Department of Energy has decided to give up the domestic component of GNEP which aimed to develop economic fuels and reprocessing in the USA.

Germany has decided to switch to direct disposal of SNF without separation of fissile material and fission products (see Tab. B.3 for overview).

An important option for spent fuel management is the reprocessing, i. e. the separation of reusable material (uranium and plutonium) and its recycling in current reactors, mainly LWR. A "western country" wishing to select this *partially closed fuel cycle option* and to reprocess its spent nuclear fuel will usually need to use either the AREVA plant at Cap la Hague in northern France or the BNFL plant at Sellafield in the UK. The wastes need to be returned to the country from which the SNF originated after electricity generation.

This option[8] has been selected by a number of countries, particularly those with large and well-established nuclear power programmes, as indicated in Tab. B.3. One of the first short-term advantages of this option is the reduction in the volume of spent fuel to be stored on site or in centralized facilities. One of the major short term disadvantages is the increased number and volume of transports of radioactive materials which are potentially hazardous and subject to strict international regulation (NIREX 2002) but also to public intervention. Another even more important disadvantage, in general, is the separation of fissile and radio-toxic material (in particular plutonium) and its storage on site; this is done for a certain period of time depending on the amount being recycled.

Reprocessing of spent uranium- or plutonium-based fuels is the term used for chemical treatment (PUREX process) of the spent fuel to enable separation of:

– unused uranium, both uranium-235 and uranium-238,
– inbred plutonium together with any original plutonium,
– fission products and higher actinides resulting from neutron capture.

The reprocessed uranium in the form or uranium oxide must be re-enriched to fabricate new LWR fuel. Since reprocessed uranium contains other, neutron-absorbing uranium isotopes such as uranium-232 or uranium-236, it is necessary to slightly increase the enrichment level when compared to use of fresh natural uranium, in order to obtain the same equivalent enrichment (NEA 2008). Significant amounts of reprocessed uranium have already been recycled in several countries such as Belgium (Doel-1 reactor), Sweden and France (900 MWe reactor cores); the remainder of the separated low-enriched uranium is kept as a reserve.

A typical commercial reactor discharges up to 32 kg of plutonium per TWhe[9] for a discharge burn-up of 45 GWd/tHM. The isotopic quality

---

[8] For further details see NEA (2007a) from which parts of the text have been excerpted or based on.
[9] A 1300 MWe commercially operated LWR produces about 10 TWhe annually on average.

of this plutonium varies with the discharge burn-up, but typically about two-thirds of the discharged plutonium consists of the fissile isotopes plutoniumium-239 and -241. To recycle this plutonium, it is possible to fabricate mixed oxide (MOX) fuel. *MOX fuel*, containing about 7% to 9% plutonium mixed with depleted uranium, is equivalent to enriched uranium fuel at a level of about 4.5% uranium-235 and fuel can be used in conventional LWR.

The currently existing MOX fuel fabrication capacities are limited: the only existing commercial plants for LWR MOX fuel are a French plant, whose licensed capacity in tonnes per year has been recently increased from 145 to 195, and a British plant for which operational difficulties limit the capacity to about 40. Japan is planning to commission a 130 tonnes plant by 2011. MOX production worldwide since 1963 accounts for the utilization of over 400 tonnes of plutonium, with between 12 and 14 tonnes being used in 2005 for production of 200 tonnes of MOX fuel.

In the past, the output of reprocessing plants has exceeded the rate of plutonium usage in MOX fuel, resulting in the build-up of inventories of civil plutonium in several countries. These stocks are expected to exceed 250 tonnes before they start to decline after 2010 as MOX fuel use increases, with MOX fuel then expected to supply 5% of world nuclear reactor fuel requirements.

At present, around 30% of SNF arisings are worldwide covered by long-term reprocessing contracts. The main goals of this "recycling" are a better utilization of the energy content of natural resources, especially for those countries which do not have abundant domestic resources, and the reduction of the volume and toxicity of the radioactive waste. Recycling of enriched uranium and plutonium would allow a saving of about 20% of the available resource (NEA 2007b).

Another significant objective of the MOX fuel concept is for dealing with the peaceful use of weapons-grade plutonium, declared as excess by both the Russian Federation and the United States. The material is fabricated into MOX fuel and used in civil LWR, thus producing energy while being irreversibly altered and made unsuitable for weapons fabrication.

1.2.3   Advanced fuel cycles

Advanced fuel cycles strive for better meeting sustainability goals, i.e. by designing a closed fuel cycle, coupled with further developed reprocessing technology and various types of advanced reactors, either with thermal or fast neutron spectrum. The latter might be critical fast reactors (FR) or sub-critical accelerator-driven systems (ADS). In particular,

- the fuel reserves should be extended by recycling irradiated fuel to recover its energy content and by converting uranuim-238 or thorium-232 to new fuel (so called breeding process); and

– the volume and radiotoxicity[10] of the waste should be reduced by more selective separation ("partitioning") of the various long-lived fission products (such as Iodine-129, Technetium-92 and minor actinides (such as Neptunium-237, Americium-241, -243, Curium-244) and by transmutation[11] into shorter lived elements.

Numerous national and international efforts are under way to develop and design advanced "reactor systems" being commercially deployable between 2030 and 2040; the Generation IV International Forum (GIF) seems to be among the most effective ones (GIF 2002).

According to the current state-of-knowledge these fully closed fuel cycles recycling both plutonium and minor actinides (long-lived fission products not yet included) will make use of fast neutron spectrum systems (FR and/ or ADS) for transmutation and multiple recycling schemes with very low losses (0.1 % for Pu and 0.5 % for minor actinides). This fuel cycle option could significantly reduce, i. e. a hundred-fold, the transuranium and long-term radiotoxicity inventories finally to be disposed of; the energy content of the final waste could also be reduced dramatically (Gonzales 2008).

This potential has been confirmed by numerous studies comparing various fuel cycle scenarios/schemes and fuel cycle parameters. Mostly, the open fuel cycle with direct disposal of spent UOX fuel serves as reference case (A1); a plutonium burning scheme (A2) with multiple recycling of plutonium only and using LWR and conventional fast reactors represents a partially closed fuel cycle with mainly disposal of MOX spent fuel. Transmutation schemes may, e. g., include burning of plutonium in LWR and of minor actinides (and plutonium after first recycling) in ADS (B3, called simple double strata). Fig. B.4 shows the development of radiotoxicity inventory over time; it is worth mentioning that fully closed fuel cycle with transmutation add a small contribution from intermediate low level waste (ILW) to the HLW radiotoxicity.

The fuel cycle schemes are also important for the development of heat load (see Fig. B.5) and by this, for the length of the needed HLW disposal galleries as the main thermal limitation for the repository concept is the maximum temperature at the gallery lining/host rock or buffer material interface.[12] This can allow further enhancement of the repository capacity (Gonzales 2008).

The effect of fuel cycle schemes including separation and burning/ transmission of plutonium and minor actinides depends proportion-

---

[10] Measure of the toxicity (health effects) after incorporation of a radionuclide including the effects of radioactive daughter products. Radiotoxicity is dependent on radiation quality (type and energy of ionising radiation) and the biokinetics of the radionuclide in the human body.

[11] Transformation of one radionuclide into another, mainly by neutron bombardment.

[12] For boom clay or bentonite buffer (granite host rock) the maximum temperature should be below 100; for salt the maximum bulk temperature should be less than 180°C.

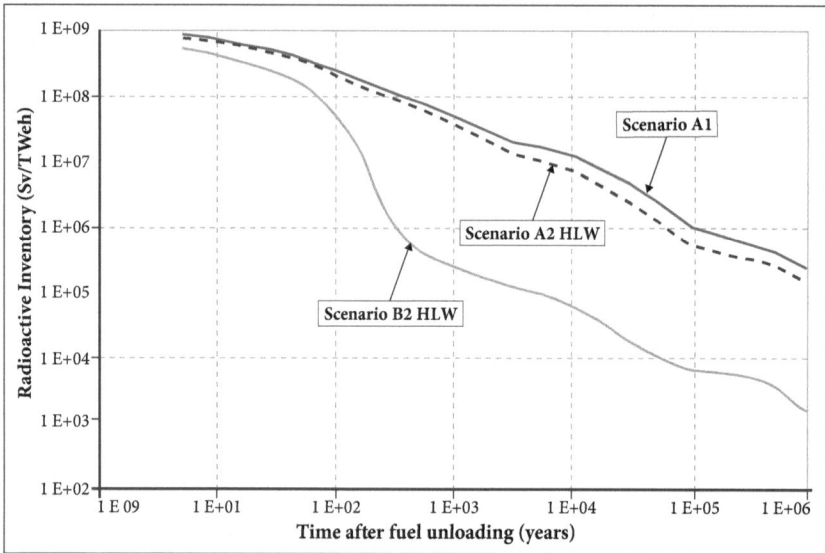

Fig. B.4:   Development of radiotoxicity inventory with time for various fuel
cycle schemes (A1 open, A2 Pu burning, B2 Pu+MA burning)
(Gonzales 2008)

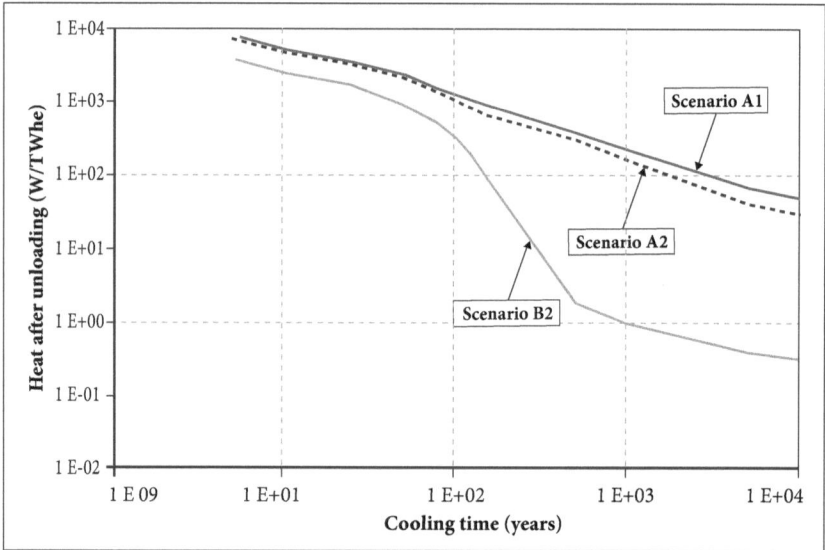

Fig. B.5:   Evolution of thermal power with time for various fuel cycle schemes
(A1 open, A2 Pu burning, B2 Pu+ma burning (Gonzales 2008)

ally on the removal factor or losses, respectively. In the referenced study
0.1 % have been assumed for plutonium losses and 0.5 % for minor actinides, in general. Pilot-scale continuous operation testing at Marcoule,

France have confirmed the feasibility of adapting the-state-of-technology PUREX process to recover 99 % of neptunium (1 % losses); for americium and curium new extractants needed to be developed which allow recovery factors of 99.9 %.

With regard to potential effective radiological doses per annum to members of a critical group within a self-sustaining community, living in an area where the radionuclides released from the repository might reach the biosphere in the very long run, analyses have been made, also to demonstrate the effect of fuel cycle options. It has been assumed that engineered and geological barriers will function as expected, i.e. container lifetime 2,000 years or longer, slow release of radionuclides due to highly delayed matrix degradation and solubility limitations as well as very slow transport through the host rock due to diffusion and sorption processes. In recent safety cases (scenario-type of analyses) the calculated annual effective doses are shown for up to 10 million years indicating for the reference fuel cycle (open cycle) that:

– releases and thereby annual effective doses depend on host rock formation and are caused by long-lived fission products, mainly iodine-129;
– no significant improvement is expected by closed fuel cycles with separation and burning/transmutation of minor actinides under normal evolution of the repository, i.e. exclusion of intrusion scenarios (see Fig. B.6).

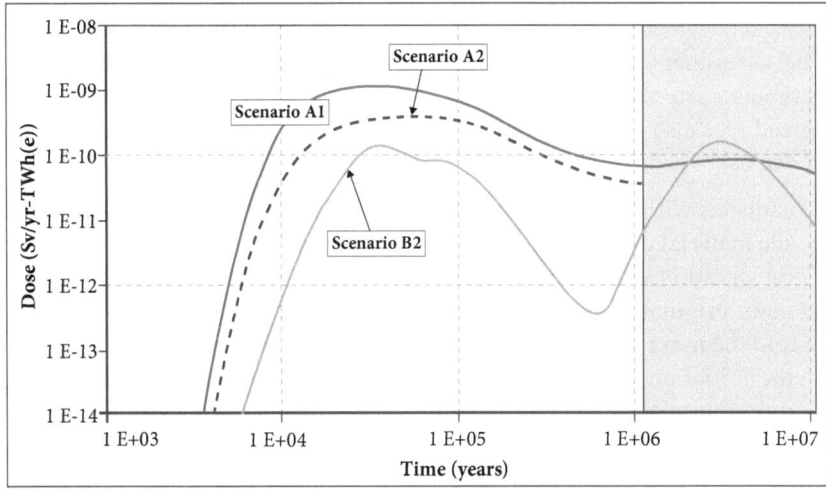

Fig. B.6: High level waste in granite: Calculated annual effective doses to the member of a critical group after release to biosphere. For assessment timeframes beyond one million years (shaded in gray in the figure), dose calculations can be performed but become increasingly meaningless due to the decreasing confidence in the underlying assumptions (Marivoet et al. 2008).

1.2.4   Ensuring subcriticality

It is a strategic objective and regulatory requirement to ensure subcriticality of wastes during all steps of the waste management chain and in particular continuing subcriticality in the repository for all kinds of SNF and wastes. Long-term estimates (up to one million years) taking decay of Pu-239/build-up of U-235 and decay of Pu-240/build-up of U-236 into account (the first increasing reactivity the latter decreasing reactivity) and ignoring fission products (conservative) and build up U-233 (optimistic) result in a minimal critical amount of uranium for water-flooded SNF of about 700 kg (GRS 1998). This translates to 1.4 UO_2-spent fuel elements of a pressurized water reactors while a cask for final repository is designed for 3 or 10 fuel elements. Therefore, criticality over an extremely long period of time cannot be ruled out, in principle (GRS 1998).

Basically, criticality accidents are regarded as prevented if analysis shows that "the effective neutron multiplication factor keff does not exceed 0.95[13] including uncertainties under consideration of all credible normal and abnormal operating conditions. Credit for fuel burn-up may be taken" (US NRC 1998). More specifically the Finnish STUK GUIDE YVL 8.5, 12/2002 claims:

> The emplaced canisters shall retain their subcriticality also in the long term, when the canisters may have lost their integrity and been subject to mechanical or corrosion induced deformations. In the criticality analyses, fuel enrichment and burn-up, the safety margin for the effective multiplication factor and other assumptions shall be selected so that a high degree of confidence in criticality safety is achieved. (STUK 2002)

The assessment of potential criticality issues have been an element of national waste management programs for many years. It is consensually agreed (see also workshop proceedings Johnson 2006) that in the context of disposal

- canisters with SNF raise largest concerns due to their high content of fissile material compared to HLW and ILW waste types;
- the question whether the canister structures and fuel bundles' shape remain original over long periods of time and whether the canisters leak and the inner volume is filled with (moderating) water is essential;
- the initial enrichment and actual depletion of uranium (burn-up) are important parameters to demonstrate compliance with formal criticality criteria.

Often, e.g. Agreius 2002, models for reference fuel assemblies (boiling water reactors (BWR) with 12, pressurized water reactors (PWR) with 4 positions) and final disposal canisters (steel, copper) are used for investigations on criticality conditions for *fresh* (unirradiated) fuel showing that

---

[13]   Criticality is given for keff=1.

as long as the canisters remain leak tight the system is indeed subcritical. If it is assumed that the canisters are leaking and water filled the reactivity increases and, at least for PWR assemblies, the subcriticality criterion is no longer met. More *realistic* assumptions would be to take credit for burn-up of the fuel and include key nuclides in depletion calculations and enrichment of the fuel for variation. In addition uncertainty calculations have been made (see Tab. B.4) leading to total uncertainties of Δk of 0.062 for BWR and 0.078 for PWR.

Tab. B.4: Summary of the uncertainties (Agrenius 2002)

Case	BWR	PWR
Fuel data	0	0
Operating history	0.002	0.002
Declared burn-up	0.005	0.008
Power density	0.001	0.001
Axial void and temperature distribution	0	0
Axial burn-up distribution	0.01	0.02
Control rods	0	0
Horizontal burn-up distribution	0.009	0.015
Isotopic prediction	0	0
Manufacturing tolerances	0.012	0.007
Calculational uncertainty	0.02	0.02
Keno uncertainty	0.004	0.004
Long term reactivity change	0	0
Sum	0.062	0.078

This means that the calculated keff without uncertainties should be smaller than 0.888 for BWR and 0.872 for PWR. One way to meet these values would be to ensure a minimum burn-up of the SNF to be disposed; Tab. B.5 provides these numbers for PWR (left) and for BWR (right), each for sets of nuclides considered in the depletion calculations.

There are some additional uncertainties related to the assumption of stable geometry of the canisters and fuel bundles made in these calculations, which may increase the number of lowest allowable discharge burn-up.

Nevertheless, if full burn-up credit and given uncertainties are accepted subcriticality can be ensured for typical SNF elements, i.e. initial enrichment of U-235 of about 3.3 % and burn-ups of 30–40 MWd/kg U. For non-open fuel cycles the issue of criticality becomes irrelevant.

Tab. B.5:   Burn-ups (MWd/kgU) giving keff=0.95 for different enrichments including uncertainties. Set 1 = [U234, U235, U238, Pu239, Pu240, Pu241, Pu242, Am241]; Set 2 = [Set 1+ Am243, Np237, Nd143, Nd145, Sm147, Sm149, Sm150, Sm151, Sm152, Eu151, Eu153, Gd155]; Set 3 = [Set 2+ Tc99, Rh103] (source: Agrenius 2002).

Pressurized Water Reactors (PWR)				Boiling Water Reactors (BWR)			
Enrichment (%U235)	Isotopics			Enrichment (%U235)	Isotopic		
	Set 1	Set 2	Set 3		Set 1	Set 2	Set 3
1.8	0.0	0.0	0.0	2.8	0.0	0.0	0.0
3.0	25.6	17.5	16.6	3.0	4.0	2.2	2.1
3.6	35.3	24.5	23.0	3.6	12.4	8.1	7.4
4.2	43.6	31.0	29.6	4.2	20.0	14.7	13.8

1.3   Potential radioactive waste management strategies and related technologies

1.3.1   Steps and building blocks

As other wastes, radioactive waste requires a strategy or plan to manage these materials. The EU COMPAS project (Dutton et al. 2004), which investigated such strategies in detail, defines a strategy as "a series of steps that each country intends to undertake in order to achieve the goal of safely managing the waste". "Safely" here means protecting human health and the environment. Although this is (or should be) the ultimate goal of each strategy, there are other issues influencing choices to be made when defining a strategy, the most prominent one perhaps being security.

For Spent Nuclear Fuel (SNF), the *first step* of its management is obviously *unloading* it from the reactor. As for any other waste which will not be recycled the *ultimate step* of a radioactive waste management (RWM) strategy is safe *disposal*. In its 2007 edition, the IAEA Safety Glossary (IAEA 2007) defines disposal as "emplacement of *waste* in an appropriate *facility* without the intention of retrieval".[14]

The interim steps between these two are *processing, transportation*, and *storage*. The exact sequence of these steps in a waste management strategy varies for different countries and for different waste types. In the case of SNF it strongly depends on the basic decisions and choices concerning the whole fuel cycle (cf. previous section).

---

[14] This definition works under the presumption that diluting and dispersing the waste is not an option for radioactive waste (although, in fact, the discharge and dispersal of radioactivity cannot completely be avoided when operating nuclear facilities).

Processing means changing the physical, chemical, and/or radiological characteristics of the waste in order to make it suitable for the next management step (and perhaps already for further steps). The goal is to "produce a waste package suitable for handling, shipment, storage and/or disposal" (IAEA 2007). In this context, the term "suitable" is meant to address one or more of the following issues:

- Provision of a physical form which can be handled or transported,
- protection of staff and the general population against direct radiation and/or contamination,
- prevention of the escape of volatile radionuclides,
- prevention of critical configurations of fissile materials,
- provision of adequate cooling,
- provision of physical and chemical characteristics favourable for disposal.

The IAEA Safety Glossary (IAEA 2007) distinguishes between processing by "pretreatment", "treatment", and "conditioning". Pretreatment activities are collection, segregation (of waste types or materials), chemical adjustment, and decontamination (i. e. removal of radioactive substances from surfaces) of the waste. Treatment activities include volume reduction (by compacting or compressing), activity removal, and change of composition. "Conditioning" covers immobilisation resulting in a solid waste form (e. g. by vitrifying high-level liquid waste from reprocessing or by encapsulation of low or intermediate level waste using cementitious grout, bitumen, or polymer), packaging of waste forms into containers, and packaging these containers into outer or secondary containers (so-called overpacks).

Transport of radioactive material (including radioactive waste and SNF) on land, water, or in the air becomes necessary when the locations at which the material arises is different from the ones at which it will be processed, stored, disposed of, or otherwise used. Obviously, the need for transport of radioactive waste strongly depends on the definition of the waste management strategy and on the choice of sites at which steps of this strategy will be undertaken. However, it is also possible that, on the contrary, issues related to transport become determining factors when defining a strategy: For example, the decision to store SNF in the vicinity of the reactor sites rather than at central storage sites in Germany (2002 amendment of the Atomic Energy Act ("Atomgesetz")) was motivated, amongst other things, by the wish to minimise SNF transports.

Transport of radioactive material is being handled as transport of other dangerous goods, but with specific account for radiation protection issues. Containment of radioactive materials, protection against direct radiation, prevention of critical configurations of fissile materials, and protection against heat effects have to be considered. According to the relevant IAEA regulations (IAEA 2009c), this is being achieved

[...] firstly by applying a graded approach to contents limits for *packages* and *convey-ances* and to performance standards applied to *package designs* depending upon the hazard of the *radioactive contents*. Secondly, [...] by imposing requirements on the *design* and operation of *packages* and on the maintenance of *packagings*, including a consideration of the nature of the *radioactive contents*. Finally, [...] by requiring administrative controls including, where appropriate, *approval* by *competent authorities*.

Storage of radioactive materials is, in contrast to disposal[15], characterised by the intention of retrieving the material (IAEA 2007). It is, thus, an interim step in a radioactive waste management strategy which can be undertaken for a variety of reasons:

(a) to allow the levels of radioactivity and heat output to decline before the next step of process of the waste management strategy can be enacted (*decay storage*); (b) to provide stock for ongoing process, transport step or immediate disposal (*buffer storage*); (c) awaiting a step for which the required facility or transportation capability are not yet available, or awaiting a decision to be made on the next step for a particular waste or material (*interim storage*); (d) for materials that, while not immediately required, have some potential future use or value and, therefore, have not been declared as a waste (*strategic storage*). (NEA 2006)

Technical solutions for waste storage vary, dependent on the materials to be stored and the purpose and anticipated duration of storage (cf. Tab. B.6).

Numerous solutions for *disposal* (or, more broadly, for long-term management especially of HLW and SNF) were discussed in the past. More recent discussions were undertaken by the British Committee on Radioactive Waste Management (CoRWM 2006) and by Canada's Nuclear Waste Management Organisation (NWMO 2005).

In the course of its work, CoRWM identified a "long list" of long-term management options which include not only disposal but also storage (Tab. B.7, 1st column).

The long list was developed using advice from UK Nirex in which the options summarised in Tab. B.7, 2nd column were considered. Similar work undertaken in order to support Canada's Nuclear Waste Management Organisation (NWMO) identifies the RWM options summarised in Tab. B.7, 3rd column (NWMO 2003).

The following can be noticed:

- Only the options at the bottom of the table represent disposal solutions in the strict sense of the word.
- The options at the top of the table ("recycling or (re-)processing") are no real endpoints of a radioactive waste managing strategy. They result in a change of waste amount and characteristics but for the remaining waste still a disposal solution needs to be found.

---

[15] Sometimes the term "final storage" is being used for something what would be, in IAEA terminology, a form of disposal. This publication, however, will use the terms "disposal" and "storage" consistent with IAEA terminology with one exception: The IAEA glossary advises NOT to use the term "interim storage" but in the NEA 2006 document this refers to a specific purpose of storage.

Tab. B.6: Examples of storage facilities for SNF and long-lived radioactive wastes in OECD member countries (from NEA 2006)

Storage facility type or concept	Examples	Expected storage time
Immediate storage for cooling after unloading from reactor	Alt SNF is cooled under water after its unloading from reactor	Months to years
Interim storage of SNF in storages at reactor sites, in central stores or at reprocessing plants	Dry or wet storage facilities for SNF at reactor sites	Months to decades
	Above ground dry storage of LWR fuel assemblies at Gorleben (Germany)	Until after 2030
	Below ground wet storage of SNF at CLAB Oskarshamn (Sweden)	At least 30 years
	Above ground storage of SNF at ZWILAG (Zentrales Zwischenlager Würenlingen, Switzerland)	40 years or more
	Storage pools for baskets with spent fuel assemblies unloaded from transport casks (La Hague)	Minimum of 2 years
Storage of vitrified HLW at reprocessing facility or (after return) in country of origin of SNF	La Hague (EE-V-SE & R7T7) Dry storage of vitrified high level waste	Up to 100 years (on the basis of studies on durability)
	Dry storage of HLW canisters at Gorleben (Germany) JNFL Vitrified Waste Storage Center (Cooling and temporary storage awaiting disposal in a future repository) (Rokkasho, Japan)	Until after 2030 30–50 years
	Dry storage of vitrified HLW at HABOG (The Netherlands)	Until 2130 (design basis ~100 years)
	Above-ground storage of vitrified HLW at ZWILAG (Zentrales Zwischenlager Würenlingen, Switzerland)	40 years or more
Geological disposal concepts that include phase(s) or step(s) of underground "storage" before closure	No existing facilities	
ILW storage awaiting disposal	Storage of medium level waste (mainly cladding from reprocessing of fuels) at ECC La Hague	Up to 100 years (on the basis of studies on durability)
	Rock cavern at Oskarshamn nuclear power plant: Storage for short-lived LILW awaiting disposal in SFR. Some long-lived LILW (e. g. core components) awaiting disposal in a future repository	License until 2010 but likely to be extended. Long-lived LILW likely to be stored for at least 30 years.
	above ground storage of LILW at ZWILAG (Zentrales Zwischenlager Würenlingen, Switzerland)	40 years or more

- The options in the middle of the table are storage measures. The technical solutions for such storage are similar to the ones for decay, buffer, interim, or strategic storage, and as these solutions they require human activities such as monitoring, control, maintenance, refurbishment etc. *over the whole storage timeframe.* They are, however, different from the storage discussed above in that they are regarded as a kind of (perhaps provisional) endpoint which is indicated by the terms "indefinite" or "long-term". Insofar, it is not completely clear to which extent the IAEA notion of having intent to retrieve the waste really applies here.
- In contrast, the disposal solutions at the bottom of the table are meant to lead to a state at which no further human intervention or follow-up activity is needed ("passive safety"). In fact, distrust concerning this idea of passive safety was one of the motivations behind promoting long-term storage solutions instead or to go for disposal approaches for which waste retrieval, although not intended, is – at least for some time after emplacement – eased by design measures.

One of the differences between the right column and the other two is that it explicitly addresses the possibility of international solutions which is presently not foreseen in most if not all national legislations and regulations. In principle, this "international option" is conceivable not only for storage and underground disposal (as in the table) but for others (e.g. borehole disposal or rock melting) as well. Obviously, international solutions (sometimes also called regional) make sense in particular for countries with smaller waste amounts and/or geographical and geological boundary conditions which make national solutions difficult to implement.

1.3.2   Reference case: direct disposal of spent nuclear fuel

In order to explain how the above mentioned elements of a RWM strategy match and to illustrate the safety and security challenges associated with these elements, in the following a rather simple strategy is described in greater detail. The strategy is based on the open "once through" fuel cycle (cf. section B 1.2): SNF is declared as waste (not as a resource or potential resource), no reprocessing takes place, and the waste is to be disposed of in a deep (geological) repository. The paragraph will focus on the SNF and not address other waste arising in the process.

Concerning the most important properties of SNF, the NAGRA safety report (NAGRA 2002) states:

> Irradiation of the fuel assemblies produces a large number of radionuclides. These include fission products, arising from fission of uranium and plutonium in the fuel pellets, and activation products, arising from neutron absorption. Some of the activation products, such as ^{14}C and ^{36}Cl, are present in both the fuel pellets and structural materials, whereas others, such as various actinide nuclides, are contained in significant amounts only within fuel pellets. Certain radionuclides are enriched at grain boundaries in the fuel, at pellet cracks and in the fuel/sheath gap as a result of thermally driven segregation during irradiation of the fuel in the reactor [...]. (Fig. B.7)

Tab. B.7: Long-term management options (CoRWM, UK Nirex, NWMO)

	CoRWM2006	UK Nirex	NWMO Background Papers 6–5
Recycling or (Re-)Processing	partitioning of wastes and transmutation of radionuclides		partitioning and transmutation
	burning of plutonium and uranium in reactors		reprocessing
	incineration to reduce waste volumes		
	melting of metals in furnaces to reduce waste volumes		
Storage	interim or indefinite storage on or below the surface	long-term above ground storage	above ground storage
		[storage below surface not mentioned in UK Nirex]	underground storage
			international storage
Disposal	near surface disposal, a few meters or tens of meters down	near-surface disposal	
	deep disposal, with the surrounding geology providing a further barrier	deep geological disposal	underground disposal
			international underground disposal
		deep boreholes	emplacement in deep boreholes
	phased deep disposal, with storage and monitoring for a period		
	direct injection of liquid wastes into rock strata	rock melting [also discussed for solid/ solidified waste]	rock melting
		direct injection	
	disposal at sea	sea disposal	disposal at sea
	sub-seabed disposal	sub-seabed disposal	sub-seabed disposal
	disposal in ice sheets	disposal in ice sheets	disposal in ice sheets
	disposal in subduction zones	disposal in subduction zones	disposal in subduction zones
	disposal in space, into high orbit, or propelled into the sun	disposal in outer space	disposal in space
	dilution and dispersal of radioactivity in the environment		dilute and disperse

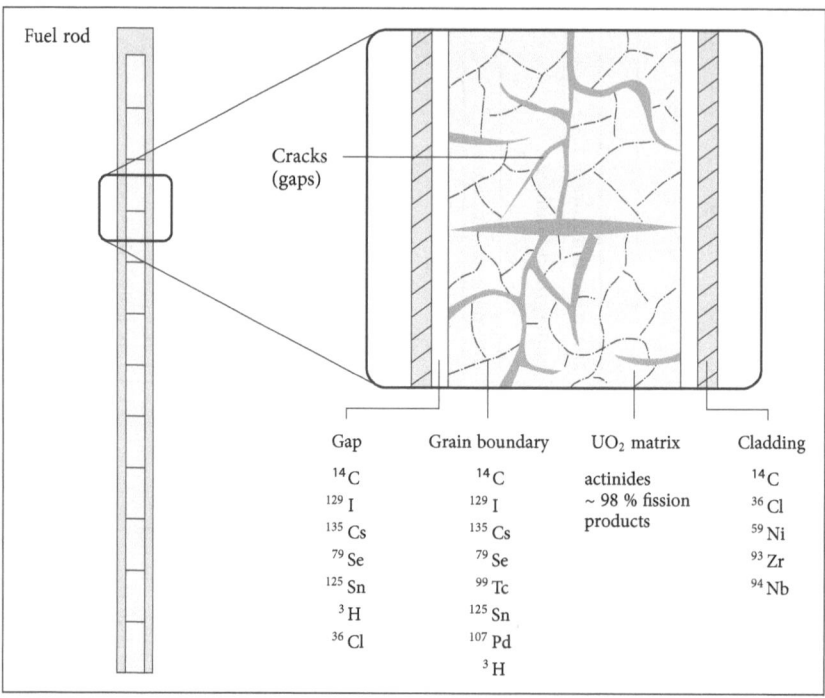

Fig. B.7:   Schematic illustration of the distribution of radionuclides within a
fuel rod (from NAGRA 2002)

After having had discharged the nuclear fuel from the reactor, some
time is needed to allow the most short-lived radionuclides to decay in
order to reduce heat generation and radioactivity ("decay storage" in the
terminology of the NEA report on "The Roles of Storage in the Management of Long-lived Radioactive Waste" NEA 2006). In Germany and
in many other countries the SNF elements are emplaced in water-filled
ponds at the reactor sites, for: cooling of the SNF and shielding purposes. The minimum time the SNF elements have to stay in the cooling ponds depends on the initial state and evolution of their heat output capacity and activity content which both depend on factors such as
the fuel element type and the burn-up. The NEA 2006 report names "at
least several months" for this time but the general practise in Germany
is to keep the SNF in the cooling ponds for several years. Fig. B.8 demonstrates that both heat output capacity and activity decrease by more
than an order of magnitude during the first year after discharge and
that another one to two orders of magnitude are gained over the next
couple of years.

As the need for cooling and shielding decreases with time transportation of the elements without permanent water cooling and shielding

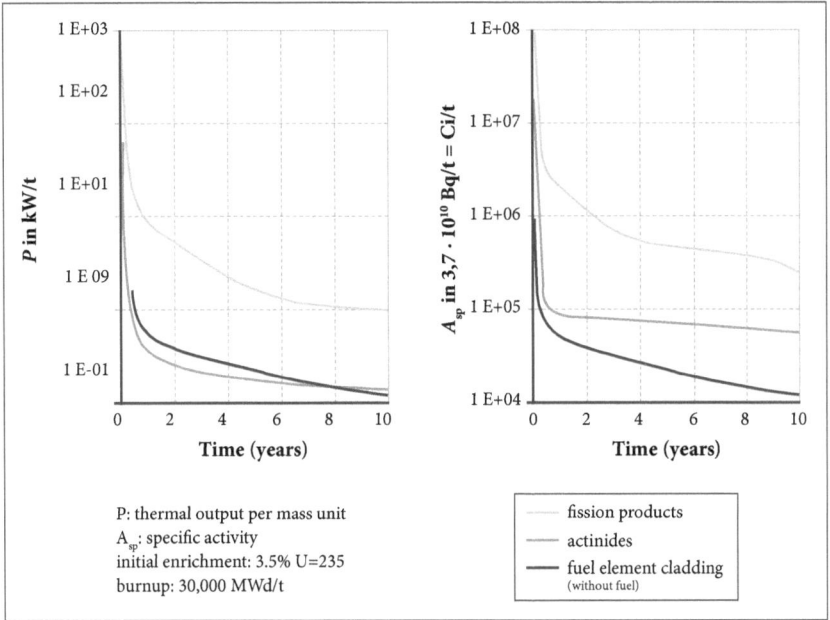

Fig. B.8: Evolution of heat generation capacity and radioactivity content of SNF (according to the online-documentation www.kernenergie.de)

over longer distances becomes possible. Nevertheless, the requirements on transport casks concerning shielding, cooling, and tightness are high and have to be maintained even in the case of major accidents. The same applies for the necessity to keep the fissile material in the casks in a non-critical configuration, leading to stringent the requirements concerning mechanic stability and integrity.

In Germany, the variety of casks used for the transport (as well as for the storage) of SNF (and of other radioactive materials) is known under the acronym CASTOR[16]. All CASTOR variants have in common that their major parts consist of ductile cast iron. Tightness is provided for by a two-lid system with pressure control. In the interior, cages keep the material in the desired configuration. The exact layout varies depending on the material the casks have to accommodate. In general, they have lengths ranging from 4 to more than 6 meters, with diameters of about 2.5 meters; they can accommodate – if designed for – SNF elements up to ten tonnes heavy metal which leads to a total mass of more than 100 tonnes per cask. The performance of the casks has to be ensured over 40 years since they are foreseen not only for transportation but also for storage.

---

[16] Cask for Storage and Transport of radioactive Materials, cf. www.gns.de.

The determining factor for any transport is the question at which location the material should be stored in advance to disposal. Such further storage may be needed as long as no disposal solution is available ("interim storage" in the terminology of the NEA 2006 report), or later to ensure a disposal procedure according to schedule ("buffer storage") or to allow further decrease of the heat production ("decay storage"). The heat production of the SNF is a determining factor when designing an underground disposal facility ("geological repository"); the more heat is being produced by each SNF element, the wider the elements have to be dispersed in the host rock which, on the other hand, will offer only a limited volume. Moreover, conditioning the waste for disposal becomes easier if its heat output and radioactivity is further reduced. Thus, a further "decay storage" of some decades considerably eases repository design, construction, and operation.

The choice of location(s) for such further storage depends on the disposal policy and the state of the disposal programme: If a disposal site is already identified it might make sense to store the material in the vicinity of this site, i. e. to establish a *central storage facility* at this site. For example, the initial plans for the Gorleben site were aimed at a "waste management centre" ("Entsorgungszentrum") which would have included facilities for storage, reprocessing, conditioning, and disposal.[17] As a consequence, the SNF (and HLW) storage facility ("TBL-G") was constructed there. This, however, turned out to become a problem because – although the Gorleben salt dome was still under investigation with regard to its suitability for disposal and no final decision was made regarding its use – the transports to the Gorleben storage facility were interpreted by some as a "creation of facts" concerning the choice of the salt dome for disposal and thus lead to the spectacular and sometimes violent protest actions against these transports.

There are, however, also examples of centralised SNF interim storage at locations which are not foreseen as disposal sites: The facility for centralised SNF storages in Ahaus ("TBL-A") and in Lubmin (Zwischenlager Nord "ZLN") in Germany and the CLAB facility near Oskarshamn in Sweden. Actually, at the time at which CLAB was planned and constructed, a site close to the facility was amongst the candidate sites for disposal and there is also an underground rock laboratory close to CLAB at which research for disposal is being undertaken. But in 2009 a decision was made to apply for the construction of a disposal facility near Forsmark (which is located more than 400 km away from Oskarshamn). Thus,

---

[17] Note, however, that these plans were developed at a time at which the reference case described here, i.e. direct disposal of SNF, was not foreseen in legislation and policy. In other words, the rationale behind the planned "Entsorgungszentrum" was somewhat different from what is being described here.

a central storage at a site does not necessarily predetermine the choice of this site for disposal.

The alternative for central SNF storage is to operate several storage facilities at different locations (de-centralised or local storage), in particular, if these locations are close to the reactor sites. Such a concept is being followed in Germany as part of the nuclear policy implemented with the 2002 amendment of the Atomic Energy Act and can be seen as motivated by the desire to avoid safety and security problems as well as protest actions associated with SNF transports. Obviously, if no decisions about the choice of a disposal site has been made so far, a policy of de-centralised SNF storage at, or close to, reactor sites requires only one long-distance transport per unit SNF (to the disposal site once it has been chosen) while a centralised storage results in two transports if the disposal site turns out to be not in the vicinity of the storage site. This gain, however, goes along with the challenge to ensure safety and security of a number of storage sites spread all over the country.

Layouts for the SNF storage facilities vary: TBL-A, TBL-G and ZNL are situated above ground and cooling is ensured by natural air convection. In contrast, CLAB is situated below the surface at a depth of about 30 m and the SNF elements are stored in storage ponds with forced water cooling. The storage facilities at the German reactor sites are situated above ground and cooling is ensured by natural air convection. An exception is the facility at Neckarwestheim at which the SNF is being stored in tunnels below the surface. There are two different types of above-ground facilities, one of them being similar to the facilities at Ahaus and Gorleben, the other with a thicker concrete structure which could in the future allow the use of storage casks other than the CASTOR containers in which case part of the safety functions normally associated to the CASTOR casks would be ensured by the storage building.

Even if a country strives for having a disposal facility available within the next one or two decades, as it is the case e. g. in France, Finland, Germany, or Sweden, design and licensing lifetimes for SNF storage facilities also range in the order of decades. In Germany, the permitted storage time (both in central and in local storage facilities) is 40 years. This goes along with the design lifetime of the CASTOR casks (cf. above).

There are, however, countries such as The Netherlands or Spain which did not yet decide on a disposal solution or which foresee such a solution for a later point in time. As a consequence, these countries have to foresee longer storage times. For example, The Netherlands plan for SNF storage for at least a century and perhaps for longer (up to three centuries). The facility for such storage is the "HABOG" (Hoogradioactief Afval Behandelings- en Opslag Gebouw, Fig. B.9) near Borssele (Province of Zeeland), an above-ground storage facility with natural air convection cooling which

Fig. B.9:   The HABOG storage facility in Zeeland (The Netherlands; photo:
K.-J. Röhlig)

is so far licensed for a century but the licensed lifetime of which might be
extended later.

The reasoning behind this policy of storage in a building is as follows:

Also it should be realised that the cumulative waste volume that is actually in
storage right now is only a few thousand m³. For such a small volume it is not
economically feasible to construct a deep geological disposal facility, as the costs
are mainly determined by the construction costs of such a facility. The waste volume collected in a period of 100years was judged as large enough to make a disposal facility viable. However, the recent decision to postpone the closure of the
Borssele NPP to 2033 implies an additional 30 years of production of high level
waste, as well as an additional 30 years of cost contribution to the disposal fund.
This means that a shorter storage period than 100years could become possible.

For this 'interim' period considered storage in buildings will be required. This
creates at least five positive effects:
  – There is a period of 100 years available to allow the money in the capital growth
    fund to grow to the desired level. This brings the financial burden for today's
    waste to an acceptable level.
  – During the next 100 years an international or regional solution may become
    available. For most countries the total volume of radioactive waste is small.
    Co-operation creates financial benefits, could result in a higher safety stand-
    ard and a more reliable control.
  – In the period of 100 years the heat-generating waste will cool down to a situa-
    tion where cooling is no longer required
  – A substantial volume of the waste will decay to a non-radioactive level in 100
    years.
  – In 100 years from now new techniques or management options can become
    available.

(Netherland's Ministry of Housing, Spatial Planning and the Environment 2009)

Long-term storage is, however, associated with a number of challenges the most important of which are summarised below (see Box B.1 from NEA 2006).

Box B.1: Some factors needed to preserve the safety and security of storage of radioactive waste from NEA 2006.

- initial conditions of stable waste form, adequate packaging or containment and good facility design
- good record keeping on waste origin, characteristics and location in store
- maintenance of store structure and all infrastructure for handling and inspecting packages and waste
- adequate control of store environment, e.g. temperature and humidity in dry stores and water chemistry in wet storage of spent fuel
- monitoring of environmental and radiological conditions and degradation, if any, of packages or waste
- security of the site and facility from malicious interference and inadvertent human events, including terrorist attacks and accidents
- protection of the site and facility from natural events, e.g. flooding, hurricane and major seismic events
- capability to assess risk from routine Operation, normal degradation and design basis accidents, including events such as above, and apply appropriate mitigation strategies
- capability to recognise when repackaging, store refurbishment or replacement of equipment is required and to perform such operations
- capability to remediate in the event of any potential failure of safety or security
- organisational capability to continue all of the above, including staff training and maintenance of safety culture, technical knowledge and records;
- appropriate regulations and independent inspections to ensure compliance with national regulations and safety requirements
- compliance with international nuclear safeguards requirements
- secure financial resources to ensure all of the above
- political and societal commitment to continue all of the above
- preparedness for the eventual implementation of an endpoint solution

Obviously, some of these challenges increase with the planned lifetime of a storage facility since they depend on societal, political, and economic stability which cannot be guaranteed forever. Others, in particular the security and safeguards issues, are more difficult to implement for many local (as opposed to few central) storage facilities.

For the (final) disposal solution (which, for the illustrative purposes of this paragraph, is assumed to be deep (geologic) disposal of the SNF in a mine-type facility), further conditioning is needed: It is either possible to dispose of entire fuel assemblies, entire fuel rods, or to cut the rods into pieces. In either case, the SNF needs to be packaged into containers or casks suitable for disposal. Dependent on the disposal concept, requirements for these containers vary. The major issues to be accounted for when designing these containers are the necessity to handle the containers during repository operation above and under ground (resulting into requirements concerning e.g. heat conduction and shielding), the physical arrangement of disposal (emplacement drifts, chambers, or vertical or horizontal boreholes of varying lengths), and requirements concerning container lifetime and performance (concerning e.g. temperature, mechanical stability, chemical issues) after emplacement. Examples of containers designed for SNF disposal are

- the POLLUX cask with a height of appr. 5.5 m and a diameter of about 2 m which consists of a ductile iron cover and a steel insert (www.gns.de) and
- the copper canister with cast iron insert (50 mm thickness of copper, length of appr. 4.8 m, diameter of 1.05 m) designed for the Swedish KBS-3 disposal concept (www.skb.se).

Normally, the conditioning should be undertaken close to the final disposal site. If undertaken elsewhere, this implies that the disposal containers or casks have to be suitable for long-distance transport.

The tables below summarise the quantities of the most important radionuclides in some SNF canisters foreseen for the Swiss programme; the inventories depend on the type of SNF elements (BWR or PWR UO_2 or PWR UO_2/MOX fuel assemblies); the burn-up is assumed 48 GWd/t_{IHM} for all variants.

> The heat output of a typical canister of SF is approximately 1500 W after 40 years of cooling (the minimum assumed pre-disposal storage period). The total α activity of MOX fuel after decay for 40 years is approximately 5 times larger than that of UO_2 fuel, while the fission product activity is approximately the same. An important consequence of this is the higher heat output of MOX fuel at longer times [...]. (NAGRA 2002)

Tab. B.8: Inventories of safety-relevant radionuclides in a reference canister containing 9 BWR UO2 fuel assemblies with a burn-up of 48 GWd/tIHM, after 40 years decay

Radio-nuclide	Fuel [Bq]	Structural materials [Bq]	Radio-nuclide	Fuel [Bq]	Structural materials [Bq]
^{3}H	7.0×10^{12}	5.4×10^9	^{228}Ra	3.2×10^1	-
^{10}Be	1.8×10^7	2.9×10^3	^{227}Ac	3.3×10^5	-
^{14}C$_{inorg}$	6.7×10^{10}	0	^{228}Th	1.9×10^8	6.8×10^1
^{14}C$_{org}$	0	5.7×10^{10}	^{229}Th	1.4×10^4	-
^{36}Cl	8.6×10^8	1.4×10^9	^{230}Th	3.2×10^7	1.6×10^1
^{41}Ca	2.2×10^8	7.0×10^7	^{232}Th	4.1×10^1	-
^{59}Ni	8.9×10^8	1.3×10^{11}	^{231}Pa	7.6×10^5	-
^{63}Ni	9.2×10^{10}	1.3×10^{13}	^{232}U	1.8×10^8	6.5×10^1
^{79}Se	1.6×10^9	8.0×10^3	^{233}U	6.7×10^6	4.6×10^0
^{90}Sr	2.2×10^{15}	6.0×10^9	^{234}U	9.7×10^{10}	8.3×10^4
^{93}Zr	1.5×10^{11}	1.8×10^{10}	^{235}U	9.1×10^8	-
^{93m}Nb	1.2×10^{11}	1.4×10^{10}	^{236}U	2.1×10^{10}	6.0×10^3
^{94}Nb	2.1×10^8	1.3×10^{10}	^{238}U	1.9×10^{10}	2.2×10^4
^{93}Mo	8.3×10^7	3.3×10^8	^{237}Np	3.2×10^{10}	4.6×10^4
^{99}Tc	1.1×10^{12}	6.7×10^7	^{238}Pu	2.6×10^{14}	6.0×10^8
^{107}Pd	1.0×10^{10}	1.2×10^5	^{239}Pu	2.2×10^{13}	2.2×10^8
^{108m}Ag	1.4×10^9	4.3×10^4	^{240}Pu	4.0×10^{13}	3.0×10^8
^{126}Sn	3.0×10^{10}	3.2×10^5	^{241}Pu	1.5×10^{15}	1.9×10^{10}
^{129}I	2.7×10^9	1.9×10^4	^{242}Pu	1.9×10^{11}	2.9×10^6
^{135}Cs	3.7×10^{10}	1.8×10^5	^{241}Am	3.0×10^{14}	3.7×10^9
^{137}Cs	3.5×10^{15}	1.9×10^{10}	^{242m}Am	5.7×10^{11}	7.1×10^6
^{151}Sm	1.6×10^{13}	1.1×10^8	^{243}Am	2.2×10^{12}	3.8×10^7
^{166m}Ho	2.1×10^9	1.0×10^4	^{243}Cm	6.7×10^{11}	1.0×10^7
^{210}Pb	9.4×10^4	-	^{244}Cm	8.0×10^{13}	1.4×10^9
^{210}Po	9.1×10^4	-	^{245}Cm	5.6×10^{10}	1.0×10^6
^{226}Ra	2.9×10^5	-	^{246}Cm	1.2×10^{10}	2.6×10^5

Tab. B.9: Inventories of safety-relevant radionuclides in a canister containing 3 PWR UO2 and 1 MOX fuel assemblies with a burn-up of 48 GWd/tIHM, after 40 years decay

Radio-nuclide	Fuel [Bq]	Structural materials [Bq]	Radio-nuclide	Fuel [Bq]	Structural materials [Bq]
^{3}H	6.6×10^{12}	3.0×10^9	^{228}Ra	2.4×10^1	-
^{10}Be	1.7×10^7	4.5×10^3	^{227}Ac	3.0×10^5	-
^{14}C$_{inorg}$	5.6×10^{10}	0	^{228}Th	1.7×10^8	5.1×10^1
^{14}C$_{org}$	0	3.2×10^{10}	^{229}Th	1.1×10^4	-
^{36}Cl	6.9×10^8	7.0×10^8	^{230}Th	3.0×10^7	1.0×10^1
^{41}Ca	1.8×10^8	3.4×10^7	^{232}Th	3.1×10^1	-
^{59}Ni	7.2×10^8	1.2×10^{11}	^{231}Pa	7.0×10^5	-
^{63}Ni	7.3×10^{10}	1.2×10^{13}	^{232}U	1.7×10^8	4.9×10^1
^{79}Se	1.4×10^9	4.4×10^3	^{233}U	5.4×10^6	3.0×10^0
^{90}Sr	1.9×10^{15}	3.4×10^9	^{234}U	9.5×10^{10}	5.1×10^4
^{93}Zr	1.3×10^{11}	9.4×10^9	^{235}U	8.4×10^8	-
^{93m}Nb	1.0×10^{11}	8.0×10^9	^{236}U	1.6×10^{10}	3.7×10^3
^{94}Nb	1.8×10^8	3.6×10^{10}	^{238}U	1.6×10^{10}	1.5×10^4
^{93}Mo	7.3×10^7	7.1×10^8	^{237}Np	2.8×10^{10}	2.9×10^4
^{99}Tc	1.0×10^{12}	1.8×10^8	^{238}Pu	3.2×10^{14}	3.6×10^8
^{107}Pd	1.1×10^{10}	6.5×10^4	^{239}Pu	2.9×10^{13}	1.5×10^8
^{108m}Ag	1.2×10^9	2.2×10^4	^{240}Pu	6.6×10^{13}	2.0×10^8
^{126}Sn	3.3×10^{10}	1.8×10^5	^{241}Pu	2.5×10^{15}	1.2×10^{10}
^{129}I	2.7×10^9	1.1×10^4	^{242}Pu	3.1×10^{11}	1.5×10^5
^{135}Cs	4.4×10^{10}	1.2×10^5	^{241}Am	4.9×10^{14}	2.4×10^9
^{137}Cs	3.3×10^{15}	1.1×10^{10}	^{242m}Am	2.3×10^{12}	5.6×10^6
^{151}Sm	1.9×10^{13}	7.0×10^7	^{243}Am	4.0×10^{12}	2.0×10^7
^{166m}Ho	2.0×10^9	5.2×10^3	^{243}Cm	2.0×10^{12}	5.5×10^6
^{210}Pb	8.3×10^4	-	^{244}Cm	1.8×10^{14}	7.1×10^8
^{210}Po	8.0×10^4	-	^{245}Cm	1.8×10^{11}	4.8×10^5
^{226}Ra	2.6×10^5	-	^{246}Cm	3.3×10^{10}	1.2×10^5

Tab. B.10: Inventories of safety-relevant radionuclides in a canister containing 4 PWR UO2 with a burn-up of 48 GWd/tIHM, after 40 years decay

Radio-nuclide	Fuel [Bq]	Structural materials [Bq]	Radio-nuclide	Fuel [Bq]	Structural materials [Bq]
^{3}H	6.9×10^{12}	3.1×10^9	^{228}Ra	3.1×10^1	-
^{10}Be	1.7×10^7	4.5×10^3	^{227}Ac	3.9×10^5	-
^{14}C$_{inorg}$	6.2×10^{10}	0	^{228}Th	2.0×10^8	4.7×10^1
^{14}C$_{org}$	0	3.6×10^{10}	^{229}Th	1.4×10^4	-
^{36}Cl	8.1×10^8	8.2×10^8	^{230}Th	3.4×10^7	1.1×10^1
^{41}Ca	2.2×10^8	4.0×10^7	^{232}Th	4.0×10^1	-
^{59}Ni	8.4×10^8	1.2×10^{11}	^{231}Pa	9.0×10^5	-
^{63}Ni	8.6×10^{10}	1.3×10^{13}	^{232}U	2.0×10^8	4.5×10^1
^{79}Se	1.6×10^9	4.8×10^3	^{233}U	6.5×10^6	3.0×10^0
^{90}Sr	2.2×10^{15}	3.7×10^9	^{234}U	1.0×10^{11}	5.4×10^4
^{93}Zr	1.5×10^{11}	1.0×10^{10}	^{235}U	1.1×10^9	-
^{93m}Nb	1.2×10^{11}	8.7×10^9	^{236}U	2.0×10^{10}	3.9×10^3
^{94}Nb	1.9×10^8	3.7×10^{10}	^{238}U	1.7×10^{10}	1.6×10^4
^{93}Mo	7.6×10^7	7.3×10^8	^{237}Np	3.1×10^{10}	2.8×10^4
^{99}Tc	1.1×10^{12}	1.9×10^8	^{238}Pu	2.3×10^{14}	3.9×10^8
^{107}Pd	9.6×10^9	7.2×10^4	^{239}Pu	2.2×10^{13}	1.4×10^8
^{108m}Ag	1.3×10^9	2.5×10^4	^{240}Pu	3.6×10^{13}	2.0×10^8
^{126}Sn	3.0×10^{10}	2.0×10^5	^{241}Pu	1.5×10^{15}	1.2×10^{10}
^{129}I	2.6×10^9	1.2×10^4	^{242}Pu	1.7×10^{11}	1.7×10^6
^{135}Cs	3.9×10^{10}	1.2×10^5	^{241}Am	2.8×10^{14}	2.3×10^9
^{137}Cs	3.4×10^{15}	1.2×10^{10}	^{242m}Am	6.2×10^{11}	5.0×10^6
^{151}Sm	1.6×10^{13}	6.7×10^7	^{243}Am	2.0×10^{12}	2.3×10^7
^{166m}Ho	1.9×10^9	6.2×10^3	^{243}Cm	6.5×10^{11}	6.1×10^6
^{210}Pb	1.0×10^5	-	^{244}Cm	7.5×10^{13}	8.7×10^8
^{210}Po	9.8×10^4	-	^{245}Cm	5.0×10^{10}	5.9×10^5
^{226}Ra	3.1×10^5	-	^{246}Cm	1.2×10^{10}	1.5×10^5

In summary, the SNF management for the most simple management option as described in this paragraph is as follows:

- discharge from reactor;
- emplacement and decay storage for several years in cooling pond at reactor site;
- emplacement in transport/storage casks;
- short-distance transport to local storage site or long-distance transport to central storage site;
- interim, buffer, and decay storage (local or central) for several decades (dependent on disposal programme);
- transport to conditioning facility (short or long distance dependent on locations of storage and conditioning);
- buffer storage, conditioning for disposal;
- buffer storage, transport to disposal facility (long or short distance dependent on locations);
- emplacement in disposal facility.

1.3.3 Other cycles for the management of spent nuclear fuel and high level waste

As described in section B 1.2, "once through" and direct disposal of SNF is by far not the only option for SNF management. This paragraph will elaborate on the implications posed by different fuel cycles to RWM.

First, the most common variant, i. e. reprocessing of SNF, is usually governed by the wish to better utilize the energy content of natural resources, to reduce the volume and toxicity of the waste, and an appropriate use of weapons-grade plutonium (cf. section B 1.2). The COMPAS report (Dutton et al. 2004) further elaborates on

> the main issues that affect the decision on whether or not to reprocess SNF which are
> - maintaining a secure supply of nuclear fuel for energy production,
> - safety and environmental considerations,
> - the prospect of a future nuclear power programme that may include advanced nuclear fuel cycles,
> - economics,
> - safeguards,
> - technical issues and
> - military requirements.

As the only operational facilities for reprocessing in Western Europe are the AREVA plant at Cap la Hague in northern France and the BNFL plant at Sellafield in the UK (reprocessing at the UK Dounreay facility ended in 1998), the SNF needs to be transported to one of these plants. For this transportation, the same boundary conditions and safety and security issues apply as for the long-distance SNF transports from reactor sites to central storage facilities explained in section B 1.3.2.

At the reprocessing plans, buffer storage as well as the processing described in section B 1.2 will take place which will then result in several types of radioactive waste:

> About 3 % of spent-fuel by mass on removal from reactor, with the exception of Magnox fuel where the percentage is somewhat less consists of highly radioactive and toxic fission-products and actinides. Following reprocessing, these are retained in concentrated nitric acid, and, after a period of storage that requires active cooling, the waste can be conditioned into glass (vitrified) blocks. After the removal of the metal parts from the fuel elements (~30–40 % of the element weight) and the uranium and plutonium from the fuel itself (~96 % of the fuel), and depending on the burn-up – one block of glass (i. e. 0.18 m³) will at present be produced for approximately 1.65 tonnes of heavy metal that is reprocessed. The Environment Council (2002) estimated that for every tonne of heavy metal that is reprocessed of the order of 0.1 m³ of vitrified high-level waste and 1 m³ of Intermediate Level Waste (ILW) is produced. (Dutton et al. 2004).

The vitrified HLW is

> [...] made up of the fission products and minor actinides separated during the recycling of the fuels. Their high β-γ activity produces a large released quantity of heat, which decreases over time, mainly with the decrease in radioactivity of the fission products with amedium physical half-life (cesium 137, strontium 90). The conditioning of these wastes consists of incorporating them in a glass matrix ; the confinement capability of this material is particularly high and durable if under favourable physical/chemical environmental conditions. (ANDRA 2005)

The radionuclide inventories of typical vitrified waste after 40 years of decay are shown in the tables below. The AREVA flasks are 1.34 m high, have a diameter of 0.43 m and accommodate 400 kg of glass coming from the reprocessing of 3 to 4 PWR fuel elements (source: BLG, online-documentation www.urananreicherung.de).

The following needs to be considered when comparing radionuclide inventory and heat production figures for vitrified HLW with figures for SNF: If the figures given for SNF relate to 4 PWR SNF elements this is approximately the amount of SNF resulting in the content of one HLW flask after reprocessing. The masses of these 4 elements are in total in the range of two tonnes (fuel and structural materials), while the corresponding glass flask has a mass of about 0.4 tonnes. The difference did, of course, not vanish – further products of reprocessing are the uranium and plutonium extracted and a considerable amount of LILW. It should also be noted that part of the [129]I contained in SNF will be released to the environment during reprocessing.

Tab. B.11: Average safety-relevant radionuclide content of a single BNFL HLW glass flask, after 40 years decay

Radionuclide	Activity [Bq]	Radionuclide	Activity [Bq]
$^{14}C_{inorg}$	7.1×10^7	^{229}Th	6.3×10^3
^{59}Ni	6.7×10^8	^{230}Th	1.1×10^7
^{63}Ni	7.1×10^{10}	^{231}Pa	2.0×10^6
^{79}Se	1.6×10^9	^{232}Th	-
^{90}Sr	1.6×10^{15}	^{232}U	2.3×10^5
^{93}Zr	1.3×10^{11}	^{233}U	3.5×10^6
^{93}Mo	5.0×10^7	^{234}U	9.9×10^7
^{93m}Nb	1.1×10^{11}	^{235}U	3.0×10^5
^{94}Nb	9.9×10^6	^{236}U	3.7×10^6
^{99}Tc	9.4×10^{11}	^{238}U	6.0×10^6
^{107}Pd	7.5×10^9	^{237}Np	2.0×10^{10}
^{108m}Ag	4.7×10^8	^{238}Pu	7.7×10^{11}
^{126}Sn	5.1×10^{10}	^{239}Pu	6.7×10^{10}
^{129}I	2.1×10^6	^{240}Pu	2.3×10^{11}
^{135}Cs	3.9×10^{10}	^{241}Pu	2.2×10^{12}
^{137}Cs	2.3×10^{15}	^{242}Pu	3.5×10^8
^{151}Sm	1.7×10^{13}	^{241}Am	1.0×10^{14}
^{166m}Ho	1.5×10^8	^{242m}Am	1.2×10^{12}
^{210}Pb	8.1×10^4	^{243}Am	8.8×10^{11}
^{210}Po	7.9×10^4	^{243}Cm	2.9×10^{11}
^{226}Ra	1.9×10^5	^{244}Cm	1.2×10^{13}
^{227}Ac	1.6×10^6	^{245}Cm	6.5×10^9
^{228}Ra	–	^{246}Cm	1.2×10^9
^{228}Th	2.3×10^5		

Tab. B.12: Average safety-relevant radionuclide content of a single COGEMA HLW glass flask, after 40 years decay

Radionuclide	Activity [Bq]	Radionuclide	Activity [Bq]
$^{14}C_{inorg}$	7.1×10^7	^{229}Th	6.3×10^3
^{59}Ni	6.7×10^8	^{230}Th	1.1×10^7
^{63}Ni	7.1×10^{10}	^{231}Pa	2.0×10^6
^{79}Se	1.6×10^9	^{232}Th	–
^{90}Sr	1.6×10^{15}	^{232}U	2.3×10^5
^{93}Zr	1.3×10^{11}	^{233}U	3.5×10^6
^{93}Mo	5.0×10^7	^{234}U	9.9×10^7
^{93m}Nb	1.1×10^{11}	^{235}U	3.0×10^5
^{94}Nb	9.9×10^6	^{236}U	3.7×10^6
^{99}Tc	9.4×10^{11}	^{238}U	6.0×10^6
^{107}Pd	7.5×10^9	^{237}Np	2.0×10^{10}
^{108m}Ag	4.7×10^8	^{238}Pu	7.7×10^{11}
^{126}Sn	5.1×10^{10}	^{239}Pu	6.7×10^{10}
^{129}I	2.1×10^6	^{240}Pu	2.3×10^{11}
^{135}Cs	3.9×10^{10}	^{241}Pu	2.2×10^{12}
^{137}Cs	2.3×10^{15}	^{242}Pu	3.5×10^8
^{151}Sm	1.7×10^{13}	^{241}Am	1.0×10^{14}
^{166m}Ho	1.5×10^8	^{242m}Am	1.2×10^{12}
^{210}Pb	8.1×10^4	^{243}Am	8.8×10^{11}
^{210}Po	7.9×10^4	^{243}Cm	2.9×10^{11}
^{226}Ra	1.9×10^5	^{244}Cm	1.2×10^{13}
^{227}Ac	1.6×10^6	^{245}Cm	6.5×10^9
^{228}Ra	–	^{246}Cm	1.2×10^9
^{228}Th	2.3×10^5		

The decay heat of an average canister of BNFL HLW glass at the time of production is ~ 3500 W, decreasing to ~ 690 W after approximately 40 years of cooling. The COGEMA reference glass has a similar composition, with a decay heat and total activity at the time of production that is, for an average flask, ~ 20 % lower than of the BNFL glass. (NAGRA 2002)

As explained earlier, it is usually intended to use the uranium and plutonium for MOX or other fuel production. To which extent or for which proportion of the existing and arising stocks this intent, however, will materi-
alise, will strongly depend on issues such as the nuclear policy of a country, safeguard considerations, or availability, and market prices of uranium.
The part of uranium and plutonium for which no re-use will take place has
later to be re-declared as (radioactive) waste. In Germany, the depleted uranium is mainly stored for future use, whereas it has to be demonstrated by
provision of corresponding plans that the plutonium will be re-used.

In principle, the materials generated by reprocessing have to be taken
back by the country from which the SNF originated. For practical reasons,
however, some compensation might take place instead of a 1:1 transfer back
to the countries of origin. For the transport back to the countries of origin,
safety and security requirements similar to the ones mentioned before for
SNF apply. The vitrified HLW from the reprocessing of German SNF will
usually be transported in a CASTOR cask designed for 28 glass flasks, or,
alternatively, in containers of French origin.

For the storage and disposal of the vitrified waste, similar issues as for SNF
to be disposed of directly have to be considered. Due to the different chemical composition and physical behaviour and the different radionuclide inventories, differences arise, e. g. in the design of disposal containers but also with
regard to the radiological considerations when assessing the safety of disposal
solutions. In addition, there are strategic differences concerning the choice of
storage sites: In Germany, vitrified HLW is being stored centrally at Gorleben
("TBL-G"). De-centralised or local storage for vitrified HLW or other materials taken back from abroad, although recently proposed in order to avoid or
reduce further problems associated with HLW transports to Gorleben, does
not make sense from a technical or logistics point of view.

As discussed before, a variety of advanced fuel cycles different from the
"traditional" or "established" ones, is presently under consideration. Implementing such cycles would have significant impact RWM issues. The most
important implications are Gonzalez (2008), Marivoet et al. (2008) (cf. also
section B 1.2):

– The radionuclide inventory of the residual HLW would be changed, e. g.
  it is estimated that the actinide inventory could be reduced by up to two
  orders of magnitude (see Fig. B.5).
– As a consequence, the heat generation of the residual HLW would decrease
  significantly faster compared to conventional fuel cycles (see Fig. B.6).
– Another consequence is the reduction of the radiotoxicity of the residual materials.
– The application of techniques such as segmentations of SNF into different components ("partitioning") and the "burning" in reactor parks
  different from the ones known today would result in ILW different in

amount and composition from the operational waste from presently implemented fuel cycles.
- As a consequence of "conventional" reprocessing and ensuing vitrification of the HLW, especially minor actinides would be no longer "available" for advanced P&T techniques.

The first and most obvious corollary for the management of radioactive waste (or, more broadly, of radioactive materials) is that all steps prior to disposal in the case of advanced fuel cycles would undergo significant changes with respect to material streams, origins and destinations of transports, and timing of management steps. Safety and security issues would be affected as well due to changes in material composition, inventory, radiation, and heat generation.

It has to be noted that there are not only potential changes in the future: Issue e) from above indicates that strategic decisions about *present* waste and material management are needed in order to provide stock of materials for potential advanced fuel cycles in the future.

The implications of the different heat characteristics (item b) from above) apply also to disposal, in particular repository layout: A design constraint for repositories in crystalline rock or clay is to avoid temperatures exceeding 100°C in order to avoid unfavourable material alterations. This constraint is accounted for by appropriate spacing of the inventory to be disposed of. In other words, less heat generation would allow for a denser disposal and the need of less host rock volume available for disposal or, for constant volume, a higher capacity of the repository. Moreover, the faster decrease of heat production might cause a temptation to delay disposal and instead to strive for prolonged decay storage (in the order of centuries). On the other hand, it must also be noted that heat generation of the waste has a positive impact on post-closure safety in the case of deep disposal in rock salt.

In any case, the impact of altered heat generation is most dramatic and visible if a country envisages continued or even increasing nuclear power production. This is illustrated in Tab. B.14 below which shows for the case of the USA that already for the existing reactor licenses two repositories would be needed in the case of direct SNF disposal. For the most extensive nuclear power production scenario, as many as 22 repositories would be required. In contrast, the most advanced fuel cycle would still result in the need for just one repository even for this most extensive nuclear power production scenario.

It is, however, less clear to what extent advanced fuel cycles have the potential to reduce the long-term radiological (or other) hazard caused by the residual materials. The above-mentioned radiotoxicity can serve as a hazard indicator for human intrusion scenarios while the radiologic consequence (annual effective dose to members of potentially exposed groups) resulting from scenarios assuming a natural evolution of a repository system will not significantly change no matter which fuel cycle is assumed (cf. section B 1.2.3).

Tab. B .13: Estimated number of geological repositories in the USA, for the different scenarios of the cumulative spent fuel in 2100 (from Gonzales 2008)

Nuclear Futures		Existing License Completion	Excluded License Completion	Continuing Level Energy Generation	Continuing Market Share Generation	Growing Market Share Generation
Cumulative spent fuel in 2100 (MTiHM)		90,000	120,000	250,000	600,000	1,500,000
		Existing Reactors Only ⟵——⟶ Existing and New Reactors				
Fuel Management Approach		Number of Repositories Needed (at 70,000 MT each)				
No Recycle	Direct Disposal (current policy)	2	2	4	9	22
No Recycle	Direct Disposal with Expanded Repository Capacity	1	1	2	5	13
Recycle	Limited Thermal Recycle with Expanded Repository Capacity	1	1	1	3	7
Recycle	Repeated Combined Thermal and Fast Recycle	(requires new reactors)		1	1	1

1.3.4 Strategic decisions on the "end point" of radioactive waste management

The disposal solutions ("end points") introduced in section B 1.3.1 (and variants thereof) have been intensively discussed in many countries, most recently in the UK (CoRWM 2006). Presently, the USA investigate options for "end points" by means of a Blue Ribbon Commission (cf. Blue Ribbon Commission 2011). So far, all of these discussions resulted in neglecting most of the solutions, the exceptions (in the case of SNF and HLW) being:

– prolonged storage in supervised, monitored and regularly refurbished engineered facilities on, or near to, the surface, and
– disposal in deep geological formations in mine-type facilities (deep boreholes sometimes considered as a variant), the long-term safety and security of which must not depend on active measures such as monitoring, surveillance, and refurbishment,
– hybrid solutions incorporating features of the aforementioned two.

Tab. B.14: Criteria applied to screen out various options (from CoRWM 2006)

Option	Criteria applied to screen out
Storage forever	– Unacceptable burden to future generations – Unacceptable risk to security of nuclear materials – Unacceptable risk to health
Direct injection	– No 'proof of concept' – Causes harm to areas of particular environmental sensitivity – Risk to security – Risk to health
Disposal at sea	– Breach of duty of care to the environment outside national boundaries – Breach of internationally recognized treaties or laws and no foreseeable likelihood of change in the future
Sub-seabed disposal	– Breach of duty of care to environment – Harm to environmentally sensitive areas – Involves a risk to future generations greater than that posed to the present generation that has enjoyed the benefits – Breach of internationally recognized treaties or laws and no foreseeable likelihood of change in the future
Disposal in ice sheets	– No 'proof of concept' – Breach of duty of care to environment – Harm to environmentally sensitive areas – Risk to future generations – Breach of internationally recognized treaties or laws and no foreseeable likelihood of change in the future
Disposal in sub-duction zones	– No 'proof of concept' – Breach of duty of care to environment – Breach of internationally recognized treaties or laws and no foreseeable likelihood of change in the future
Disposal in space	– Breach of duty of care to environment – Harm to environmentally sensitive areas – Risk to security – Risk to health – Cost disproportionate to benefits received
Dilution and dispersal	– No 'proof of concept' – Breach of duty of care to environment – Breach of internationally recognized treaties or laws and no foreseeable likelihood of change in the future

As an illustration of the reasoning behind neglecting other disposal solutions, the results of the Committee on Radioactive Waste Management (CoRWM 2006), an advisory body in the UK with members representing a "diversity of viewpoints, experience and knowledge" are presented here. CoRWM first developed a "long list" of options which were then evaluated against a set of screening criteria (cf. Tab. B.14).

After the screening and an ensuing discussion process, the following remaining options and variants were shortlisted for HLW/SNF disposal (Tab. B.15):

Tab. B.15:   Options and variants shortlisted for HLW/SNF

Long term interim storage	1. Interim stores, above ground, at or near current locations of waste and protected to current standards 2. Interim stores, above ground, centralised and protected to current standards 3. Interim stores, above ground, at or near current locations of waste and protected 4. Interim stores, above ground, centralised and protected 5. Interim stores, underground, at or near current location of wastes and protected by ground cover 6. Interim stores, underground, centralised and protected by ground cover
Geological disposal	7. Geological disposal 8. Deep borehole disposal
Phased geological disposal	9. Phased geological disposal

As CoRWM also investigated the management of short-lived wastes, several variants of near-surface disposal for this waste were also shortlisted.

A similar reasoning for not further considering other solutions is reported in the COMPAS report:

> The uncertainties associated with some options, notably the disposal of SNF into outer space, disposal in ice sheets and disposal in subduction zones in ocean trenches are considered to be so great [...] that all countries have discounted them. Although such options were previously discounted in the UK, a rigorous and public review by the Government of all options (except where they have been ruled out by international agreements), started in 2003[18].

> The options of sea dumping and disposal in deep-sea sediments are politically and socially unacceptable in many countries and this has been reflected in international treaties such as the London Dumping (International Maritime Organisation 1991) and OSPAR (OSPAR 1992) conventions, so that the signatories are precluded from these options.

> Thus the long-term (i.e. greater than a few hundred years) options that are being considered are land-based geological disposal and indefinite storage [...].

---

[18]   This refers to the CoRWM process described above which was at an early stage at the time the COMPAS report was produced.

Note that the CoRWM screened out "storage forever" (which is, strictly spoken, in the IAEA terminology introduced earlier not storage but disposal since no retrieval is foreseen) but allows for a number of storage solutions the duration of which is not further defined. For further work, however, CoRWM defined "a 'reference case' of storage for a maximum of 300 years".

Apparently, the two most remarkable differences between the "storage" and the "disposal" options are

- the degree to which their safety and security in the long run depends on active measures (which involve radiological exposures of workers), and
- the degree to which they are reversible (and, in particular, the waste is retrievable) at different future points in time.

Preferences concerning these issues depend on the confidence one has into the safety and security of the different options and on the interpretation of the principle of cross-generational equity.

The Joint Convention on the Safety of Spent Fuel Management and on the Safety of Radioactive Waste Management phrases the principle of cross-generational equity as follows

> […] to ensure that during all stages of spent fuel and radioactive waste management there are effective defences against potential hazards so that individuals, society and the environment are protected from harmful effects of ionizing radiation, now and in the future, in such a way that the needs and aspirations of the present generation are met without compromising the ability of future generations to meet their needs and aspirations […]. (IAEA 1997)

A question very much debated amongst ethicists is whether the "ability of future generations to meet their needs and aspirations" would require a maximum degree of flexibility for these generations or rather a minimum burden or responsibility (COMPAS, see section B.3 below). The former would result in storage solutions for which retrievability is ensured for the case that future generations have the desire to get the material back, be it in order to re-use it or to apply a disposal solution they consider more appropriate. These solutions, however, go along with an economical and perhaps radiological burden due to the necessity of ensuring long-term monitoring, surveillance, and refurbishment of the storage facility. This burden would, on the contrary, vanish if a deep disposal facility is once closed but one would have to pay the price of less (or no) retrievability and flexibility in that case.

This target conflict as well as sometimes diverging views about the safety and security of solutions depending on long-term societal and economic stability and societal commitment (storage) or on passive safety based on geological and engineering features (deep disposal) some try to resolve by compromising resulting in hybrid concepts. Such concepts can, e. g., foresee deep disposal going along with engineered features allowing or at least easing waste retrieval for a certain timeframe and/or with a schedule in which irreversible decisions are left for later generations, e. g. by leaving the facility open for prolonged timeframes. CoRWM arguments concerning this target conflict are presented in Box B.2:

Box B.2:   CoRWM arguments in favour of 'early' closure as well as in favour of a prolonged open period.

[...] arguments in favour of 'early' closure (in practice at least a century from now), assuming sufficient confidence in the long-term safety of the concept, are as follows

- It minimises the burdens of cost effort and worker dose transferred to future generations.
- It recognises that future generations may lack the skills or motivation to deal with the wastes.
- It places less reliance on maintaining institutional controls.
- It provides greater safety in the near term, which is what concerns people most.
- It provides greater security from terrorist attacks and the problem of nuclear proliferation.

Keeping a repository open even for a few centuries will not add materially to existing knowledge of its probable long-term behaviour. The arguments that support early closure also support early backfilling vault by vault as the waste is emplaced. [...] The case for leaving the repository open after it is filled with waste (for up to 300 years or possibly more) rests on the following arguments:

- It provides flexibility for future generations to take decisions.
- It allows for the lack of trust and confidence in the long-term safety case for disposal.
- It enables future generations to have access to a potential resource.
- It leaves open the possibility of alternative or improved methods for management of wastes.
- It seeks to maintain flexibility whilst making progress in reducing the burden.

More generally, approaches allowing for flexibility are covered by the concept of *reversibility*[19]. This concept accounts for the fact that disposal programmes, especially in the case of HLW/SNF, last over decades which in itself allows for flexibility provided that the development process is designed accordingly – note that the definition of reversibility as given by NEA (2001) presupposes that repository development takes place in a step-wise process with well-defined decision points. According to the IAEA Safety Requirements WS-R-4 (IAEA 2006b), such steps might be

- site characterization,
- geological disposal facility design,
- geological disposal facility construction,
- geological disposal facility operation, and
- geological disposal facility closure.

---

[19] "[T]he possibility of reversing one or a series of steps in repository planning or development at any stage of the programme" (NEA 2001).

The European Pilot Study (Bodenez/Röhlig et al. 2008) adds a "conceptualisation" step in the beginning.

In practice, the choice of steps will depend on the boundary conditions of each national programme. In particular, there will certainly be interaction and iteration amongst the activities described by these rather general and schematic headlines, e.g. construction of parts of the repository might proceed at a time at which operation (i.e. waste emplacement) takes place in others.

The above mentioned concept of *reversibility*

[...] implies the review and, if necessary, re-evaluation of earlier decisions, as well as the means (technical, financial, etc.) to reverse a step. Reversibility denotes the fact that fall-back positions are incorporated in the disposal policy and in the actual technical programme. Reversibility may be facilitated, for example, by adopting small steps and frequent reviews in the programme, as well as by incorporating engineering measures. In the early stages of a programme, reversal of a decision regarding site selection or the adoption of a particular design option may be considered. At later stages, during construction and operation, or following emplacement of the waste, reversal may involve the modification of one or more components of the facility, or even the retrieval of waste packages from parts of the facility.

Retrievability denotes the possibility of reversing the action of waste emplacement. It is thus a special case of reversibility. *Retrieval* is the action of recovery of the waste or waste packages. Retrievability, the potential for retrieval, may need to be considered at various stages after emplacement, including after final sealing and closure. In discussing retrievability and retrieval, it is important to specify what is to be retrieved, since this affects the implementation and technical feasibility. Retrievability could, for example, refer to: retrieval of individual waste packages which are identified as faulty or damaged, even as emplacement of other packages continues; retrieval of some or all of the waste packages at some time after emplacement; or retrieval of the waste materials if the packages are no longer intact. Retrievability may be facilitated by the repository design and operational strategies, for example, by leaving underground access ways open and emplacement/retrieval systems in place until a late stage, and through the development and use of durable containers and easily excavated backfill. (NEA 2001)[20]

A further discussion of retrievability issues can be found in section B 1.3.7 below.

Other approaches, e.g. the Dutch one, attempt to strive for flexibility as well rather with a long-term (indefinite?) storage solution (instead of retrievable disposal) as the starting point. Presently, SNF and HLW are being stored in the "HABOG" (cf. section B 1.3.2) which is so far licensed for a century but the licensed lifetime of which might be extended later. The

---

[20] Note that terminology varies: In its "Safety Requirements Governing the Final Disposal of Heat-Generating Radioactive Waste" (BMU 2010a), the German Federal Ministry for the Environment, Nature Conservation and Nuclear Safety (BMU) defines retrievability as the "planned technical possibility for removing the waste containers out of the repository mine" while recovery is defined as "retrieval of the radioactive wastes as an emergency measure".

idea of providing flexibility for future generations by doing so is illustrated in Fig. B 10, for the reasoning behind cf. section B 1.3.2.

1.3.5 Safety and security issues

The ultimate goal of radioactive waste management (RWM) is to ensure safety and security in the long run. According to the IAEA (2007) glossary, (nuclear) security means "prevention and detection of, and response

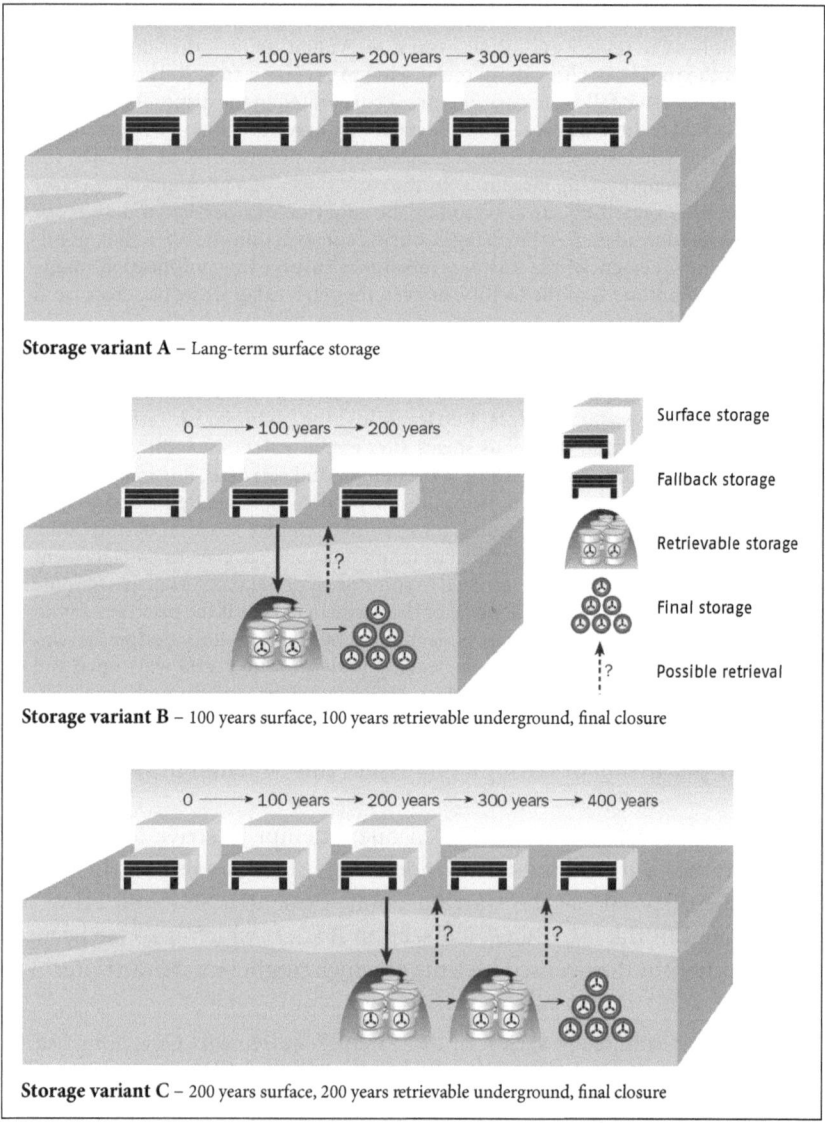

Fig. B.10: Schematic of long-term storage variants considered in the Netherlands (from NEA 2006)

to, theft, *sabotage*, unauthorized access, illegal transfer or other *malicious* acts involving *nuclear material*, other *radioactive substances* or their associated *facilities*". By "safety", the Joint Convention on the Safety of Spent Fuel Management and on the Safety of Radioactive Waste Management (IAEA 1997) understands that "individuals, society and the environment are adequately protected against radiological and other hazards", the latter being "biological, chemical and other hazards". In the long run, achieving safety means "to avoid actions that impose reasonably predictable impacts on future generations greater than those permitted for the current generation". Security and safety have to be ensured for a broad spectrum of hazards and threats, i.e. natural, technical and human, and have to include "conventional" (such as construction building safety, traffic and mining safety, potential release of chemo toxic substances) and nuclear issues. Most of these safety and security issues are well-known and are (or should be) well-addressed in daily practice. In contrast, safety and security issues for disposal facilities are outstanding for two reasons: (i) They relate to nuclear facilities to be constructed at several hundred meters depth and (ii) safety and security has to be ensured and demonstrated over long timeframes.

Opinions vary, however, concerning the question over which timeframes safety and security have to be achieved and demonstrated.

> These range from societally relevant time frames of at most several generations or a few hundred years, to the much longer time frames relevant to large-scale geological (e.g. tectonic) changes, with the time frames over which safety assessments are considered to be meaningful and relevant generally lying somewhere in between these two extremes. (NEA 2007c)

Some approaches try to derive demonstration timeframes based on radiotoxicity decrease and to define cut-off times (in the order of some 100,000 years) at which the radiotoxicity of the waste can be compared to natural circumstances.

> It has been shown, for example, that after about 100,000 years the radiological toxicity of one tonne of Swedish spent fuel is on a par with the radiological toxicity of the natural uranium from which it was derived. (NEA 2004)

Such approaches, however, do not account for the facts that the waste has other compositions and concentrations than materials containing natural uranium, or that radiotoxicity is not necessarily a measure of hazard, or that natural materials are not *per se* harmless (NEA 2004). Other approaches are based on the request to perform consequence calculations for at least as long as the calculated indicators (e.g. annual effective dose to a potentially exposed group) are increasing. It is, however, clear that there are practical limitations for the timeframes for which safety can be achieved and demonstrated. Model calculations describing e.g. radionuclide migration to the biosphere become increasingly meaningless since the underlying assumptions lose their justification with increasing problem time. If the calculated consequence curves are still rising at the end of the timeframes of predict-ability, one might perform the calculations for longer timeframes, e. g. until peaks becomes obvious, in order to get an impression of the time pattern but must then be aware of the limited information which can be derived from such calculations.

The rationale behind the often quoted one million years timeframe applicable in Germany (BMU 2010a) is the ability to find sites and geologi-cal environments which are, according to present-day geoscientific knowl-edge, believed to conserve their favourable features for a time of that order of magnitude.

Conventionally, there is a close connection between the concepts of safety and of protection or control, the latter being associated with active measures:

> *Safety (as addressed before)* is primarily concerned with maintaining *control* over *sources*, whereas *(radiation) protection* is primarily concerned with controlling *exposure* to *radiation* and its effects. Clearly the two are closely connected: *radi-ation protection* (or *radiological protection*) is very much simpler if the *source* in question is under *control*, so *safety* necessarily contributes towards *protection*. (IAEA 2007)

Safety in this sense, as well as security, is conventionally achieved in reg-ularly monitored, controlled, maintained and refurbished storage facil-ities. The longer the lifetime of such facilities lasts, however, the greater are the challenges associated with ensuring safety and security (cf. section B 1.3.2). In contrast, the concept of "passive safety" in the long run has to be achieved by a combination of geogenic and anthropogenic components the performance of which would not depend on active measures. This does not exclude activities such as monitoring which might, dependent on reg-ulations, even be obligatory, but the idea is that safety or security must not depend on such measures. According to international regulation (IAEA 2006b), the following features are important pillars of long-term safety and security of geological disposal:

- To contain the waste until most of the radioactivity, and especially that associated with shorter lived radionuclides, has decayed;
- To isolate the waste from the biosphere and to substantially reduce the likelihood of inadvertent human intrusion into the waste;
- To delay any significant migration of radionuclides to the biosphere un-til a time in the far future when much of the radioactivity will have de-cayed;
- To ensure that any levels of radionuclides eventually reaching the bio-sphere are such that possible radiological impacts in the future are acceptably low.

In contrast, the safety and security to be ensured during the operation of disposal facilities does depend on active measures. In that, safety and security issues are similar to the ones to be faced in other nuclear installations. However, in the case of deep (mine-type) repositories additional challenges come from the needs

- to ensure "conventional" nuclear safety in an *underground* installation, and
- to ensure, in addition, "conventional" mining safety.

An example for the specific issues to be addressed is the design requirements for ventilation: First, appropriate ventilation as in every other mine needs to be ensured during normal operation. But second, in the case of an accident associated with radionuclide release (e. g. if a waste canister is dropped and damaged) undue dispersal of the released radionuclides by ventilation (and resulting contamination of the environment and exposure of the population) has to be prevented e. g. by under pressure.

More generally, it has to be kept in mind that several aspects of safety and security have to be ensured and, if conflicting, to be balanced: short term issues vs. long-term issues and, for the former, issues concerning the staff of the facility vs. those concerning the general population or the environment.

For example, sophisticated emplacement and closure measures for ensuring long-term safety might result in longer exposure duration for repository workers who have to implement these measures. Also, prolonged open periods (cf. previous section) of the repository might result in longer exposure duration and/or higher mining risks for the staff which has to maintain the facility over this period.

1.3.6 Deep (geologic) disposal: potential host rocks and associated repository concepts

Conceivable and debated "end points" of HLW/SNF management include (cf. section B 1.3.4) variants of long term interim storage, geological disposal in deep boreholes or mine-type facilities, and phased geological disposal. Technical solutions for interim storage and the implications of running storage-type facilities over long timeframes have been discussed in sections B 1.3.2 and B 1.3.4.

Disposal in (several 1000 m) deep boreholes (Fig. B.17) has been investigated especially in Sweden, although the investigations certainly lacked the detail and the emphasis put on the concept of disposal in mine-type facilities (namely the so-called KBS-3 concept, cf. below).

Reasons in favour of the deep borehole option include the expectation to find stable groundwater conditions at these depths and a lack of contact to surface waters as well as the higher security of this option compared to disposal in mines at some 100 m depth. The former, however, remains to be shown by extensive geoscientific research. There are also open questions concerning the feasibility of the necessary drilling and emplacement techniques. A contractor of the Swedish waste management company SKB con-

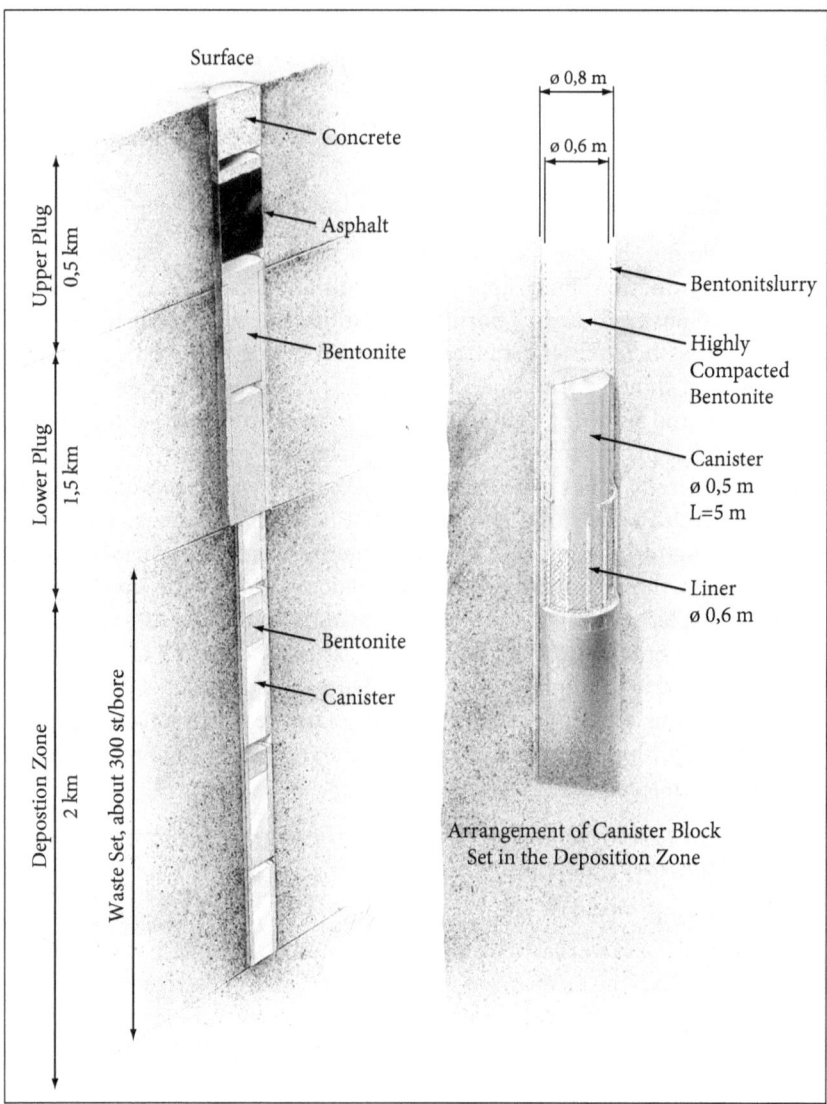

Fig. B.11:   Deep borehole disposal (Svensk Kärnbränslehantering AB 2000)

cludes " [...] that it is possible to drill the well with currently existing technology, although it represents one of the biggest challenges to be presented
to the drilling industry." (Harrison 2000)

The detailed investigation of the deep geology as well as of drilling and
emplacement techniques represents an effort that SKB apparently considers
unjustified given the effort already invested into, and the advanced state of,
the "mine-type" concept KBS-3. Some, however, consider this position as
biased (Åhäll 2006).

In the following, some (representative) technical solutions for the deep (geologic) SNF/HLW disposal in mine-type facilities are briefly described. They can also be considered as starting point for any "phased" disposal solution. When referring to the facilities as "mine-type", this refers to the fact that there are several similarities to the design of excavation mines: The waste will be emplaced in vaults (vertical or horizontal boreholes of varying lengths, tunnels/drifts, caverns/chambers) the access to which will be provided via drifts and one or more shafts (by hoists) or ramps (by vehicles). The materials for the containers or casks in which the waste is emplaced vary, dependent on the functions the containers have to fulfil (cf. below). So do the techniques and materials applied for surrounding the containers ("buffers") as well as for backfilling and/or closing the repository vaults. Baldwin/Chapman/Neall (2008) distinguishes the concepts given in table Tab. B.16. Note that the table also includes the above discussed emplacement in deep boreholes and that combinations of more than one of the "mine-type" concepts (e. g. for emplacing different waste forms) are under consideration as well. The ensuing figures illustrate the main features of some of the concepts.

Tab. B.16: Key features and variants leading to the disposal concepts (from Baldwin/Chapman/Neall 2008)

Key Feature	Variants	Concept No.
In-tunnel (borehole)	Vertical borehole	1
	Horizontal borehole	2
In-tunnel (axial)	Short-lived canister	3
	Long-lived canister	4
In-tunnel (axial) with super-container	Small working annulus	5
	Small annulus + concrete buffer	6
	Large working annulus	7
Caverns with cooling, delayed backfilling	Steel MPC + bentonite backfill	8
	Steel or concrete/DUCRETE Container + cement backfill	9
Mined deep borehole matrix		10
Hydraulic cage	Around a cavern repository	11
Very deep boreholes		12

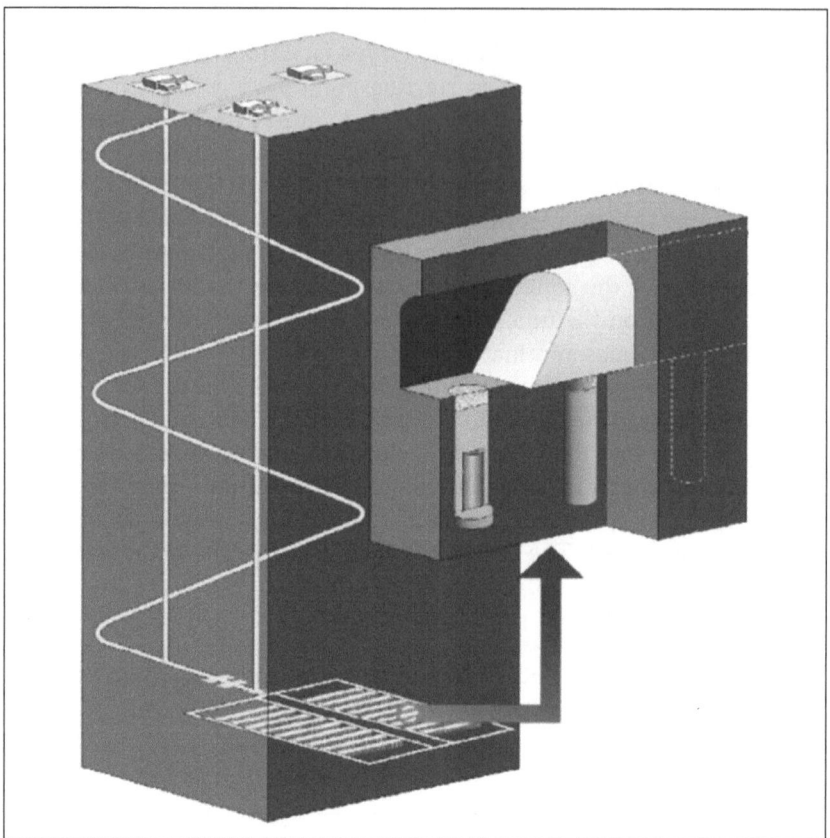

Fig. B.12:   Schematic illustration of Concept 1, in-tunnel (vertical borehole) with long- or short-lived canister. In this illustration, 'panels' of disposal tunnels are shown incomplete and the extent of the repository and additional ventilation shaft(s) is not indicated. In all the concepts, construction (in some panels) and waste emplacement (in others) could be carried out in parallel. This is an operational and strategic decision. The remaining concept illustrations for tunnel and cavern repositories show only a small, nominal disposal area.

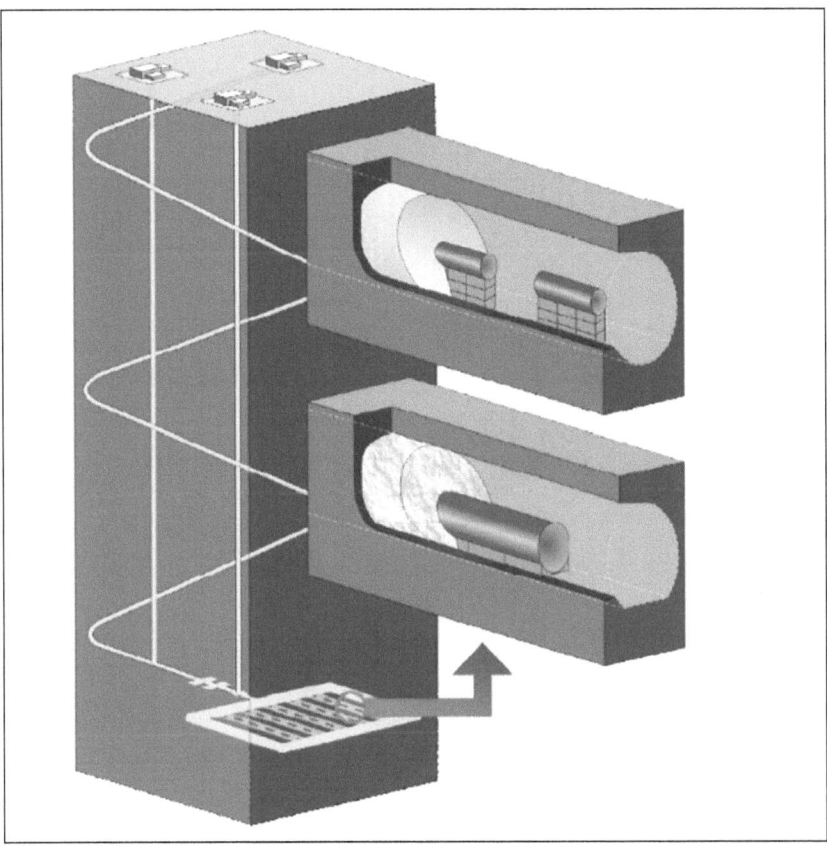

Fig. B.13: Schematic illustration of concepts 3 and 4, in-tunnel (axial) with long- and short-lived canister and buffer (the lower figure illustrates disposal in salt, with a salt backfill) (from Baldwin/Chapman/Neall 2008)

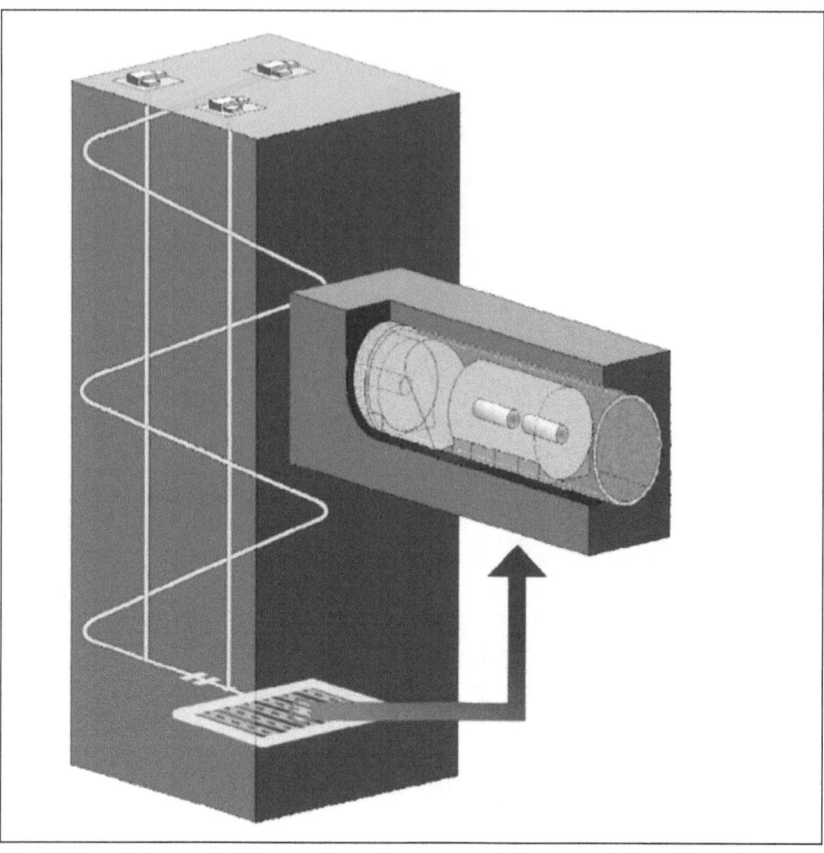

Fig. B.14:   Schematic illustration of concept 6, in-tunnel (axial) with supercontainer (concrete buffer) (from Baldwin/Chapman/Neall 2008)

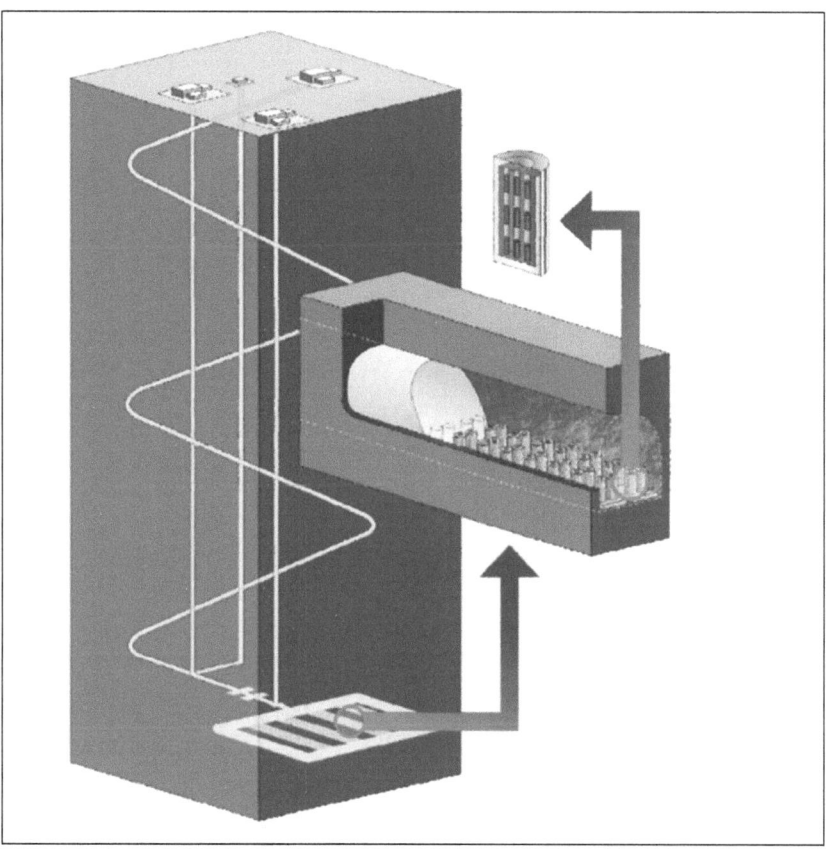

Fig. B.15:   Schematic illustration of concept 8, caverns with steel MPC (bentonite backfill). Note that there is scale change on the blow-up diagram: the caverns are of the order of 10–20 m wide and high, compared to the previous tunnel concepts, where the tunnels were only a few meters in diameter (from Baldwin/Chapman/Neall 2008)

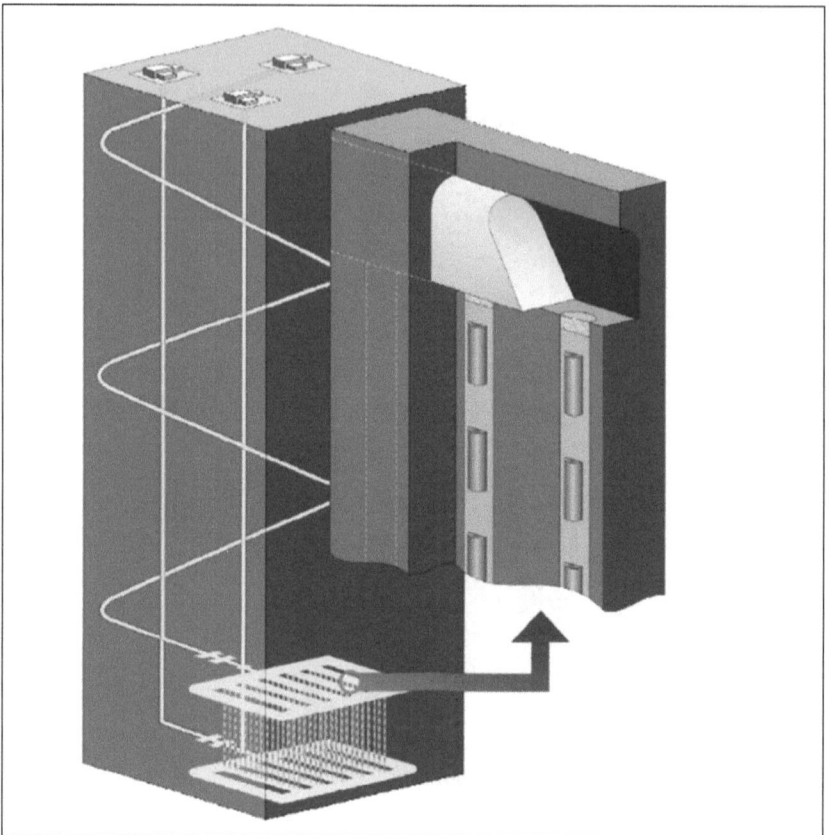

Fig. B.16: Schematic illustration of concept 10, mined deep borehole matrix
(from Baldwin/Chapman/Neall 2008)

Although the ultimate goal of disposal is the same for all concepts, there
is still considerable variation, the reasons for which include the following:

– In different countries, different geological formations potentially hosting the repository ("host rocks") are available. Their characteristics important for erection, operation, and post-closure performance as well as
   their geometric configuration and extension vary, and accordingly affect
   the repository concepts.
– Although the pillars of safety and security quoted in the previous section (containment, isolation, migration delay, limitation of impacts) are
   the same, the weight to be put on each of them (depending on the point
   in time after closure) as well as the ways they are achieved vary. Technically spoken, designs and (geo-)technical measures have to be adjusted
   in order to take advantage of the favourable features of the host rock and
   to compensate for the less favourable ones.

– Attitudes towards the confidence in the performance of anthropogenic/
  engineered vs. geogenic/natural components over long times may vary.
– Forms and amounts of waste to be disposed of vary.
– Expectations concerning the remaining degree of flexibility after waste
  emplacement (cf. above) vary.
– National legislative and regulatory backgrounds as well as tradition and experiences (e.g. concerning mining activities in different environments) vary.

In any case, a host rock and site will be chosen and a repository will be
designed in a way that the pillars of safety and security will be achieved by a
system of possibly diverse or redundant functions the repository components
(geogenic and anthropogenic) have to fulfil. Although these components are
often referred to as a "multi-barrier system", this package of barriers is somewhat different from the multi-barrier ("defense-in-depth") principle in other
sectors, e.g. in NPP safety, due to different conditions and phenomena and,
in particular, to longer timeframes in which components have to fulfil safety
functions varying over time. Fig. B.17 "[…] illustrates how safety rests on different safety functions or combinations of safety functions in different time
frames" for the Belgian disposal concept (Ondraf/Niras 2009).

   Fig. B.18 from the French disposal programme takes this a bit further by
providing a more detailed breakdown of safety functions and their acting over

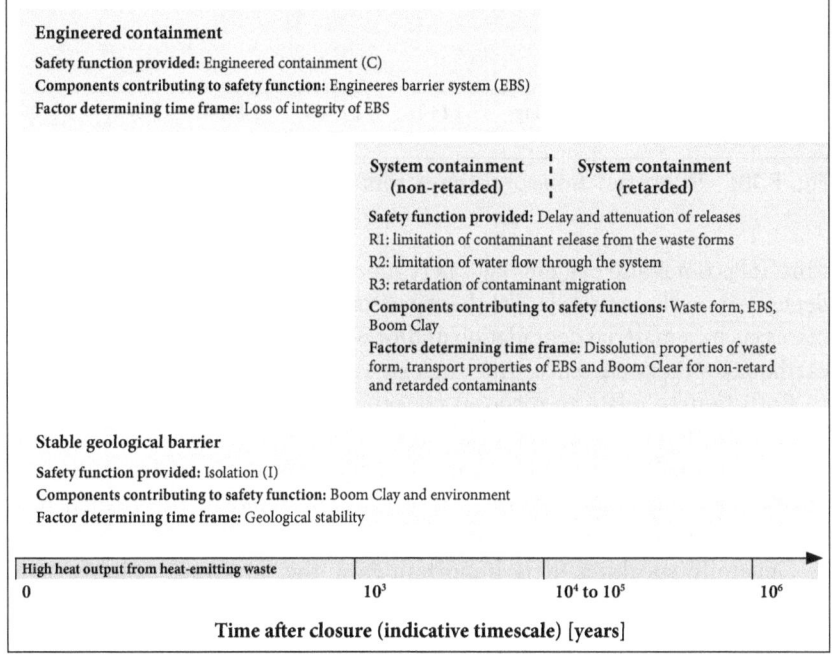

Fig. B.17:   Illustration of the safety concept in the case of geological disposal
                of heat-emitting wastes (from Ondraf/Niras 2009)

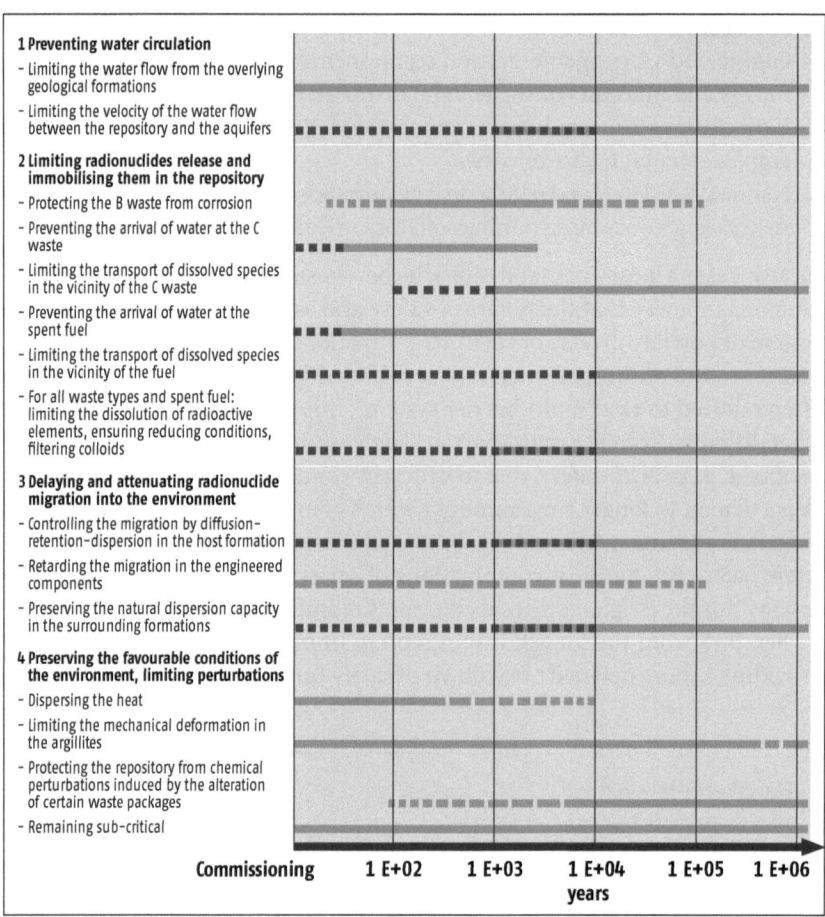

Fig. B.18:   Safety functions over time (from ANDRA 2005)

time. It becomes also evident that safety functions might either act or be latent dependent on the actual state of the repository system due to continuous processes such as material degradation and/or potential disruptive events such as earthquakes the occurrence, time and extension of which is unknown.

Both figures relate to disposal concepts in clay formations which are, however, different from each other in terms of their properties: While the formation considered as a host rock in Belgium is a plastic clay with a relatively high water content, the host rock considered in the French programme is an indurated claystone.

Generally speaking, it is important how the host rock behaves with regard to the desired safety functions, how this behaviour will be preserved over time and to which extent and for which time this can be demonstrated. It has also to be accounted for that the host rock will not act "as given" but react to the disturbance caused by the repository: First, the erection of the facility and the excavation of the underground vaults represent mechanical disturbances which will also affect the hydraulic regime. Second, the heat emitted by the emplaced waste causes a thermal disturbance. Third, the emplaced materials (waste, containers, material for buffering, backfilling and closing) alter the chemical conditions. Finally, the feasibility of erection and operation is important: It must be possible to construct a repository mine both in terms of mining feasibility and in terms of available volumes.

Given this variety of desirable features on one hand and the high variation of geological properties on the other, it is clear that there will always be a need for compromise – no host rock will provide only favourable conditions with regard to all of these aspects. The lack of features not or not sufficiently well provided by the geology has to be compensated by appropriate (geo) technical measures, where appropriate.

Tab. B.17: Relationship of a selection of host rock properties to fundamental requirements for waste disposal

	Mining feasibility	Long-term stability and predictability	Reaction to disturbances by disposal facility	Containment and migration delay
Vertical and lateral extension	+			
Depth	+	+		
Thermal conductivity			+	
Hydraulic conductivity			+	+
Mechanical Strength	+	+	+	
Deformation behaviour	+	+	+	
Vault stability	+	+	+	
in situ stresses	+	+	+	
Solubility		+		+
Sorption behaviour				+
Heat resistance		+	+	

Tab. B.17 schematically relates a selection of host rock properties to the fundamental requirements of mining feasibility, long-term stability and predictability, reaction to disturbances, and the "classical" safety functions containment and migration delay.

The properties accounted for in the table are the ones considered in a comparison of the most-often investigated host rocks undertaken by the German Federal Geological Survey BGR (Tab. B.18) which further substantiates the point made above about the need for compromising and compensating.

An example for a compensation for less favourable host rock properties is the Swedish KBS-3 concept for SNF disposal in granite. Despite of several other favourable properties, granites and other crystalline host rocks are

Tab. B.18: Features of potential host rocks relevant for disposal

Feature	Rock Salt	Clay/Claystone	Crystalline (e.g. granite)
Thermal Conductivity	high	low	medium
Hydraulic Conductivity	practically impermeable	very high to high	very high (not fractured) to conductive (fractured)
Mechanical Strength	medium	low to medium	high
Deformation behaviour	viscous (creep)	plastic to brittle	brittle
Vault stability	inherent stability	lining necessary	high (not fractured) to low (highly fractured)
In Situ Stresses	lithostatically isotropic	anisotropic	anisotropic
Solubility	high	very low	very low
Sorption behaviour	very low	very high	medium to high
Heat resistance	high	low	high

favourable feature   unfavourable feature   medium

often fractured and thus provide pathways for the migration of toxic substances with the flow of underground waters. In particular, this is the case for the host rock formations available in Sweden and in Finland, where the KBS-3 concept (or a variant thereof) is considered as well.

The most striking component of the concept is the so-called copper canister, a disposal canister with a copper surface and a cast iron insert. While the former is meant to provide containment of the emplaced SNF for several hundred thousand years, the latter should grant mechanical stability. Contributing to mechanical stability is also one of the roles of the surrounding bentonite buffer: The bentonite (an industrial clay material) will be saturated with water, swell, and compensate possible mechanical disturbances from outside. In addition, the swollen bentonite will hinder advective migration of toxic substances in the case the canister fails and also provide a man-made chemical barrier. The configuration of the concept is shown in figure Fig. B.19. Obviously, it belongs to the category 1 of Tab. B.16 (vertical in-tunnel boreholes, here with long-lived canisters). A variant with horizontal boreholes ("KBS-3H", in contrast to "KBS-3V") is also being studied, with Finland showing preferences for the latter.

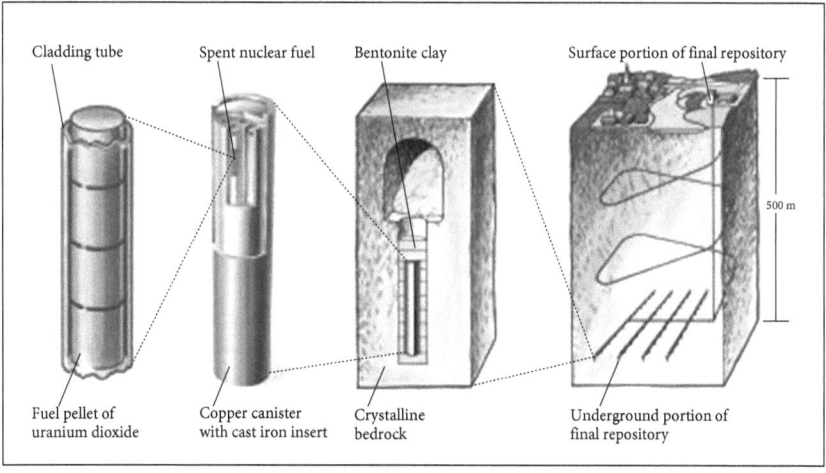

Fig. B.19: The KBS-3 concept for disposal of spent nuclear fuel (from Svensk Kärnbränslehantering AB 2006)

The detailed safety functions to be performed by the components are shown in figure Fig. B.20. From the colour coding the outstanding role of the copper canister becomes evident – it has to perform all "primary" safety functions (except for retardation – indicated by black frames) while the other components have to support the fulfilment of these primary func-

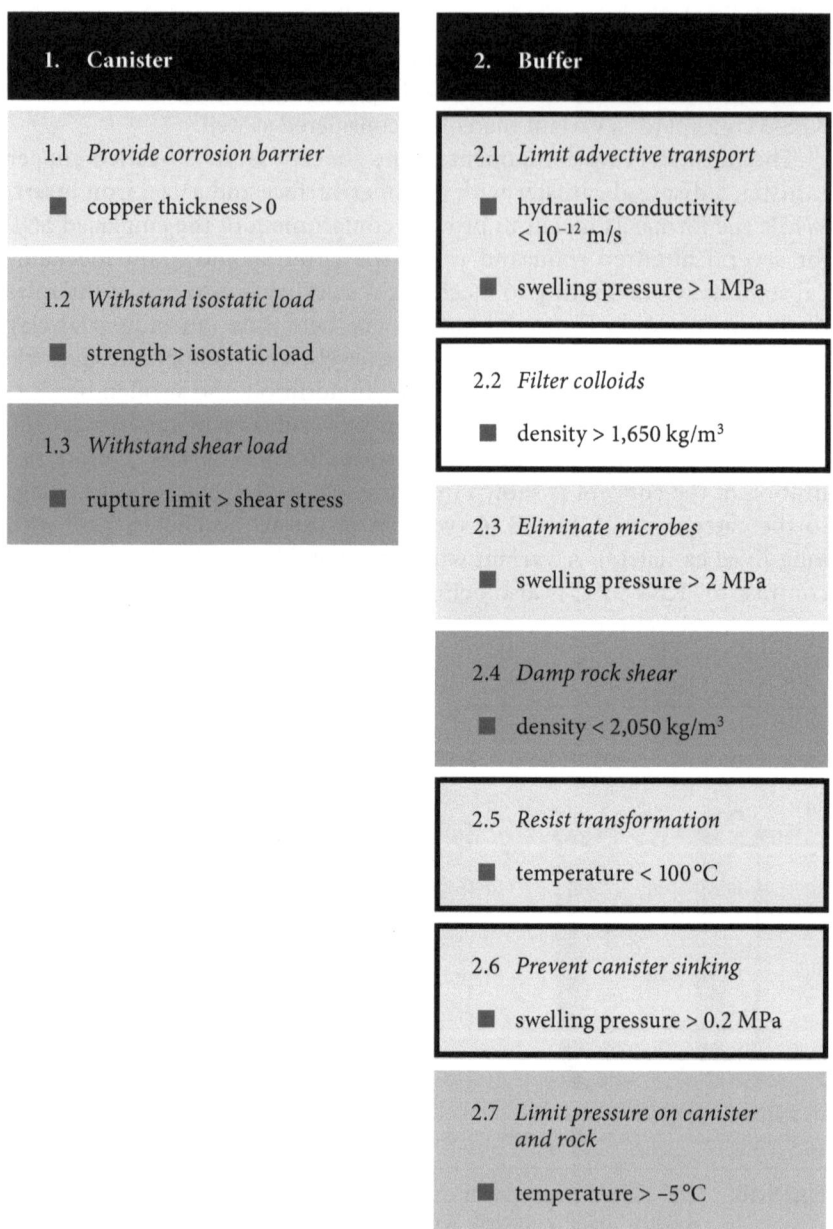

Fig. B.20: Safety functions (italics), safety function indicators and safety function indicator criteria. When quantitative criteria cannot be given, terms like "high", "low" and "limited" are used to indicate favourable values of the safety function indicators. The colour coding shows how the functions contribute to the canister safety functions 1.1 (light), 1.2 (medium), 1.3 (dark) or to retardation (black frame). Many functions contribute to both 1.1 and retardation (light box with black frame) (according to Svensk Kärnbränslehantering AB 2006)

3. Deposition Tunnel Backfill	4. Geosphere

3.1 *Limit advective transport*

- hydraulic conductivity $< 10^{-10}$ m/s

- swelling pressure > 0.1 MPa

- temperature $> 0\,°C$

4.1 *Provide chemically favourable conditions*

- Reducing conditions; Eh limited

- Salinity; TDS limited

- ionic strength; $[M^{2+}] > 1mM$

- concentrations of K, HS$^-$, Fe; limited

- pH; pH < 11

- Avoid chloride corrosion; pH > 4 or $[Cl^-] < 3M$

4.2 *Provide favourable hydrologic and transport conditions*

- transport resistance; high

- fracture transmissivity; limited

- hydraulic gradients; limited

- Kd, De; high

- Colloid concentration; low

4.3a *Provide mechanically stable conditions*

- shear movements at deposition holes < 0.1 m

4.3b *Provide mechanically stable conditions*

- GW pressure; limitedness

4.4 *Provide thermally favourable conditions*

- temperature $>$ buffer freezing temperature

tions. 1.1 is meant to ensure containment, 1.2 and 1.3 to ensure mechanic stability.

In contrast, disposal concepts in clays (both plastic and indurated) heavily rely on the low permeability of the clay materials which result in a very slow diffusion-dominated (if any) migration of toxic substances with effective diffusion coefficients in the order of magnitude of 10^{11} m^2/s (NEA 2005). Sorption by e.g. ion exclusion or surface complexation further decreases the migration velocity of many migrants. This is reflected in the "high-level" safety functions of the French disposal concept:

– Resisting the circulation of water;
– Limiting the release of radioactive nuclides and immobilising them within the repository;
– Delaying and reducing the migration of radioactive nuclides.

(Referred to in Fig. B.18 and further illustrated in Fig. B.21–Fig. B.23.)

Concepts of SNF/HLW disposal in clay host rocks vary: The French ANDRA envisages a "type 2" disposal according to Tab. B.16 (horizontal in-tunnel boreholes with short-lived canisters which should provide containment for the "early transient" or "thermal" phase of a few centuries after disposal closure).

> Andra note that the use of horizontal boreholes, up to a few tens of metres in length, makes better use of the laterally (rather than vertically) extensive argillaceous formations present in their proposed repository siting area, which outweighs the extra difficulties of handling packages in horizontal rather than vertical deposition holes.

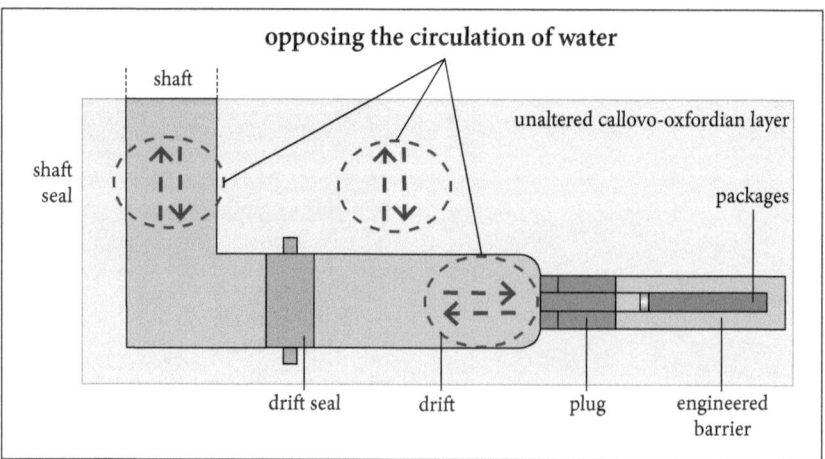

Fig. B.21:   Safety function "Resisting the circulation of water" (from ANDRA 2005)

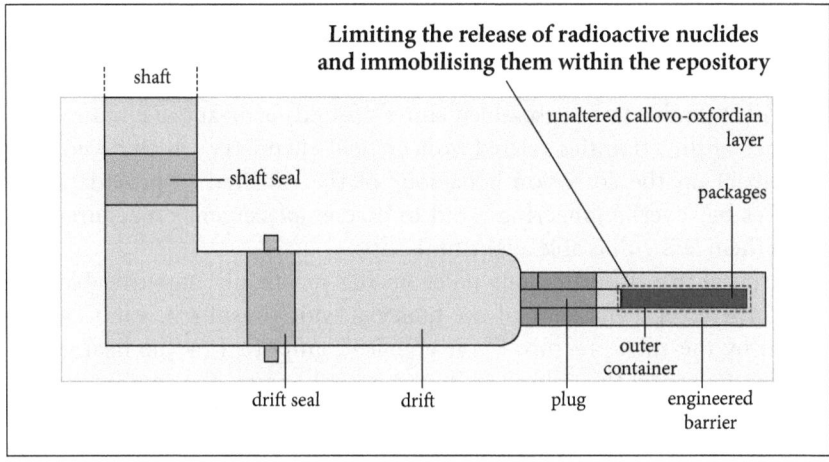

Fig. B.22: Safety function "Limiting the release of radioactive nuclides and immobilising them in the repository" (from ANDRA 2005)

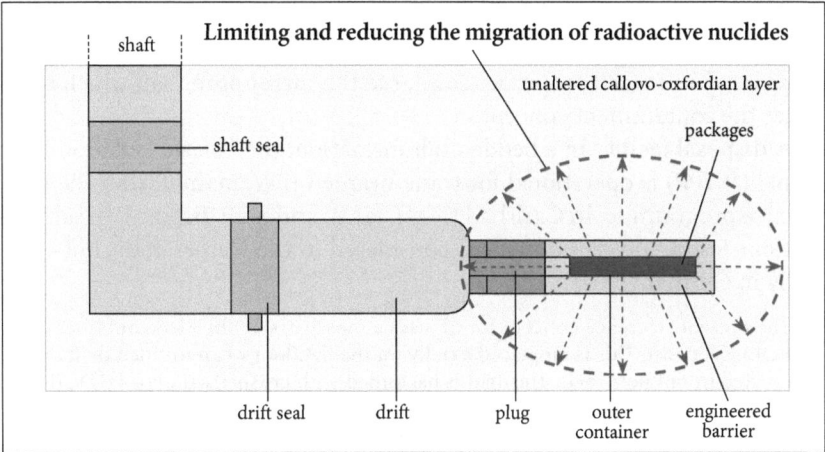

Fig. B.23: Safety function "Limiting and reducing the migration of radioactive nuclides" (from ANDRA 2005)

Andra also prefers to use metal-lined deposition holes with no buffer material as this provides adequate containment in the host rocks identified whilst reducing hole size, use of concrete and the volume of clay disturbed by the excavation. The use of a substantial borehole liner also allows retrieval of waste emplacement for a significant period after emplacement, which is a requirement of the Andra disposal programme (Baldwin/Chapman/Neall 2008).

In Switzerland, NAGRA investigates a concept of type 3 according to Tab. B.16 (in-tunnel (axial) with short-lived canister and buffer) for HLW and SNF emplacement in carbon steel containers, preferably in an indurated clay formation.

The Belgian ONDRAF.NIRAS plans to implement a concept of type 6 "in-tunnel (axial) with supercontainer (concrete buffer)" in a plastic clay formation. The use of the "supercontainer" (basically a container-buffer-backfill ensemble to be assembled above ground) is meant to circumvent a number of uncertainties related to near-field chemistry which could have an impact on the corrosion behaviour of the canister, to provide safety reserves by "over-engineering", and to ease emplacement procedures and make them less vulnerable against mishaps.

Disposal in salt formations relies on the practically impermeable salt. Given the creep behaviour of the material which results in vault closure driven by the rock pressure ("convergence", amplified by the heat generated by the waste) this allows for a concept relying on the containment to be provided by the geological component. The containment properties of the waste containers are only needed for a limited time, especially as long as the repository is still operational. Migration delay and attenuation by surrounding or overlying strata can be considered a latent function coming only into play if containment fails. To prevent this, geotechnical barriers have to preclude water or brine inflow especially in the early post-closure phase. Later, the crushed salt backfill will, driven by convergence, be compacted and reach properties similar to the surrounding salt which completes the containment concept.

A disposal facility in a bedded salt formation, the "Waste Isolation Pilot Plant" (WIPP) is operational for trans-uranic LILW, mainly from the USA defence programme, in Carlsbad (NM). HLW and SNF disposal on salt formations (especially salt domes) is considered in the Netherlands, but especially in Germany:

> The present reference concept for SF uses a massive, self-shielded container (a POLLUX cask). This is emplaced axially on the flat floor of an unlined drift excavated in salt host rock. The drift is backfilled with crushed salt. For HLW, the reference concept comprises the emplacement of HLW containers (without any overpack) in long (possibly up to 300 m) vertical boreholes drilled from the floor of disposal drifts. The same method is also being considered as an alternative for emplacement of thin-walled steel SF containers of the same diameter. The boreholes are unlined and the space between containers and in the narrow annulus around them is filled with crushed salt. (Baldwin/Chapman/Neall 2008)

More precisely, the "containers" intended for borehole disposal are flasks (height of about 1.3 m, diameter of about 0.4 m) with vitrified HLW from reprocessing ("glass flasks"), compacted structural parts sleeves and solid waste from SNF reprocessing (CSD-C – Colis de déchets compacté) and medium-active glass products (CSD-B – Colis de déchets boues).

The concept is thus a combination of "type 3" (in-tunnel (axial), but without overpack) for SNF and "type 10" (Mined deep borehole matrix) for HLW (Tab. B.16). Variant concepts considered include using long boreholes (as opposed to POLLUX emplacement in drifts) for the SNF as well. For this variant, the use of thin-walled SNF containers ("BSK-3"), which would have a profile shape similar to the one of the HLW flasks but a height of about 5 m, is being studied. Advantages of borehole disposal include a better thermo-conducting connection between the canister and the rock salt and thus an increased salt creep, vault closure and thus an earlier state of confinement as well as less potential gas generation from the metal materials emplaced, and a better utilisation of the rock salt volume available. Further modification of both the POLLUX and the BSK concepts are conceivable, in particular when addressing retrievability requirements (cf. section B.1.3.7).

In a very simplifying and schematic manner, Tab. B.20 below illustrates how the different primary safety functions are assigned to engineered anthropogenic and natural geogenic parts of the repository system in different host rocks, respectively.

Tab. B.19: Primary safety functions assigned to engineered anthropogenic and natural geogenic parts of the repository system

Function	Isolation from biosphere, reduce likelihood of human intrusion	Stability (mechanical, hydraulic, chemical)	Contain waste	Attenuate / delay migration
Chrystal-line, e.g. granite	Host rock	Canister, buffer	Canister	
	Overburden	Host rock		Host rock
Indurated clay	Host rock	Host rock	[Canister] – limited in time!	Host rock [Overburden]
	Overburden			Seals
Rock salt	Host rock	Host rock	Host rock	Host rock [Overburden]
	Overburden		Seals	Seals

geologic   engineered

In Germany, strong reliance on the geological barrier is preferred. This is evidenced in the idea of a "confining rock zone" ("einschlusswirksamer Gebirgs-bereich") the basis of which was developed by the Committee on a Site Selection Procedure for Repository Sites (AKEnd 2002) and which, in a modified form, found its way into the recently published "Safety Requirements Governing the Final Disposal of Heat-Generating Radioactive Waste" (BMU

2010a). It is unlikely that such a "confining rock zone" can be found in German crystalline formations as host rock. In contrast, rock salt sites and indurated clays might have this potential. In 2007, the German Federal Institute for Geosciences and Natural Resources (Bundesanstalt für Geowissenschaften und Rohstoffe BGR) published, based on earlier studies, a map showing potentially feasible salt domes and clay formations in Germany (Fig. B.24).

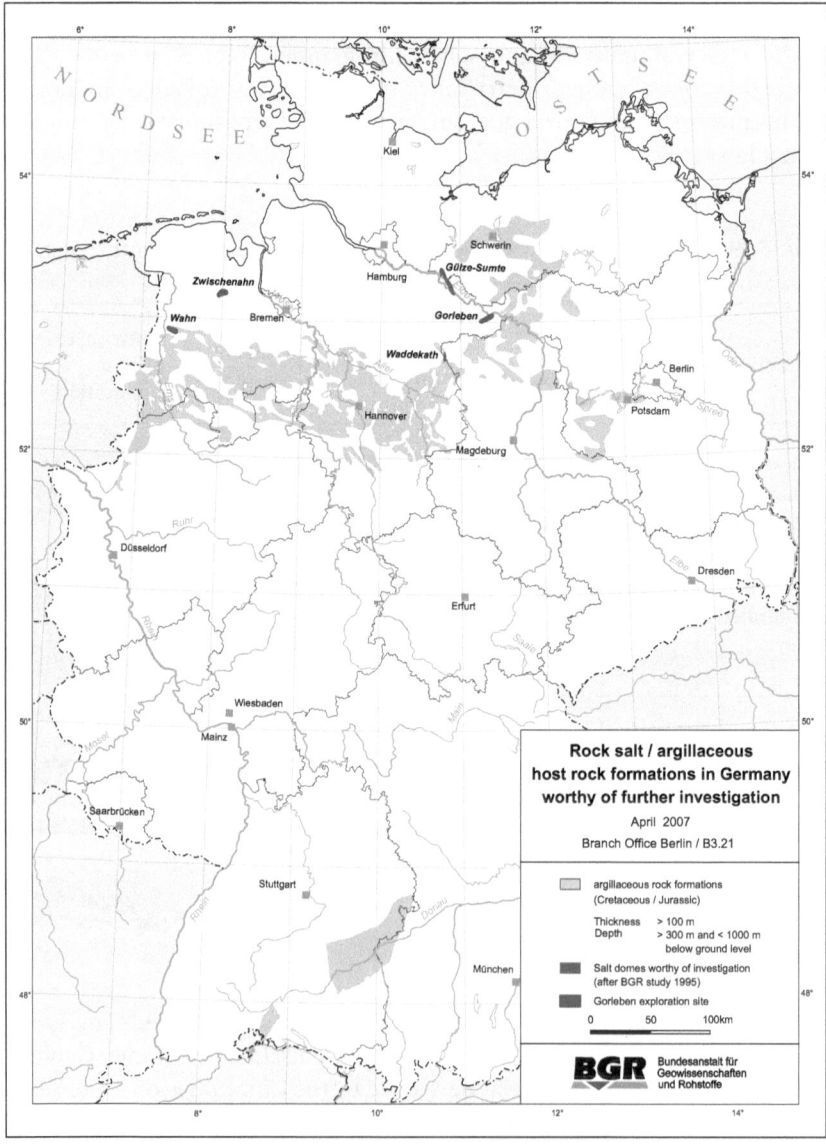

Fig. B.24: Map of rock salt and argillaceous rock formations in Germany worthy of further investigation (BGR 2007)

1.3.7   Retrievability issues

As explained in section B 1.3.4, retrievability of the waste from the disposal facility for a certain period of time might be desired for a number of reasons including the following:

- It could be argued that radioactive materials today considered as waste might be considered a resource in the future, especially if new fuel cycle technologies become available (cf. section B 1.2.3).
- It could also be imagined that in the future better or "safer" options for managing the waste might become available.
- Waste retrieval might be seen as an option in the case of emergencies, i. e. the disposal facility as originally constructed turns out to be less safe than planned or in cases in which standards about what is considered safe change.

Sometimes it is also argued that safety of deep disposal in principle cannot be achieved or demonstrated and that therefore the emplacement in deep geological formations has to be seen as a temporary measure (i. e. a storage solution) in any case.

It is also conceivable to require retrievable emplacement of a small waste amount in a so-called pilot repository (a part of the whole facility) which will be monitored during a certain period in order to finally support the decision to close the whole facility (cf. online-documentation http://www.nagra.ch/g3.cms/s_page/80760/s_name/tlhaa).

In the following, some options for implementing retrievability and their technical implications are described and discussed.

The UK Nirex concept of "phased geological disposal" (NIREX 2002), which was developed for LILW management, strives for flexibility during the stages (or phases) of waste management. It foresees the phases

1. Packaging the waste,
2. Interim surface storage,
3. Transport,
4. Waste emplacement,
5. Monitored underground storage,
6. Backfilling – "when sufficient confidence is obtained, or society desires it" (NIREX 2002),
7. Repository closure – "at an appropriate time" (NIREX 2002) – and
8. After closure.

During the phase of monitored underground storage

> [...] the waste packages could remain with no need for intervention for as long as desired by future generations. This gives future generations a wide range of options. They could proceed to the next phase of the plan when desired, or they could even retrieve the packages for some other type of long-term management. (NIREX 2002)

The French approach to "reversible disposal" provides an example of a concept developed for HLW disposal specified for a particular host rock (claystone) and more concretely accounting for technical boundary conditions and timeframes (ANDRA 2005), cf. Fig. B.10.

– "Post package-emplacement" stage

The "Post package emplacement" stage corresponds to the period in which one or more cells are filled with packages but not sealed. During this stage, the cell is made safe from a radiological point of view by a totally reversible shielded physical barrier, which protects the operators present in the access drifts. The network of connecting and access drifts and shafts remains fully accessible. [...] At all times, the physical disposal cell closing device can be opened so that the packages can be retrieved without further delay. This stage is comparable with a storage configuration.

– "Post cell-sealing" stage

The "Post cell-sealing" stage starts once the cells have been sealed by a swelling clay plug. [...] During this stage, it is technically possible to retrieve the packages. It requires preparatory work consisting of removing the clay plug and restoring the head of the cell.

– "Post module-closing" stage

The "Post module-closing" stage starts once all the components of a module have been sealed and back-filled. [...] The module's connecting drifts remain ventilated and accessible. At this stage, it is also technically possible to retrieve the packages. The preparatory work for this operation is longer than that in the previous case. It consists of removing the back-filling from the drifts using excavation techniques similar to those used during construction the, as during the previous stage, renewing the tops of the cells. [...]

– "Post repository zone-closing" stage

The "Post repository zone-closing" stage starts when the repository zone's internal connecting drifts are back-filled and sealed. At this stage, the main connecting drifts giving access to the repository zone in question remain ventilated and accessible. The transition to this stage [...] only increases the lengths of the drifts to be re-excavated in order to gain access to the packages in the event of their retrieval.

– "Post-closure" stage

The "post-closure"-stage starts after sealing and back-filling the shafts. It corresponds to the end of the disposal process. The repository is then in the "post-closing" configuration. However, it is possible to envisage a monitoring period for the repository and its environment. Although more complex, it remains technically possible to retrieve the packages in this configuration [...].

In order to maintain the geometry of the repository vaults and to limit deformations of the argillite host rock in advance to and during waste emplacement, but also afterwards (cf. above), the described concept in particular relies on an elaborate system of liners. Other technological measures are in place in order to ensure the protection of workers (in particular radiological shielding) during the possible waste retrieval.

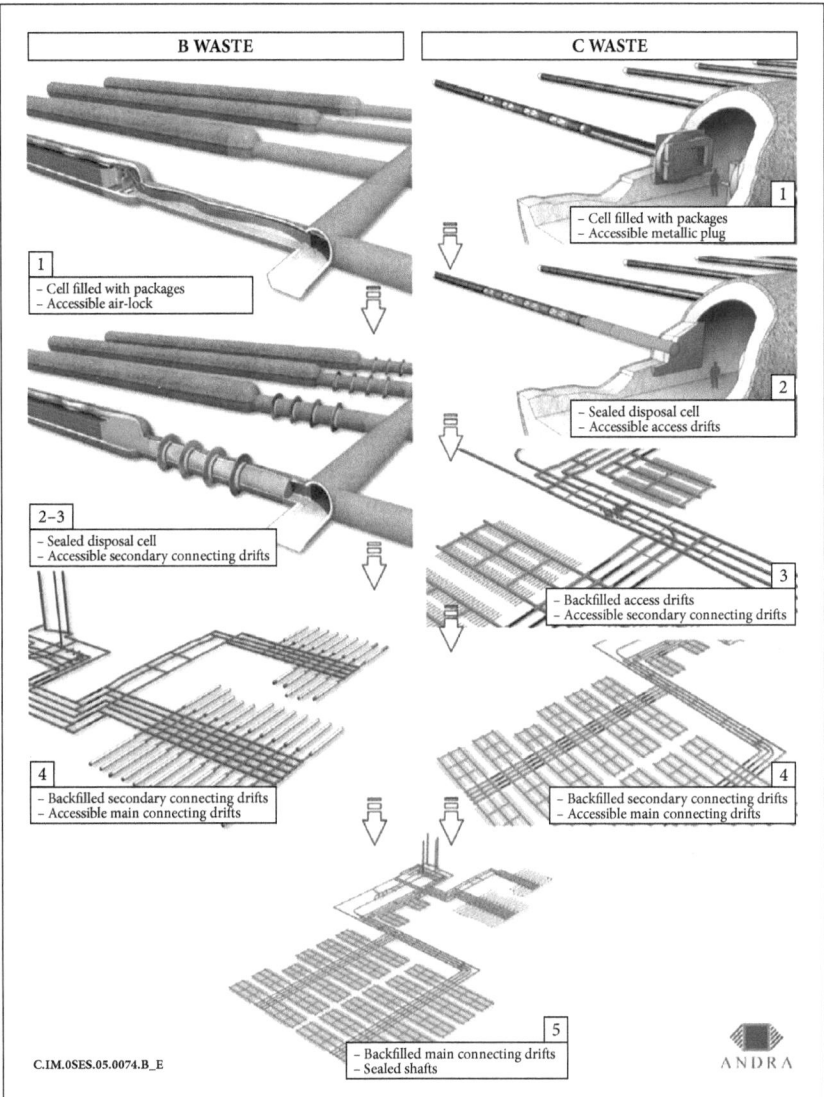

Fig. B.25: Key repository operating and closing stages (from ANDRA 2005)

It becomes evident,

a) that implementing reversibility or retrievability is a complex undertaking,

b) that there is no simple "yes or no" answer to the question whether disposed waste is retrievable but rather a varying "degree of retrievability"[21] can be considered, and

---

[21] In more recent discussions, sometimes between retrieval (of whole intact waste packages) and recovery (of the waste *per se*) is being distinguished.

c) that this degree of retrievability strongly depends on the timeframes considered which, however, are considerably shorter (in the order of decades up to three centuries) compared to the timeframes post-closure safety has to be ensured (up to hundred thousands of years).

It might also be the case that measures for retrievability jeopardise long-term (passive) safety which is, however, considered unacceptable:

> No relaxation of safety standards or requirements could be allowed on the grounds that waste retrieval may be possible or facilitated by a particular provision. It would have to be ensured that any such provision would not have an unacceptable adverse effect on safety or performance. (IAEA 2006b)

In practise, this is not trivial as the intent for retrievability might compete with the wish to efficiently confine the waste as illustrated by the following example: In its "Safety Requirements Governing the Final Disposal of Heat-Generating Radioactive Waste" (BMU 2010a), the German Federal Ministry of the Environment (BMU) requires retrievability of the waste containers for the duration of the operation of the repository as well as the possibility to handle the containers in the case of recovery out of the decommissioned and closed facility for a period of 500 years (likely system evolutions to be considered).

The safety concepts so far developed for disposal in rock salt (cf. section B 1.3.6) do not foresee measures to address such requirements. In particular the idea of disposal in long boreholes, but potentially also the POLLUX drift emplacement concept, might need to be revised in order to address the retrievability/recoverability requests of the 2010 BMU Safety Requirements. Possible revisions might imply larger vaults and prolonged periods over which drifts need to be kept open. Recalling the fact that the rock salt itself is intended to be the main barrier the performance of which will be dependent on preferably negligible disturbance of the host rock and (fast) vault closure by salt creep ("convergence") illustrates that retrievability measures might compete with safety-relevant features and processes. The same would apply if such measures would imply emplacement of additional large amounts of metal. Details need to be investigated in the near future.

Although the examples given above concern only two host rocks (claystone and rock salt), they illustrate that measures for retrievability need to ensure, amongst other things, accessibility of the emplaced waste which results in requirements concerning the dimensioning and lifetime of repository vaults. The more the safety concept relies on properties of the (undisturbed) host rock and the more "dynamic" the geological environment is (crystalline rock being the least and salt being the most "dynamic" environment), the likelier an (adverse) impact on passive safety will be. Therefore, it can also be concluded that

a) retrievability and passive safety might become competing targets, and that
b) such potential competition depends on the host rock, geological environment and related safety concept.

1.4 Long term safety assessment and the safety case

1.4.1 Security against intrusion

Security issues related to disposal are consensually focused on security against human intrusion, which could be either inadvertent or intentional. Human intrusion is considered as inadvertent if the awareness of the repository and the knowledge of the hazard potential of the waste emplaced have been lost. In the case of intentional intrusion, society is still aware of the repository and its hazard potential, but starts e. g. drilling through geological formation, e. g. because the waste became an attractive material or a resource. It is widely agreed that it is exclusively inadvertent intrusion that the safety case has to deal with. Intentional intrusion can only be placed in the responsibility of the respective acting society.

For both intentional and inadvertant human intrusion, the radio toxicity, depending on the fuel cycle, can be used as metrics.

A comprehensive study of human intrusion on the basis of a systematic scenario development would require the prediction of human actions as well as of the state of the art in science and technology of future generations. Such a prediction is regarded highly speculative or even impossible as expressed by a Working Group "Scenario Development" composed of representatives of German organisations (Working Group on "Scenario Development" 2008).

The working group holds the view that suitable and appropriate measures have to be taken upon the planning and construction of a licensed repository in the future that hinder or prevent inadvertent human intrusion and/or reduce the consequences. These measures must not impair the safety of the repository. The most effective measures against inadvertent intrusion consist of establishing the repository in deep geological formations and providing knowledge maintenance in the long run. With the decision for the concept of concentrating and isolating the radioactive waste in a repository, the possibility inevitably has to be accepted that radiation exposure limits may be exceeded in the event of intrusion into the repository. In addition, it is not possible to quantify appropriately the consequences associated with human intrusion due to the lack of predictability of the boundary conditions and other parameters to be assumed. Therefore the working Group holds the view that it is not reasonable to evaluate consequences of human intrusion by means of radiological limit values.

In the BMU Safety Requirements (BMU 2010a), a slightly different position is taken by requiring to analyse likelihoods and potential radiological consequences when optimising the repository system. The consideration of human intrusion should rank, however, secondary compared to other optimisation targets such as radiation protection during repository operation, long-term safety, operational safety, reliability and quality of long-term confinement, safety management, and technical and financial feasibility.

1.4.2   Challenges to demonstration of long-term safety

In the previous section, several concepts for *achieving long-term safety* of disposal facilities were discussed. Safety has, however, not only to be achieved but also to be *assessed* in order to demonstrate compliance with legal requirements or, more generally, with the principles underlying radioactive waste management.

However, *assessing safety* can be seen as an important means of *achieving* it – the necessity to demonstrate how a system performs is an important driver and a source of knowledge when designing it. Thus, safety assessment, described in (IAEA 2006) as

> the process of systematically analysing the hazards associated with the facility and the ability of the site and the design of the facility to provide for the safety functions and to meet technical requirements,

involves "phenomenological analysis of the system, providing quantitative information on its behaviour" (ibd.) and serves two purposes: First, it informs siting, research, and repository development by "evaluating the prevailing level of understanding of the disposal system and assessing the associated uncertainties" (ibd.). Moreover, it provides a basis for licensing situations in that it contributes to compliance demonstration.

In radiation protection (cf. section B 2), safety assessment is a well-established tool for demonstrating that a facility or activity does not result in undue exposure and hazard. Such safety assessment deals with relatively short timeframes (up to decades) and with systems and activities well amenable to prediction. For disposal facilities it is also necessary to carry out such assessments in order to demonstrate that the activities connected with construction, waste emplacement, and closure do not present an undue hazard for the facilities' staff or the public ("operational" or "short-term" safety). For this purpose, it is natural (and legally required) to perform safety assessments comparable to the ones for other nuclear facilities. The challenge connected with achieving and assessing operational safety lies in the fact that radiological protection has to be ensured in an underground facility. This results in technical boundary conditions (e. g. in connection with the definition of controlled or monitoring areas underground, with the handling of heavy waste packages in hoists, or with potential nuclide transfer by the ventilation system) as well as in the necessity to account for legal requirements from different areas (in particular nuclear and mining regulations).

On top of that, "long-term" or "post-closure" safety has to be achieved and demonstrated. In fact, achieving safety and security in the long run is the very and only purpose of constructing disposal facilities (in contrast to other nuclear facilities).

Long-term safety includes "conventional" issues such as mining safety (e. g. prevention of cave-ins). Moreover, and more specifically for disposal facilities, it includes the protection of man and the environment against exposure with radiotoxic or chemotoxic substances contained in the emplaced waste. Historically, the idea to demonstrate radiologic safety was to transfer radiation protection criteria for the present to the long-term (mostly in terms of annual individual effective dose or risk). This is coherent with the principle that future generations should not be exposed to hazards greater than the ones accepted today.

The idea was then to derive so-called scenarios (descriptions of potential future evolutions of the system), model the system behaviour for these scenarios, in particular potential radionuclide migration, and to compare calculated releases and resulting exposures with these criteria. Fig. B.26 is a coarse illustration of how an assessment model chain and its compartments for the numerical calculation of radionuclide migration and its potential consequences might look like: Assuming a scenario in which a fluid (e. g. a liquid such as water or brine) gets access to the emplaced waste, radiotoxic substances could be dissolved and the fluid could carry them from the repository mine ("near field") via the host rock, its overburden and/or surrounding strata ("far field", "geosphere") to the accessible groundwater. Gas resulting e. g. from container corrosion might, dependent on the scenario, serve as transport medium as well. Once in the accessible environment, the nuclides could enter the food chain and be ingested by humans (e. g. via irrigation, cattle grazing contaminated plants, and humans consuming milk or beef). Another potential pathway would be inhalation of dust stemming from soil which once before came into contact with contaminated groundwater. There are established "biosphere models" (cf. section B 2) modelling such exposure modes and being used for assessing radiolog-

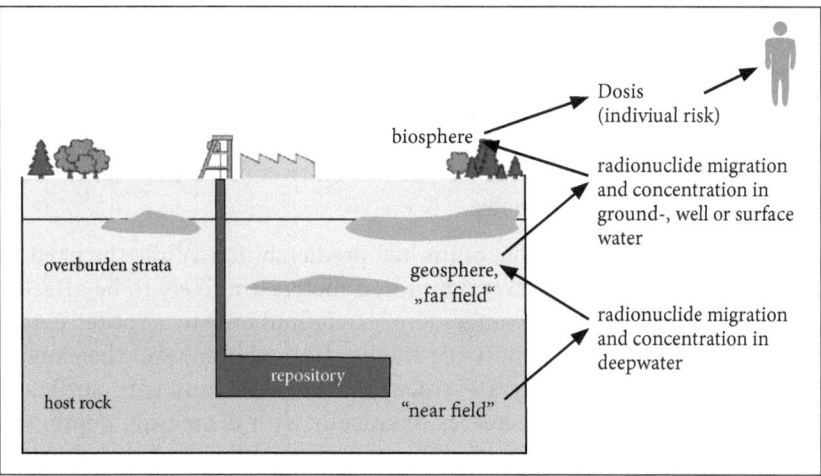

Fig. B.26: Typical model compartments and chain for assessing potential long-term consequences assuming a release scenario

ical safety of nuclear facilities. These, however, need to be modified when addressing long term issues: First, long-term changes e.g. of the climate might drastically change the biosphere and human behaviour which would result in potential pathways different from the ones considered in the short run. And second, certain radionuclides tend to be built up over long time-frames which will result in radiological consequences different from, and often higher than, short-term consequences.

Potential consequences of the release of chemotoxic substances can be addressed using a similar methodology. In addition, it has to be ensured and demonstrated that the waste disposal will not result in critical configurations of the fissile materials emplaced (cf. section B 1.2.4). There are, however, some peculiarities of disposal systems resulting in differences to other nuclear facilities when assessing that they do not create unacceptable radiological or chemical hazards or end up in critical configurations:

– The space scale to be considered lies in the range of some 100 m to several km.
– The systems to be considered are geological and thus heterogeneous. As a consequence, their characterization will always remain incomplete. Moreover, some characterisation methods such as exploration drillings might jeopardise the confinement potential of the geological features.
– It is necessary to account for complex interacting processes which are hardly ever amenable to a fully satisfying scientific characterisation. For example, the geochemical phenomenon of surface uptake of radionuclides (sorption) is obviously highly relevant. It depends, however, on complex thermodynamic effects in which soil and groundwater composition (which both change in space and time) play an important role and which are sometimes hard to analyse and quantify even in the laboratory, to say nothing of field conditions.
– Each repository system is unique. Therefore, the possibilities of conventional testing, comparison and validation as usual for engineered systems are very limited.
– In regulations, often demonstration timescales up to one million years are required. As a consequence, the reliability of predictions for the subsystems is often questionable.

Fig. B.27 illustrates this issue of limited predictability: While the parts of the system at a depth of several hundred meters are likely to be affected by climate and geological changes (e. g. glaciation) only to a limited extent (provided that a sufficiently stable site has been chosen) and their major features relevant for safety are amenable to predictions with sufficient certainty, this certainty decreases drastically with decreasing depth. On the other extreme, ecologic conditions and resulting human habits (in particular food patterns) might change very soon and are practically not predictable over timescales exceeding some years or decades. These lat-

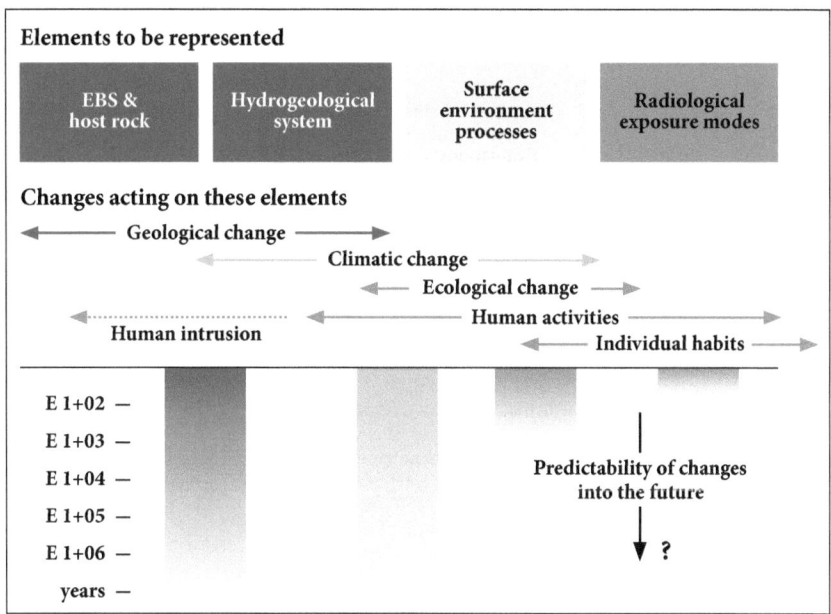

Fig. B.27: Timescales and limited reliability of predictions for systems relevant in safety assessments (from NEA 1999)

ter issues are, however, decisive when calculating entities such as effective doses.

As a consequence, and also as a consequence of the uncertainties
depicted in section B 2.5, *annual effective doses or individual risks calculated in assessments can never be seen as predictions of consequences from
real exposures to be happen. Rather, they are indicators for the performance
of the multiple barrier system the most important part of which indeed should
perform predictably well over the timeframe of concern. It must, however, be
admitted that scientific proof or verification in the strong sense of the word
about compliance with e. g. a dose or risk limit or target will never be achievable. It is, however, achievable (and necessary) to demonstrate system and barrier performance beyond reasonable doubt.*

1.4.3 Safety case concept

The concept of the so-called "safety case" (Fig. B.28) is a means for such
demonstration. It evolved over the past two decades in several countries
and at international organisations. A *Safety Case* is created by bringing together safety-related elements and arguments from site exploration, research, disposal site development and construction, safety analysis, etc.

This concept, which has now made its way into international regulations
(IAEA 2006b), is generally based on two elements:

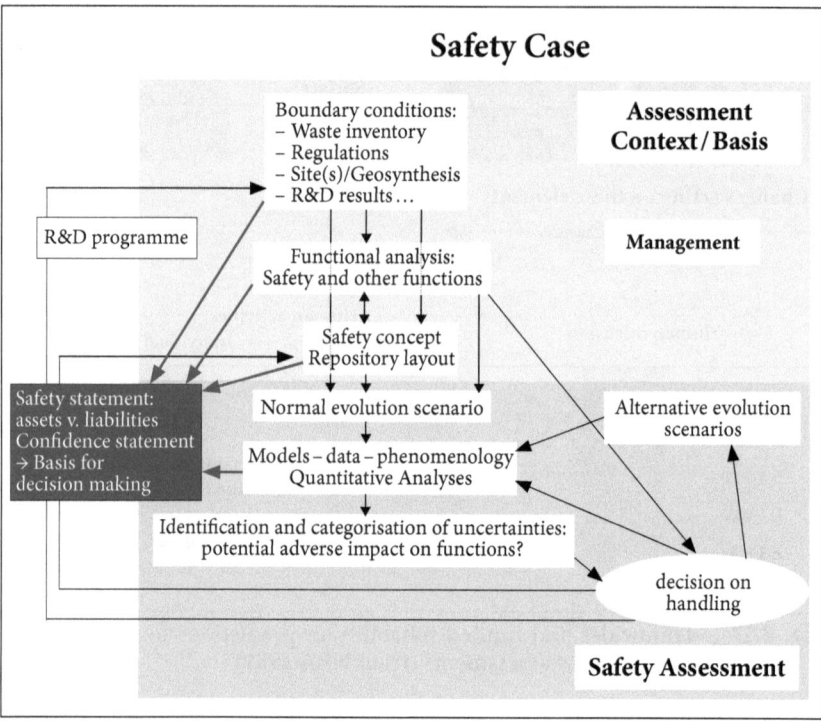

Fig. B.28:   Structure and evolution of a Safety Case

(1) The integral element:

The Safety Case accounts for the idea that safety is not a concept that can be standardised or calculated in the narrow sense; its evidence only comes about through the total of safety principles, scientific results, concepts, analyses and the confidence that is inherent to those aspects.

The starting point for the development of the Safety Case is the legal framework, the respective set of rules and the national waste disposal strategy. It is from the latter that the waste inventory to be disposed of can be derived. If a Safety Case is created for a final disposal project at a particular site, the results of the site exploration and characterisation form an essential component. In their early phase, when the decision about a site has yet to be made, final disposal programmes rely on generic site data. In either case it is important that in the compilation of geoscientific basics ("geosynthesis" – often a separate volume of the overall documentation of the Safety Case) it is made very clear what is already known, what must still be explored and which assumptions (to be verified later) were used as a basis. For such assumptions the plausibility must be elaborated. It has to be discussed what impacts on safety can be expected if results diverging from the assumptions occur in the later course of the exploration.

It is on this basis that the development of a safety concept for final disposal sites is carried out. The necessary safety functions must be demonstrated (e. g. mechanical stability, containment of waste, attenuation or prevention of fluid movements, chemical retention of radionuclides, etc.) along with the way in which they can be ensured in the interplay between the site (i. e. the mechanical, hydraulic and chemical properties of the host rock and the overburden) and the technical components (e. g. waste containers, backfillings, dam constructions, drift and shaft seals). Accordingly, the development and demonstration of a final disposal concept must be carried out by determining the repository mine layout, the definition of containers, backfills, plugs and seals, the compilation of the required properties of these components (e. g. permeability, chemical environment, mechanical properties) and by illustrating on the basis of material science and engineering how these properties can be ensured. In the safety assessment then, first of all that particular potential future evolution of the final disposal system ("scenario") is described and analysed by means of numerical simulation models for which the safety functions are performing as planned ("normal development"). The demonstration must make evident that this particular scenario is the expected or most likely scenario. This is followed by a systematic identification and evaluation of imponderables, ambiguities and uncertainties which may have a safety-related impact on the functioning of the final disposal system. Depending on the situation, these uncertainties can be addressed within the safety analysis, e. g. by varying parameters and models or by taking into account scenarios which deviate from the normal evolution ("alternative scenarios"). Safety related uncertainties for which this is not possible or feasible must be considered and resolved in later steps of repository development, i. e. after developing the Safety Case, in the context of research, site investigation and repository development (cf. the following paragraph). The Safety Case is completed by a safety and confidence statement in which the results of the analysis are put into context with further lines of argument (e. g. based on ground water age or tracer profiles at the site) and in which the particular question of uncertainties and their resolution is addressed. Apart from these scientific and technical components it is also essential for applicants/operators to prove their ability for managing the final disposal project (which will last over several decades) by demonstrating the relevant managerial structures and procedures.

(2) The dynamic element:

The Safety Case forms the basis upon which decisions must be made at particular breakpoints in a step-by-step development and optimisation process (e. g. the transition from above ground to below ground site exploration, determination of site suitability, the beginning of the construction work or of the repository operation). For this, it is necessary to further develop and update the Safety Case step-by-step and to document what has been achieved so far and what must be achieved in the future (e. g. unanswered scientific questions and plans for addressing them). In accordance with each national situation, its rules and standards, concepts and current stage of the final disposal programme, these decisions must be made either by the operator/ developer, in the form of an authorisation or license submission or, if necessary, as a political decision, e. g. in the form of a legislative act.

In France, for instance, Safety Cases with different objectives were prepared in 1996, 2001 and 2005: In 1996 an application for the permission of an underground laboratory in Bure was made; in 2001, the so-called Dossier 2001 was used to test the methodology for the subsequent Dossier 2005 (ANDRA 2005). The Dossier 2005 finally formed the basis for the 2006 legislation regarding the disposal of nuclear waste (République française 2006). This legislation complied with the strict time and action schedule of the so-called "Loi Bataille" (République française 1991) from 15 years earlier.

As mentioned before, the handling of uncertainties – which are unavoidable due to the given time scales, the heterogeneity of the systems considered and the complexity of relevant processes– is pivotal to the Safety Case and essential for the further step-by-step development process of a final disposal programme. The Safety Case provides relevant information and explains how to act in individual cases.

The decisive factor in assessing uncertainties is the question of their relevance regarding safety functions. Unanswered questions and uncertainties must be catalogued systematically in the Safety Case and evaluated in terms of their safety relevance (Vigfusson et al. 2007). This can be done either in a qualitative or – in the context of the safety assessment – quantitative way (cf. the subsequent examples 1–3). Safety-related unanswered questions must be resolved to such an extent that in the further course of the program a readiness for license application can be reached. This can be achieved in two ways:

– By reducing existing uncertainties through research and exploration (cf. the following example 4),
– by avoiding or attenuating the impacts of the uncertainties in the context of the development of the final repository by way of a robust repository construction (cf. the following example 5).

Such possible ways of dealing with uncertainties are demonstrated with the help of a few examples:

1. Quantitative approach: In the French safety concept the properties of the host rock (the Callovo-Oxfordian clay stone) play a significant role: Low permeability hinders the groundwater flow which results in a diffusion-dominated (and thus slow) migration of released radionuclides. Furthermore, most nuclides are retarded in the rock by sorption. Therefore it is very important to confirm these and other relevant properties during site exploration. In safety assessment the significance of such uncertainties was examined, e. g. by means of model calculations using varied parame-

ter values (increased permeability, lower sorption parameters, higher solubility limits, higher diffusion coefficients) (ANDRA 2005).

2. Alternative scenario approach: Commensurate with the significance of the aforementioned processes, cases were examined in which the related safety functions partly or fully failed: For instance, an alternative scenario was considered in which the failure of tunnel plugs and shaft seals effected a through-flow of the final disposal cavities, thus avoiding the claystone-barrier described above.

3. Confidence building regarding the models used: The models used in dealing with the geochemical behaviour of the host rock were substantiated by data collections, experiments and quality control measures. Furthermore, for those cases in which data could be explained on the basis of different models, several model variants were used in parallel for the safety assessment.

4. Research: In the context of the development of the French 1996 safety report, the question of the dominance of diffusion as a transport process for radionuclides in claystone arose, particularly regarding the possibility of faster transport processes in undiscovered fractures. Accordingly, as part of the research programme, seismic measurements and the exploration via drifts in the underground laboratory as well as via drillings as means for an improved site characterisation were planned. The results of this and other, e.g. paleohydrogeological and -geochemical research, made it possible for the 2001 Dossier to determine the host rock claystone as the main barrier.

5. Avoiding uncertainties through a robust repository construction: In the Belgian safety report SAFIR 2 (ONDRAF/NIRAS 2001), uncertainties in the near-field chemistry, especially in the area of backfillings and their possible effects on the corrosion behaviour of waste containers, were identified. The then-planned repository technology raised concerns as to whether operational procedures might in certain cases not be carried out failure-free and without interferences to the planned barrier system. These concerns led to the development of the so-called "supercontainer" in which the waste containers are enclosed by backfilling material and enclosed with a stainless steel cover already above ground in order to simplify the disposal process and to help control the geochemical environment later on, thus creating safety reserves ("overengineering").

6. Stylisation: There are, however, areas for which the above mentioned approaches for addressing uncertainties fail due to the high degree of speculation inherent to any attempt for prognoses. Examples for such areas are issues related to future society evolution or human behaviour which are important when addressing potential future human actions (scenarios) jeopardising repository performance, e.g. inadvertent human intrusion into the repository (cf. above). Future human behaviour is also an important factor when addressing potential radiological impact on future human beings (cf. section B 2). Generally, such uncertainties are addressed by agreeing on basic assumptions the analyses should be built upon ("stylised" or "conventional"

approaches and scenarios). E.g., the applicant and the licensing authority could agree upon a well-defined limited set of human intrusion scenarios to be addressed in the safety case or on conventions for the radionuclide transfer models to be used. Although the necessity for such approaches clearly demonstrates the limitations of any prediction, it is important to note that such stylisation will not be applied when assessing the performance of the repository with regard to central safety functions such as confinement or attenuation/delay of radionuclide migration. These functions have to be provided by robust features of the site and the engineered features amenable to prognoses over the timeframe of interest (cf. Fig. B.27).

1.5   Timescales and potential roadmap

The intent of this paragraph is to explore which evolutions in time (scenarios) the German disposal programme might take and which potential impacts it might have regarding the ability to solve the problem in due time as well as the flexibility of the German society, e.g. future uses of the fissile material, alternative (including today unknown) options to manage the waste, etc.

As explained earlier, the emplacement of the radioactive materials in a repository results, on one hand, in a solution which aims at reducing or lifting the burden of responsibility on future generations and increasing safety by making it less or not dependent on the (uncertain) future state of society and economics. Thus, an "early" implementation of disposal would serve these aims. On the other hand, such a solution would result in a loss or (in the case of retrievable disposal) decrease of flexibility for future generations, e.g. regarding future uses of the material or application of waste management solutions considered better by future generations. It is therefore important to assess possible timeframes for repository implementation.

The starting point consists of two bounding scenarios which are extreme in terms of time they need for implementation. Given that the present government wishes to pursue the Gorleben option, the variables with the highest impact on this time are:

– the pace at which the Gorleben project will be implemented,
– the question whether the project will succeed or fail,
– and, in case of its failure for technical, legal or political reasons:
    – the time of failure,
    – the point in time at which the consideration of alternatives will start,
    – the pace and seriousness with which alternative options are pursued, and
    – the success (or otherwise) of an alternative option.

The *point in time at which the consideration of alternatives will start* could be right now (parallel consideration of alternatives), at the time of the (postulated) Gorleben failure, somewhere in between, or even considerably later after a "wait and see" period. The text will not address the possibility of such a "wait and see" period since its duration would be subject to pure speculation.

For the first bounding scenario "Gorleben implementation at highest possible pace (fast GIHP)" it is assumed that, from now on, each step of the implementation will be taken successfully and as soon as possible. This includes the assumption that, although of course a lawsuit might be filed against a planning permission, such a lawsuit would not have any suspensive effect.

The second bounding scenario "Gorleben late failure, late consideration of alternatives (slow GLFLA)" defines the other extreme: The Gorleben project would fail at the latest possible point in time, i.e. as a result of a lawsuit successfully filed against a planning permission. In other words, up to this point this slow scenario is identical to the fast GIHP scenario (no other delays in the Gorleben implementation are considered). The other underlying assumption is that during the whole Gorleben project no alternative options are considered which would result in a complete restart of the programme after the Gorleben failure. For this restart it is assumed that a thorough consideration of waste management options, e.g., similar to the UK CoRWM process, would be initiated. It is further assumed that, again similar to the UK, this process results in the decision that deep disposal is still the preferred option (if another option would be the outcome it is hardly possible to speculate on any time schedule connected with such an option). The assumption is then that a site will be selected in a process according to the schedule of AKEnd proposal. Activities such as preliminary safety analyses and peer reviews are assumed to take place in parallel.

However for a number of reasons this second bounding scenario "slow GLFLA" is *not* the "slowest conceivable option": First, it is postulated that *in advance to the Gorleben failure* the project proceeds without delay. Second, as explained above, no "wait and see" is assumed. Third, it is assumed that societal decision processes would take place according to ideas considered today (CoRWM, AKEnd). It is, however, known that the attitude of society or parts of it to such processes (and thus their possible duration) changed considerably over the last decades. One could, therefore, argue that such a process initiated ten or twenty years from now might have a format and duration different from the ones envisaged today. Fourth, as explained above, the consideration of management options other than deep disposal might lead to different schedules. And finally, of course the consideration of alternative options might fail as well which would then lead to another restart (or "wait and see") – a process which could theoretically continue *ad infinitum*. All these possibilities are, however, subject to highly speculative considerations and were therefore not considered in the slow GLFLA scenario. Possible scenarios "in between" might result from a delay of the Gorleben project, from its failure at an earlier point in time, and from an earlier consideration of alternatives parallel to the Gorleben project.

The table below shows that the difference concerning the implementation times between the scenarios amounts to 37 years.[22] 23 + 3 of these years

---

[22] Note that figures such as "37 years" might suggest a precision which is of course unrealistic due to the uncertainties involved in estimating the duration of each step or phase. Nevertheless, it has been abstained from using rounded numbers in order to avoid cumulating rounding errors.

are a result of the CoRWM (three years) and AKEnd (23 years) procedures assumed for the slow GLFLA scenario. The remaining eleven years difference comes from the fact that certain time periods have to be absolved twice (underground exploration first for Gorleben, then for the two candidate sites; planning permission procedure) or can be considered "lost" (time from planning permission to rejection of Gorleben). The reasons that these time periods do not amount to the full period of 18 years between now and the assumed high court decision are the facts that part of these 18 years (preliminary safety analysis, peer review) are not assumed to "happen again" and that parts of the Gorleben underground exploration were already carried out by now.

The duration of repository operation (i.e. waste emplacement) is uncertain for a number of reasons: As explained in sections B 1.3.6 and B 1.3.7, there are several disposal concepts (POLLUX and BSK-3) considered for direct disposal of SNF which might undergo further modifications, in particular when accounting for retrievability requirements. Moreover, for the borehole emplacement of flasks with waste from reprocessing and BSK-3 containers with SNF it is not yet clear which type of transfer container will be used when transporting the material into the repository mine. However, the by far most sensitive and still uncertain parameter with regard to the duration of emplacement is the capacity of the conditioning facilities. For the purpose of this estimate, facilities capable of conditioning $500\,t_{HM}$ SNF per annum are postulated. Under the assumption that the presently envisaged lifetime increases result in $15,000\,t_{HM}$ SNF directly to be disposed of[23] this would imply 30 years of SNF conditioning and ensuing emplacement in the repository.

It is further postulated that emplacement of HLW and LLW from reprocessing would result in additional ten years of repository operation. Since we are interested in the loss of flexibility with time when implementing "early" disposal, one could assume that HLW/ILW from reprocessing will be emplaced "as early as feasible" while the emplacement of SNF (which might be considered as a resource in the future) could be delayed for as long as feasible. There are, however, several boundary conditions governing waste emplacement (appropriate distribution of heat sources in the repository system, logistics of conditioning and buffer storage, considerations w.r.t. repository engineering, repository operation etc.) which suggest that HLW/ILW and SNF emplacement will take place alternately, thus resulting in a gradual loss of flexibility over the assumed total of 40 years repository operation.

Seven more years for repository closure were estimated under the assumption that certain backfill and closure works already are to be carried out in parallel to waste emplacement. If, however, a policy orientated towards retrievability would require that repository vaults have to be kept open for some time, this would later result in an increase of the time needed for repository closure.

---

[23] $10,500\,t_{HM}$ SNF from the 32 years lifetime under the Atomic Energy Act from 2002 (www.bfs.de, last visited on December 5, 2010) plus $4,500\,t_{HM}$ SNF from lifetime extensions.

	Fast GIHP	Slow GLFLA	Other possible developments related to power generation*
2010	Restart of Gorleben site investigation, development of preliminary safety analysis[a] International peer review of preliminary safety analysis[a]		
2015	Underground investigation completed[b] Federal government declares Gorleben suitable[a]		"certainty" about overcoming of "show stoppers" within Gen IV nuclear systems development
2020	Planning permission[b], lawsuit(s) filed		"certainly" about realistic potential of new renewables, CCS technologies
2025	Court decision: Gorleben in accordance with law[c]	Court decision: Gorleben unlawful[c] CoRWM-type process initiated	"certainty" about overcoming of "show stoppers" within Gen IV nuclear systems development – "clarity" about commercial deployability role of CCS – "clarity" about commercial deplorability/role of future unclear systems
2030	Shaft construction completed[b]	CoRWM-type body recommends deep disposal and site selection according to AKEnd procedure, Federal government accepts proposal[d] Potentially suitable regions identified[e]	– "clarity" about commercial deplorability/role of future unclear systems

* Focused on foreseeable political and techno-logical developements, i. e. time horizon limited to next 20 to 30 years.

	Fast GIHP	Slow GLFLA	Other possible developments related to power generation*
2035	Repository construction completed, start of operation (emplacement)[b]	5 areas identified[e]	
2040		Site exploration from above ground[e] 2 sites for exploration from underground identified[e]	advanced reactor systems commercially deployable[f,h]
2045			
2050		Exploration from underground completed[e]	Fusion reactor systems commercially deployable[g]
2055		Site decision by Federal government[e]	
2060		Planning permission[b]	
2065		Shaft construction completed[b]	
2070		Repository construction completed, start of operation (emplacement)[b]	

	Fast GIHP	Slow GLFLA	Other possible developments related to power generation*
2075	Emplacement completed, start of repository closure		
2080	Repository closure completed		
2085			
2090			
2095			
2100			
2105			
2110		Emplacement completed, start of repository closure	
2120		Repository closure completed	
2125			
2130			
2135			

* Focused on foreseeable political and techno-logical developements, i. e. time horizon limited to next 20 to 30 years.

Basis/sources for the steps and their timing assumed in the table:

(a) focused on foreseeable political and technological developements, i. e. time horizon limited to next 20 to 30 years;
(b) BMU planning;
(c) Thomauske (2004);
(d) Duration of procedures at court estimated from Konrad experience;
(e) CoRWM experience (CoRWM 2006);
(f) Thomauske (2004) (estimate based on procedures proposed by AKEnd). The AKEnd proposal requires:

> Only those regions declaring their willingness to be involved in the performance of surface investigations on their territory remain in the procedure. [...] If less than three site regions declare their willingness to participate regarding exploration from the surface, the selection procedure is halted, and further action will have to be reconsidered. (AKEnd 2002 [translation Röhlig])

If a procedure indeed followed this recommendation, this would thus be another potential for delay.
(g) cf. section B 1.2.3;
(h) Hasinger (2010);
(i) www.gen-4.org.

In summary, it can be concluded that even for the fast GIHP scenario a gradual loss of flexibility with regard to the SNF (due to its emplacement underground) will only arise from the 30s of this century on, a point in time at which according to today's understanding and visions advanced reactor systems may get close to be operated on a commercial basis. For the less optimistic (slow) GLFLA scenario this would happen about 37 years later. In other words, even the fast GIHP scenario cannot be considered as based on a schedule unduly decreasing flexibility for the next generation – in any case, there will be sufficient time to take or revise decisions in view of changing societal attitudes and/or technology developments such as the ones depicted in the right column of the table. Consequently, even if the highest priority is assigned to flexibility there is no need for intentionally delaying the process.

It can also be concluded that reaching passive repository safety (i. e. repository closure) would, at the earliest, take place in the late 21st century. It should also be noted that, as explained above, the slower GLFLA scenario (closure in the first half of the 22nd century) is by no means the slowest conceivable option. In contrast, in particular a late failure of the Gorleben project with no reserve option or plan of action in place might lead to a considerable delay in addition to the decades derived for the GLFLA scenario.

Obviously, the consideration of reserve options could significantly reduce the difference between the two scenarios in terms of time of repository closure: The earlier and the more thoroughly reserve options are considered, the earlier repository closure could become possible even if Gorleben failed. The resulting in-between scenario would be – provided the consideration of alternatives would lead to success – slower than the GIHP scenario but faster than the GLFLA scenario.

Moreover, there are additional arguments in favour of such a "mixed approach":

– The chances of such an approach being politically accepted would increase (compared to the situation today) if the approach was based on principles such as fairness, legitimacy, clarity of decision-making, openness, transparency, and reversibility.

– Such an approach would maximise chances to implement disposal: On one hand, so far the Gorleben site investigation did not result in scientific findings precluding the construction of a safe repository, but it is still possible that such findings arise later in the process and/or that the Gorleben project will fail due to political or legal reasons. On the other hand, there is no guarantee that even a well-designed site selection and characterisation process will finally result in finding a site considered suitable since bad surprises might arise even at relatively late stages of any site characterisation and repository development process. Thus, excluding Gorleben would result in losing a potentially suitable site without guarantee that an alternative will be found.

2 Radiation risk and radiological protection

2.1 Introduction

With respect to radioactive waste health risks are caused by ionising radiations which occur in conjunction with the decay of radionuclides. Biological risks of ionising radiation were already identified a few years after the discovery of X-rays by Röntgen (1895) and of radioactivity by Becquerel (1896). A very extensive knowledge has been achieved about estimating radiation doses (see section B 2.2) and radiation effects by clinical studies as well as by experimental investigations. For risk evaluations, the shape of the dose effect curves is of eminent significance (Fig. B.29). In radiation research as well as in toxicology of chemicals in general two principal categories of dose effect relationships have been described (ICRP 1977). The shape of these dose response curves is based on manifold experimental studies of radiation effects which have been experimentally investigated by studying molecular structures, living cells and animals after radiation exposures. Further data have been obtained from clinical experiences and epidemiological studies which have been observed after the exposure to ionising radiation in humans in connection with radiotherapy, diagnostics, atomic bomb explosions in Japan and further accidents (UNSCEAR 1993; 1994; 2000; 2006 BEIR 1990; 2005).

Two principle classes of dose effect curves have been formulated for the action of ionising radiation and are used for risk evaluation in radiological protection. The principal and fundamental difference of these dose effect relationships is very important in the low dose range. The most significant feature is whether a threshold dose does exist – these effects are defined as deterministic or non-stochastic effects – or it is assumed that no threshold dose exists; these effects are defined as stochastic effects – (Fig. B.29) (ICRP 1977). The so-called Linear-No-Threshold (LNT) concept has been proposed and used for risk evaluation in the low dose range (ICRP 1991; 2007). As no health effects have been measured after radiation exposures in the low dose range (doses <100 mSv[24]) until now the radiation risk can only be quantified by extrapolation from radiation effects observed in higher dose ranges. Therefore the LNT concept has been seriously criticised as an overestimate of radiation effects (Tubiana et al. 2005). However, there are indications from experimental studies that certain radiation effects follow the LNT concept especially after exposures to high LET radiation like exposure to α-particles which are very relevant for the considerations of risk from radioactive waste as the long living radionuclides quite often are α-emitters. Therefore the International Commission on Radiological Protection (ICRP) came to the conclusion that it is "prudent" to stay with the LNT concept (ICRP 2007).

---

[24] Sievert (Sv) is the unit for the equivalent dose which is obtained from the absorbed dose multiplied with the radiation weighting factor (w_R).

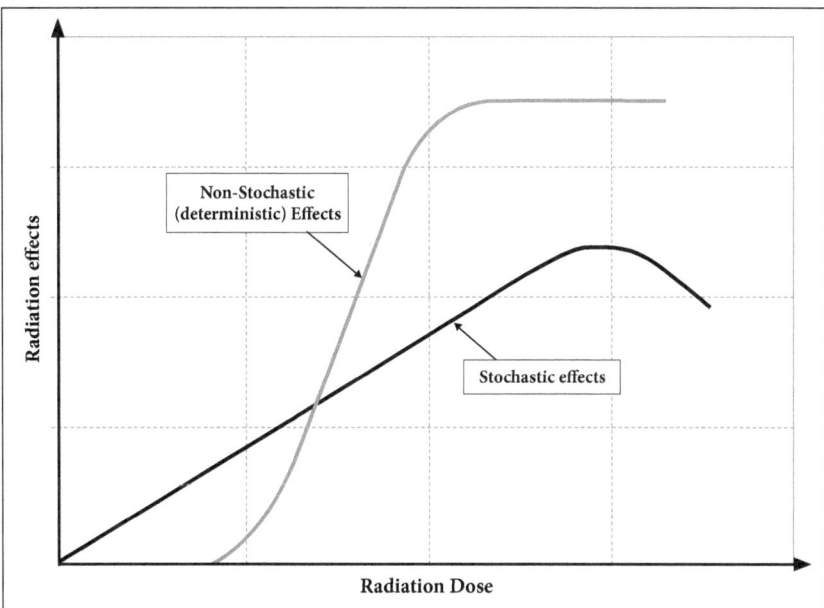

Fig. B.29: Classical dose response curves for radiation effects.

In the case of deterministic radiation effects health effects are only observed when the radiation doses are higher than the corresponding threshold doses. Above the threshold the number of the deterministic effects rises rapidly with increasing radiation doses but also the severity of the radiation effects becomes stronger with increasing dose (ICRP 1991; Streffer et al. 2004). Especially the acute radiation effects, which are mainly caused by cell killing in the corresponding organs and tissues and which can lead to acute lethality, fall into this category. Acute radiation effects are therefore preferentially observed in such organ systems in which a high cell renewal takes place. The renewal of blood cells in the bone marrow, the cell renewal in epithelia of the skin and the intestine for example are inhibited or reduced by radiation doses higher than 1 Gy.[25] The acute radiation effects occur within days or a few weeks after the exposure (Scherer et al. 1991). But there are also some other somatic, deterministic effects which become manifest only after months or years, cf. the induction of cataracts in the eye lens, fibrotic changes in almost all tissues, damage of the blood vessels, necrotic processes in the skin and other epithelia, fall also into the class of deterministic effects (Scherer et al. 1991).

For the development of most deterministic radiation effects, cell death is the most significant process. In all cases not only one cell but many cells

---

[25] Gray (Gy) is the unit for the absorbed energy dose. 1 Gy corresponds to the absorbed energy of 1 Joule per kg mass (tissue).

have to be damaged or changed in order to cause these effects (Scherer et al. 1991). Based on the knowledge of the damaging mechanisms, which are involved in the development of these radiation damages, the dose ranges for the threshold can be described. They are found for almost all effects above acute radiation doses of 0.5 Gy low LET radiation.[26] After chronic or fractionated exposures to low LET radiation, these threshold doses are higher than after acute exposures (Scherer et al. 1991; Hall 1994; ICRP 1991; UNSCEAR 1993). The dose limits as well as the reference doses for workers in radiation facilities, as well as for individuals of the general population, have been fixed below the threshold doses so that these effects can be avoided. These are valid for the work places as well as for the population in or near planned waste facilities during the operative activities and especially in the following later phases after closure where the reference doses should be in the range of 0.1 mSv per year.

For the second type of dose effect relationship, it is assumed that a threshold dose does not exist (Fig. B.29). The following effects are called stochastic effects (genetic effects, the induction of cancer) (ICRP 1991; 2007) as well as some developmental changes after prenatal radiation. For the development of these radiation effects, non-lethal damages of the genome in the corresponding cell nuclei are of significance (mutations, cell transformations). For the description of the mechanisms it is assumed that genetic changes in a single cell are sufficient in order to induce this type of damage (UNSCEAR 2000; Streffer et al. 2004; ICRP 2007). For risk evaluation, it is assumed that stochastic effects increase proportionally with radiation dose in the low and medium dose range up to about 2 Gy low LET radiation. In the higher dose range the frequency of stochastic radiation effects may decrease when cell killing occurs and this can dominate then the dose effect curve (ICRP 1991; Streffer et al. 2004).

Such a unicellular mechanism is obvious for genetic. If the genome is damaged in one single germ cell and if in case of a female this germ cell becomes fertilised or in case of a male fertilises a female mature germ cell, organisms will develop which carry the genetic mutation. Similarly, it is assumed that cancers can develop from one damaged cell after a radiation exposure.

For the induction of cancer, this mechanism is not as obvious as for genetic mutations. With experimental studies on animals and epidemiological studies on irradiated humans, it has been clearly shown that ionising radiation can cause an increase of cancer rates. The data with statistically significant

---

[26] The Linear Energy Transfer (LET) describes the deposited energy of an ionising particle along the track of the particle (eV/µm). This represents a measure for the radiation quality which includes the type of radiation (α-, β-, γ-rays etc.) and the energy of the radiation. β-, γ-rays and photons are low LET radiations with a value for the radiation quality (Q) of 1, α-rays, neutrons are high LET radiations with values for Q of 20 (α-rays) and 5–20 for neutrons depending on their energy. For protons a Q value of 5 was defined (ICRP 1991).

effects of this kind are only found in the range of acute doses of 100 mSv and higher but could not be observed significantly in a dose range below 100 mSv for a general population of all age groups (UNSCEAR 1988; BEIR 2005; ICRP 2007; Streffer 2009). The dose limits which are considered in radiological protection for the effective dose (see below) are in the range of 20 mSv for workers including workplaces during the operative phase of a waste depository and of 1 mSv for individuals of the public during the operative phase and lower at the phase of closure. For the described reasons the dose estimation is of high importance in order to get an information about the radiation risk.

Therefore it can be concluded that no deterministic radiation effects will occur by possible releases from depositories of high level radioactive waste. The expected radiation dose will be far too low and the levels of the threshold doses will not be reached. On the other hand stochastic effects cannot be excluded as it is assumed for the causation of these effects that no threshold doses exist but they are in a range where no significant measurement of such effects can be achieved. The possible risk can only be estimated by extrapolation using the LNT model.

2.2 System of dose quantities in radiological protection

Radiological protection has the general aim of protecting humans and the environment from harm caused by ionising radiation after external as well as internal radiation exposures. This requires a quantitative description of the radiation exposures, e.g. of humans, from external radiation sources as well as from the internal exposures which are caused by the decay of incorporated radionuclides within the human body. While the exposures from radiation sources external to the body can be well described by physical quantities and comparatively exactly measured, the exposures following the intake of radionuclides can only be roughly estimated with the aid of models. These exposures depend upon the metabolism and biokinetics of the radionuclides as well as on the anatomical and physiological parameters of the human body.

In radiological protection the absorbed dose, D, is the basic physical dose quantity. It is used for all types of ionizing radiation and any irradiation geometry. D is defined as the quotient of $d\varepsilon$, by dm, where $d\varepsilon$ is the mean energy transferred to the matter of mass dm by ionising radiation (ICRP 2007). The SI unit is $J\ kg^{-1}$ and its special name is gray (Gy). As radiation effects depend not only on the absorbed dose but also on the type of radiation, on the distribution of energy, absorption of energy in time and space within the human body, and on the radiosensitivity of the exposed tissues or organs, some weightings of the absorbed dose has to be performed in order to express and judge on the radiation effect. As adopted by ICRP D is used as the fundamental physical quantity and this quantity is averaged over specified organs and tissues. The resulting "Mean absorbed dose", $D_{T,R}$ in an organ or tissue (T) by a radiation (R) is obtained by this averaging procedure (Fig. B.30). Thereafter suitable weighting factors are chosen to take account of differences in the biological effectiveness

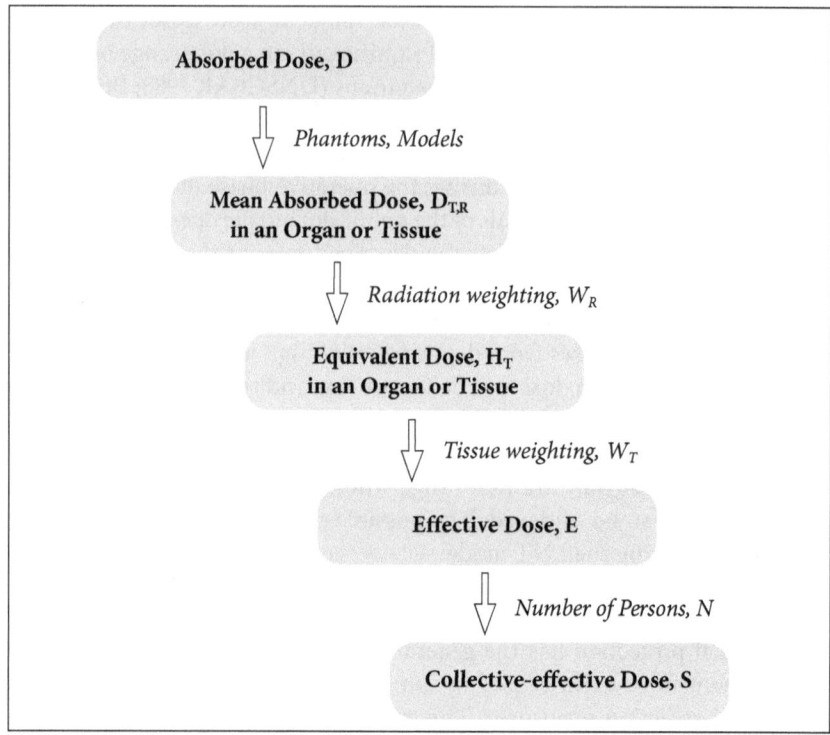

Fig. B.30: System of dose quantities for use in radiological protection (ICRP 2007)

of different radiations (w_R) (Tab. B.21) resulting in the equivalent dose, H_T, in an organ or tissue (T). These radiation weighting factors (w_T) are selected on the basis of experimentally obtained RBE-values which are determined by measuring the relative biological effectiveness (RBE) of a radiation quality in certain organisms or other biological systems and by biophysical considerations.

Tab. B.20: Radiation weighting, factors in ICRP recommendation (2007). All values relate to the radiation incident on the body or, for internal sources, emitted from the source.

Radiation type	Radiation weighting factor, w_R
Photons	1
Electrons and muons	1
Protons and charged pions	2
Alpha particles, fission fragments, heavy ions	20
Neutrons	A continuous function of energy

A further weighting is performed in order to account for differences in sensitivities of organs and tissues to the stochastic health effects (w_T) (Tab. B.22) in order to obtain the effective dose, E, which is used as a general unit for radiological protection. The tissue weighting factors (w_T) are averaged for cancer risk and hereditary effects over all ages and both sexes. They are fixed for a reference person (ICRP 2007). Effective dose (Fig. B.30) is therefore a quantity based on the internal and external radiation fields and the primary physical interactions in human tissues as well as on judgements about the relative efficiency of the radiation qualities and the biological organ and tissue sensitivities resulting in stochastic health effects (ICRP 2007). Effective dose is generally used for setting of dose limits in legal and regulatory systems.

Tab. B.21: Tissue Weighting Factors, wT in ICRP Recommendations (2007) Notes: (1.) The wT for gonads is applied to the mean of the doses to testes and ovaries. (2) The dose to the colon is taken to be the mass-weighted mean of ULI and LLI doses, as in the publication 60 formulation. (3.) The specified remainder tissues (14 in total, 13 in each sex) are: adrenals, extra thoracic tissue (ET), gall bladder, heart, kidneys, lymphatic nodes, muscle, oral mucosa, pancreas, prostate (σ), small intestine (SI), spleen, thymus, uterus/cervix (φ).

Organ/Tissue	Number	w_T	Total Contribution
Lung, Stomach, Colon, Bone marrow, Breast, Remainder	6	0.12	0.72
Gonads	1	0.08	0.08
Thyroid, Oesophagus, Bladder, Liver	4	0.04	0.16
Bone surface, Skin, Brain, Salivary glands	4	0.01	0.04

The radiation doses differ between various organs and tissues in most exposure situations. They are also dependent on the exposure pathways. This is especially the case for internal exposures from incorporated radionuclides due to the biokinetics behaviour and retention in the specified tissues which are different after ingestion, inhalation or uptake through wounds. Effective dose has the advantage that the possible risks in certain organs and tissues especially caused by the heterogeneous exposures and the variability of radiosensitivity can be taken into account and are expressed in one distinct value. For instance the incorporation of radioactive iodine leads to a very high radiation dose in the thyroid due to a specific metabolic uptake of iodine into this organ in comparison to other organs and tissues in which the exposure is by a factor of more than 1,000 smaller than in the thyroid. On the other hand after an uptake (ingestion) of a number of α-particle emitting radionuclides with long physical half time like plutonium, radium, uranium the uptake and in consequence radiation dose is much higher to the skeletal structures (by a factor of around 10 to 100) than to other organs and tissues, as the retention of these radionuclides is high in the bone. On the other hand the incorporation of tritium in the chemical form of water or of the radionuclides of caesium leads to a comparatively homogeneous exposure of all important organs and tissues.

Radionuclides incorporated in the human body often irradiate tissues over long time periods determined by both their physical half-life and their biological retention within the body. Radionuclides with long physical half time become dominating during the long-term phase of a repository after closure, as radionuclides with shorter physical half-life's have decayed to great deal at these late periods. However, for a long effective half-life a long biological half-life of the corresponding radionuclide is also necessary. Thus the decay of the radionuclides may give rise to doses in body tissues either over very short periods after incorporation or throughout life. For example, in the case of intakes of triturated water, as a consequence of its short biological half-time (around 100 days, this biological half-life and a physical half-life of 12.3 years result in an effective half-life time of only ten days), essentially the total dose is delivered within two to three months after intake. For ^{239}Pu, however, both biological retention times especially in the skeleton and the physical half time (around 24,000 years) are very long, and doses will accumulate over the remaining lifespan of the individual. Another interesting example is iodine-129 (^{129}I) with a physical half-life time of 15.7 million years but a biological half-life of iodine of around 120 days the effective half time for ^{129}I is about 100 days.

The need to regulate exposures by incorporated long-living radionuclides and the accumulation of radiation doses over extended periods of time led to the definition of "committed" dose quantities (ICRP 2007), which present the total dose expected to be delivered within a specified time period. The committed equivalent dose, $H_{T(\tau)}$, in a tissue or organ T is defined by

$$H_T(\tau) = \int_{t_0}^{t_0+\tau} \dot{H}_T(t)\mathrm{d}t$$

where τ is the integration time following the intake at time t_0. The quantity committed effective dose, $E(\tau)$, is then given by

$$E(\tau) = \sum_T w_T H_T(\tau)$$

For workers, the committed dose is normally evaluated over the 50-year period following the intake, i. e. a rounded value considered by (ICRP 2007) to be the life expectancy of a young person entering the workforce. The committed effective dose from intakes is also used in prospective dose estimates for members of the public. This would then also be valid for the planning of the late phase of the repository. In these cases a commitment period of 50 years is considered for adults. For infants and children the dose is evaluated up to age 70 years (ICRP 2007).

For practical purposes the values from both external and internal radiation exposures should be combined in the assessment of the value of total effective dose for demonstrating compliance with dose limits and constraints (ICRP 2007). In most situations of occupational exposure the effective dose, E, can be derived from operational quantities using the following formula:

$$E \cong H_p(10) + E(50)$$

where Hp(10) is the personal dose equivalent from external exposure and E(50) the committed effective dose from internal exposure, assessed by:

$$E(50) = \sum_j e_{j,inh}(50) \cdot I_{j,inh} + \sum_j e_{j,ing}(50) \cdot I_{j,ing}$$

where $e_{j;inh}(50)$ is the committed effective dose coefficient for activity intakes by inhalation of a radionuclide j, $I_{j;inh}$ is the activity intake of a radionuclide j by inhalation, $e_{j;ing}(50)$ is the committed effective dose coefficient for activity intakes of a radionuclide j by ingestion, and $I_{j;ing}$ is the activity intake of a radionuclide j by ingestion (ICRP 2007).

The dose coefficients used in the above described equations are those specified by ICRP for all organs and tissues through anatomical, physiological, and biokinetic parameters given for the Reference Male and Reference Female (ICRP 2002). As mentioned before for waste management the radiation exposures to the workers and to the public must be considered for the operation of the waste repository and to public only for the later periods after closure. During the operation of the waste repository the regulatory measures are certainly the same as with other nuclear installations. The dose limits for the effective dose in representative persons of workers are 20 mSv per year and for persons of the public are in these cases 1 mSv per year and lower as recommended by ICRP (2007). However, for the later time periods after closure ICRP (1998) has proposed a dose constraint of 0.3 mSv per year and today in most regulatory proposals a radiation dose of 0.1 to 0.3 mSv is foreseen.

Recently the German Federal Ministry for Environment published a document for the safety requirements of repositories in which the reference dose for a single person of the public was reduced to 10 μSv per year during the late phase after closure under the condition that this exposure takes during the whole life span (BMU 2010). In this connection the BMU refers to the ICRP publication 104 (2007) which is interpreted by the BMU in that way that 10μSv is defined as a "trivial dose". This interpretation is a misunderstanding and an "over-interpretation" of the LNT risk model. In ICRP publication 104 it is stated:

> The conclusion was that an individual radiation dose, regardless of its origin, is likely to be regarded as trivial if it is of the order of some tens of microsieverts per year. It was noted that this level of dose corresponds to a few per cent of the annual dose limit for members of the public recommended by ICRP and is much smaller than any upper bound set by competent authorities for practices subject to regulatory control.

Such a radiation dose of the "order of *some tens* of microsieverts per year" is trivial and can be exempted from the radiological control. In ICRP publication 104 (2007) it is also stated:

> The fact that the Commission's recommendations consider all levels of radiation exposure, however small, does not mean that all radiation exposure situations can be or need to be formally regulated and controlled. On the basis of the principles of justification and optimisation of protection, account may be taken of the amenability of controlling the exposure situation and of whether or not control is warranted.

The proposed dose of 10 μSv per year should be reconsidered. Such a radiation dose is much less than a per cent of the average radiation dose which originates from natural sources. It is around one per cent of the average α-radiation dose from radon and its radioactive daughter products. The radiation dose of 10 μSv per year lies within the scatter of exposures from natural sources. A change of the living place will usually lead to an alteration of the radiation dose from natural sources which is larger than 10 μSv per year. Such comparisons with the radiation from natural sources is valid as in both cases (natural sources and repository for HLW) α-, β- and γ-radiations observed with the same radiation quality. This radiation dose is several orders (around a factor of 10^5) lower than radiation doses which can cause a measurable health effect. The radiation dose of 10 μSv per year is by a factor of 10 lower than the present international consensus of 0.1 mSv for probable situations during the late phase of repositories after closure.

Then the annual effective dose is the sum of the effective dose obtained within one year from external exposure and the committed effective dose from incorporated radionuclides within this year. For the dose assessment of the worker during the operational phase the radiation exposure is monitored individually. The dose of persons of the public is usually not obtained by individual monitoring as for occupational exposure but is mainly determined by environmental measurements, habit data, and modeling. It can be estimated from:

- Simulation and prediction of radionuclide levels in effluents from the technical installation or source during the design period;

- Effluent and stray radiation monitoring during the operational period; and
- Radioecological modeling (pathway analysis of environmental transport, e.g., from the release of radionuclides to the geological phase, to the biosphere and then transport through the soil to plants to animals to humans). A typical pathway would be: Transport of radioactive material from soil of the biosphere to grass to cow to milk to human.

In Germany it is hypothetically assumed that the total consumed food (e.g. milk) of a person living in the neighbourhood of a nuclear installation is produced at the place with the highest concentration of radioactivity. External exposures of individuals may occur from radionuclides released from the waste repository during the operational phase.

For the later phase of waste repositories after closure it is assumed that no release of radioactive material occurs from the deposited material for a certain time period due to the confinement and retention capability of the technical and natural barriers. In case of much later release the radionuclides must first migrate through the geosphere, can enter then into the biosphere which is assumed to occur through the groundwater or via gaseous pathways and to reach humans living in the concerning region through exposure pathways like the grass – cow – milk pathway as stated before. In these cases a possible external radiation exposure from such radioactive material appears negligible but an uptake of the radionuclides through the food chain is then the most relevant pathway for exposure. The modalities of dose assessment vary from country to country. In Germany very hypothetical models are used as explained before. Effective doses can be estimated by modeling and further calculation. Besides the uncertainties which are connected with the choice of the underlying scenarios, models and parameters for these calculations in the far future further uncertainties arise related to questions such as: What will be the population living in the potentially affected region several thousands or even millions of years later? What will be the habits and lifestyle of these populations? Etc.

2.3 Application of effective dose

As has been stated before the effective dose has the great advantage that a judgement of the radiation dose and the possibly related risk can be assessed in cases of a very heterogeneous distribution of the radiation exposures between various tissues and organs of the human body. Such a heterogeneous distribution occurs especially after the incorporation of radioactive substances. However, it has always to be considered that the effective dose is calculated for a reference person with reference parameters and it can generally not be used for risk estimation in individual cases although risk constraints are given in a general way for a system. The main and primary use of effective dose is to provide a means of demonstrating compliance with dose limits. It is an instrument for rough assessments. In this sense effective dose is used for regulatory purposes worldwide (ICRP 2007) and can also be used for the planning processes of waste repositories.

Effective dose is used to limit the occurrence of stochastic effects (cancer and hereditary effects) and is not applicable to the assessment of the possibility of deterministic tissue reactions. In the dose range below the annual effective dose limit tissue reactions will not occur. Such radiation effects will be avoided by this concept. This will also be valid for all phases of a waste depository.

The calculation of reference dose coefficients for intakes of radionuclides and dose conversion factors for external exposures are based on reference anatomical and physiological data for the organs and tissues of the human body together with defined biokinetic and dosimetric models. – It can be assumed that the anatomical and physiological parameters used for the reference dose coefficients will not change too dramatically even during several thousand years or even a million years. The general approach is to monitor individuals or the environment during the short-term phase and from these measurement data to assess the external exposure or radionuclide intake can be derived. For the long-term phase the amounts of radionuclides in the environment have to be obtained via appropriate models. It should be reminded that the weighting factors used in the calculation of reference dose coefficients and conversion factors are selected values that apply to a population of both sexes and all ages. Thus dose coefficients and the reference models and weighting factors used in their calculation are not individual specific but apply to a reference person for the purposes of regulatory control. Conversion coefficients or dose coefficients are calculated for a reference adult worker or a reference member of the public of a defined age group (ICRP 2007).

Particularly in retrospective dose assessments for occupational exposures information may be available that differs from the reference parameter values used in the calculation of dose conversion factors and dose coefficients. In such situations it may be appropriate, depending on the level of exposure, to use specific data in the assessment of exposure or the intake and calculation of doses. Such a situation may occur at workplaces during the operational phase of waste depositories.

Effective dose is a risk related quantity based upon the consequences of whole body exposure. The w_T values are selected values that are chosen to take account of the contribution of individual organs and tissues to total radiation detriment from stochastic effects, in terms of cancer and hereditary effects, on the basis of current epidemiological evidence. Therefore the present situation of cancer frequencies, the ratio of morbidity and mortality by these diseases including the efficiency of therapy is included in the present tissue weighting factors w_T. One has to expect that these parameters will change in the long term of consideration with respect to depositories for high level radioactive waste.

In summary, effective dose should be used for assessing exposure and controlling stochastic effects for regulatory purposes. It can be used to demonstrate compliance with dose limits, reference doses and for dose records. In situations approaching or even exceeding the dose limits, effective dose provides a convenient quantity for the assessment of overall radiation exposure, taking account of all exposure pathways, internal and external, for dose record keeping and regulatory purposes. Used in this way effective dose is a valuable quantity for practical radiological protection purposes especially in planning phases but it applies to a reference person. In retrospective situations the assessment of effective dose gives an insight into the quality of radiological protection and gives information on whether the dose limits or the reference dose might exceed.

2.4 Collective dose

The dosimetric quantities for radiological protection discussed above refer to single reference persons. However, the task of radiological protection is not only to protect single persons but also to optimise and reduce the radiation exposure of groups of occupationally exposed persons or of the public. This does not mean minimisation of the radiation dose. Societal and economic aspects have to be taken into consideration. This is relevant for the various phases including the planning of waste depositories. For these purposes of optimisation ICRP has introduced the collective dose quantities (ICRP 1977; 1991; 2007) which should be used and seen as an instrument to achieve optimisation. It is not a suitable instrument and should not be used for risk estimation. These quantities take account of the number of persons exposed to radiation from a source and the time period of exposure. Collective dose is obtained in principle by multiplying the number of exposed persons with the average dose to the exposed persons over a specified time period from a source. The specified quantities have been defined as the collective equivalent dose, S_T, which relates to a tissue or an organ T, and the collective effective dose, S (ICRP 1991). The special name of the unit of these collective dose quantities is the man Sievert (man Sv) (ICRP 2007).

The definition of collective dose quantities as described above has led people to use collective effective doses incorrectly for summing up radiation exposures for a wide range of doses over very long time periods and over large geographical regions and to calculate on these basis radiation-related detriments. However, this is only meaningful if there is sufficient knowledge of the risk coefficients for the detrimental radiation effects in the dose ranges which contribute to the collective dose (Kaul et al. 1987).

In this context it has to be realised that the risk factors e. g. for carcinogenesis at low doses, are obtained from the extrapolation of epidemiological data observed in much higher dose ranges. As described the extrapolation is based on the assumption of a linear dose effect relationship without a threshold dose (LNT model). The ICRP considers that in the low dose range the risk factors have a high degree of uncertainty. This is particularly the case for very low individual doses which are small fractions of the radiation doses in comparison with doses received from natural sources. The use of collective dose is not at all a valid procedure under such conditions for detailed risk estimates (ICRP 2007).

To avoid the aggregation of low individual doses over extended time periods and geographical regions limiting conditions need to be set. If such calculations are performed in connection with waste depositories such considerations are especially valid and should be regarded. Especially for longer time periods the use of collective dose which is obtained from the summation of small individual doses with high uncertainties is not reasonable (ICRP 2007, Annex B).

The collective effective dose due to individual effective dose values between E_1 and E_2 is defined as

$$S(E_1, E_2, \Delta T) = \int_{E_1}^{E_2} E \frac{dN}{dE} \, dE \ ,$$

where $\dfrac{dN}{dE}$ denotes the number of individuals who are exposed to an effective dose between E and $E + dE$ and ΔT specifies the time period within which the effective doses are summed.

In the calculation of collective effective dose the following aspects could be considered:

– number of exposed individuals;
– age and sex of exposed persons;
– range of individual doses;
– dose distribution in time;
– geographical distribution of exposed individuals.

2.5 Radiotoxicity of safety–relevant radionuclides for waste repositories

The radiotoxicity of radionuclides is determined by the degree to which health effects are caused by the ionising radiation which originates from the decay of the radioactive nuclides. Most important certainly is the resulting radiation dose, which is determined by several physical and biological parameters like

– the type of ionising particles which originate from the decay of the radioactive material,
– the energy of these particles,
– the physical half life time of the radionuclide,

- the biokinetics and the metabolism of the radioactive material which determines the distribution between the organs and tissues within the organism,
- the location of the radioactive material within the tissues and in the cells, the biological half-life which results from these biological parameters.

With respect to health effects after radiation exposures to low doses caused by released radionuclides from waste repositories at the post-closure phase the possible induction of stochastic effects especially of cancer is most relevant. As has been pointed out before no threshold dose is assumed for these effects and therefore even very small doses may cause these radiation-induced health effects although they will not be measurable as they will be covered in a population by the scattering of the spontaneous cancers with respect to life style of the population, time (year by year), region of the population etc.

With respect to long term risk estimates of repositories for high level radioactive waste a number of radionuclides with a long physical half-life are relevant for the assessment of possible radiation exposures and thereby relevant for safety analyses during the late phases of several thousand years and beyond after closure of the repository. As has already pointed out before, radiation doses from the external sources will be negligible. The exposure of humans will result from the incorporation of radioactive material. Under these conditions the dose coefficients for committed effective dose (CED, given in Sv/Bq) are the relevant parameters to determine radiotoxicity. As has been pointed out the effective dose is risk related and therefore the corresponding coefficient per Bq describes best the degree of radiotoxicity. In some cases where the incorporated radioactive material is distributed very heterogeneously in the human body the dose coefficient for the committed equivalent dose in an organ or tissue with the highest accumulation of radioactivity (also given in Sv/Bq) may be more appropriate.

On the first sight one would expect that the α-particle emitting radionuclides like the actinides are most relevant for risk estimates of repository for HLW as these radionuclides usually have a long physical half-time and the α-particles have a high RBE-value. Further the inventory of these isotopes in the HLW is comparatively high. However, it turns out that besides these physical and biological factors the mobility of the radioactive material is most important for the risk assessment of a deeply located repository.

2.6 Assessment of potential radiation doses from repositories

The analysis of inventories of safety-relevant radionuclides in a canister containing 4 PWR UO$_2$ with a burn-up of 48 GWd/t$_{IHM}$, after 40 years decay before final storage (see Tab. B.10) results in the radionuclides which are listed in Tab. B.22 with the corresponding amounts of radioactivity in Bq. In this table as well as in Tab. B.23 and Tab. B.24 the inventories of radionuclides in a canister containing 4 PWR UO_2 with a burn-up of 48 GWd/t$_{(HM)}$ after 40 years decay are given (column 4, data are from Tab. B.10 of section B 1.3.2. Column 2 contains the physical half-life times of the corresponding radionuclides (column 1) taken from ICRP publication 107 (2009). In column 3 the committed effective dose coefficients in Sv/Bq (see section B 2.2 above) are presented which are taken from ICRP publication 68 (1996) These 51 radionuclides are the most relevant radioactive isotopes from which a radiation exposure to humans and the environment can occur and therefore a safe storage or disposal of this waste is necessary. There are 17 of these 51 radionuclides which also appear in nature without any nuclear technology and therefore the behaviour in the biosphere as well as the biokinetics in humans has been studied. – The mass of fuel in a canister is 1.556 t$_{IHM}$; the total mass of structural materials 0.589 t.

In the list of these radionuclides the dose coefficients are also given with which the resulting committed effective dose (CED) (column 5) can be calculated under the condition that the total radioactivity is ingested by humans. There are 35 of the listed 51 radionuclides which contribute with more than 1 Sv to the total radiation dose. The by far highest contributions to the radiation dose with more than 1,000 Sv per radionuclide would come at this stage of waste treatment from the fission and activation nuclides, strontium (^{90}Sr), caesium (^{137}Cs) and samarium (^{151}Sm), as well as from a number of isotopes of the actinides (uranium, neptunium, plutonium, americium and curium) (Tab. B.22).

However, the two fission products ^{90}Sr and ^{137}Cs have a physical half-life of only around 30 years and ^{151}Sm a physical half-life of 90 years. Thus within a period of 10,000 years these radionuclides have decayed over this time period for more than 300 and 100 half-life times respectively. Therefore these radionuclides have been reduced by the decay to such a degree that they are no longer relevant for the consideration of the long-time safety of a waste repository (Tab. B.23). After a time period of 10,000 years there are 21 radionuclides left which would contribute with a radiation dose of more than 1 Sv per radionuclide when the total radioactivity would be ingested by humans (Tab. B.23). With a contribution of more than 1,000 Sv under these conditions only some isotopes of the above mentioned actinides are left. After a time period of one million years there would be still eleven radionuclides (five fission and activation products, six actinides) of the deposited radioactive waste which would contribute with an ingestion dose of more than one Sv per radionuclide to the total possible dose through the whole inventory (Tab. B.24).

For the actual situation with respect to potential radiation exposures from a waste repository the potential release of the radionuclides from the canisters and the mobility of the specific radionuclide in the geological material covering the repository (host rock) is decisive. Such analyses have been performed by several groups, e.g. as parts of safety studies (cf. section B 1.3.5). One study has been performed by Marivoet et al. (2008). An example of calculated radiation doses by this group is demonstrated in Fig. B.31. It was assumed that the radioactive material (HLW) started to be released from the canisters after 1,300 to 10,000 years and that the "waste matrix lifetimes are 72,000 years for HLW" (Marivoet et al. 2008). The reference fuel came from a "once through" cycle in a pressurised light water reactor (PWR) plant with uranium oxide fuel. The migration of the radioactive material took place through granite (the assumed host rock) in these modeling procedures. The migration with the water movement is most important.

Tab. B.22: Coefficients for committed effective dose (Sv/Bq) and committed effective dose (Sv) for radionuclides in a reference canister after storage of 40 years

1	2	3	4	5
Radio-nuclide	Phys. Half-life $T_{1/2}$ (y)	Committed Effective Dose (CED) Coeff. Adult (Sv/Bq)	Inventories (4 PWR UO_2 48 GWd/t 40 y) (Bq)	CED (Adult) (Sv)
^{3}H	1.23E+01	1.80E-11	6.90E+12	1.24E+02
^{10}Be	1.51E+06	1.10E-09	1.70E+07	1.87E-02
^{14}C	5.70E+03	5.80E-10	6.20E+10	3.60E+01
^{36}Cl	3.01E+05	9.30E-10	8.10E+08	7.53E-01
^{41}Ca	1.02E+05	1.90E-10	2.20E+08	4.18E-02
^{59}Ni	1.01E+05	6.30E-11	8.40E+08	5.29E-02
^{63}Ni	1.00E+02	1.50E-10	8.60E+10	1.29E+01
^{79}Se	2.95E+05	2.90E-09	1.60E+09	4.64E+00
^{90}Sr	2.88E+01	2.80E-08	2.20E+15	6.16E+07
^{93}Zr	1.53E+06	1.10E-09	1.50E+11	1.65E+02
^{93m}Nb	1.61E+01	1.20E-10	1.20E+11	1.44E+01
^{94}Nb	2.03E+04	1.70E-09	1.90E+08	3.23E-01
^{93}Mo	4.00E+03	3.10E-09	7.60E+07	2.36E-01
^{99}Tc	2.11E+05	5.50E-11	1.10E+12	6.05E+01
^{107}Pd	6.50E+06	3.70E-11	9.60E+09	3.55E-01
^{108m}Ag	4.18E+02	2.30E-09	1.30E+09	2.99E+00
^{126}Sn	2.30E+05	4.70E-09	3.00E+10	1.41E+02
^{129}I	1.57E+07	1.10E-07	2.60E+09	2.86E+02
^{135}Cs	2.30E+06	2.00E-09	3.90E+10	7.80E+01
^{137}Cs	3.02E+01	1.30E-08	3.40E+15	4.42E+07

1	2	3	4	5
Radio-nuclide	Phys. Half-life $T_{1/2}$ (y)	Committed Effective Dose (CED) Coeff. Adult (Sv/Bq)	Inventories (4 PWR UO_2 48 GWd/t 40 y) (Bq)	CED (Adult) (Sv)
^{151}Sm	9.00E+01	9.80E-09	1.60E+13	1.57E+05
^{166m}Ho	1.20E+03	2.00E-09	1.90E+09	3.80E+00
^{210}Pb	2.22E+01	6.90E-07	1.00E+05	6.90E-02
^{210}Po	3.78E-01	1.20E-06	9.80E+04	1.18E-01
^{226}Ra	1.60E+03	2.80E-07	3.10E+05	8.68E-02
^{228}Ra	5.75E+00	6.90E-07	3.10E+01	2.14E-05
^{227}Ac	2.18E+01	1.10E-06	3.90E+05	4.29E-01
^{228}Th	1.91E+00	2.40E-08	2.00E+08	4.80E+00
^{229}Th	7.34E+03	4.90E-07	1.40E+04	6.86E-03
^{230}Th	7.54E+04	2.10E-07	3.40E+07	7.14E+00
^{232}Th	1.41E+10	2.30E-07	4.00E+01	9.20E-06
^{231}Pa	3.27E+04	7.10E-07	9.00E+05	6.39E-01
^{232}U	6.89E+01	8.30E-07	2.00E+08	1.66E+02
^{233}U	1.59E+05	5.10E-08	6.50E+06	3.32E-01
^{234}U	2.46E+05	4.90E-08	1.00E+11	4.90E+03
^{235}U	7.04E+08	4.70E-08	1.10E+09	5.17E+01
^{236}U	2.34E+07	4.70E-08	2.00E+10	9.40E+02
^{238}U	4.47E+09	4.50E-08	1.70E+10	7.65E+02
^{237}Np	2.14E+06	1.10E-07	3.10E+10	3.41E+03
^{238}Pu	8.77E+01	2.30E-07	2.30E+14	5.29E+07
^{239}Pu	2.41E+04	2.50E-07	2.20E+13	5.50E+06
^{240}Pu	6.56E+03	2.50E-07	3.60E+13	9.00E+06
^{241}Pu	1.44E+01	4.80E-09	1.50E+15	7.20E+06
^{242}Pu	3.75E+05	2.40E-07	1.70E+11	4.08E+04
^{241}Am	4.32E+02	2.00E-07	2.80E+14	5.60E+07
^{242m}Am	1.41E+02	1.90E-07	6.20E+11	1.18E+05
^{243}Am	7.37E+03	2.00E-07	2.00E+12	4.00E+05
^{243}Cm	2.91E+01	1.50E-07	6.50E+11	9.75E+04
^{244}Cm	1.81E+01	1.20E-07	7.50E+13	9.00E+06
^{245}Cm	8.50E+03	2.10E-07	5.00E+10	1.05E+04
^{246}Cm	4.76E+03	2.10E-07	1.20E+10	2.52E+03

Tab. B.23: Coefficients for committed effective dose (Sv/Bq) and committed effective dose (Sv) for radionuclides in a reference canister after storage of 40 years and 10,000 years storage in a repository. Only those radionuclides are listed which would yield a total dose >1 Sv.

Radio-nuclide	Phys. Half-life $T_{1/2}$ (y)	Committed Effective Dose (CED) Coeff. Adult (Sv/Bq)	Inventories (4 PWR UO_2 48 GWd/t 40 y) (Bq)	CED (Adult) (Sv)	Inventories after 10^4 y (Bq)	Resulting CED after 10^4 y (Sv)
^{14}C	5.70E+03	5.80E-10	6.20E+10	3.60E+01	1.84E+10	1.07E+01
^{36}Cl	3.01E+05	9.30E-10	8.10E+08	7.53E-01	7.92E+08	7.36E-01
^{79}Se	2.95E+05	2.90E-09	1.60E+09	4.64E+00	1.56E+09	4.53E+00
^{93}Zr	1.53E+06	1.10E-09	1.50E+11	1.65E+02	1.49E+11	1.64E+02
^{99}Tc	2.11E+05	5.50E-11	1.10E+12	6.05E+01	1.06E+12	5.85E+01
^{126}Sn	2.30E+05	4.70E-09	3.00E+10	1.41E+02	2.91E+10	1.37E+02
^{129}I	1.57E+07	1.10E-07	2.60E+09	2.86E+02	2.60E+09	2.86E+02
^{135}Cs	2.30E+06	2.00E-09	3.90E+10	7.80E+01	3.89E+10	7.78E+01
^{230}Th	7.54E+04	2.10E-07	3.40E+07	7.14E+00	3.10E+07	6.51E+00
^{234}U	2.46E+05	4.90E-08	1.00E+11	4.90E+03	9.72E+10	4.76E+03
^{235}U	7.04E+08	4.70E-08	1.10E+09	5.17E+01	1.10E+09	5.17E+01
^{236}U	2.34E+07	4.70E-08	2.00E+10	9.40E+02	2.00E+10	9.40E+02
^{238}U	4.47E+09	4.50E-08	1.70E+10	7.65E+02	1.70E+10	7.65E+02
^{237}Np	2.14E+06	1.10E-07	3.10E+10	3.41E+03	3.09E+10	3.40E+03
^{239}Pu	2.41E+04	2.50E-07	2.20E+13	5.50E+06	1.65E+13	4.13E+06
^{240}Pu	6.56E+03	2.50E-07	3.60E+13	9.00E+06	1.25E+13	3.13E+06
^{242}Pu	3.75E+05	2.40E-07	1.70E+11	4.08E+04	1.67E+11	4.01E+04
^{241}Am	4.32E+02	2.00E-07	2.80E+14	5.60E+07	3.01E+07	6.02E+00
^{243}Am	7.37E+03	2.00E-07	2.00E+12	4.00E+05	7.81E+11	1.56E+05
^{245}Cm	8.50E+03	2.10E-07	5.00E+10	1.05E+04	2.21E+10	4.65E+03
^{246}Cm	4.76E+03	2.10E-07	1.20E+10	2.52E+03	2.80E+09	5.87E+02

Tab. B.24: Coefficients for committed effective dose (Sv/Bq) and committed effective dose (Sv) for radionuclides in a reference canister after storage of 40 years and one million years storage in a repository. Only those radionuclides are listed which would yield a total dose >1 Sv.

Radio-nuclide	Phys. Half-life $T_{1/2}$ (y)	Commit-ted Effec-tive Dose (CED) Coeff. Adult (Sv/Bq)	Inventories (4 PWR UO$_2$ 48 GWd/t 40 y) (Bq)	CED (Adult) (Sv)	Invento-ries after 10^6 y (Bq)	Resulting CED after 10^6 y (Sv)
^{93}Zr	1.53E+06	1.10E-09	1.50E+11	1.65E+02	9.54E+10	1.05E+02
^{99}Tc	2.11E+05	5.50E-11	1.10E+12	6.05E+01	4.12E+10	2.27E+00
^{126}Sn	2.30E+05	4.70E-09	3.00E+10	1.41E+02	1.47E+09	6.92E+00
^{129}I	1.57E+07	1.10E-07	2.60E+09	2.86E+02	2.49E+09	2.74E+02
^{135}Cs	2.30E+06	2.00E-09	3.90E+10	7.80E+01	2.89E+10	5.77E+01
^{234}U	2.46E+05	4.90E-08	1.00E+11	4.90E+03	5.97E+09	2.93E+02
^{235}U	7.04E+08	4.70E-08	1.10E+09	5.17E+01	1.10E+09	5.16E+01
^{236}U	2.34E+07	4.70E-08	2.00E+10	9.40E+02	1.94E+10	9.13E+02
^{238}U	4.47E+09	4.50E-08	1.70E+10	7.65E+02	1.70E+10	7.65E+02
^{237}Np	2.14E+06	1.10E-07	3.10E+10	3.41E+03	2.24E+10	2.47E+03
^{242}Pu	3.75E+05	2.40E-07	1.70E+11	4.08E+04	2.68E+10	6.43E+03

The radiation dose was calculated for a reference person of the critical group over a time period of 1,000 to 10,000,000 years after migration through the geosphere reaching the biosphere and passing then through exposure pathways like the grass-cow-milk-human pathway as mentioned earlier or the radioactivity reaches the human through groundwater directly. As can be seen from Fig. B.31 the total dose reaches a value of around 1 nSv per year and per TeraWatt-hour(e) with a maximum at the time period of around 20,000 to 100,000 years. In former times the safety modelling for waste repositories was performed over 10,000 years whereas today the time cut-off is usually one million years. Marivoet et al. (2008) compared the amount of radioactivity released over a time period of ten million years with the original amount of radioactivity of the inventory and found a reduction by a factor of 10^{-6} to 10^{-7} for the amount of radioactivity. The radiation dose in the environment originates dominantly from ^{129}I. This is due to the high mobility of iodine in general and especially in wet material when iodine appears as the very soluble iodine-ion and alkali-metals like Na or K are the kations. Further iodine has a high vapour pressure in the chemical form of elementary iodine. It has been described before

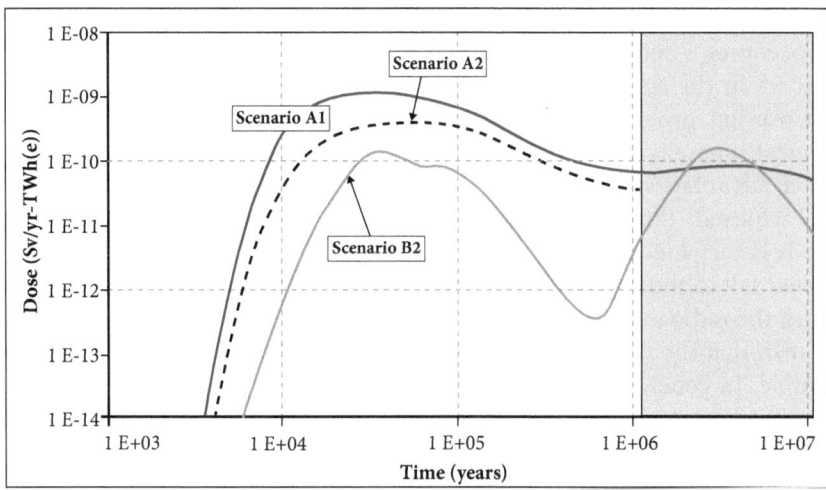

Fig. B.31: Estimated radiation doses caused by release of radioactive material from the canisters with high level waste (HLW) in granite. For assessment timeframes beyond one million years (shaded in gray in the figure), dose calculations can be performed but become increasingly meaningless due to the decreasing confidence in the underlying assumptions (Marivoet et al. 2008).

that after the incorporation of iodine isotopes the thyroid dose is by a factor of around 1,000 higher than in other tissues due to the selective uptake of iodine into the thyroid. Therefore the effective dose is dominated by the thyroid dose and the dose coefficient for the thyroid dose is by a factor of about 20 higher than for the effective dose (ICRP 1995). This would mean that at the level of 1 nSv for the effective dose, the time period of the maximal effective dose in Fig. B.31, the thyroid dose has a value of around 20 nSv, as ^{129}I is the dominating radionuclide.

On the other hand it is interesting and plausible that the actinides with very high amounts of radioactivity in the repository even 10,000 or 1 million years after closure do not contribute significantly the radiation dose in humans and apparently in the biosphere. This is due to the very low mobility of the actinides which are mainly in the chemical form of oxides and apparently only small parts are released from the repository and then can migrate through the geosphere. Therefore it appears that the transmutation of actinides which is often discussed for the management of high level radioactive waste is not very reasonable for such aims. The management of iodine-129 and other smaller radionuclides which appear as ions could be a successful procedure for reducing the exposure to humans in the late phase after closure.

Such possibilities of management may be achieved by physical, chemical or biological means: It could be possible to immobilize the iodine in stable waste matrices (Marivoet et al. 2008). With respect to chemical procedures a better knowledge of the chemical form in which the iodine occurs in the spent fuel elements. In case of the iodine anion chemical absorption processes during the migration through the covering geological sphere is a possibility. For such processes a repository in salt may have advantages as an exchange of the iodine and chlorine anions may be achieved. Finally the uptake and retention of iodine in the human body is very much dependent on the daily supply with iodine e. g. by iodinated salt to those persons as the inactive stable iodine will then compete with the radioactive iodine and the retention will be reduced. It has been shown that the use of iodinated salt can reduce the uptake of radioactive iodine. In general the management of the radioactive material has to be focussed more on the mobile fission and activation products than on the less mobile actinides.

2.7 What is a low radiation dose?[27]

It has already been pointed out that the distribution of radiation dose in the tissue is very heterogeneous in the low dose range. This distribution is dependent on the physical processes which occur when the energy of ionising radiation is absorbed by interaction with atoms or molecules of the living organism. The spatial and temporal distributions of these events and the following biological processes are important for the development of radiation effects. Therefore, physical and biological considerations must be discussed for the decision about the question "What is a low dose?".

2.7.1 Microdosimetric considerations

In the medium to high dose ranges of low LET radiation (100 mGy and higher) a relatively homogeneous exposure of cells and tissues occurs with low LET radiation (β- and γ-rays). This is different with the low dose range when the effects of single ionising particles have to be considered. When a tissue with many cells (several hundred millions cells per g tissue) receives an averaged dose of 1 mGy only 63.2 % of the cells in the tissue will be hit, 36.8 % of the cells experience the track of one particle. Further cells will be hit by several particles (Tab. B.25). With a further decrease of the radiation dose the number of cell without a hit will increase. Under the assumption that the sensitive target is to be seen the DNA located in the cell nucleus, these processes of dose distribution are of high significance.

On the other hand an increased expression of the protein p21 was observed in unhit cells (Little 2000). Such an effect is called "bystander effect". Until now it is not clear what will be the impact for health effects

---

[27] For more detailed information see Annex C 1.

Tab. B.25: Proportions of a cell population traversed by tracks for various average doses from γ-rays and α-particles (approximately 1 mGy for γ-rays and 370 mGy for α-particles per track passing through a cell nucleus on average) (UNSCEAR 2000)

Mean tracks per cell	Percentage of cells in population suffering					
	0 track	1 track	2 tracks	3 tracks	4 tracks	>5 tracks
0.1	90.5	9	0.5	0.015	–	–
0.2	81.9	16.4	1.6	0.1	–	–
0.5	60.7	30.3	7.6	1.3	0.2	–
1	36.8	36.8	18.4	6.1	1.5	0.4
2	13.5	27.1	27.1	18	9	5.3
5	0.7	3.4	8.4	14	17.5	56
10	0.005	0.05	0.2	0.8	1.0	97.1

after radiation exposures. The situation with respect to dose distribution is completely different for the exposure to densely ionising radiation with high LET. α-Rays have a very short range in tissue that is dependent on the energy of the α-particles which are formed through the radioactive decay of the corresponding radioactive isotopes. Thus for α-particles which are formed through the radioactive decay of ^{226}Ra, ^{238}U, ^{239}Pu and others with energies of up to about 8 MeV a maximal range of around 40–80 μm is observed in mammalian tissues. If one considers that the diameter of cell nuclei of human cells is in the range of 5 to 10 μm and the diameter of the cells in the range of 10 to 30 μm this demonstrates that α-radiation can reach in average around 1 to 2 and maximally up to 5 cell layers from their place of origin. The energy dose in the cell nucleus can vary from small doses (in the range of mGy) up to more than one Gy. These considerations demonstrate clearly that the definition of average tissue doses is an oversimplification for energy deposition especially of high LET radiation. It is very important how the α-emitting radioactive isotopes are distributed within the tissue.

Thus in the low dose ranges (average tissue doses of 1 mGy or smaller) ionisation events will occur only in a small percentage of the cells and the number of hit cells depends significantly on the radiation quality (radiation energy and type of radiation). This means that small doses cannot really be defined based on these microdosimetric considerations and they are very heterogeneously distributed on the cellular and sub-cellular level.

For depositories of high level radioactive waste a dose limit or constraint of 0.1 to 0.3 mSv per year is foreseen for the late phase after closure by many national and international proposals as well as recommendations. This means for low LET radiation that such a radiation dose will cause a hit in about 10 per cent of the cells per year. In the case of 1 nSv, as calculated in Fig. B.31, only around 1 cell in a million of cells will be hit.

2.7.2 Biological considerations

Another possibility exists in order to describe the low dose range based on biological effects. After the exposure to low LET radiation (γ-rays, β-rays) the extent of radiation effects can be described by dose effect relations with a linear and a quadratic term of the dose. It has to be regarded that biological effects, which are observed after radiation exposures, like chromosomal aberrations, mutations or cancer, already occur without any radiation (spontaneous effects). For this reason, a constant term "C" has to be considered in possible equations. A dose effect curve can then be written in the form:

$E(D) = αD + βD^2 + C.$

In this formula α and γ are constant coefficients for the linear and quadratic term of the dose respectively. These coefficients vary for different endpoints and possibly also for various defined radiation conditions. Frequently α/β-ratios of around 200 mGy have been observed for Co-60 γ-radiation. This corresponds to a medium radiation dose after which the linear and the quadratic terms contribute to the radiation effects to about the same extent. From such a value it results by calculation that the action of radiation increases in a linear way in the low dose range with radiation dose up to around 20 mGy, as the contribution of the quadratic term is low in this dose range (UNSCEAR 2000). On this basis and convention a radiation dose in the range of 20 to 40 mGy has been called a low dose (UNSCEAR 2000). The UNSCEAR committee concluded further that a dose rate of 0.05 mGy per minute could be considered as a low dose rate.

With respect to waste depositories radiation doses in the range of 1 mSv and lower are discussed for a person of the public during the operational phase and of 0.1 to 0.3 mSv for the later phase. Further we have seen from the estimates of Marivoet et al. (2008) that radiation doses in the range of nSv per year and Terawatt-hour(e) have been calculated. These doses range in all cases in the very low dose region on biophysical as well as on biological considerations.

2.8 Radiation exposures from natural and man-made sources today

Since life exists on earth it is exposed to ionising radiation from natural sources. Although a widespread regional variability exists the average values in larger countries is worldwide very similar. There are only few exceptions. (UNSCEAR 2000). The components with the average annual dose values for Germany are listed in Tab. B.26. The cosmic radiation consists mainly of muons and γ-radiation from solar activities. The solar radiation exposure is lowest at sea level and increases with altitude. An increase of 1,500 m leads to about a doubling of the dose. It also depends on the geomagnetic latitude with a lower value at the equator and reaches maximal levels with a plateau at around 55 degree (UNSCEAR 2000). It results in an almost homogeneous, external whole body radiation exposure. Therefore flying above 10,000 m can enhance the dose rate appreciable.

The terrestrial radiation is caused by external exposures to γ-radiation originating from the decay of radionuclides which are contained in the soil and especially in stones. The main sources of this component are the γ-emitting radionuclides in the ^{238}U and ^{232}Th decay chains as well as ^{40}K (UNSCEAR 2000). In Germany ranges have been measured 40-1.340 Bq kg^{-1} for ^{40}K, 11-330 Bq kg^{-1} ^{238}U, 5-200 Bq kg^{-1} for ^{226}Ra and 7-134 Bq kg^{-1} for ^{232}Th. The total radioactivity concentrations lead out-doors to absorbed dose rates in air of 50 nGy h^{-1} (average value) with the range of 4–350 nGy h^{-1}.

Tab. B.26: Radiation exposures from natural sources (average values per year)

Cosmic Rays (whole body)	~ 0.3 mSv
Terrestric Exposure (whole body)	~ 0.5 mSv
Internal Exposure (almost whole body)	~ 0.3 mSv
Inhalation of Rn and daughters (Lung)	~ 10 mSv

The internal exposure is caused by radionuclides in the human body which are taken up with food including drinking water. The main source is ^{40}K which is contained in all types of food. Around 4,400 Bq ^{40}K and 4,000 ^{14}C are measured per person in Germany on average. Potassium and therefore also ^{40}K is present in all living cells, the same is the case for ^{14}C and therefore an almost whole body radiation exposure is caused by these radionuclides whereas some other radionuclides of Pb, Po, Ra and U are taken up specifically in the bone and lead only to smaller exposures in other tissues.

The highest and most critical radiation exposure from natural sources comes from the inhalation of radon and its radioactive daughter products. This radionuclide is a noble gas and it is found everywhere, it is a product of the decay chain of ^{238}U and originates directly from the decay of ^{226}Ra by release of an α-particle. Extensive measurements of this radioactivity have been performed in many countries. In German houses an average value of around 50 Bq m^{-3} has been observed. Again a wide variation has been found with areas in Bavaria, Saxony and Thuringia in which several hundred or even thousand Bq m^{-3} have been measured. It depends very much on the content of ^{238}U and ^{226}Ra in the soil as well as on the quality of the soil in the region where the houses are located.

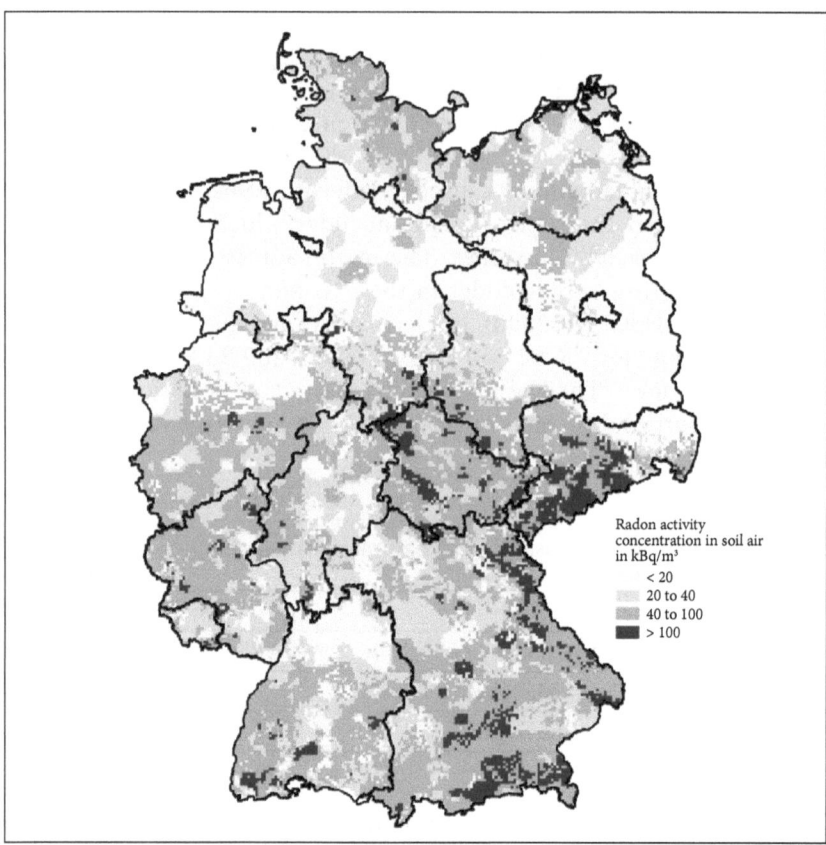

Fig. B.32:   Concentration of radon in soil air one meter of depth in Germany
              (BfS 2008)

The exposure takes place predominantly in the respiratory tract and
therefore it leads to a very local average dose of around 10 mSv in the lung
of the German population (Tab. B.26). Here the effective dose is a useful
instrument to cope with this heterogeneous exposure. The resulting effective dose sums up to around 2.23 mSv per year and about 50 per cent of this
dose is coming from the inhalation of radon and its radioactive daughter
products in the lung. Effective doses are observed for humans worldwide
in this range when large regions are considered. However, it can be much
higher when smaller regions are studied. In Germany, it can be three to four
times higher in smaller regions than the average value (Tab. B.27). In some
countries like India, regions (parts of Kerala) with even much higher exposures have been found.

The risk of radon has been studied worldwide in a number of epidemiological investigations and it is the only natural source where a significant risk
for cancer (lung cancer) could be observed not only in miners with expo-

Tab. B.27:   Radiation exposure of humans in different areas of life in Germany

1. Medical use of ionizing radiation (dose per treatment)	
Therapy	several 10,000 mSv (mGy)
Diagnostics, local, regional	1–50 mSv
2. Exposures at workplaces (effective dose per year)	
Staff in control areas in 2007 (BfS 2008)	Average 0.79 mSv
Flying staff in average in 2007 (BfS 2008) —Exclusively flying North Atlantic route	2.2 mSv 6–8 mSv
Welders (electrodes with 232Th)	6–20 mSv
Workplaces with very high Rn-concentrations (e. g. water industry, Fichtelgebirge, Erzgebirge)	6–20 mSv
3. Environmental exposures (effective dose per year)	
Average of natural exposure in Germany	2.23 mSv
High regional natural exposure in Germany	8–10 mSv
High regional natural exposure in India	15–70 mSv
Nuclear facilities Germany (BfS 2008)	<0.01 mSv
Exposures from the Chernobyl accident in Germany 2007 (BfS 2008)	0.01 mSv

sures to very high radon concentrations but recently also the risk for lung cancer has been shown in houses for the normal population. An increase of lung cancer could be demonstrated with increasing radon concentration in houses which was significant especially with cigarette smokers (more than additive) whereas the risk in non-smokers was much smaller. An increase of lung cancer risk of 10 % was obtained in houses when the radon concentration increased in the houses by 100 Bqm^{-3} (Darby et al. 2006).

Besides the described radiation exposures from natural sources a number of other radiation exposures occur in medicine, at workplaces and in the environment to the general population (Tab. B.27). Very high radiation doses are necessary in radiotherapy for the therapy of cancer. In these situations it is necessary to kill the cancer cells. Therefore high radiation doses are needed. These doses are given mostly to small regions, namely preferably as limited as possible to the cancer. Ionizing radiation is also used for diagnostic procedures in medicine, mainly X-ray diagnostics, again these are frequently local or regional exposures. It is difficult under these circumstances to calculate effective dose. However, during recent years computer tomography is used with increasing frequencies where wider regions of the human body are exposed. Similar is the situation with the diagnostic application of radionuclides in nuclear medicine and the increased number of PET studies. This leads to the situation that the radiation exposures in medicine reaches almost the exposures from natural sources when the average effective dose per person is calculated (BfS 2008).

At workplaces around 320,000 persons were registered as occupationally exposed persons of whom about 57,000 really got an exposure at the workplace with an average dose of 0.79 mSv in Germany in the year 2007. The dose at workplaces should be limited to 20 mSv per year by legal regulations. This limit was exceeded by 15 individuals in Germany in 2007 (BfS 2008). Interestingly the flying personal has received an average radiation dose of 2.2 mSv in 2007. Of the 35,000 individuals around 19,000 persons were exposed with doses in the range of 2.01 to 6.00 mSv. The highest exposures occurred at flights at a latitude of 55 degree and higher. Comparatively high effective doses can be achieved in water work with high concentrations of radium and radon. In 84 workers the calculated average effective dose was 6.3 mSv in 2007, one worker even reaches an effective dose of 41 mSv (BfS 2008). These data of realistic exposures today including the exposure to each person from the natural sources may be of interest for comparison with the potential radiation dose which is expected from repositories during the late phases after closure. In Fig. B.33 the annual effective

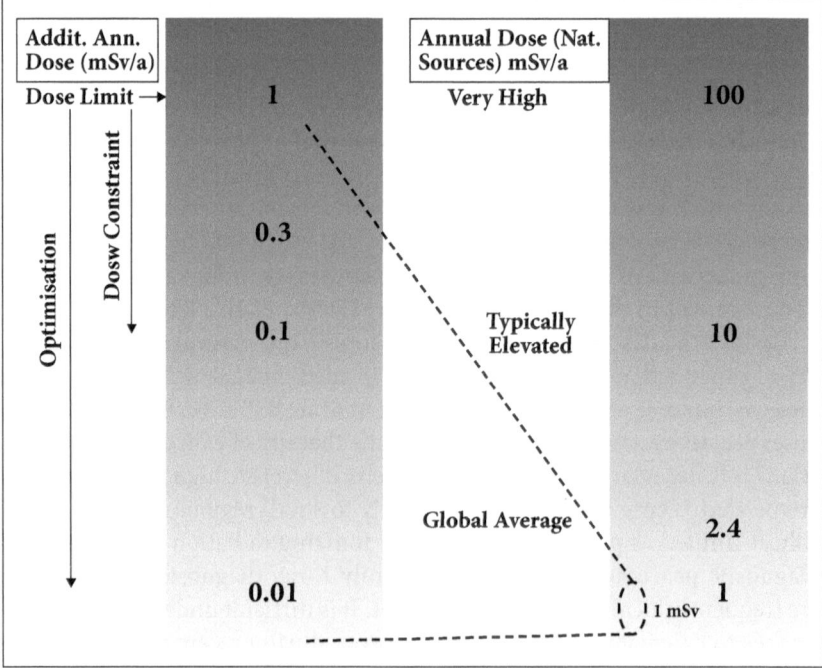

Fig. B.33: Range of effective doses per person and year from natural sources worldwide and potential annual additional doses from nuclear installations including repositories for radioactive high level waste for persons of the public (ICRP 1999)

doses from natural sources which have been observed are schematically summarized. Further additional doses from releases from nuclear installations including waste repositories are schematically presented. As stated earlier, ICRP (1998) has proposed a dose constraint of 0.3 mSv per year for waste repositories. In the more recent proposals from OECD and IAEA a dose constraint of 0.1 mSv per year has been recommended and Marivoet et al. (2008) estimated for their inventory of radioactive material and repository modela radiation dose of around 1 nSv per year for a certain release scenario.

2.9 Development of health effects after radiation exposure[28]

It has already been described that two principal dose response curves have been described. For the radiation effects in the low dose range (<100 mSv) which are relevant for all phases of waste repositories only stochastic effects with a linear dose response without a threshold (the LNT model) have to be considered. The effects under these conditions are the possible induction of hereditary effects and of cancer.

For hereditary effects radiation effects can only be extrapolated from animal experiments. With these studies the so called doubling dose has repeatedly been estimated. For the doubling dose a value of 1 Gy has been estimated for low LET radiation and chronic exposure (UNSCEAR 2001).

The situation is more open for the induction of cancer as it has already been pointed out. There are critical discussions whether the LNT model is also valid for the induction of cancer (BEIR 2005; Tubiana et al. 2005; ICRP 2007; Streffer 2009). There is no scientific proof for the dose response of cancer induction in the dose range below 100 mSv as the possible effects are too small for being measured. During recent years a number of biological processes have been studied which may modulate the dose response especially in the low dose range which will be mentioned. The most extensive epidemiological studies after exposure to ionising radiation are the investigations of cancer incidence and mortality of the survivors of the atomic bombing in Hiroshima and Nagasaki. With the recent data cohorts of 86,572 survivors with 9,335 cancer deaths and 105,427 survivors with 17,448 primary cancer diseases were analysed (Preston et al. 2003; Preston et al. 2007).

These studies with some other studies on nuclear workers and further exposed populations which are for various reasons not so conclusive are the basis from which ICRP derived the risk factor of 5×10^{-2} per Sv for stochastic effects after exposure to low LET radiation in the low dose range with low dose rates and of 10^{-1} per Sv for high LET radiation (ICRP 2007). Under the assumption of a linear dose response the risk for stochastic effects can be estimated by extrapolation. This would mean that after a radiation dose of 1 mSv the risk factor is 5×10^{-5} for low LET or 10^{-4} for high LET radiation.

---

[28] For more detailed information see the Annex C 1.

With radiation doses of 1 μSv or 1nSv the risk factor is three or six orders respectively smaller. Today around 40 % of our population will suffer from cancer and the scatter of this figure with respect to regions, time factor etc. is so large that the above estimated risk figures will be covered by this scatter (Streffer 2009).

An individual cancer which may have been caused by ionising radiation can by no means be distinguished from cancers which originate from endogenous or other unknown causes ("spontaneous" cancer or background). It appears that a better knowledge of the mechanism for cancer development after radiation exposure can help to solve the open question whether really no threshold exists. From experimental and clinical studies it has been shown that cancer develops over a long period, in humans very often over decades. It is a multistep process and several mutations as well as changes in cell proliferation are most important (UNSCEAR 2000; Streffer 2009). A number of biological processes like DNA repair, cell killing by apoptosis, adaptive response, bystander effects, increase of genomic instability, genetic disposition and immunological processes can modulate the development of cancer and by this the dose response. Until now it is unclear during which steps of cancer development the listed biological processes can interfere and have an impact on the development of these health effects (Streffer 2009).

2.10 Uncertainties and variability in dose and risk assessment

The dose and risk assessments are fundamental for radiological protection and are associated with large uncertainties in the low dose range of less than 1 mSv. Additionally due to the time scales under the consideration for repositories for radioactive high level waste ranging over thousands or even up to one million years these uncertainties increase tremendously. The calculated effective doses are uncertain with respect to a number of conditions typically connected to repositories (cf. chapter 2). Furthermore, the radiological considerations underlying the dose estimates are subject to uncertainty and variability as described in the following.

The possible exposures by ionising radiation can only be expected from radioactive material which is released from the deposited canisters. In order to cause an exposure to humans or the environment with living systems this radioactive material has to leave the cave where the canisters are stored and then the material has to migrate through the geological phase in order to reach the biological phase.

It can be expected that the exposure will originate from internal exposures of incorporated radionuclides mainly with the uptake of food and water and much less from external exposures. From studies after the release of radioactive material into the environment (e. g. radioactive fallout from nuclear tests in the atmosphere, releases from the Chernobyl accident) it is well known that the uncertainties are much higher with internal than with external exposures (UNSCEAR 2000; CERRIE 2004; ICRP 2007). These internal exposures cannot be measured directly but only estimated through modelling of biokinetic systems and of exposure pathways in the case of environmental contaminations.

A number of such exposure pathways have been developed. For radioactive iodine the most relevant pathway takes place through the consumption of milk as explained before. This exposure pathway causes a relatively high radiation dose in the thyroid and it is important that the risk factor for thyroid cancer is comparatively high in children. A number of transfer factors are involved in these processes and these differ very much. For example the transfer of iodine into plants is very much dependent on the type of the soil. Additionally all these parameters have a wide range of variability and also uncertainties. The uncertainty of the dose and risk assessment will increase if the parameters in the described chains are selected under considering always conservative criteria.

One has to differentiate between variability and uncertainty. The first term is inherent to the system, e. g. a population with many individuals has a wider range of genetic disposition. Variability is further dependent on e. g. different types of soil with respect to transfer factors, individual differences of biokinetic behaviour of radioactive substances etc. Uncertainties are due to imprecision of measurements, lack of knowledge, statistical behaviour of the system etc. This is especially valid in the low dose range where e. g. radiation risk cannot be measured. The variability and uncertainties of radioecological models are large and will be in the order of several magnitudes. The variability and uncertainties of the dose and risk assessment, after the radioactive material has reached the human sphere and body, are usually better known and can be better judged, as many clinical experiences and results from experimental studies have been obtained. However, it is certainly only the case under the presumption that these parameters do not change drastically over thousands up to one million years. Studies of metabolic chains or of biological systems e. g. DNA-, protein-synthesis and other processes and their similarities as well as stability over millions of years during biological evolution demonstrate some evidence that drastic changes cannot be expected. It has to be assumed that the food and water consumption of the human population will not change drastically in the region of the repository during the large periods discussed with respect to waste repositories. The dose assessment is performed under the condition that all the food of these humans will be produced in this region in the same quality and variety. In the following some problems and open questions will be summarized with respect to dose and risk assessment (ICRP 2007):

The heterogeneity of energy deposition within tissues has been described earlier in the low dose ranges of external as well as of internal exposures. For the evaluation of these doses models are necessary to simulate the geometry of the external exposure, the biokinetics of the intake and retention of radionuclides in the human body, and the human anatomy as well as physiology. Dosimetric considerations with respect to its methodology and practical use are also of great importance. These models and their parameter values have been developed in many cases from experimental investigations and human studies in order to derive "best estimates" of model parameter values. It is recognized that there may be large uncertainties in the values of some of the parameters and in the formulation or structures of the models themselves (ICRP 2007). Some of these uncertainties have been addressed in publications (Leggett et al. 1998; ICRP 2002; Harrison et al. 2003; Likhtarev et al. 2003) and estimates of the illustrated variability of parameter values e. g. for physiological and anatomical characteristics have been demonstrated (ICRP 2002). Such variations of parameter values are of particular significance with respect to the models necessary for dose assessments from internal exposure. From different experimental situations with a broad range of values the necessary parameters are selected by judgments in order to evaluate weighting factors (wR-factors for different radiation qualities; wT-factors for different radiosensitivities of organs and tissues) and other parameters for the dose assessment.

Thus the radiation weighting factors (Table 1) are selected on the basis of measurements of the relative biological effectiveness (RBE) of radiations with different linear energy transfer (LET) through experimental studies and clinical experience as well as on the basis of biophysical considerations. There exists a broad range of values (ICRP 2004) from which the w_R-values are selected by judgment (ICRP 2007). Similar procedures have been used for the selection of the tissue weighting factors w_T (Table 2). These latter values represent the radiosensitivity of the specified tissues and organs for the development of stochastic effects (cancer and for the gonads additionally hereditary effects) which determines the radiotoxicity of a certain radionuclide.

Uncertainty refers to the level of confidence that can be placed in a given parameter value or prediction of a model or estimate of the central value of dose for a population. Uncertainties of the measurements in the low dose ranges of the determined parameters become larger. These are important factors in all extrapolation procedures and particularly in assessing radiation doses and their effects in the low dose range. It is also determined by the imprecision and sensitivity of measurements. Statistical problems play a role.

Variability (strictly, biological variability) refers to quantitative differences between different members of the population in question e. g. with respect to their physiological and metabolic parameters. For example, two healthy persons of the same age and sex with identical diets may exhibit substantially different rates of transit of material through the colon. Similarly individual members of a population will show substantial variation in the uptake of radioiodine by the thyroid for the same initial intake. Variability will be an important source of uncertainty in the estimate of a central value when the estimate is based on a few, highly variable observations. Such a variability is certainly of utmost importance when biokinetic parameters and habits are considered over thousands of years. What will be the food intake in later populations? What will be the environmental conditions etc.?

Risk factors for stochastic effects, from which w_R and w_T values are derived, have been obtained from epidemiological and experimental radiobiological data in the medium and higher dose ranges. The risk factors for the lower dose ranges that are important for radiological protection as well as the concept of effective dose, are based on extrapolation from the measured data in the higher dose ranges using the linear no threshold model (LNT). This model is an assumption which has not been scientifically validated until now. It is considered to be the most appropriate interpretation of current experimental and epidemiological data and is consistent with current understanding of stochastic radiation effects. Under certain conditions aspects of the precautionary principle are certainly included. However, its use also introduces a high degree of uncertainty especially in relation to exposures at low doses and low dose rates (UNSCEAR 2000; BEIR 2004; ICRP 2007). On the other hand risk factors for stochastic effects are mainly based on cancer incidence and mortality in the moment. In this connection the question is valid: What will be the cancer risk and success of cancer treatment many generations later?

The uncertainties which are associated with the assessment of radiation doses and health detriments have been discussed. Some of the more important factors considered are (ICRP 2007):

- The heterogeneity of energy deposition within tissues occurs in the low dose ranges of external as well as of internal exposures.
- The heterogeneous distribution of radionuclides has been described in the body and in tissues which is especially significant when considering ionising particles with short ranges such as alpha-particles.
- For dose assessments from internal exposures the biokinetic models and their parameter values are variable and dependent on the specific conditions of exposure. Frequently animal data have been used in the present estimations and have been extrapolated to humans.
- Human populations vary worldwide on ethnic grounds with respect to physiological and other parameters (ICRP 2002). Variability can become large when radioecological models are used to assess concentrations of radionuclides in food, and hence intakes from habit data as the parameters are frequently very uncertain, biological variability is large, and measured activities are frequently in the low range. These phenomena and their variability will change tremendously when considered over thousands of years.

- The target cells for the induction of cancer and their location in tissues are unclear. The dose response in the low dose range for stochastic effects, the mode of extrapolation and the LNT concept are uncertain until now.
- For the estimation of parameters connected to the assessment of health detriments sex and age averaging is performed which causes uncertainty.

The degree of uncertainty varies for the various parameters and the circumstances in defined situations. Therefore it is not possible to give general values of uncertainties but considerations of this kind should be and have been made for special cases and should be included in comprehensive evaluations (e. g. CERRIE 2004; ICRP 2006). The degree of uncertainty differs between various radionuclides (ICRP 2007). These aspects will be of very high significance for the estimation of radiotoxicity in connection with the later phase of waste depositories.

It should be noted that the dosimetric models, conversion coefficients and other parameters recommended by ICRP have been developed principally and primarily for planning and assessing normal occupational exposures, for planning for discharges into the environment and for generic assessments of doses. They are needed to demonstrate compliance with dose limits. These are circumstances in which doses are low (ICRP 2007).

In conclusion: The reference models and their parameter values have been developed primarily for use in prospective radiological protection. These models and parameter values can be used for demonstrating compliance with dose limits when exposures are low but in general should not be used for risk estimates. This limitation of usage applies particularly to effective dose.

Variability can become extremely large when radioecological models are used to assess concentrations of radionuclides in food, and hence intakes from habit data as the parameters are frequently very uncertain, biological variability is large, and measured activity values are frequently low. The RBE values which are important for the choice of w_R values vary with the end point considered and the experimental design. Frequently the values rely on experimental animal and in-vitro data, therefore the target cells for the induction of cancer and their location in tissues are unclear. The dose response in the low dose range for stochastic effects, the mode of extrapolation and the LNT model are uncertain. For the estimation of parameters connected to the assessment of health detriments, sex averaging is performed which causes uncertainty.

In general there is no doubt that the uncertainties are very high when prospective statements have to be made into the far future of 10,000 or even 1 million years. The knowledge about the occurrence of ionizing radiation from natural sources and the behaviour of natural radionuclides in the environment as well as the biokinetics in humans is certainly very helpful for these estimates. In a number of steps of this chain judgments have been made with tendencies to the conservative side. Therefore there is some confidence that those potential radiation exposures will stay in regions where no significant risk can be measured. Besides these aspects which are supported by scientific knowledge and background it can certainly not be foreseen how much regions where repositories are located will be inhabited by how many people and what will be the habits.

3 Management of high level radioactive waste with reference to long-term responsibility

3.1 Ethics as rational conflict resolution

3.1.1 Rational conflict resolution

Judging by the prolonged protests and debates in Germany and other countries over the management of radioactive waste, the issue of finding an adequate waste management strategy seems to be afflicted with a considerable potential for conflict. The easily raised tempers and the emphasis and passion with which arguments are put forward are just an external indicator of this – and not a particularly reliable one. Clearly, the heated mode in which the conflict is sometimes carried on in public is a characteristic trait of the debate and deserves closer attention, especially if one is interested not only in the theoretical development of strategies, but in the practical resolution of the conflict. And if a solution is to be brought about by resolving factual issues rather than issues of power, the conflict must also be analysed in terms of its social and political dimensions, in order to test how the partisan insistence on a particular position can be transformed into a constructive discourse about rational strategies acceptable to all parties (see below, section B 3.3). This is the only way to find a legitimate as well as legitimized solution.

But in order to develop proposals and make recommendations as to which problem-solving strategies could prove acceptable (serving as a preparation of such a discourse), it is helpful to first consider the problem causing the conflict, independent of its resonance in society and politics and beyond the definitions which often already bear a partisan stamp. This does not mean, however, that the problem should be reconstructed as a merely technical issue – on the contrary. Such a reconstruction would rightly be in danger of being classified by some of those involved in the discourse as biased by a technicistic and reductive interpretation of the problem, and therefore ignored.

From the fact that two persons may be in perfect agreement over the available data (e. g. certain courses of events and the likelihood of their occurrence) and yet differ in the confidence they put in a technology, it becomes evident that the decision for or against a particular technical installation cannot be settled merely by solving technical problems. Unlike the estimated likelihood of an event, which is proven right or wrong by the actual course of events, confidence in a technology manifests itself in people's decision-making, and is thus a practical category: decision-making is always partly influenced by individual experiences which affect the way the decision-maker interprets the available data. Decision-making is shaped by short-term stimuli and emotions as well as long-term dispositions, formed by cultural factors to which belong, among so many others, the marketing strategies of manufacturers and providers of a technique.

Subjective decision-making behaviour of this kind is unproblematic as long as the decision to use a technology is made individually and nobody else is affected by the consequences of this decision. For large technical structures, however, this is no longer a given; for example when the decision for or against an investment in a technical structure must be made collectively and members of the collective differ in their evaluation of its reliability. If the differences in the evaluations lead to different decisions, but the technical installation calls for a joint investment, a conflict arises: the parties involved do not merely have different views or assessments; they *want* different things and cannot achieve them without one another. Technical data may make an important contribution to settling such a conflict, but the conflict is ultimately produced by the evaluations of the data, which are partly shaped by other factors.

Due to the complex and often inconsistent volumes of information and to the different conditions in which the parties involved develop the criteria for their respective evaluations, collective decisions about large technical structures often have a high potential for conflicts. The individual willingness to invest in a large-scale technical structure often depends crucially on the assessment of the benefit to be gained from its proper functioning on the one hand, and the expected harm its malfunctioning might cause on the other hand. Such differences in evaluation cannot be settled by methods which help demonstrate that one of them is the "right" or "appropriate" evaluation, even if methods can be established which make it possible to reject some evaluations as incoherent or inconsistent. Since these differences occur in the first place because some people *want* something and others *want* anything else, and because some see the technology as an appropriate means to achieve what they want, while others do not. For example, consider a situation where a technology is used to produce a state that is desirable and hence beneficial from the point of view of party A, but which requires an investment by party B, although B does not benefit from it, or benefits less than A. Conflicts arising from such unequal distributions of costs and benefits, or opportunities and risks, are a characteristic trait of technically developed cultures. Their ubiquity becomes apparent when the talk of investments is not inappropriately restricted to monetary resources, but also comprises the imposition of risks and harmful emissions of all kinds. At this point normative issues can no longer be ignored. This concerns the justification of impositions and demands, and ultimately reaches into the realms of law and justice.

Therefore, to sum up, for a promising approach to the development and justification of recommendations for how to deal with radioactive waste, (1) the problem should be considered independent of its social effects and independent of the existing, often biased, definitions, (2) the problem should not be reduced to its technical aspects, and the various conflict-generating differences in the evaluations and justifications should be taken into account. Thus the starting point should be the existing conflict, and not the technical problem. The conflict should, however, be clarified (3) with regard to both the available options and the existing normative restrictions.

3.1.2   Ethics and morality

Traditionally, questions arising with regard to the desirability and admissibility of states resulting as planned and unplanned consequences of an action (with a more or less quantifiable probability), and with regard to the reasonableness of costs and/or risks and their distribution, and to the appropriateness of the distribution of proceeds and/or opportunities – in short, questions about the acceptability of actions and their consequences – have fallen into the domain of ethics. It would be a misunderstanding, however, to deduce from this that the task of ethics is to find or develop criteria which, with a little practice and skill, might enable one to determine the admissibility, reasonableness or acceptability of any action. This misunderstanding is suggested both by the question "What ethical principles must necessarily be considered when dealing with the issue of the disposal of nuclear waste?" (Boetsch 2003:39) and by the definition of ethics given by the Committee on Radioactive Waste Management (CORWM 2006:38) in its influential study "Managing our Radioactive Waste Safely":

> Ethics are sets of principles or standards concerned with behaviour and well-being. They act as a guide to what is acceptable or unacceptable, what we should do, what is right or wrong, good or bad. Ethics are about how we ought to act in contexts that have significant implications for human and non-human lives and well-being.

It is easy to see that this – admittedly widespread[29] – conception of ethics as a provider of behaviour-regulating principles is distorted, if one focuses on an essential presupposition of all ethical endeavours and bears in mind that it is the very criteria for reasonableness, legitimacy, and acceptability that are controversial in the public debates. It might seem reasonable to apply the well-established scientific approach to *ethical* questions about the acceptability of actions and their consequences, as if the aim were to go beyond the many opinions being voiced and test what is "really" true, and which of the conflicting parties is actually "right". Such an approach, however – regardless of how suitable it is in other scientific contexts – would misconceive the fundamental structure of such debates: these are not differences of opinion or disagreements over matters of fact that might be settled in favour of one party or the other – or against both – by methodically checking the facts. On the contrary, they are *conflicts*, in which the plans

---

[29] Some further examples out of the area of RWM are NWMO Roundtable on Ethics 2005:1 (Ethical questions "aim to identify basic values, principles, and issues"), Damveld/van den Berg 2000:14: "Ethics deals with fundamental values, rights and concepts"; Serre 2004:84: "Pour l'éthique, une valeur est un 'bien en soi'".

of one party fail due to the plans of others, and in which one party cannot attain what it needs or wants because others are acting in favour for *their* own needs and wants. Ethics is as a part of the effort to develop and provide tools allowing for a non-violent resolution of such conflicts, and as a scientific endeavour in that it takes into account neither authority nor privileges, nor any hierarchical relationships, but takes it as a methodological presupposition that all parties to the disputes are equal and have equal rights.

This conception of ethics as a scientific endeavour entails methodological predeterminations that motivate a closer look at, and a comparison to, the natural sciences. A fundamental presupposition of all ethical endeavours is that parties to a dispute who are engaged in conflict with one another are to be considered fundamentally equal in their role as participants in a practical discourse aimed at conflict resolution. This principle is a prerequisite for the success of all ethical endeavours, just as assumptions of continuity such as "natura non facit saltus" are prerequisites for scientific endeavours. An explanation of an event qualifies as scientific only if it does not violate this principle of continuity, for example by ascribing an unusual event to a spontaneous "freak of nature" or divine intervention. Basing one's methods on the assumption of continuity is virtually standard practice in scientific explanation. There may be other types of explanation, but an explanation is called "scientific" only insofar as it endeavours to respect this assumption. Focusing on this model of explanation rather than a different model, however, presupposes a decision based on considerations of expedience. Taking the assumed continuity of events as a given fact of nature, and trying to prove it with scientific methods, would constitute a category mistake. That is why I. Kant (AA III KrV:194 f) considers the sentence "nothing happens by blind chance (in mundo non datur casus)" a condition of the possibility of sensual experience – a principle that cannot be vindicated by experience but is presupposed by it. In the same way, the assumption of the fundamental equality of all parties in a discourse targeted at conflict resolution is standard practice for ethical endeavours. Attempts at conflict resolution which do not conform to this *methodical* presupposition are not ethical, whatever else they may be.[30] Again, any attempt to prove the fundamental equality of the conflicting parties in the ethical discourse using "ethical" means would be a category mistake. To quote Kant again: wherever our actions are meant to be guided by ethical ("sittliche") considerations, no one is available as a "mere means" for other people's ends, but each person must always be considered "an end in himself" at the same time (AA IV GMS:433). Every rational being, as a member of the (discursive) community, the "kingdom" of these ends, has a "share in the general law-mak-

---

[30] This is Kantian as well: here "ethical" is used in a traditional sense, as opposed to the increasingly widespread usage in recent times, to mean a particular type of rational qualification of claims, not a morally qualified type of action.

ing [...] which determines all value", and is, in this, "giving universal laws, [...] but also himself subject to these laws" (AA IV GMS:433). If a rational being wants to act ethically ("sittlich"), rather than following his inclinations without regard for the concerns of others, he must therefore

> act as if he were [...] always a legislating member in the universal kingdom of ends. (AA IV GMS:438)

To put it in a less theory-laden way:

> If my wish is to be morally good [...] I ask myself only: can you wish for your maxim to become a universal law? (AA IV GMS:403)

If this question receives a positive answer for the commission of an action, but not for its omission, it yields a moral ("sittliche") obligation. The source of this obligation, however, is obviously not the ethical principle, but the claims of the agent and all those affected by the action. The principle merely establishes the appropriate and well-considered integration of these claims as a criterion.

The above-quoted criterion for the acceptability of an action and its consequences for any rational being, and the implied equality of all people in the "kingdom of ends", is the categorical imperative Kant elaborates in his "Grundlegung zur Metaphysik der Sitten" ("Groundwork of the Metaphysics of Morals"). This categorical imperative underlies or precedes moral deliberation and determines it, but cannot be revealed or imposed as an obligation by means of moral deliberation.

Despite the structural similarities between ethics and the natural sciences, the difference is clearly marked: parties to a conflict, unlike parties to a disagreement over matters of fact, do not simply state different opinions which can be interpreted as diverging hypotheses. A conflict is a clash, not of opinions, but of volitions: one party does not want to allow what the other party wants to do, does not want to accept something imposed by the other party, or rejects a distribution of a resource that grants it only this or that particular share of it. The question of whether any of the parties are right can be posed legitimately if it is intended to mean whether they are mistaken in their assumptions, and for example overestimate the probability or the dimension of certain consequences which they regard as undesirable. The question can also be considered valid if there is evidence that one party is inappropriately focusing on the short-term satisfaction of needs over its long-term interests. Above and beyond such questions that concern the rationality of factual connections or the correct self-attribution of volitions, however, the application of standards of correctness is inappropriate. Individuals can be assumed to have different desires, needs, preferences, interests, intentions, purposes etc. It is clashes between these which cause conflict. From an ethical point of view, however, based on the principle mentioned above, all the parties have an equal right to make these desires, needs etc. the basis of the claims put forward in the conflict. Thus following the decision-making model used in the natural sciences to identify which of these disparate opinions are "correct" would require standards that go beyond the realm of ethics. Hence anybody – be they a party to the conflict or an outsider – who wishes to contribute to the resolution of a conflict while fulfilling ethical standards cannot lay claim to privileged knowledge or "deeper" insights, but at best a certain advantage of experience or professional expertise. These advantages can be used to prepare the discourse, to make recommendations, and to give advice, but not to bring about decisions.

Nevertheless claims must not be raised completely arbitrarily: as a minimum, we demand consistency of any subject wishing to interact with us, and evaluations given must be independent of the situation and the respective interlocutor, in other words: universalizable. In case of apparent discrepancies, especially those which favour the subject himself, we demand reasons that explain the perceived exception as an application of the rule under particular circumstances, hence re-establishing consistency or excusing the discrepancy. For example, if somebody claims a greater share of a certain resource according to some rule of distribution, he must apply the same rule even if, due to a change in circumstances, it puts him at a disadvantage in an otherwise similar situation. Anyone who claims that three situations s_1, s_2, s_3 can be brought about at roughly the same cost, but declares that he prefers s_1 over s_2 and s_2 over s_3, in every respect, cannot refuse a decision to invest in s_1 and instead vote to invest the resources in s_3. Anyone who defies principles like these in practically relevant situations faces at the very least a loss of reputation, often the withdrawal of others' willingness to cooperate, and sometimes even more severe social sanctions which will interfere with his plans and actions. Anybody planning for the long term, wishing not just to impose a particular argument, but to prevail in conflict, and not just to prevail in a particular conflict, but to assert himself in his social environment and to lead (in Aristotelian terms) a good life in his society, will try to adhere to these principles, and at the same time demand the same adherence from others. It is conceivable that different practices of evaluation may emerge in different societies or parts of societies, and that they may remain consistent for long phases at least. It is equally conceivable (and is closer to the given conditions) that different practices of admitting and denying entitlements may exist side by side. In any event, these practices must at least be largely consistent and structured by principles in order to grant a certain amount of predictability, which is indispensable for planning. Thus when choosing practices for evaluating and granting entitlements one can assume a plurality of options on the one hand, but on the other hand ineluctable restrictions which make this choice non-arbitrary.

The concept underlying (as one example among many others) the above-cited study by the CoRWM (2006), that "ethics" is a decision-making resource in conflict situations, and the popular demand that decisions of

importance for society should be guided by "ethical standards", are ultimately based on an inappropriate definition of the relationship between
"ethics" and "morals".[31] Unlike ethics, which merely states formal criteria and procedural rules for practical discourses, morals, which have their
roots in tradition and conventions, supply material rules and standards
that serve as guidelines for our own plans, inform our expectations of others' behaviour, coordinate the actions of community members and can be
evoked as grounds for legitimation in cases of conflict. Hence moral rules
and standards constitute highly efficient instruments of conflict resolution in the countless situations where agents have to coordinate their volitions and actions with those of others. They tend to fail, however in cross-
cultural conflicts, since different cultures and subcultures have established
different morals. Indeed they often exacerbate the conflict, if the expectations individuals develop based on their morals are disappointed by members of other moral communities, or the legitimation of an action based
on one morality is disputed on the grounds of another morality. The critical analysis of morals and their scope, and the investigation of moral-tran-
scending means of conflict resolution, is therefore another crucial task of
ethics. The rules and standards applied by ethics in this critical analysis
are neither identical with those of morality, nor are they applied in order
to substantiate the claims of a particular morality. An ethical examination
of the debate over the management of radioactive waste should therefore
examine individual arguments to see to what extent they uphold the principle of equality between all parties to the conflict, and to what extent they
merely express the attachment of individual parties to their own morality.

3.1.3 Ethical analysis of conflict

Not only are divergent opinions on the effects and side effects of the technologies employed to be expected in the conflict on the management of
high level radioactive waste, but also divergent goals and interests, differing prognoses about the consequences linked with particular methods, and
divergent moral convictions. This will also be connected to differing understandings of who is obliged to legitimize their actions towards whom, what
actions require legitimation, and what counts as a basis for legitimation.
Consequently, a rational resolution of this conflict requires not only agreement on the cause-effect-relationships involved, but also a critical reflection
on the desired ends and the grounds for legitimation put forward for the
choice of methods ("waste management strategies").

An ethical analysis of conflict requires a concept of conflict that remains
neutral towards morality-specific presuppositions. This analysis will there-

---

[31] Therefore Cotton 2009 is not justified in his scepticism about ethics although
he is right in his sceptical analysis of the "top-down" approach of the CoRWM-
Process.

fore be based on a concept of conflict which merely assumes the incompatibility of the respective aims and plans, but does not for example identify any one party as the originator of the conflict, "troublemaker" or guilty party. When actions which can be described as preserving the status quo clash with actions bringing about change, the former enjoy an advantage – according to widespread moral models for the evaluation of conflicts which are operative in many situations – insofar as the entire burden of justification falls on those who wish to deviate from the status quo. This distribution of the burden of justification, well established in moral practice, may in some cases pass ethical examination, in particular if the "old" is also "tried and tested", while a change would amount to trying out an utterly untested alternative. Generally speaking, however, conservative choices – which include avoidance as well as preservation – can prevent others from achieving their goals just as much as actions bringing about change. The very existence of a conflict between agents wishing to preserve the status quo and those wishing to change it shows that different evaluations of the status quo and its alternatives are possible. A form of ethics aimed at the independent development of conflict resolution strategies therefore requires a definition of the conflict that is independent of the normative expectations which are – particularly in the context of environmental protection and technologization – attached to descriptive expressions such as "conserving", "preserving", "avoiding" or "preventing", "ending" or "destroying", and the associations they evoke.

Ethical conflict analysis also remains neutral with respect to the modes of action "commission" and "omission", thereby diverging somewhat from everyday preconceptions. For pragmatic reasons, the mode of action is an important criterion for the evaluation of actions in everyday morality: in our moral practice, someone who causes harm to another person by committing an action is sanctioned more rigorously than someone who brings about the same state through an omission, allowing it to happen although he or she could have prevented it. This unequal treatment is not based on a difference in the consequences of the actions or in the actions themselves, but rather on the interest of those who are applying the sanctions to influence the future behaviour of the agent: somebody who makes an active effort to injure others can be assumed to have a different disposition to somebody who would have had to make an effort to prevent this injury, and who prefers to tolerate harm to others than the trouble it would take to prevent it. Since sanctions are (among other things, at least) supposed to have an effect on the future dispositions of the agent and his future actions, a different allocation of sanctions and a difference in the moral evaluation of omissions and commissions seems justified. Thus the unequal moral treatment of the modes of action makes sense, pragmatically speaking, and is also in principle justifiable from an ethical point of view. Firstly, however, it is always necessary to check whether the conflict in question is suitable for such treatment: if the question to be settled is whether the present generation should take up certain options and impose the associated physical risks on future generations, or whether they should refrain from this action and thus potentially deprive future generations of a benefit, then according to ethical standards the resulting conflict should not be evaluated as a function of the unequal moral treatment of omission and commission. This is firstly because the agents in this scenario are collectives, to which dispositions cannot be ascribed, or at least not in any straightforward way. Secondly, due to the temporal relationships and particularly the singularity of the action in question, unequal treatment would be pointless from the point of view of moral praxis. The distinction between modes of action should therefore not play a part in the ethical consideration of the conflict, and thus in the question of how the present generation should, following ethical criteria, reach its decision.

Neither the temporal order in which the plans were devised by the parties nor the kind of actions intended to be employed in the implementation of a plan can in themselves substantiate the claim that a certain action must not be taken. In order to achieve conflict resolution which is sustainable in the long term, we must draw on a concept of conflict that is neutral towards the characteristics of the parties and their relations to one another (in particular, their temporal, spatial or social proximity). This means that the ethical analysis must also be neutral towards those standards which parties apply in judging their own goals or goals conflicting with these.

This fundamental and radical approach to conflicts would overtax individual agents in their everyday lives, and would be more a hindrance than a help in the achievement of their goals. If humans had unlimited cognitive resources, a completely rational form of conflict resolution would probably be superior to these instruments and strategies. Given the limited resources we actually have, however, the use of established morals is in most cases not just good enough, but superior to the resource-intensive rational strategy. The latter will be reserved for those conflicts that cannot be adequately resolved by the established instruments of a moral community, e. g. because the scope of the conflicts requires particular reflection, because the established instruments fail, or because the conflicts transcend the boundaries of individual moral communities. Since such reflections make high demands on cognitive and thus temporal resources, they should be organized in such a way that they can be carried out when pressure to act does not require an immediate decision. In more complex societies there is also the option of delegating this task to specialists who, equipped with appropriate resources and released from the pressure to act, can build up a stock of rational strategies and develop proposals for actual configurations of conflict. In modern societies this is the task of academic ethics, as well as other disciplines and consultatory institutions. Thus considered, ethics has no decision-making or admonishing function, but a service function.

Significant as traditional morals are for everyday action, with their ability to relieve pressure and facilitate action, it would be a category mistake to presuppose, without further testing, criteria such as everyday moral intuition in problems like the management of high level waste – if the pressure to act still leaves sufficient (temporal) resources to develop proposals which are as rational as possible and consider as many aspects of the problem as possible, and to change the conditions of action and decision-making to allow for an optimized resolution of conflicts.

Three category mistakes of this kind – closely linked with one another in theme – pervade the debate on radioactive waste disposal:

(a) Immediate action

The obvious requirement for a solution to the problem of radioactive waste management is widely interpreted as including a requirement for immediate action. This demand is usually justified with a reference to future generations, derived from an intuitive idea of justice; it is argued that we cannot simply leave the waste behind for them or "dump it on them". This demand, which has deeper implications and is by no means self-evident, comes with a considerable burden of justification, however. This burden can only be discharged when one interprets the situative conditions of action as an emergency situation requiring an immediate response, where the short reaction times limit the number of optimization strategies available. In such emergencies it is impossible to first inspect and compare the available options, to invest, if necessary, in the creation of more favourable conditions in order to open up further options, to evaluate how the solution to the problem at hand can possibly be combined with other tasks, in order to produce synergies etc. Thus emergencies of this kind justify impositions on third parties that would otherwise be unacceptable in longer-term planning. In particular, the rights of disposal and participation of those affected may be overridden with reference to emergencies. For that reason alone, demands for immediate action must be interpreted very restrictively from an ethical perspective: limiting the available options to what is possible or is perceived to be possible here and now means foregoing opportunities to resolve conflicts in a fair trade-off of interests and to do justice to the claims of all those involved in the best way possible. This is particularly true of the interests of future generations (see section B 3.2 below). Consideration for the latter seems a particularly strong argument for keeping options open for as long as it is possible without exposing others to unacceptable risks, and for making use of the available time to adopt, in a structured process, an optimized waste management strategy that does justice to all parties concerned, whether present or future. The general conditions of waste management described in the preceding chapters and the time scales given in section B 1.5 above show that there is still ample opportunity for this.

Since – as mentioned at the outset – the problem obviously has to be solved, and since the future always contains risks, an emergency plan should also be elaborated and kept at the ready, allowing the adoption, at short notice, of a solution that may be less than ideal but is acceptable and superior to inaction.

(b) Fairness

It is a commonplace demand in the debate on the management of radioactive waste that the solution should be *fair* – by cross-generational as well as intragenerational standards. Even in everyday questions about the distribution of resources, however, traditional concepts of fairness do not always provide sustainable foundations for a successful resolution of conflicts. In the same situation, one person might claim a larger share of a resource due to his or her greater need for it, a second person may claim a greater share because of his or her greater output, while a third person might declare that his or her ability to use the commodity productively and multiply it justifies a share considerably larger than one third. Fairness, according to common usage, does not necessarily mean equality, it only means that exceptions to the rule of equality require justification. What kind of justification individuals are willing to accept often depends on their specific situation. One thing is evident, however: anybody who, in a dispute of this kind, demands a distribution favouring him or herself, and tries to base this claim on a particular rule of distribution, must also accept this rule if a similar situation arises in which, due to changes which have occurred in the meantime, the same rule now disadvantages him/her.

Given that there is no immediate pressure to decide how to fairly distribute the burdens of radioactive waste management, however, this is not just a matter of applying established rules of fairness to a given situation with a fixed "volume to be distributed". Neither the costs of disposal nor the risks resulting from the presence of radioactive waste are constant factors, since both can in principle be influenced by the investigation and implementation of waste management strategies (cf. section B 1.2). Hence a plan must be devised that, allowing for justified exceptions to the rule of equality, is able to bring about the strategic optimization of the distribution situation and the "volume to be distributed". The conflicting claims that are made with reference to moral principles can thus be transformed into considerations (or negotiations) about conflict-pacifying allocations of all possible burdens and benefits which are at stake. The bartering of burdens and benefits in a way that is acceptable for all parties is always one option among others. This option can be chosen if e. g. transferring the tasks of research and implementation to future generations seems likely to produce an optimized overall distribution. This of course presupposes that appropriate compensation will be provided by the present generation.

In the long-term perspective, if direct interaction is impossible, and no exchange can take place, compensation – which is problematic in view of the uncertain interests of future generations – should be weighed up against preventive and curative strategies. The precedence of prevention is established in the German constitution, but can in principle be weighed up against other principles.[32]

(c) "Polluter pays" principle

It is a commonplace argument that those who have "made the mess" and caused the burden must also step in and take care of its proper disposal. On the one hand fairness demands this, on the other hand it is based on the moral-pragmatic aim of guiding future behaviour: this is how parents train their children to be more careful, and how states encourage producers into resource-efficient behaviour by the internalization of costs. As far as long-term problems related to radioactive waste management are concerned, however, there is only one configuration of conflict between the members of present and future generations: past actions have resulted in waste that entails risks for future generations and therefore needs to be disposed of. The moral-pragmatic justification for the "polluter pays" principle thus ceases to apply, leaving only the principle of fairness. In the light of this, we should be searching for an optimal resolution to this cross-generational conflict, rather than general strategies for an overall best solution in recurring instances of this type of conflict.

Given the vast number of people who will be affected by the risks in the future, however, it may be in the interests of fairness not to entrust the management of the risks to those who have *created* them only because they *have* created them. If there are others who are (due to greater competence or more favourable circumstances) better equipped to ensure a form of waste management that meets the needs of future generations, it may even be advisable, from an ethical point of view, to transfer the task of waste management to them, in exchange for a compensation they are willing to accept without coercion. This holds on both a cross-generational and a cross-national scale: the availability of skills and resources, rather than the boundaries between nations or generations, must be considered as the relevant factors for the adequate fulfilment of long-term obligations. In the light of the time-scales at issue such boundaries seem mere historical contingencies.

If better opportunities for conflict resolution are offered by a division of labour across the boundaries of presumed "waste-producing generations" and of "waste-producing communities" understood in terms of the nation state, the polluter-pays principle would at any rate not be an obstacle to this, provided verifiably reliable mechanisms of exchange across these bounda-

---

[32] For ethical requirements to RWM with respect to the problem of justice cf. Grunwald 2010:77ff.

ries could be established, and responsibility for waste management were not transferred at the expense of third parties.[33]

* * *

The arguments put forward so far, rejecting the demand for immediate action, widening the concept of fairness, and relativizing the polluter-pays principle, all make reference to the rights and demands of future generations. In actual fact, consideration for remote future generations is not a relevant factor in our established moral practice, the foundation for the expectations rebutted with the above arguments. Yet the incorporation of the claims of remote future generations into our planning, and the resulting long-term obligation, are of major relevance for the management of radioactive waste, and must thus be considered separately, using rational assessment rather than traditional moral principles.

3.2   Ethics and morals

3.2.1   Long-term obligations as a topic of ethics

As a normative discipline, ethics has always dealt with the future. Historically, however, the debate has been limited to the near future, inasmuch as this has been affected by problems of conflict resolution, and that primarily means conflicts that occur within direct interaction. There was no need, at least there was no recognisable need for trans-generational conflicts to which ethical debates could have reacted. It is not simply that for long periods of time the history of mankind has been largely interpreted as an (ascending or declining) process of salvation, with the belief being that the future of coming generations depended on the Lord's benevolence and the good conduct of the members of these generations. Even when, in the Enlightenment, what man was and would become came to be perceived as his own responsibility and dependent on his own choice, there was no option to actions that, in the eyes of the actor, could affect the wellbeing of future beings. On the contrary, the productivity of the older generations was always perceived as benefiting later generations as well, so that Kant for instance could only recognize a question of "trans-generational justice" in the opposite sense to that discussed today:

> Befremdend bleibt es immer hiebei: daß die ältern Generationen nur scheinen um der späteren Willen ihr mühseliges Geschäfte zu treiben, um nämlich diesen eine Stufe zu bereiten, von der diese das Bauwerk, welches die Natur zur Absicht hat, höher bringen könnten; und dass doch nur die spätesten das Glück haben sollen,

---

[33] Certainly, Shrader-Frechette (2000:773) is right in stating: "just because A is better able to deal with B's problems than B is, does not mean B has the right to impose his problems on A". But that does not rule out the possibility that obligations to another C might include an obligation to pass over the problem to A – which might again include B's obligation to compensate A for the imposed burdens.

in dem Gebäude zu wohnen, woran eine lange Reihe ihrer Vorfahren (zwar frei-
lich ohne ihre Absicht) gearbeitet hatten, ohne doch selbst an dem Glück, das sie
vorbereiteten, Antheil nehmen zu können. (AA VIII Idee:20)[34]

Furthermore, not only did any cross-generational comparison seem only to
favour future generations, there also seemed to be no limits as long as there
was no scarcity recognisable, or at least none which could not be avoided
simply by moving on to another resource. But if due to the availability of
external inflows there is no scarcity of a commodity or at least no scarcity
can be perceived, then there will be no management, no economical and no
ethical administration of that commodity.

Very much in contrast to this, the modern debate is characterized by the
opposite perception, that the legacy of the present generations will be ecological and economic burdens. Following J. Mittelstraß (2008:7) this modern debate actually began with the problem of radioactive waste.

3.2.2 Long-term obligations and "intergenerational justice"

Today, the concern for future generations is mostly formulated in terms
of "intergenerational justice"[35] and – especially in the context of what is
called sustainable development – focuses on the problem of fair distribution
of scarce resources between generations as it was first globally recognized
when oil reserves were perceived to be scarce in the 1970s (from today's
perspective this perception was skewed).[36] Furthermore the problem was
addressed theoretically in the book *A Theory of Justice* by J. Rawls (1971),
which has been very influential for the actual debates on ethics and political philosophy: his so called "difference principle" requires that social institutions be arranged so that inequalities of wealth and income work to the
advantage of those who will be worst off. This arrangement of a fair intragenerational distribution immediately leads to a problem of trans-generational justice: what is to be distributed must be fixed so that just distribution in the first generation is not established at the expense of later generations (Rawls 1971:72).

The related term "intergenerational justice", prevalent in the English-
language literature on the subject, implies a set of problems evoking mis-
leading connotations and producing pseudo problems. The prefix "inter"
points to the idea of a mutual relationship that has to be brought into some

---

[34] "It remains strange that the earlier generations appear to carry through their
toilsome labour only for the sake of the later ones, to prepare for them a foundation on which the later generations could erect the higher edifice which was Nature's goal, and yet that only the latest of the generations should have the good
fortune to inhabit the building on which a long line of their ancestors had (un-
intentionally) laboured without being permitted to partake of the fortune they
had prepared."

[35] For a survey on concepts and questions of intergenerational justice cf. Tremmel
2009 and, with a special respect to radiological protection, Gosseries 2008.

[36] Cf. above all, the popular *The Limits of Growth* (Meadows et al. 1972).

kind of balance ("justice"). Such issues of justice between generations do exist. The universal use of the expression "intergenerational justice", however, suggests that the ethical problems concerning remote future generations are basically of the same kind. And yet in the contexts in which long-term obligations are investigated, *inter*generational relationships very soon cease to exist (almost certainly from generation G_{0+4} on). The future generations at issue here have no mutually interactive relationship with us: true, future generations will be confronted with the consequences of our technology-related actions, but conversely the actions of future generations are not relevant for us. Relationships between generations of the "inter" type must therefore be clearly distinguished from those relationships to be dealt with in terms of long-term responsibility.

But even if one does not misunderstand the problems at issue as problems of *inter*generational justice, and speaks more cautiously of, for example, problems of "cross-generational justice", it is not self-evident that, from an ethical perspective, those are problems of justice at all. Contrary to the tacit assumption often made in the debate following Rawls' *Theory of Justice*, not all problems of moral obligation can be reformulated as problems of distributive justice. The converse does apply, however: all problems of distributive justice can be reformulated as problems of moral obligation. For example, we can consider whether the present generation is under an obligation to safeguard some quantity of some resource for remote future generations, or to spare them some particular risk. But problems of obligation can be discussed even where there are no issues of distribution regarding future generations. One example is the elementary, non-trivial question of whether the present generation is obliged to ensure the survival of the species *homo sapiens* with its reproductive behaviour, or whether a refusal to conceive is, in moral terms, at least permissible. It would be strangely artificial, however, to reconstruct this as the question of whether we are obliged to grant future generations the same access to a resource called "life" as we demand for ourselves. Put in a much simpler, more general way, the question is whether the members of a generation g_0 (e.g. our own generation) are under any obligation towards remote future generations.

There is a whole range of ethical questions relating to the future which cannot always be (easily) reconstructed as problems of inter- or cross-generational distributive justice, e.g. those concerning a shrinking biodiversity (extinction of species), large-scale geological formation changes above and below ground which often go hand in hand with ground water lowering (*Ewigkeitslasten*, literally "eternity burdens"), genetic changes in the human, animal or plant genome and so on. And it seems as if – at least when the long-term-dimension of this problem is at stake – the question of what to do with toxic or radioactive waste is another example of this kind.

3.2.3 Long-term obligations vs. Long-term responsibility

The great success of Hans Jonas's book *The Imperative of Responsibility* (1984), particularly among a non-specialist public with an interest in ethical and political issues, has reinforced a tendency, initiated by Max Weber's distinction between *Gesinnungsethik* (ethics of conviction) and *Verantwortungsethik* (ethics of responsibility) (Weber 1919), to explicate all moral problems as problems of responsibility. The overuse of the word "responsibility" ("Verantwortung") in the social sciences and in politics has additionally obscured the terminology. This is not the place to try to bring about a comprehensive terminological clarification.[37] The reflections set out here are based on the assumption that obligations and entitlements arise from moral discourses about solutions to action-related conflicts.[38] If the outcome of such a discourse is that remote future generations enjoy an entitlement, the corresponding obligations are binding for all members of the present generation. Responsibility, on the other hand, falls to only a few of them, through the delegation of obligations.[39] Obligations give rise to responsibility through the moral division of labour, with which a society organizes the implementation of obligations, just as in economics the division of productive labour and in science the division of cognitive labour help to organize the relevant tasks.

Moral division of labour is particularly useful in the treatment of moral problems which emerge when one assesses the risks and opportunities connected to the possibilities (such as the production of electricity through nuclear fission) and necessities (such as the management of radioactive waste) which have arisen due to modern technology. Due to the highly complex, sometimes "overly complex" nature of these problems, modern technicized societies often have no choice but to delegate the collective obligation towards future generations, making it the responsibility of individuals who deal "professionally" with the issues involved in ascertaining and fulfilling these obligations. This societal delegation of obligations poses its own specific problems, however: naturally, the more society delegates obligations to people in positions of responsibility, the greater the uncertainty about whether they are assuming their responsibility in the desired manner. One of the fundamental problems of modern technicized civilizations is organizing the societal assumption of responsibility in such a way that on the one hand the delegation of obligations helps to avoid "moral overload",

---

[37] For criticism of the concept of responsibility cf. Wieland 1999. For an analysis of what Jonas' conception can contribute to RWM cf. Löfquist 2008:173–197.

[38] For more detail cf. Gethmann 1982; Gethmann 1991; Gethmann/Sander 1999.

[39] This suggestion follows Kamlah 1973:110ff. To avoid misunderstandings let it be noted that this terminological rule does not supply a method for distinguishing an ethics of conviction from an ethics of responsibility. Whatever the origins of moral obligations may be – they must exist already in order to be delegated "into responsibility".

and that on the other hand the moral competence of the decision-makers remains transparent.

According to the explanations given, the question of whether there are actually obligations toward future generations comes first, methodically, while questions of distribution and responsibility are methodically secondary. An objection often raised against deontological approaches of this kind, however, is that this focus on reconstructing the 'ought' neglects the consequences of actions. Based on the Anglo-Saxon tradition, a distinction is made between deontological and teleological conceptions: while the former deduce imperatives from principles, the latter attempt to evaluate actions based on their consequences alone. The question of whether obligations can be justified with recourse to consequences alone does not need to be settled here.[40] A deontological ethics that disregards consequences entirely seems at best a didactic exaggeration, at worst enemy propaganda from the utilitarian camp. Actually it is hard to think of a deontological conception in which examining the consequences of action is not an essential element of ethical evaluation (Kamp 2010). If for example the question is whether a maxim (e.g. breaking promises if it is to one's advantage) can serve as a universal norm, the answer must draw on the consequences which this would have (namely, for this example, the collapse of the promise as an institution of human interaction).[41] Deontological conceptions differ from purely consequentialist conceptions in that they determine what should be done on the basis of a combined evaluation of reasons for acting and its consequences, which can take various forms, depending on the precise composition of the ethical conception.[42] Thus a deontological approach is entirely compatible with the concept of a moral weighing-up. Moral obligations always have an aspect of inalienability (which cannot be "traded off"), and an aspect of alienability which can be weighed up in terms of 'more' or 'less'. This is reflected in the law: in the framework of legally canonized norm systems, it is forbidden to appropriate the property of others – not more or less, not sometimes for some people, but always and for everyone (the corresponding norm is unconditionally valid in this respect). And yet if you were to let someone drown rather than picking up a swim ring found lying on the beach and throwing it to them, you would be liable to prosecution (thus the corresponding norm is conditionally valid in this respect). In the same way, moral obligations are inalienable, but can at the same time be weighed up in terms of 'more' or 'less', insofar as they are embedded in a context of obligations. Nevertheless, such a weighing-up must be guided by methods

---

[40] This was already questioned by Frankena (1963). Cf. the comprehensive study in Nida-Rümelin (1993).

[41] The example is formulated following Kant (AA IV GMS:422).

[42] The conceptions of Hartmann (1925) and Scheler (1913) are prominent for giving moral qualification solely on the basis of reasons, disregarding the consequences of action.

and principles – and the principles must be subjected to weighing-up in the same sense. Thus the fact that a moral weighing-up is possible in the framework of a deontological argument based on reasons and consequences does not mean that everything can be converted into a single currency, so to speak, and then compared.

3.2.4 Long-term obligation – fundamental considerations

The enquiry into obligation and responsibility to remote generations touches on a number of non-trivial ethical questions that need to be answered if such long-term responsibility is to be justified on its merits and in terms of its type. In particular, the three following questions are of paramount importance:[43]

(1) Do we have obligations only towards those who are interacting with us directly, in our time?

A negative answer to this question is a prerequisite for the concept of obligations that reach beyond the agent's immediate present into the future. The question would have to be answered in the affirmative by defenders of ethical approaches that base obligations solely on claims or preferences that have actually been voiced (e.g. preference utilitarianism). These conceptions give rise to an apparent paradox: members of a generation g_0 can never have any obligations towards members of any future generation g_n, unless the future beings are capable of actually giving voice to claims against members of g_0 – which, given sufficient temporal distance (say, $n > 3$), is never the case. To counter conceptions which reduce obligations to actually expressed claims, it is necessary to distinguish between *having* (being entitled to) a claim and *making* (voicing) a claim. If there are grounds for suggesting that future generations have claims against earlier generations, even if they cannot yet assert them, then for the same reasons the earlier generations also have corresponding obligations towards the later ones.

(2) Does obligation only extend to generations of some degree $k > i$ (e.g. $i = 3$)?

Everyday experience seems to support the stance taken by many economists, namely that obligation has to end with some greater or lesser degree k, because knowledge about the needs of future generations decreases in proportion to temporal distance. If one tries to express this intuition more precisely, however, it appears far less plausible. If all obligations end with some degree k, this would mean that there are no obligations at all towards the members of generation g_{i+1}, as opposed to g_i, just because they are members of a generation of a degree $k>i$. Restricting obligations like this would be incompatible with the methodological principle of ethical

---

[43] The issues dealt with in summary in this section are discussed in more detail in: Gethmann 1993.

universalism. Temporal egoism ("only members of the present generation have rights") or temporal particularism ("only the next three generations have rights") can of course be argued with as much consistency as any other kind of particularism, linking moral rights to tribal affiliation, religion, race, class or gender. If one grants entitlements to moral subjects at all, however, it seems quite arbitrary to set a temporal, spatial, racial or other limit on them.

There may be pragmatic reasons – such as the need for an efficiently functioning practice – for drawing such a line and making it binding for certain decisions about the future consequences of action (for example, determining that the age of legal majority begins with one's 18th birthday, and thus imposing a binding limit onto the continuum of increasing maturity). It must be possible, however, to justify the line towards all those concerned, including in particular those who are disadvantaged by it. The very fact that we are willing to accept this duty of justification confirms that we do not dispute the entitlements of those who are "behind the border" once it is drawn.

The decisive argument for not limiting the scope of obligation is not a universalistic ethical conception, but the ethical irrelevance of all reasons for such a limitation. Purposes such as the efficient organization of actions and the means necessary to achieve them can certainly be of ethical relevance, insofar as they may potentially cause conflicts. That is why the maxims that are subject to ethical evaluation will, for example, read "in situations of type t I want to assign greater priority to the efficiency of my decisions than to the avoidance of dangers for members of generation g_{i+1}, and thus want to assign rights only as far as generation g_i". The justification of this maxim, however, cannot be based on pragmatic reasons. Such reasons are irrelevant for the ethical issue of the acceptability of this maxim, regardless of whether the person posing the ethical question does so from an egoistic, particularist, or universalist point of view.

The limitations considered for the design of permanent disposal sites with regard to technical feasibility or controllability are pragmatic in nature. If the time frames resulting from such demarcations reached further than the potential of the consequences of the action to cause conflict (for example due to a decrease in the radiotoxic hazard potential of the waste), these limitations would become ethically irrelevant: they are incompatible with ethical universalism, but cause no conflicts. Limits which are set for pragmatic planning-related reasons, however, and which do not extend beyond the potential of the consequences to cause conflict (because of the continuing radiotoxic or chemotoxic hazard potential of the waste), might then indirectly cause conflicts. Demarcations of this kind thus require an acceptable justification, i.e. a justification which is likely to be accepted, after rational examination, even by those who do not benefit from it.

(3) Are we obligated to members of the kth generation to the same degree as we are to those of the first generation after us?

An affirmative answer to question (2) often leads to the conclusion that we must also answer question (3) in the affirmative. Either there is an obligation towards members of the generation g_k, just as for members of the generations g_{k-1}, g_{k-2}, ... g_1. This consequence does not necessarily follow. On the contrary, it is always essential to distinguish – regardless of the temporal distance between the obligated and the entitled parties – between the existence of an *obligation* on the one hand and the degree to which it is *binding* on the other (in the choice of words, if not in substance, the distinction follows Kant; cf. AA VI MS:390ff).

If one neglects to distinguish between obligation and binding force, and assumes an equally binding obligation to exist towards everyone, however spatially, socially, or temporally distant they may be, then ethical universalism leads directly to an overtaxing of the individual, which in turn results in a complete inability to act or make decisions for every action where one can expect a sequence space that is not clearly and narrowly defined, and thus corresponding potential for conflict. For issues of long-term obligation, the assumption of an unchanging obligation for the individual leads to an utterly paralysing pragmatic paradox: what if we had obligations towards the 10,000[th] generation which were just as binding as those towards our children's generation? While we are sufficiently familiar with the "life-world" of our children to determine the circumstances and consequences of actions, in material terms, with a fair degree of certainty, we can describe these only in the framework of an imagined "life-world" for the 10,000[th] generation after us. Thus while we can anticipate the "life-world" of our children enough to reach a moral judgement, for the 10,000[th] generation after us we would have to be prepared for all possibilities. These possibilities could be varied *ad libitum* in such a way that both the commission and the omission of any action would have disastrous consequences for the members of the 10,000[th] generation after us. This would mean that, taking long-term obligation into consideration, it would be wrong for us to either commit or omit an action H which is currently under debate – even if not committing (not omitting) H were to have disastrous consequences for our children's generation, and hence committing (omitting) H were an urgent necessity in terms of their needs.[44]

3.2.5   Long-term obligation in the absence of knowledge

Given the consequences that would arise if we based our actions on the assumption of an equally binding obligation towards all future generations,

---

[44]   Thus, according to principles from the theory of decision-making, when faced with complete uncertainty the appropriate attitude is one of indifference towards the decision-making options, cf. e.g. etwa Kern/Nida-Rümelin 1994.

regardless of how little we know about them, it is necessary for *ethical* reasons to distinguish between the obligation which exists universalistically, i.e. indefinitely, and a degree of binding force which decreases with spatial, social and temporal distance. If, however, it is impossible on the one hand to imagine a future generation towards which there is no obligation, and on the other hand the assumption of an obligation that remains equally binding for all futures, however distant, must be rejected, then the question arises: how is the obligation towards future generations to be qualified?

The obvious answer is to apply a sort of "constructive" procedure: the obligation towards distant future generations ("future long-distance obligation") is different to the obligation towards our children and grandchildren ("future local obligation"), but cases of local obligation allow us to reconstruct the rules upon which long-distance obligation is also constructed. The difference between local and long-distance obligation is that, in the case of long-distance obligation, we have obligations towards people (or conversely, they have claims on us) who will never, not even potentially, face us as partners in interaction and thus in conflict. Our rules for conflict resolution ("peace strategies") can thus only be "projected" onto these generations.[45]

The starting point for this is the perception that the gradation of inalienable obligations according to degrees of binding force is entirely in keeping with everyday moral experience, as can easily be demonstrated with examples:

(a) Let us suppose that every parent has an obligation to ensure the well-being of his or her children. Then this obligation exists – or it does not exist. It does not exist "a little", "in parts" or "largely". The binding force of this obligation can, however, differ greatly in degree, relating to different parameters: what a parent has to do so that it can be said that he/she has fulfilled his/her obligation can differ, i.e. increase or decrease, relating to the parent's economic resources, the parent's other obligations (e.g. towards other children who demonstrate greater need of help), the child's financial status, the child's age and other parameters. For two different situations of parenthood one can say that the obligation exists in both cases, but not necessarily that it is equally binding.

(b) Let us suppose that there is a categorical obligation to help anyone in need. If for example someone faints during a flight, this obligation applies to everyone present. The degree of binding force is, however, higher for professional flight attendants or for doctors who are present by chance than for other people present. In the case of two random witnesses to the fainting attack, both have the obligation to help, but it is not necessarily equally binding for both. It is more binding for the flight

---

[45] Cf. Gethmann 1991 on the ethically elementary situation of mediating an argument.

attendants and any doctors who may be present because of the delegation of obligation into professional hands.

The examples already give some indications of important dimensions of differentiation in degrees of binding force. These include interactive proximity and distance, both synchronous and diachronic, the moral division of labour by means of occupational competence, delegation (responsibility), representation (community) or anticipation (guardianship). The moral division of labour is not something predetermined, but is organized precisely for the efficient fulfilment of obligations, for example by distinguishing corresponding skills and responsibilities. Such a division of labour can be thought of both synchronously as a division of labour between those living at the same time, and diachronically, across generational boundaries.

The particular requirements of long-term obligation become clear when one bears in mind that the moral paradox developed above (cf. section B 3.2.4) regarding an obligation spanning 10,000 generations, makes considerable use of the fact that our relationship to the future is characterized by decreasing levels of knowledge. If the ethical relevance of our lack of knowledge about the future were not acknowledged, the qualification of moral obligations would be separated from important factors of action and its consequences – with roughly counterintuitive results. This should not, however, be misunderstood: one cannot first assume a constantly existing correlation between the temporal distance and the degree of knowledge/ignorance with regard to any developments and then – in the manner of a fixed discount rate – allow the binding force to diminish along the time line.

Thus the degree of ignorance about temporally remote world conditions is no more a constant than knowledge about spatially remote conditions (e.g. in as yet unexplored continents) has been, throughout human history. In some areas at least, research and exploration make it possible, despite variance, to identify constants, formulate parameter dependencies and the laws governing courses of events, propose more or less proven hypotheses, and reject some hypotheses as impossible.

For this reason, the "exploration of the consequences of scientific and technological developments" represents a core moral task for a scientific-technological civilization. If we wish to act technologically, there is an obligation to procure knowledge about the consequences. Here there is no strict correlation between the degree to which such knowledge can be attained and the resources which are employed for this, but the volume of the resources used can change the degree of knowledge attainable.

As a general rule, obligations are not only universalist, insofar as they exist towards every person who may in the future be affected by the consequences of decisions. They are also universalist in the sense that every person is the addressee of the obligation. If – as might be the case for long-distance obligations – individual actors do not assume their obligation, or do not do so adequately, because they do not possess the requisite cognitive or material resources, then in many cases the collective organization of the assumption of obligations allows a greater scope. Thus the increase in the scope of our actions through technologization and collectivization comes with the obligation to form organizations and equip them with the requisite resources so that they can, on the basis of the knowledge gained, or of rationally founded suppositions, organize society's long-term obligations, without being dependent on the resource constraints of the individual actors. The obligation towards future generations thus also includes the obligation of the individual actors to try to ensure an adequate supply of the resources necessary for this. In the process responsibility is transferred to organizations, by means of the delegation of obligations, and the binding force applying to different people becomes differentiated. Nonetheless, the individual as the actual bearer of the obligation cannot completely release him or herself from responsibility – on the contrary, in delegating the fulfilment of his or her obligations, the individual is at the same time assuming responsibility for monitoring the responsible organizations. This can in turn be partly organized through the division of labour, by creating further organized responsibilities ("checks and balances"), but ultimately it is up to the individual actor as moral subject to supervise and demand the exercise of delegated responsibility as part of the fulfilment of his duties. Participation of the actual bearers of obligation in consultation processes within and outside the organization is thus a key prerequisite for the adequate fulfilment of duties.

However, the fulfilment of duties towards the members of future generations responds to demands which cannot be directly raised. In relation to the time spans which are at stake in the management of high level waste, the claims actually made by contemporaries in fact represent only a very narrow strand of all of the claims which must be taken into consideration. First of all, we must make well-founded imputations about the claims of these future generations. Here we cannot assume a systematic coherence between the claim profiles of present-day parties to the conflict and the future generations who are to be included as parties to the conflict. The certainty with which such imputations can be made tends to decrease with temporal distance, as does the justifiability of the assertion that one can competently represent the claims of future generations. Thus for issues with a long-term perspective the decision-making burdens must be deferred to procedures: the rights to represent future generations must be transferred in a legitimized manner and require in particular the procedural organization of measures for avoiding contingency. This transfer then also develops self-binding powers, with which the participant himself limits his rights of participation and control in order to guarantee the legitimacy of the procedure. A lack of acceptance for decisions then no longer means that the process is inadequate (no more than widespread acceptance means that the process is adequate). On the contrary, only the correct execution of a process which is in actual fact accepted as producing legitimation does in actual fact produce legitimation (Luhmann 1969; Grunwald/Hocke 2006). The responsible representation of future claims thus requires a mandate which is legitimized by procedural organization, anchored in institutions, and controlled by society – rights of representation cannot simply be claimed by declaring oneself to be competent, responsible, or personally affected.

3.3 Legitimation and participation

3.3.1 Preliminaries

Reacting to the lasting political problems and the continuing lack of public acceptance, most recent studies on the topic of radioactive waste management suggest that only participative strategies will be able to cut the Gordian knot.[46] The progress that some, especially the Scandinavian, countries have been able to make in recent years has given further support to this view.

It is obvious that, in designing a complex plan, the success of which involves a high degree of responsibility, the participation of all those with local and technical expertise is desirable for optimal planning, regardless of actual certification and profession. To a certain extent, an imperative to activate these optimisation resources can be derived from the obligation towards the members of future generations and from the demand for a just distribution of burdens (cf. section B 3.1.3). The only limits placed on this participation arise from balancing it against the requirement for a certain decision-making efficiency, on the one hand because processes of optimization are always subject to the law of diminishing expectation of marginal utility (the benefits of further consultation become simultaneously smaller and more expensive), on the other hand because the risk (that further postponing the implementation of the plan will narrow the freedom to manoeuvre) will at some point exceed the opportunities which can be expected from further consultation.

As well as aiding plan optimization, the inclusion of those affected in decisions about restructuring measures or the establishment of large technical facilities has proven its worth as a strategy which can help to counteract prejudices and misunderstandings, to ease fears, and to build trust. Participation can make a substantial contribution towards making transparent what impositions for the individual are connected with the measures, how he should be compensated for these, what social need the measure is responding to, what risks one is prepared to take for its sake, and how opportunities and risks are distributed. Such an inclusion of those affected, which focuses on the one hand on allowing people to have their say and help to shape events, and on the other hand on disseminating information,

---

[46] E.g. AkEnd 2002, CoRWM 2006, NEA 2010b to mention just three prominent ones.

can contribute substantially to introducing a more factual tone into debates which are otherwise conducted on an emotional level. This more factual tone is likely to be beneficial in terms of both efficiency and justice (cf. section B 5 for more detail on this).

From an ethical perspective, there is another expectation linked with the reference to participatory procedures which also merits sceptical attention, an expectation found particularly in concepts of so-called participatory technology assessments (TA), namely that participation will give planning decisions a higher degree of legitimacy. Since concepts of this type increasingly form the backdrop for the debate on the management of radioactive waste, the risks of misconceptions about participation will be pointed out here, before we turn to the question (as done in section B 5) of how the opportunities involved in participation can be used.

According to the concepts of participative TA, the participants are lay persons, citizens, directly concerned persons, consumers, stake-holders, etc. on the one hand, as compared with (technical) experts, scientists, institutional decision-makers, producers, share-holders, etc. on the other. Fundamentally, we are currently experiencing a thrust in the direction of allotting certain areas of decision-making competence (e.g. to the citizens) and withdrawing certain areas of competence (e.g. from the experts). According to the concepts under debate in this process, cognitive, more precisely scientific competence is certainly involved, too. For this reason not simply plebiscitary procedures alone but rather a multitudinous variety of combined procedures are put forward. This is why the many different variations of participatory TA do not merely presuppose a shift of competence (e.g. from scientist to lay person); but rather something like a competence schism along the dividing line between cognitive versus evaluative competence. Thus, it is ultimately a question of allocating competence.

In the allocation of competence a number of premises and pre-suppositions reveal themselves. All in all, these premises and pre-suppositions lead to the creation of basic, more or less democratic decision-making procedures, or at least of symbols (like consensus conferences), which are in some way reflective of basic democratic decision-making processes. These premises and pre-suppositions can be developed more precisely within three theses:

Thesis 1   The sciences are not endowed with any primarily evaluative competence. This premise pre-supposes scientific descriptivism and moral non-cognitivism (*scientific premise*).

Thesis II   The citizen, in contrast to the scientist, is primarily endowed with this evaluative competence (*ethical premise*).

Thesis III The democratic institutions, constituted by delegation and representation of competence, are unsuitable (at any rate in a number of important cases) to execute the will of the citizens (*political premise*).

In the following these three premises will be addressed critically, but in order to avoid misunderstanding, before the theses themselves, which are certainly controversial, will be discussed, there firstly seem to be the need to make it clear which hypotheses are *not* denied:

- It will not be disputed that in a democratic society all power and thus all decisions on technology policy must ultimately stem from the people (and not from technical, scientific or economic elite).
- It will not be disputed that it is right that the citizens should have a part in community decision-making processes involving consequences for the people concerned. This is especially valid for political decisions which *immediately* affect the citizens themselves.
- And finally, it will not be disputed that it is desirable that the people, on the basis of a good scientific education, are able to understand scientific and technological processes.

3.3.2 The tribalisation of science

To provide misunderstandings there are, furthermore, some remarks necessary on the premise that the sciences are not primarily endowed with evaluative competence. In accordance with this they are then not primarily responsible when it is a matter of technology and scientific policy decision-making on questions with considerable bearing on the people at large or at least on major parts of the population. This view is often characterised as scientific descriptivism and moral non-cognitivism. The questions to be asked here may be discussed from two different points of view:

- Is science really exclusively or primarily, a purely descriptive, explanatory undertaking?
- Do scientists, as such, really possess no evaluative competence?

The *first* question leads into the field of general philosophy of science and is widely discussed there. Expounding upon this topic would lead too far here.[47] Therefore just a few brief observations: Contrary to the dictum of Max Weber, there are genuinely normative sciences, e. g. economics, jurisprudence and ethics, to name but a few. In addition, the so-called descriptive sciences are riddled with norms of a methodological nature, which in turn must be justified with regard to their rational expedience. These norms belong to the field of competence of the scientists (and not lay persons).

The *second* question first involves conceding that it is not possible to conclude directly from the methodological status of a given science the competence of those engaged in that science. It is *not* valid to assume a priori that economists behave more economically, lawyers more lawfully and

---

[47] For a more detailed discussion cf. Janich/Kambartel/Mittelstraß 1974, Carrier 2006.

moral scientists more morally than other people. Although it *is* valid that economists *know* better what is economically rational, jurists *know* better what is in conformity with the law and moral scientists *know* better what is moral. If this is applicable, then it is not permissible to place the scientists (experts, ...) from the point of view of competence as a group alongside other groups (groups of citizens, ...) and to regard them as equal with respect to their competence[48], as an ethnologist would do in placing tribes alongside other tribes, or as a theologist does in setting confessions alongside other confessions. Such a strategy of pluralistic juxtaposition is what could be called the "tribalisation" or "confessionalisation" of scientific competence.

The tribalisation of science reveals itself, for example, in the following phenomenon. When scientists deem a toxic risk low and a citizens' action group high, it would be a misconception to equal this constellation with that one in which a tribe thinks rain could be brought about by sacrificing hens, while another tribe were to assert that it must be geese. In general: Such an equalization would take scientists as a social group among others, without special competences being ascribed to them, overlooking the fact that the very inherent defining attribute of the sciences is exactly the self-imposed duty to uphold certain standards of rational reason, which under certain (albeit non-trivial) circumstances permits the claim that a particular assertion of the scientists is true and a demand correct. Though scien*tists* are only humans like everyone else, the scien*ces*, on the other hand, provided that certain conditions are fulfilled, can rightly claim acceptability. As long as they speak as representatives of the sciences they are responsible for a special contribution to the debate that is not just their opinion, not even the opinion of "the sciences", but is distinguished by having passed a procedure that is designed to remove or at least reduce personal and situational biases. When participation is conceptualized without adequate allocation of the different functional roles citizens and scientists have in participatory processes, and when the different statements of experts and lay person are merely treated as a "variance of opinions", one has fallen victim to the tribalistic fallacy.

3.3.3   The overtaxing of the citizens' competence

The above quoted "ethical premise" does assign different functional roles to citizen and scientists. And so do participatory concepts that are built on the assumption that the citizen, in contrast to the scientist, is primar-

---

[48]   Cf: "Participative TA is characterised by a procedure in which, in addition to the scientific disciplines, non-scientists, i.e. decision-makers and those directly or indirectly affected by decisions are deliberately involved in the analysis and evaluation process, and are permitted to influence the investigation process, the determination of options for action and/or also the resolving of conflicts by finding solutions through negotiation." (Baron 1997:148 [translation Gethmann]).

ily endowed with the *evaluative* competence necessary as a basis for making the "right" decisions (whereas the sciences are perceived as an endeavour that is completely non-judgemental). In this sense e. g. L. Hennen emphasises, that participation in TA

> is implicitly committed to a republican model of democracy in which the citizens themselves are the ones who should make the decisions on those questions which concern them (Hennen 1999:569 [translation Gethmann])

and similarly G. Kass (2000) understands participation as a "direct involvement of citizens in political decision-making" (p. 20). But this at the same time overtaxis the citizens' competences.

First of all, critical attention must be drawn to an inconsistency in the resulting TA strategies, which can be traced back to unclear interpretations of the word "discourse". If it is primarily a matter of evaluative, prescriptive competence, then why all the effort of "discourse", in which the citizens must be taught the basics of energy technology, molecular biology or reproductive medicine? Programmes in the U. S. A. and other countries including Germany which are combined under the heading "education" reflect the misunderstanding existing here (e. g. Schell/Mohr 1994). The citizens do not doubt that the biologists are masters of their field and know their business, on the contrary, they are all too afraid that they do; neither do they wish to evaluate the cognitive achievements and performance of the scientists but rather the latter's evaluative prescriptive competence, their reliability and trustworthiness and other character attributes. Trust also has something to do with familiarity with the given topic field, but not with quasi-scientific training. What is to be understood here by "discourse" is urgently in need of precise definition (cf. Gethmann/Sander 1999:124ff).

But now to the core question: Is the citizen himself endowed with this competence? Is it really appropriate to presuppose that everyone really knows "what is best for himself"? Or isn't it more a matter of postulating that I may expect others to recognise that this is applicable in my particular case? The hypothesis that each individual in regard to his own needs is the best expert to decide what is good for himself, will be called the "self-competence thesis" ("Eigenkompetenzthese").

Contrary to this thesis stand, first of all, some primary phenomena of life experience, experience gained through the nature of life in its various phases and biological changes from the needs of the child for upbringing and education to the needs of old people for care and attention. In addition, experience is gained from the manipulation and the strategic creation of needs.

Theoretical analysis and interpretation of such experience offers a number of anthropological concepts concerning the factual immaturity of the individual and his inability to determine his own existence. Kant's demand for enlightenment as a way out of self-caused disability (AA VIII

Aufklärung:33), Marx's concept of false consciousness, widespread over the works of Marx and Engels (cf. especially MEW 3 Ideologie), the concepts of Freud and other psychoanalysts concerning the repression of needs, those theories which have to do with the character of goods as merchandise (e.g. Adorno 1980, Horkheimer/Adorno 1988) and the power-impregnated nature of all discourses (e.g. Foucault 1971) could be mentioned here.

Naturally, it is not intended here to fundamentally dispute the thesis of self-competence would be justified, not only for the reason that this would mean strategies in whatever political form to place the individual under disability, withdrawing his rights to self-determination. The thesis of self-competence can a priori neither be founded nor refuted. Rather it is a "regulative idea" (in the sense of Kant), which a community must presuppose to be fulfilled, which, however, in the individual case is always only more or less fulfilled. That the citizens, lay persons, persons concerned, ... are endowed with self-competence is therefore not a fact but rather a regulative concept, which must guide us in the organisation of the community but which we must not regard as factually realised.

On the other hand the thesis of factual self-competence constitutes an overtaxing of the evaluative competence of the citizen, from which in many cases he or she rightly would like to see himself or herself freed by the political organisation of the community. By means of delegation and representation procedures, he elects those fellow citizens to whom he wishes to hand over his competence in order to avoid being overburdened by the obligation to exercise his competence in all the decision-making processes. Thus, the idea of grass roots democracy is not at all in the interests of the citizens, who, in fact, for the most part are seeking to fend off the burdens of individual responsibility, but rather in the interest of those political actors who simply want materially different decisions or even a different political system.

3.3.4 *The plebiscitarism of the will of the people*

The thesis described above as the political premise contends that the democratic institutions, constituted by delegation and representation of competence, are (in part) unsuitable to execute the will of the people in technical and scientific policy issues.

A particularly conspicuous example is given by the following passage of W. Baron:

> Within a normatively oriented, theoretically democratic framework, participation is designed to afford possibilities for actively taking part in and influencing political policy-making over and above the act of voting in the elections of a representative democracy, which is increasingly being experienced as insufficient.... participation constitutes the central principle of action of modern societies, i.e. without opportunities to participate, society would no longer be capable of consensus and thus be incapable of survival. (Baron 1997:147 [translation Gethmann])

The denunciatory undertone against representative democracy ("no longer capable of survival") is unignorable. Such a concept of participation as "the central principle of action of modern societies", closely related to the ideas of a grass-roots democracy will imply a substantial dismantling of the political institutions, above all of the parliament. It thereby pre-supposes that, to a certain extent, it was possible to rescind procedures of delegation and representation and the accompanying processes of professionalised decision-making.

It may be helpful to outline, in a somewhat caricature-like manner, a phenomenon best described as a "competence allocation cycle". This cycle works as follows:

- Citizens elect parliaments
- parliaments call upon the advice of experts
- experts demand citizens' decisions
- citizens elect parliaments
- ...

In this highly simplified stylisation it becomes plain to see that the substantial dismantling of political representation will result in the substantial dismantling of the very scientific expertise which is needed by the political institutions.

A democratic society is one in which the procedures for forming political will be established in such a way that the filtered will of the people may find its path.[49] Here everything depends on a closer scrutiny of this filter function. The expression "filtered will" is in contrast to "immediate will" of the people. In the long and indeed, in part bloody history of the development of political institutions, European societies have learned that it is not the immediate will of the people which is a guarantor against despotism. Such filter functions are, for example, the division of power, by which means the organs of state power exercise reciprocal control over one another, subsidiarity among the various levels of regional corporations (local, state and federal government bodies) or "time dilatation", i.e. delaying procedures, such as hearings, several readings, etc. So these are the elements which, in the European democracies, belong to the essential structure of the institutions albeit in a wide variety of forms. The structure of democratic institutions in non-plebiscitary democracies expresses, on the one hand, that the power is held by no-one else but the people. On the other hand, the *immediate* will of the people warrants suspicion. Therefore, the institutions must be "forged" in such a way that the people are protected against their own will by means of institutional filters.

---

[49] This filtering is an effect of the principle of representation that, following Kant, is an essential of each republican form of government (AA IV GMS:339–342; AA VIII Frieden:349–353). Within theory of constitutional law the task of these representative processes is called the "increasing and improvement" of the people's will (cf. Krüger 1966:232–253, esp. 233).

But are not these considerations in contradiction to the immediate evidence contained in the demand for participation in cases involving far-reaching decisions, such as the deposition of radioactive waste? As will be shown in section B5 below participation of those directly affected is an essential instrument for improving rational planning in the case of planning procedures for large-scale technical plant. The participation of the citizens in such cases improves communication and by doing so (albeit without being certain) might increase public acceptance, too. But is the legitimisation of the decision itself also improved? This question may also be answered in the affirmative provided that those involved are really directly concerned and affected. But what about the deposition of radioactive waste where we have to consider technical and scientific policy decisions in which the degree of being affected by the consequences is only loosely connected to the degree in which participative involvement is claimed and in which participative involvement is possible? Here there is no deductive connection between the communicative aspect of participation and the aspect of legitimisation.

Therefore, in conclusion, there certainly should be participation of those being affected by the consequences in any relevant decision. But it is decisive to conceptualise participation as a means to optimise the outcome with respect to the best available knowledge, especially the local knowledge of those directly affected, and in consideration of its communicative functions. Understanding participation as a means of legitimisation would lead – not least because of the demands of future generations – to inadequate measures. Decision procedures are to be designed in a way that assures, that the choice behaviour of representatives is not biased by implicit or explicit representation of interests of the present generations. It should be oriented to the demands of the future generations as well.

4 Legal questions of managing high level radioactive waste

4.1 Basic legal issues

4.1.1 The responsibility of the state

In the 1980s, there was a controversial debate in Germany about the constitutionality of the peaceful use of nuclear energy which also addressed the problem of radioactive waste disposal. Hasso Hofmann, a specialist of constitutional law and philosophy of law with high reputation, took the position that the use of nuclear energy was ethically irresponsible and unlawful from a constitutional law point of view (Hofmann 1981:286), relying to a major degree on arguments related to the problem of final disposal of high level radioactive waste. Hasso Hofmann was particularly concerned about the fact that, due to the long half-time of high level nuclear wastes, these wastes had to be deposited under strictly controlled conditions over a period of time the dimensions of which exceed human imagination, at least human calculation. Leaving such a nuclear legacy to hundred thousands of future generations whose needs, values, technological capabilities and political-administrative governance structures could not be predicted, was in Hasso Hofmann's eyes ethically irresponsible and unconstitutional. Therefore he strongly pleaded for abandoning the use of nuclear energy. Other authors[50] strongly opposed this view, arguing that the waste problem was technically and socially manageable.

Almost 30 years later, the problems present themselves somewhat differently. For more than 50 years the nuclear power plants in Germany and in increasing numbers of states all over the world have been generating high level radioactive wastes that need to be deposited in a repository. Even assuming an immediate abandonment of nuclear energy, only a certain quantitative reduction of the problem would ensue while the problem of high level radioactive waste as such would remain. The volumes of high level radioactive wastes are not very high and the future generation of high level radioactive waste over the potential remaining lifetime of German nuclear power plants would only add a fraction to what already exists. The quantitative reduction of waste volumes that would be achievable by immediate abandonment of nuclear energy does not really mitigate the problem that has to be tackled.

Presently, in accordance with Section 9a(1b) German Atomic Energy Act, spent nuclear fuel in Germany is stored in interim storage facilities at or near the premises of nuclear power plants. Both for reasons of safety and security this is not a state that can be tolerated for a long period of time; the design and licensing lifetime of these interim storage facilities is 40 years. Irrespective of whether one advocates the present German pol-

---

[50] Lawrence 1989:177ff; Stober 1983:585; Ossenbühl 1983:674–675.

icy of abandoning nuclear energy by the year 2020 as envisaged in Section 7(1), 2nd sentence and Section 7(1a) of the Atomic Energy Act or favours at least a prolongation of the operating times of existing nuclear power plants under strict safety and security precautions as a temporal solution in order to reach the objectives of climate change policy, the long-term problem of final disposal of high level radioactive waste must be solved. That the selection and development of an appropriate repository for high level radioactive waste may have certain side effects such as encouraging the future use of nuclear energy, should not serve as a pretext to postpone the timely solution of a problem that is serious.

Final disposal of high level nuclear waste inevitably involves the state as the main player. However the allocation of responsibility between the state and the waste generators may be shaped – the latter may be directly or indirectly responsible for waste management or only have to bear its financial burdens –, in view of the long-term risks associated with high level radioactive waste disposal there will always be an ultimate responsibility of the state. This responsibility may be based on the national constitution or, as regards the EU, on the EU Treaty or the Euratom Treaty. It may also be laid down in the national laws that regulate nuclear energy in general or nuclear waste management in particular.

In Germany, Article 20a of the Federal Constitution obliges the state to protect the environment (the "natural bases of life"), also in responsibility for future generations. Moreover, the German Federal Constitutional Court has derived a state duty to protect from the fundamental right to life and health set out in Article 2(2) Federal Constitution. In developing and specifying this duty to protect, the hazards and risks associated with nuclear energy have played a particular role. It is doubtful whether the subjective foundation of this duty to protect allows for extending its scope of protection beyond living generations. However, the specifications elaborated by the Constitutional Court and constitutional law doctrine can be transferred to the state duty to protect the environment set forth by Article 20a Federal Constitution. The duty is not limited to preventing harm to human health and the environment. It reaches out to preventing or at least adequately reducing significant risks and even low risks where the possible harm may be high. At EU level, Article 37 of the Charter of Fundamental Rights established by the Treaty of Lisbon (Article 6 EU Treaty) likewise contains an obligation of the Union to protect the environment. It provides that a high level of environmental protection and the improvement of environmental quality must be integrated into the policies of the Union and ensured in accordance with the principle of sustainable development. Furthermore, Articles 2 and 3 of the Charter contain a fundamental right to life and physical integrity. Since the Euratom Community formally remains outside the EU, the Charter does not apply to it. The Euratom Treaty establishes a duty to protect in its provisions on basic health standards and the notification procedure relating to transboundary discharges (Articles 30–38).

The constitutional duty to protect normally only provides for a broad framework within which political discretion can be exercised. Therefore the levels of legislation, sub-delegated legislation and administrative rule-making are more relevant than that of the constitution for defining the scope and content of the duty to protect. Such crucial questions as the basic options of waste management, the technical strategies to be employed including the requisite level of protection and the selection and development of a repository for high level radioactive waste normally are addressed by the legislature or the executive.

In implementing the duty to protect at the level of the executive, the administrative responsibilities must be allocated. A certain problem in this respect is that the executive often both is involved in promoting nuclear energy and managing radioactive waste and is responsible for regulation and supervision. This creates a potential conflict of interests that may endanger the appropriate performance of the duty to protect or at least impair the trust in the executive. There are various organisational arrangements for ensuring the effective independence of the regulatory functions from other functions where the relevant government or governmental agencies are involved in both promotion of nuclear energy and radioactive waste management and in their regulation. These range from privatisation of waste management to the establishment of separate agencies to intra-organisational devices that ensure independence of the regulatory and supervisory functions.

4.1.2 Principles of radioactive waste management

In the course of its development over the last 40 years environmental law has shaped a number of principles that provide this fairly new branch of law particular features. Historically speaking, the precautionary principle has been considered as the most important principle. Nuclear and radiation protection law has developed specifications of the precautionary principle such as the principles of ALARA ("as low as reasonably achievable"), ALARP ("as low as reasonably practicable") and more generally optimisation.[51] Closely related is the principle of justification insofar as, in view of the potential risks and hazards associated with ionising radiation, it prohibits any new or increased exposure that is not justified by net benefits to individuals or society derived from it. However, the principle of justification is also applied in deciding on the degree of reducing exposure, which constitutes an application of the principle of proportionality in the framework of precautionary action (see, e.g., ICRP 2007:89–90). More recently, the principle of sustainable development that entails a special considera-

---

[51] See, e.g., ICRP 2007:88ff; ICRP 2006:81ff.

tion for future generations has emerged as a kind of competitor, at least as a complimentary principle. In view of the long-term dimensions of high level radioactive waste management, the concern for future generation is now enshrined in the relevant international conventions as well as in modern national laws.

However, both principles do not unequivocally answer the crucial questions raised in the field of radioactive waste management. It is not clear when, in case of scientific uncertainty, the precautionary principle ends and tolerable risk ("residual risk") that has to be borne by the community as a burden of civilisation, begins. It is not clear how the burdens caused by the present use of nuclear energy have to be allocated to the present generations that benefit from it and to future generations that do not. The language used in the Joint Convention (see section B 4.2.1) whereby future generations should only be exposed to risks that are considered tolerable regarding the present generations at its face's value provides for intertemporal equity, plainly disregards our ignorance about the needs, values, technological and economic capabilities and political-administrative governance structures of hundreds of thousands of future generations that are confronted with high level radioactive waste generated by the present generations. Even if one defined the temporal scope of the duty to protect in the light of the principle of sustainability as to encompass all generations affected, this problem of asymmetry remains and the answer to it, if any, does not lie in the principle itself but rather at best in its responsible specification.

Another important principle of environmental law that relates to the allocation of responsibilities between state and waste generators is the polluter pays principle. Notwithstanding the state duty to protect, under the polluter pays principle the generators of radioactive waste are responsible for managing this waste. It seems clear that this entails at least financial responsibility. Beyond this core the contours of the polluter pays principle get blurred. It is commonly held that the polluter pays principle is subject to shaping and specification by the state and therefore its essential function only is to justify – rather than requiring – the imposition of physical or operational responsibility on the waste generators. Such a responsibility is common as regards interim storage but may also be extended to final disposal in repositories in the form of joint private management institutions. Also, research and development obligations of nuclear waste generators may be imposed in the name of the polluter pays principle. This possible extension of waste generator responsibility shows that there is no clear-cut borderline between the polluter pays principle and the state duty to protect. In any case the state bears a residual responsibility to ensure the effectiveness of this kind of "delegation".

4.1.3 Degree of legalisation of waste management

Any juridical analysis of high level radioactive waste management policy must ask the questions as to whether and to what degree the relevant solutions are to be laid down in statutory or other legally binding
texts (are to be "legalised") or can be left to discretionary political or
administrative decision. This question does not only concern substantive
solutions such as options, strategies, protection levels and site selection,
it also relates to procedures. In Germany, the Federal Constitutional
Court and the Federal Administrative Court[52] in a series of decisions on
nuclear power plants have developed the "essential questions" doctrine.
This doctrine requires that, even beyond the need to base restrictions of
fundamental rights on a statute, essential questions of society must be
settled by an act of parliament. However, the courts recognise the right of
parliament to only establish a statutory framework by recourse to broad
statutory terms and leave the requisite specification to the executive.
Therefore the legal power of the essential question doctrine is bound to
remain slight. Thus it is more or less a political question whether, beyond
parliamentary framework regulation or specific statutory provisions as
mandated to restrict particular fundamental rights, questions such as
options, strategies, level of protection and procedures of site selection
are decided by the legislature (are "legalised") or more or less left to the
executive.

In favour of legislation, one can sustain that there is a greater transparency, stability and continuity of substantive solutions and procedures.
Moreover, the high scientific and technological complexity may suggest a
legally structured procedure for generating the requisite knowledge base for
taking decisions on tolerability of risk. Perhaps most important, the parliament can exert a greater influence on decisions that are important to vot-
ers and show them that their interests, values and anxieties have a chance
of being considered in the decision-making process. In other words, legalisation increases the sensed legitimacy of regulation although it will not
appease conscientious objectors against nuclear energy.

On the other side of the coin, the choice of strategies and site selection
primarily involve scientific and technological decisions that need to be
translated in order to be understood by the decision-makers. This feature
of the decision-making process provides a powerful argument for entrust-
ing the executive and its experts with the necessary specification of the parliamentary framework. In real political practice the weights of legalisation
and political decision will vary from country to country. However one will
rarely find pure systems.

---

[52] BVerfGE 49, 89, 129 – Kalkar I; BVerwGE 72, 300, 316 – Wyhl; BVerwG, Neue
Zeitschrift für Verwaltungsrecht 2007:837 para. 7 – Schacht Konrad II.

4.1.4   Decision-making levels

Decisions on high level radioactive waste management take place at different levels or, in other words, in different decision blocks. One can distinguish between strategies including the basic options and the level of protection to be achieved, site selection, construction and operation of the relevant facility. At all levels the questions of allocation of responsibilities and competencies as well as procedure are relevant. The levels of decision-making are analytical tools; they do not possess an ontological value. Moreover, it would be erroneous to conceive the decision-making levels as strict temporal sequences. Finally, their delimitation from one another is not clear-cut.

For example, decisions on strategies, site selection and construction may go hand in hand as developing strategies, the concrete design and construction also depend on the characterisation of the bedrock and the geological and hydrological features of the site; they may have to be modified following more recent knowledge acquired in the siting process. Moreover, while site selection may be concluded with an administrative decision on the site, it may also be comprised in the administrative decision on construction so that the two levels of decision-making are blurred.

Nevertheless, the distinction between the different levels of decision-making provides an adequate tool for analysing the relevant legal problems if one keeps these limitations in mind.

4.2   International conventions and recommendations

4.2.1   The Joint Convention on the Safety of Spent Nuclear Fuel Management and on the Safety of Radioactive Waste Management

The Joint Convention on the Safety of Spent Nuclear Fuel Management and on the Safety of Radioactive Waste Management of 1997 (in short: Joint Convention) is the first legal instrument that addresses the issue of radioactive waste management on a global level.[53] It entered into force on 18 January 2001 and presently comprises 41 parties. The International Atomic Energy Agency (IAEA) functions as the Secretariat of the Convention. The Convention comprises spent nuclear fuel and radioactive waste in the strict sense under a common regulatory umbrella, although essentially dealing with these two subjects separately. The notion of waste includes spent nuclear fuel if no further use is intended and the fuel is designated for disposal (Articles 2 lit. h and 10). Military nuclear material is only encompassed under narrow circumstances.

The objective of the Joint Convention is to achieve and maintain a high level of nuclear safety and ensure that at all stages of spent fuel and radio-

---

[53]   It is a sister convention of the Convention on Nuclear Safety of 1994. As to both conventions see Schärf 2008:65–69.

active waste management there are effective defences against potential hazards so that individuals, society and the environment are protected from harmful effects of ionising radiation; these objectives shall be pursued in the interest of both the present and future generations (Article 1 lit. i and ii). Finally, the Convention aims to prevent accidents with radioactive consequences and mitigate, where they should occur, their consequences (Article 1 lit. iii). In essence, the duty to protect set forth in the Convention is based on the precautionary principle and the principle of sustainable development.

Chapter 3 (Articles 11–17) contains the regulation of radioactive waste management. The relevant rules do not distinguish between various activity levels of radioactive waste and in this respect leave the contracting parties quite some discretion which, however, must be exercised in keeping with the principles laid down in the Convention. It should also be noted that the Convention does not address the issue of radioactive waste management options including that of having an international repository. However, in the latter respect it leaves the door open in permitting international transports of radioactive wastes also for managing them abroad (Article 27(1) lit. iii, iv) (see Boutellier/McCombie 2006:138).

In the first place, the Joint Convention establishes general safety requirements that specify the duty to protect individuals, society and the environment against radiation and other hazards associated with radioactive waste management (Article 11). These are in particular measures that adequately address criticality and removal of residual heat during radioactive waste management, keep radioactive waste to the minimum practicable, effectively protect individuals, society and the environment by applying suitable protective methods having regard to internationally endorsed criteria and standards and consider non-radiological hazards that may be associated with radioactive waste management (Art. 11 lit. i, ii, iv and v). The provision also addresses the protection of future generations against risks and other hazards as well as other burdens associated with radioactive waste management (Article 11 lit. vi, vii).

The specific requirements set out in the Joint Convention address the siting, design, construction and operation of radioactive waste management facilities in a rather detailed manner (Articles 13–16). All waste management activities shall be subject to the requirement of a licence (Article 19(2) lit. ii and with respect to operation Article 16 lit. i). As regards the siting of facilities, Article 13 of the Joint Convention requires the competent authorities to evaluate all site-related factors likely to affect the safety of the facility during operations and after closure, evaluate the likely safety impact of the facility, taking into account possible changes of site conditions after closure, and provide sufficient information on the safety of the facility to the public (Article 13 lit. i, ii and iii). Before constructing the facility a systematic safety assessment and an environmental impact assessment relating both to operations and the post-closure period shall be carried out (Article 15 lit. i, ii). The design and construction of the facility shall provide for suitable measures to limit possible radiological impacts, including those from discharges or uncontrolled releases and the technologies incorporated in the design and construction must be supported by experience, testing or analysis (Article 14 lit. i and iv). Moreover, technical provisions for the closure of the facility shall be prepared (Article 14 lit. iii). Before starting operations, the safety assessment and the environmental assessment shall be updated or further detailed as necessary (Article 15 lit. iii). Compliance with all relevant assessments made previously as well as conformity of the facility, as constructed, with all design and safety requirements shall be demonstrated (Article 16 lit. i). Moreover Articles 16 and 17 of the Joint Convention list a number of technical safety requirements relating to operational limits and conditions, maintenance, monitoring, inspection and reporting, and engineering and technical support. In addition, in the operation of a radioactive waste disposal facility, the requirements for operational radiation protection under Article 24 of the Joint Convention and emergency preparedness under Article 25 of the Joint Convention must be complied with. There are also special post-closure requirements (Articles 16 lit. ix, 17). In principle, the Convention leaves the contracting parties the option between active and passive controls (Article 17 lit. ii).

The Joint Convention further provides that the primary responsibility for the safety of radioactive waste management rests with the licensee (Article 21). As regards financial responsibility, the Convention requires that adequate financial means are available during the operational life-time of the disposal facility as well as for post-closure period (Article 22). However, the Convention abstains from prescribing that the license holder shall bear the financial responsibility.

An important aspect of the regulatory strategy of the Joint Convention concerns the improvement of the national regulatory and institutional framework (Articles 19–20). In this regard the Convention requires the contracting parties to establish appropriate substantive regulation and licensing and to institutionalise control, inspection, documentation, reporting and enforcement mechanisms. In particular, the Convention mandates

> a clear allocation of responsibilities of the bodies involved in the different steps of spent fuel and radioactive waste management (Article 19(2) lit. vi)

and an assurance of

> the effective independence of the regulatory functions from other functions where organisations are involved in both spent fuel or radioactive waste management and in their regulation (Article 20(2)).

The exact meaning of the requirement of independence of regulatory functions is not entirely clear. The reference to "waste management" could imply that only storage and disposal activities shall be separated from regulatory ones. However, the requirement can also be understood as to mean that "management" includes promotional activities.[54] Finally, the Joint Convention (Articles 29–36) establishes a peer review process of the effectiveness of national regulation of radioactive waste management (see Tonhauser 2006:131–136; Warnecke/Bonne 1995:26–29). The idea underlying the peer review process is twofold: The process compels the national authorities to review their own activities and draw conclusions on future improvements. Moreover states benefit from the experience of other states through an open and frank discussion even on technically difficult matters. The process consists of reporting obligations (Article 32) and periodic review of the national reports through preparatory and peer review meetings of the parties supported by implementation base documents. The parties are required to attend these meetings. Subjects of the review process are in particular the national strategies, the development of the regulatory and institutional framework and the engagement with stakeholders and the public regarding consultation and achievement of public acceptance.

4.2.2 Recommendations of international organisations and other bodies

4.2.2.1 International Atomic Energy Agency

Apart from various other tasks, the International Atomic Energy Agency (IAEA) is also active in the field of the peaceful use of nuclear energy in general and protection against risks and hazards originating from the handling of nuclear material in particular. It tries to improve nuclear safety by publishing recommendations that are based on a broad consensus of the international scientific and technological community.

In 2006 the IAEA published the "Fundamental Safety Principles – Safety Fundamentals"[55] that are applicable to all activities that involve the handling of radioactive material including radioactive waste management. The document understands "safety" in the broadest possible sense, meaning protection of people and the environment against radiation risks, safety of facilities, risks of normal operations, incidents and loss of control over a nuclear facility. With respect to radioactive waste management the Fundamental Safety Principles supersede the earlier "Principles of Radioactive Waste Management" of 1995 (IAEA 1995). The Principles have the character of a recommendation. The document sets out ten fundamental safety principles that often also contain some specification regarding radioactive waste: (1) the primary responsibility of persons or organisations that give rise to radiation risks; (2) the requirement of an effective legal and gov-

---

54 In this sense various IAEA recommendations; see section B 4.2.2.1; to the same extent IAEA 1994, Article 8(2), which, however, is limited to nuclear power plants.

55 IAEA 2006; see also Zaiss 2009:189–93; Schärf 2008:8–15.

ernmental framework for safety, including a regulatory body that is effectively independent of the operator of the disposal facility and of any other body, so that it is free from undue pressure from interested parties; (3) the requirement of effective leadership and management for safety; (4) the need for justification of facilities and activities (positive risk/benefit balance); (5) the requirement of optimisation of protection (highest level of safety that can reasonably be achieved); (6) limitation of risks to individuals (dose limitation); (7) protection of people and the environment also in remote areas (transboundary risks) and in the future (protection of future generations); with respect to radioactive waste: avoidance of undue burdens being imposed on future generations, such as the need to take significant protective actions, and minimisation of radioactive waste generation; (8) prevention of accidents; (9) emergency preparedness and response for nuclear or radiation incidents; and (10) protective actions to reduce existing or unregulated risks, especially natural radiation risks.

Under the umbrella of the previous and present principles, IAEA has adopted a large number of more specific "Safety Standards" (safety requirements), among others, relating to radioactive waste management. The most important document in the field of radioactive waste management is the Safety Standard "Geological Disposal of Radioactive Waste" of 2006 (IAEA 2006b). The document contains objectives and principles (Section 2) and detailed safety requirements (Section 3), distinguishing between the operational period and the post-closure period. It still refers to the old principles of 1995 but seems to be also consistent with the new principles. According to the document, geological disposal facilities are to be sited, designed, constructed, operated and closed so that protection in the post-closure period is optimised and a reasonable assurance is provided that doses or risks to members of the public will not exceed the relevant dose or risk level (No. 2.12). The effective dose of 1mSv from all practices in a year is also considered a criterion not to be exceeded in the post-closure period; likewise, the individual dose originating with the repository alone shall be not more that 0.3 mSv in a year or the risk of death or serious illness not exceed 1 : 100,000 (IAEA 2006b:10–11).

As regards the institutional arrangements, the Safety Requirement "Legal and Governmental Infrastructure for Nuclear, Radiation, Radioactive Waste and Transport Safety" of 2000[56] is of particular importance. It recommends that the regulatory body shall be effectively independent of organisations and bodies charged with the promotion of nuclear technologies or responsible for facilities or activities. This is motivated by the objective to enable the taking of regulatory judgements and enforcement action without pressure from interests that may conflict with safety. Both the interest in effectiveness and in credibility have guided this recommendation (see IAEA 2002:para. 2.3).

---

[56] IAEA 2000, para 2.2(2); to the same extent IAEA 2002, para. 2.2, 2.3, 2.6.

4.2.2.2 International Commission on Radiological Protection

The International Commission on Radiological Protection (ICRP) is a non-governmental scientific organisation (non-profit organisation) that was founded in 1929 and restructured in 1950 (see Lindell/Dunster/Valentin, 1998). Its objective is to advance for the public benefit the science of radiological protection, in particular by providing recommendations and guidance on all aspects of protection against ionising radiation. In making recommendations, the ICRP aims at providing an appropriate standard of protection without unduly limiting the benefits derived by the practices that give rise to radioactive exposure. The composition and organisation of ICRP ensure independence, impartiality and high scientific reputation. Its recommendations are often the basis of regulatory action all over the world.

ICRP has published some reports that focus on radioactive waste disposal, in particular in 1997 the report on "Radiological Protection Policy for the Disposal of Radioactive Waste" (ICRP 1997) and in 1998 the report "Radiation Protection Recommendations as Applied to the Disposal of Long Lived Solid Radioactive Waste" (ICRP 1998). Moreover, the general recommendations of ICRP regarding radiation protection such as the 2006 report on optimisation and the 2007 recommendations (ICRP2006; ICRP 2007) are relevant. ICRP recommends that the control of exposure should be achieved by the use of "constrained" optimisation, meaning that below the dose limits source-related constraints and reference levels limit the optimisation process. ICRP proposes as a constraint of optimisation an individual exposure standard for a single source under normal circumstances of 0.3 mSv or its risk equivalent of 1:100,000 per year.[57] On the other hand, more recently the economic limitations of the optimisation concept are emphasized (ICRP 2006, para. 35, A6–A18). Moreover, ICRP demands the reduction of long-term exposure and the establishment of standards for inadvertent human intrusion (ICRP 1998, para. 64). Although with less emphasis, the principle of optimisation is also applied to the planning of activities, the geological features of the site and the design of facilities.[58]

4.2.2.3 Nuclear Energy Agency

The Nuclear Energy Agency (NEA) is a specialised agency within the Organisation for Economic Cooperation and Development (OECD), an intergovernmental organisation of 28 industrialised countries from North America, Europe and the Asia-Pacific Region. NEA has a non-political, technical mission. It tries to pool the member states' experience, provide a forum for the exchange of views, identify areas of future work, carry out joint research and develop common positions on key issues ("consensus opinions"). One of NEA's areas of work is radioactive waste management.

---

[57] ICRP 1997, para. 48; ICRP 2007, para. 240, Table 5.
[58] ICRP 2006, para 29; ICRP 2007, para. 216, 253–254; ICRP 1998, para. 51.

In this field, NEA assists member countries in the management of radioactive waste and in the development of strategies. On request, NEA also conducts peer reviews of nation radioactive waste management programmes and regulations. A relevant report published in 2009 deals with the issue of timescales in the post-closure safety assessment (NEA 2009).

4.3 European regulation

4.3.1 Euratom Treaty and European Directives

The Treaty for the Establishment of a European Atomic Energy Community (Euratom Treaty) established a common European regime for trade with, and the use of, nuclear energy for civilian purposes. Among others, the Treaty contains harmonised regulation for the protection of health and safety of workers and the general public against the dangers arising from ionising radiation. Although focusing on nuclear power plants, these rules also apply to radiation originating from radioactive, including high level, waste management. Article 30 Euratom Treaty empowers the Community to set basic health standards in the form of maximum permissible doses, maximum permissible levels of exposure and contamination and fundamental principles of workers' protection. The member states are obliged to ensure compliance with these standards (Article 33) and carry out continuous monitoring (Article 35). The notion of health standards is interpreted broadly. It does not only encompass radiological protection in the conventional meaning but according to the jurisprudence of the European Court of Justice[59] also the safety of facilities.

There are Treaty provisions that explicitly apply to projects. Article 37 of the Euratom Treaty establishes a consultation procedure with respect to potential transboundary radioactive pollution. Any plan for the disposal of radioactive waste must be submitted to the Commission in order to determine whether the implementation of such plan is liable to result in the radioactive contamination of the water soil or airspace in another member state. The notion of disposal in the meaning of Article 37 Euratom Treaty is broad. It comprises any intentional or accidental discharge of radioactive substances from a facility that handles radioactive material, including a waste disposal facility (Haedrich 1993:1042), arguably also in the post-closure phase since any unforeseen radioactive discharge is covered. Therefore it also includes radiation originating from high level radioactive waste management in a repository, provided it is to be constructed near the border of another EU member state. The Commission is to deliver an opinion on such projects before they are implemented (Article 37). For that purpose it must be provided on time with extensive information about the site, construction, planned or possible discharges and

---

[59] European Court of Justice, 2002 ECR I-11221 para. 82 – Commission/Council; 2009 ECR I-10265 para. 105 – Land Oberösterreich/ČEZ as.

their evaluation, and controls in due time before the implementation of the project.[60] Although the project does not require the formal consent of the Commission, the Treaty assumes that the member states concerned will normally follow the opinion of the Commission. Moreover, Article 38 Euratom Treaty empowers the Commission to issue a non-binding recommendation to the member states, in urgent cases also a binding directive requiring the member state concerned to take all necessary measures to prevent infringement of the basic standards and to ensure compliance with Euratom regulations.

The basic health standards referred to in Article 30 Euratom Treaty have been laid down in the Directive 96/29/Euratom (see Schärf 1998:344–364). Although they are intended to provide a harmonised level of protection for the whole European Atomic Energy Community that is sufficiently protective of human health, they only constitute a minimum harmonisation so that the member states can adopt more stringent regulation.[61] The standards also apply to the storage and disposal of radioactive substances (Article 2(1) (a) of the Directive). It is not clearly set out that they extend to the post-closure phase although the caput of Article 2 refers to all practices that merely involve a risk from ionizing radiation. The Directive lays down general safety principles, especially the requirement of best possible protection entailing the need for justification of exposure and risk/benefit analysis, keeping exposures as low as reasonably achievable and consideration of the combined effects of doses from all relevant sources (Article 6). The Directive sets dose limit values and principles of occupational health protection for workers (Articles 8–12, 17–39), dose limit values and principles governing operational protection for the general population (Articles 13, 14, 43–47) and provisions regarding emergencies (Articles 48–53).

Any operation and decommissioning of a facility in the nuclear fuel cycle requires an authorisation by the competent national authorities (Article 4(1) lit. a of the Directive). The same is in principle true for the disposal, recycling and reuse of radioactive materials arising from such facilities, except where certain clearance levels are not exceeded (Article 5(1) and (2) of the Directive). The prerequisites for granting the authorisation primarily consist in compliance with the general safety principles set forth by Article 6 and additional fundamental principles governing operational protection of the population as set forth in Articles 44 to 47 (Article 43). These lat-

---

[60] European Court of Justice, 1988 ECR 5013 para. 14–20 – Land de Sarre/Ministère de l'Industrie, des P et T et du Tourisme and others; European Court of Justice, Case C-115/08 – Land Oberösterreich/ČEZ as, supra note 60, para. 104; Haedrich 1993.

[61] European Court of Justice, 1992 ECR I-6153 – Commission/Belgium. The case concerns the predecessor directive but is also valid for Directive 96/29; see Jans/Vedder 2008:449; see also Haedrich 1993:1040–1042.

ter principles relate to the siting, design, construction and operation of the facility without always making a clear distinction between these four different phases. The siting, design and construction must be in conformity with radiation protection. The acceptance into service is subject to the requirement of providing adequate protection to the population against exposure or radioactive contamination outside the perimeter, taking into account, among others, geological, hydrological and ecological conditions. Moreover, the authorisation shall be based on dose estimates for the population and vulnerable groups the details of which are laid down (Article 45) and it shall prescribe inspections and surveillance (Article 46). Enterprises that operate relevant facilities are obliged to achieve and maintain an optimal level of protection of the environment and the population, check the effectiveness of protective devices within the facility and exercise an appropriate degree of surveillance of radiation protection including the assessment of exposure and radioactive contamination (Article 47). Finally, the Directive contains provision relating to emergencies (Articles 48–53). The Directive has been transposed into the national law of the member states, normally in the manner that more specific and sometimes also more stringent regulation is provided.

In the framework of the political programme of a further Europeanization of nuclear Energy of 2002 (so called "Nuclear package") (EU Commission 2002), the Commission also proposed a Directive on the Safe Management of Spent Nuclear Fuel and Radioactive Waste[62] that would have entailed the option of constructing a common repository for, or offering an existing national repository to, member states having small volumes of radioactive waste (Article 4(4) of the amended proposal). According to the definition of a radioactive waste disposal facility (Article 2(2)), the Directive would apply to the post-operational phase. However, the proposal did not find the agreement of the Council.[63] Other parts of the "Nuclear package" have passed the legislative process but are not directly relevant to high level radioactive waste management. This is true for the Directive 2003/1227/ Euratom on the Control of High Activity Sealed Radiation Sources and Orphan Sources and the Directive 2009/71/Euratom establishing a Community Framework for the Nuclear Safety of Nuclear Facilities that concern stages of the waste management cycle that precede disposal in a repository. There are also special rules relating to transboundary transport of nuclear waste and spent nuclear fuel for reprocessing (Directive 2006/117/ Euratom). These rules apply both to internal and extra-Community transports. Apart from reprocessing, they would be relevant in our context if the

---

[62] EU Commission 2003 amended in view of the proposals by Parliament: EU Commission 2004; as to an internal draft of 2002 see Boutellier/McCombie2006:137.

[63] See the Resolution of 8 May 2007 where preference is given to coordinated action.

option of having a common repository for high level radioactive waste were provided for in a Euratom legislative text.

The failure of the proposal on waste disposal did not mean that the Commission entirely abandoned this issue.[64] In November 2010, the new Commission resumed the issue of harmonisation and proposed a new version of a EU Directive on the management of spent nuclear fuel and radioactive waste that is designed to bring EU law in line with the Joint Convention but also aims at consistency with the Directive on the safety of nuclear power plants (EU Commission 2010). The draft establishes, mostly in accordance with the Joint Convention, general safety principles such as the requirement of long-term safety, specified by provisions on the safety case that is to include all steps from siting to the post-closure phase, moreover the responsibility for future generations and the principle of domestic disposal (unless otherwise agreed by the member states (Articles 4 and 8). It mandates the establishment of an appropriate regulatory, legislative and organisational framework (Article 5). Specifying the relevant rules of the Joint Convention, the draft provides that the regulatory authorities shall be functionally separate from any other body involved in the promotion or exploitation of nuclear energy, including management of radioactive waste, in order to ensure effective independence from undue influence on its regulatory functions (Article 6). Further, the draft requires transparency and effective public participation in the decision-making process on radioactive waste management (Article 12). While the prime responsibility for radioactive waste is vested in the operators (Article 7), the member states bear the ultimate responsibility. The major strategic options of radioactive waste management including the choice between direct disposal and reprocessing are left to the decision of member states (Recital 25). Of particular importance are the proposed rules on national programmes for the management of spent nuclear fuel and radioactive waste (Articles 13–15).[65] Among others, the national programmes shall contain concepts, plans and technical solutions for radioactive waste disposal, concepts and plans for the post-closure phase including the relevant time-frames and the major milestones, time-frames and responsibilities for implementation as well as financial decisions (Article 14). Although not specifically addressed, the national programmes include the siting of facilities for final disposal. The programmes must indicate when, where and how these facilities shall be constructed and operated.

---

[64] See the communications, EU Commission 2006 and EU Commission 2008; EU Commission 2007. There have also been some research projects on joint final disposal, the so called SAPIERR projects; see Boutellier/McCombie 2006:136–141.

[65] The programmes do not need to consist of a single reference document (Recital 33).

4.3.2 Western European Regulator's Association and European Nuclear Safety Regulators Group

The Western European Regulator's Association (WENRA) is a non-governmental organisation established in 1999 that comprises the heads and senior staff members of nuclear regulatory authorities of European countries with nuclear power plants. Except for Switzerland, all of its members come from the EU. The original objective of WENRA was to accompany the accession process of candidate countries to the EU. Nowadays, harmonisation of approaches to nuclear safety and the exchange of experience on significant safety issues have become more important. WENRA develops and publishes "reference levels" that serve as benchmarks for assessing and adapting national regulation and implementation (see Drábová 2009:183–186; Zaiss 2009:193–197). It recognises the IAEA safety standards while the reference levels are considered to represent good practice in Europe. WENRA has established a Working Group on Waste and Decommissioning the work of which started in 2002. This body has published two reports on common reference levels in the area of its competence, among them in 2006 a Reference Levels Report on radioactive waste management (WENRA 2006). However, the problems of geological disposal of radioactive waste have as yet not been addressed.

WENRA's tasks have to some extent been taken over by the European Nuclear Safety Regulators Group (ENSREG) that was set up in 2007 and comprises high level regulators from all EU member states (Commission Decision 2007/530: o. J. 1970 No. L 195, 44). ENSREG focuses on management of spent nuclear fuel and radioactive waste. It submitted a first report in 2009 (ENSREG 2009). The recent Commission proposal for a Directive on radioactive waste management rests to a major degree on inputs form ENSREG.

4.4 Comparative experience

4.4.1 General remarks

The following section presents a summary of the regulatory experience in selected nuclear countries (United States, France, United Kingdom, Switzerland, Sweden, Finland, Japan and Spain). It is based on a more detailed comparative review which will be presented in the Annex. Although the political-administrative systems and cultures of the countries included in the comparison, the geological conditions, the objectives of site selection (suitable vs. optimal site) and the attitudes of their population toward nuclear energy in general and the disposal of radioactive waste in particular are quite different, the internal logic of the problem of radioactive waste management is more or less the same and there is a limited reservoir of solutions. Moreover, international regulation, recommendations by international scientific organisations as well as quite extensive international exchanges between regulators have led to a certain convergence of solutions that makes comparison easier and more fruitful. Therefore, failures and successes of substantive solutions and procedural approaches abroad constitute lessons that may be highly relevant for Germany with respect to the various aspects of high level radioactive waste management. However, in view of the differences just mentioned one can rule it out that successful solutions achieved abroad could be simply transferred to Germany.

The selection of the countries included in the comparison is inspired by their importance as nuclear countries and hence also their involvement in radioactive waste disposal, the expectation that their experience could make interesting contributions to problem-solving in Germany, both at substantive and procedural level, and the accessibility of national regulation.

4.4.2 Comparative evaluation

The comparison of the law of major nuclear countries regarding high level radioactive waste management reveals a number of common features but also important differences. In some areas, the influence of the Joint Convention and international exchanges and peer review on gradual harmonisation is evident. The remaining differences can be mainly explained by differences of political-administrative systems or cultures, geological and economic conditions, objectives of site selection (suitable vs. optimal site) and public attitudes toward nuclear energy in general and radioactive waste disposal in particular, especially the strength of an anti-nuclear-power movement.[66]

At the level of statutory regulation, there are either special statutes on radioactive waste disposal or the general laws on nuclear energy that also apply to waste disposal. In a number of countries (Switzerland, Sweden, Finland and Japan) radiation protection is regulated in a special statute that also covers radioactive waste management. Where general laws on nuclear activities apply, specialisation of regulation normally occurs at the level of sub-delegated legislation and administrative rule-making. In addition to these laws, often general statutes on spatial planning, environmental impact assessment and public participation apply. The overall degree of legalisation is quite varied. It is high in the United States, France, Switzerland, Sweden and Finland and very low in the United Kingdom and Spain. But even in countries having a high degree of legalisation, some basic policy questions of radioactive waste management and procedure may be left to purely political decision-making.

In all countries under comparison, the ultimate responsibility for radioactive waste disposal is vested in the government. In more or less detailed pronouncements, the applicable laws impose on the government a rather strict duty to protect that reflects the precautionary principle and the principle of inter-generational justice. Subject to some qualification regard-

---

[66] See also AkEnd 2002:40–48; Hörnschemeier 2008:600.

ing Finland and Switzerland the executive bears the major responsibility for specifying the duty to protect. By contrast, the responsibility of industry is quite varied. On one side of the scale there are clearly state-oriented systems where the responsibility of industry, that is radioactive waste generators, is limited to financing waste management (United States, United Kingdom and Spain). On the other side we find regulatory systems where the responsibility of the nuclear industry is very strong (Switzerland, Sweden, Finland and Japan). In these latter countries, organisations established by industry are responsible for the management of high level radioactive waste including site development, construction and operation of the repository. The state limits itself to the steering of the procedure or, as in the case of Sweden, even only the approval of site selection; in all cases it retains responsibility for regulation and supervision of construction and operation. A special case is presented by France where waste management is vested in a public enterprise that is composed of radioactive waste generators but supervised by the state. In Sweden and Finland, industry also has specific research and development obligations.

As regards the institutional arrangements, the typical pattern is the existence of one or more ministries that bear the political and basic regulatory responsibility and one or more separate agencies that are competent for regulation and supervision, in case of state-oriented systems including the construction and operation of the repository. The requirements of separation of management of radioactive waste disposal and regulation/supervision set forth in the Joint Convention have been respected in most countries if one interprets the term "management" in a narrow fashion, at least as the result of certain reorganisations that took place more recently (France and the United Kingdom). However, if management is deemed to include any promotional and political activities in the field of nuclear power, the picture becomes much more varied.[67] A number of countries (France, Switzerland, Sweden, Finland and Japan) have opted for a government-integrated regulation and vested the authority to grant the basic permissions for HLW repositories in the Government, sometimes subject to a Parliamentary vote. They thereby accept a major degree of politicisation of the relevant decisions and to this extent practically reduce the existing independent or separate regulatory bodies to the exercise of advisory functions.[68] However, as long as the government is not directly involved in operational disposal activities, one could argue that the requirements of the Convention are met. The Joint Convention does not require the establishment of an independent regulatory body. Moreover, it should be underlined that the more or less technocratic model of an independent regulatory agency may not be appropriate

---

[67] See also Bredimas/Nuttall 2008:1346–1347, 1351–1352; Michanek/Söderholm 2009:4092–4093 (both limited to nuclear power plants).

[68] An exception is presented by Spain where the Government formally grants the permissions, but acting on a binding report by an independent regulatory body.

where decisions are at stake that are deemed to be of crucial interest to society as a whole such as the siting of a HLW repository.[69] However, the intentions underlying the farther-going IAEA recommendations are missed. In order to secure effective safety and credibility these recommendations call for the establishment of a regulatory agency that is independent of management and promotional functions.

The basic strategic options of high level radioactive waste management are normally set out in the relevant statutes and not just based on political decisions. This is in particular true for the decision in favour of geological disposal and the choice between direct disposal (United States, Finland, temporarily also Switzerland) and reprocessing (France, United Kingdom, Japan). However, in the United Kingdom and Spain geological disposal is only based on a political decision and in Sweden the rejection of reprocessing is not set out by statute. Moreover, the principle of domestic waste disposal is normally established by statute. Only Switzerland has retained the option of a joint international solution. As regards the requirement of retrievability or even reversibility, the statutory pronouncements are less frequent. It is laid down in the legislation of the United States, France and Switzerland. In a number of countries, the question is deliberately left open for future decision (United Kingdom, Sweden, Spain, Japan and – after removal of the requirement from its legislation – also Finland). A special regarding strategy is presented by Spain. This country has opted for a long period of central interim storage of high level radioactive waste, hoping that it thus can profit from technological development in the next decades.

Other, especially the safety-related and technical aspects of the final disposal strategy for high level radioactive waste are more often set out in regulations, administrative rules or plans. To a large extent they still have to be developed. As a means of securing public acceptance, in many countries new forms of public participation in developing the waste disposal strategy, both in the form of pluralistic advisory bodies and open public debates, have been introduced (especially in France and the United Kingdom). It is clear that during active operations the repository must comply with the normal radiation protection, safety and security requirements that are applicable to other nuclear facilities, while the technical details are normally still to be worked out pending further development of the repository. However, due to the uncertainties associated with estimates that reach out to several 100,000 years, the post-closure period causes major problems and here the differences between the countries under comparison are the greatest. The relevant time-scales range from 100,000 years to several 100,000 years to one million years. Japan has entirely renounced to setting any time-frame, limiting itself to the period of time where robust estimates are still possible.

---

[69] See also Bredimas/Nuttall 2008:1351, who, however, favour the approach taken by the IAEA recommendations.

A similar assessment can be made with respect to France where only a first period of proof of 10,000 years has been established while the length of the total time-frame is left open. In some countries such as the United States, Sweden and Finland, there are different radiation protection standards for the first 10,000 or 100,000 years after closure and the following period of time (in the United States 0.1/1 mSv per year, in Sweden and Finland 0.1 mSv/lower than natural radiation). A standard of individual exposure in the post-closure period of 0.1 mSv and/or a tolerable risk of 1:1 million are quite common, but there are also countries that have opted for 0.15 or 0.25 mSv (United States, France).

Site selection for a repository has proven to be the stumble-stone of high level radioactive waste disposal. Several factors coincide here to burden or even threaten the selection process. These are in particular the strong inter-relationship between the general attitudes towards nuclear energy and radioactive waste disposal, the uncertainties and ambiguities associated with the time-scale of high level radioactive waste disposal, the normal aversion of people against undesirable development in their neighbourhood, conflicts between central government and the regions, and the prospects of future technological advances that may suggest postponing the solution. As a response to wide-spread lack of public acceptance and the failure of prior technocratic site selection attempts, most of the countries under comparison rely on new, more legitimate, i.e. transparent, comprehensible and inclusive forms of procedure. The major elements of the new procedure are a criteria-led and staged decision-making process that includes alternatives, broad public information and participation and voluntary participation of the affected municipalities in the selection process. This is in particular true for France and the United Kingdom, but also for Sweden, Finland and Japan and in a qualified form even for the United States and Switzerland. However, it is to be noted that except for Finland, voluntarism is accepted as a political principle only so that the state in case of need can resume its competences to impose a decision. In Sweden and Switzerland the prerequisites for overriding a municipal or cantonal veto are formally laid down in the applicable statute and the decision is subject to judicial review. In the United States, the new administration's renouncement to the early congressional site determination at Yucca Mountain has also resulted in a new, more open and criteria-led site selection process.

Looking at the success of the new approach, one can state that the site selection decision has already been taken in Sweden and Finland. In both countries the decision was facilitated by the fact that the repository could be built on or very near the site of an existing nuclear power plant. Such conditions do not exist in most other countries under review and certainly not in Germany. In Finland, the success of voluntarism is somewhat flawed by the later decision by government to construct a new nuclear power plant while the willingness of the relevant municipality to host the repository was based on the draft framework decision that had limited the waste volume to be accepted. By and large the same is true for Sweden because here, too, the decision to permit new reactors that replace existing ones followed the siting decision. By contrast, the new approach still is on probation in France, the United Kingdom, Switzerland and Japan.

The legal status of the site selection procedure is quite varied. The degree of legalisation of the process is different. In many countries, site selection precedes a formal siting decision that is taken under nuclear energy or waste disposal law or under general planning law (United States, United Kingdom, Switzerland, Sweden, Finland, and Spain). In other countries, it is only at the level of construction permission (for the underground laboratory or even the final repository) that the site becomes part of a formal decision-making process (France, Japan). The normal requirement to carry out one or more staged environmental impact assessments is normally geared to the formal permission process (except for the United States) and therefore may be relatively late.

In all countries under review, as an expression of the polluter-pays principle, nuclear industry must bear the costs of high level radioactive waste disposal. This includes the costs of developing, selecting, constructing and operating the repository, but not necessarily the post-closure period. However, the legal mechanisms used for protection against other creditors, especially in case of insolvency, and against the consequences of cessation of operations and dissolution of the operator of a nuclear power plant are quite different. Some countries only require the establishment of reserves which provide limited protection (United Kingdom). More frequent are fund solutions and obligations to feed the fund on time in order to ensure timely coverage of the predictable costs of final waste disposal. Here, the separate administration of the fund shall ensure that the money collected is exclusively available for final waste disposal. Where the state administers the fund, protection against its use for the general budget is not formally provided. In Sweden, the waste generators must also provide guarantees for their future ability to pay into the fund.

4.5 German law

4.5.1 Sources of legal regulation

In Germany, the most important source of legal regulation of radioactive waste disposal is the Atomic Energy Act of 1959 in its consolidated version of 1985. This act does not only regulate the construction and operation of nuclear power plants and other forms of handling nuclear fuel and radioactive substances. It also encompasses the disposal of radioactive waste. After 1985, the Atomic Energy Act was fundamentally amended by the Law of 22 April 2002 that provided for the gradual phasing-out of nuclear energy and also changed the waste disposal regime. The recent 11th Amendment that is not yet law would extend the operation times of nuclear power plants but does not put the phase-out policy as such into question. As regards radiation protection, the Radiation Protection Regulation of 2001 is applicable. This regulation covers all kinds of handling of radioactive substances including waste disposal (Sections 2(1) No. 1 lit. c, 3(1) No. 34). However, planning permits for repositories granted under Section 9b Nuclear Energy Act are paramount. According to Section 14 Water Resources Management Act, the planning permission does not include water pollution; to this extent, a separate authorisation is required. It is also to be noted that in contrast to previous law the Federal Mining Act of 1980, as last amended in 2006, applies to the authorisation of underground repositories for radioactive wastes in parallel to the Nuclear Energy Act (Section 9b(4) No. 3 Atomic Energy Act).

The planning permission procedure for repositories is governed by various legal rules. The basic regulation is set out in Section 9b Atomic Energy Act. Most of the details are contained in the Administrative Procedure Act's rules on planning permits. Pursuant to a reference by Section 9b (5) Nuclear Energy Act also certain provisions of the Nuclear Law Procedure Regulation of 1977 (in its consolidated version of 1995) are applicable. Finally the Environmental Impact Assessment Act of 1990 (in its consolidated version of 2005) is applicable to the planning permission procedure (Section 9b (2) Atomic Energy Act). As regards the permissions required under water and mining law, the relevant provisions of the Water Resources Management Act and the Federal Mining Act are applicable. The requisite permissions are also granted by the authority competent for the planning permission.

Although one could conclude from this list of legal sources that the degree of legalisation of nuclear waste disposal is relatively high in Germany, this is only partially correct. First of all, important substantive questions are addressed by administrative rules, guidelines and recommendations of the executive or advisory bodies only. Secondly, the site selection process for radioactive waste repositories is essentially a political process with few, if any, legal framework rules. This is in strong contrast to many other nuclear states where there have been, at least recently, clear trends towards legalisation of the selection procedure.

4.5.2   Responsibilities

The major responsibility for the short- and long-term safety and security of radioactive waste disposal in Germany is vested in the state although the Nuclear Energy Act provides for some forms of privatisation of radioactive waste disposal. Moreover, the Act also allocates major responsibilities to the radioactive waste generators.

4.5.2.1   The constitutional duty to protect

The responsibility of the state for, in particular high level, radioactive waste disposal has strong roots in German constitutional law. Already at the turn of the 1970s to the 1980s, the Federal Constitutional Court interpreted the fundamental right to life and physical integrity laid down in Article 2(2) Federal Constitution to the extent that is also conferred a protective duty on the state to secure a high degree of nuclear safety.[70] In view of the particular hazards and risks presented by the peaceful use of nuclear energy, the Federal Constitutional Court required that the state must provide best possible prevention of dangers (significant risks) and precaution against potential (or low) risks. The causation of harm had to be excluded according to the standard of "practical reason", that is, in conformity with the most advanced state of science and technology. Potential risks that remained when all these measures had been taken ("residual risks") had to be borne by the population as a common burden of civilisation. As more clearly spelt out by later decisions of the Federal Administrative Court,[71] it is not necessary that there is a sufficient probability of future harm occurring as the consequence of nuclear activities. Rather, already a "potential of concern", that is, a plausible reason to believe that future harm might occur, triggers the duty to protect.

Moreover, Article 20a Federal Constitution obliges the state to protect the natural bases of life, also in consideration of future generations. This provision establishes a "state goal" to protect the environment that has a purely objective character and does not confer subjective rights on individuals. Its particular importance lies in the fact that it explicitly encompasses long-term safety that surpasses existing generations. It is an expression of the principle of sustainable development (in its environmental dimension). The fundamental rights, due to their individualistic concept, are in principle limited to protecting interests of living persons. It is doubtful whether, insofar as they establish an objective system of values, the duty to protect derived from them extends to future generations. By contrast, the state duty to protect established by Article 20a Federal Constitution clearly covers all aspects of long-term safety of radioactive waste repositories that go beyond existing generations.[72] Therefore, this constitutional provision constitutes the core elements of state responsibility regarding high level radioactive waste management. Parallel to the state duty to protect derived from Article 2(2) Federal Constitution, Article 20a Federal Constitution is to be interpreted to the extent that future harm to the environment, including human health affected by environmental factors, must be practically excluded and that a mere potential of concern triggers the state duty – and this at a long-term time scale.

However strong these formulations may appear, the constitutional state duty to protect has a fundamental weakness. The German courts normally

---

[70] BVerfGE 49, 89, 143 – Kalkar I; BVerfGE 53, 30, 38–39 – Mülheim-Kärlich; recently confirmed in BVerfG, Neue Zeitschrift für Verwaltungsrecht 2009:171.
[71] BVerwGE 72, 300, 316 – Wyhl; BVerwGE 104, 36, 44ff – Obrigheim.
[72] BVerwG, NVwZ 2007:833 para. 60 – Schacht Konrad I; see also BVerwGE 105, 6, 20 – Morsleben.

defer to the judgement of the legislature as to the need for protection and the measures to be taken and grant it a broad margin of political discretion. It is only when the state remains entirely inactive or its regulation is evidently insufficient for an effective protection that the courts consider themselves as entitled to disregard the political prerogative of the legislature.

4.5.2.2   *The statutory duty to protect*

For practical purposes the statutory duty to protect is more important. This duty is inherent in the planning permission requirements for high level radioactive waste repositories. Under Section 9(3) of the Atomic Energy Act, the Federal State is responsible for establishing and operating facilities for the final disposal of radioactive waste. The construction and operation of such facilities require a planning authorisation. The applicable Section 9b(4) of the Atomic Energy Act partly refers to the permit requirements for nuclear power plants under Section 7(2) of the Act, in particular its Nos. 3 and 5. In accordance with these latter provisions, the planning permission may only be granted where precautions against harm caused by the construction and operation of the facility, as are necessary according to the state of science and technology, have been taken and the necessary protection against disturbances and other interventions by third person is ensured. Moreover, Section 9b(4) Nos. 1 and 2 of the Act requires that no impairment of the public interest is to be expected from the facility and no other public law provisions, especially as regards the compatibility of the facility with the protection of the environment, rule out its authorisation. In contrast to the constitutional duty to protect these requirements are not at the disposition of the competent authority. Rather, they must be fulfilled before a planning permit is granted. On the other hand, their content is understood in parallel to the constitutional duty to protect. This means that the competent authority must ensure best possible prevention of significant risks and precaution against low risks and that harm associated with the activity must be practically excluded. Moreover, already a "potential of concern", that is, a scientifically plausible reason to believe that future harm might occur, triggers the duty to protect. This also applies to long-term risk.[73]

A further expression of the duty to protect is to be found in the principles of radiation protection under the Radiation Protection Regulation, namely justification of exposure, avoiding unnecessary exposure, minimisation of exposure and dose limitation (Sections 4–6). These principles also apply to the authorisation of repositories under Section 9b Atomic Energy Act as far as protection from exposures (in contrast to safety) during the operation phase is concerned. Risks that remain when all these measures

---

[73] See BVerwG, NVwZ 2007:833 para. 59 – Schacht Konrad I; BVerwG, Neue Zeitschrift für Verwaltungsrecht 2007:837, para. 19 – Schacht Konrad II; BVerwGE 105, 6, 20 – Morsleben.

have been taken ("residual" or tolerable risks) are in principle to be borne by the population as a common burden of civilisation.

If one goes more into detail, one discovers a number of further refinements inherent in the duty to protect. As regards precaution, there is a distinction between individual risk and collective risk (to which the minimisation principle belongs) that is important for judicial review.[74] The residual risk may be further reduced at the discretion of the executive. The limits between mandatory precaution against potential, including long-term, risk and discretionary reduction of residual risk are not clear-cut and subject to some divergences of opinion, even among the courts.[75] Finally, the ultimate responsibility for determining and evaluating the relevant (potential) risks and deciding on their tolerability is vested in the executive that is accorded a broad margin of appreciation (non-technically speaking: discretion).[76]

4.5.2.3 Responsibility of waste generators and privatisation of state responsibilities

Sections 21a and 21b Atomic Energy Act provide for financial responsibility of waste generators for all waste-related activities that are carried out by the state. This concerns the utilisation of the repository for which the waste generators have to pay fees or market prices depending on the organisation of the final disposal activities (Section 21a Atomic Energy Act). It also relates to the necessary costs for the planning, acquisition of land, facility-oriented research and development, and construction of the repository (Section 21b(1) Atomic Energy Act). Moreover, in accordance with the polluter-pays principle as understood as a principle that also establishes material responsibility, the waste generators are responsible for interim storage of high level radioactive waste on or near their premises (Section 9a(1b), (2), 3rd sentence Atomic Energy Act). For securing better public acceptance by avoiding transports of high level radioactive waste, the 2002 amendment of the Atomic Energy Act has given up the previous possibility of interim storage in central storage facilities operated by private enterprises and introduced this form of decentralised interim storage;[77] it is only under narrow conditions, especially in case of taking back residual waste from foreign reprocessing plants, that high level radioactive waste can still be delivered to

---

[74] BVerwGE 61, 256, 267; BVerwGE 72, 300, 318–319 – Wyhl; BVerwGE 104, 36, 46 – Obrigheim; BVerwG, Neue Zeitschrift für Verwaltungsrecht 2008:1012 para. 20; Sellner/Hennenhöfer 2007, part 12 No. 107.

[75] See with respect to external interventions BVerwGE 104, 36, 44ff – Obrigheim, on the one hand, and BVerwGE 131, 129 para. 30–33 – Standortzwischenlager Brunsbüttel, on the other.

[76] BVerwGE 72, 300, 316–317 – Wyhl; BVerwGE 81, 185, 190ff – Werkschutz; BVerwG. Neue Zeitschrift für Verwaltungsrecht 2007:83, para. 7; BVerwGE 131, 129 para. 19. The restriction of this discretion regarding fact-finding in BVerwGE 106, 115, 128 – Mülheim-Kärlich has been given up again.

[77] Critical: Huber 2002:246; see also Kloepfer 2004, § 15 No. 130.

central storage facilities (Section 9a(2), 2nd and 4th sentence Atomic Energy Act). The state retains a residual responsibility for ensuring compliance with all applicable requirements (so called "guarantee responsibility").[78]

The responsibility of the Federal State for establishing and operating a HLW repository is not absolute. Section 9b(3), 2nd sentence of the Atomic Energy Act provides that the state can charge third persons with performing its duties. Under Section 9b(3) 3rd sentence, for discharging its duties, the Federal Government can even entirely assign the task of waste management and confer the sovereign powers necessary for fulfilling it to third (also private) persons if these provide all assurances for an appropriate fulfilment of the task; the assignee is subject to supervision by the state. For such purposes, the German Company for Construction and Operation of Final Repositories for Wastes (Deutsche Gesellschaft für den Bau und Betrieb von Endlagern für Abfallstoffe – DBE) has been founded by the Federal state and state-owned enterprises; it is now an entirely private enterprise. This entity would be a natural candidate for constructing and operating the high level radioactive waste repository in the future.

The transfer of official waste management duties to private entities relies on the German administrative law model of assignment of official duties coupled with the conferral of sovereign powers to the assignee ("Beleihung").[79] The assignee in principle exercises a public duty as a matter of responsibility of his own, although in the field of nuclear waste disposal he is subject to normal supervision of nuclear activities by the state (Section 19(5) Atomic Energy Act) and, if organised as a private corporation owned by the state, may be subject to directions given by the shareholders. In the field of managing HLW the advantages of assignment lie in a greater expertise and efficiency through specialisation and synergy effects. Nevertheless, in view of the very long-term nature of the associated risks and hazards the model of assignment is controversial here.[80] Moreover, the misuse of the underground research facility in the former mine of Asse as a sort of disposal facility by the German Research Centre for Environment and Health (Deutsches Forschungszentrum für Umwelt und Gesundheit – GSF) in the public opinion casts some doubts on the trustworthiness of outside operators of radioactive waste disposal facilities. Although the GSF is a public research centre, the mere fact that it also worked for private clients associates it with "private" waste disposal and is liable to diminish public trust in the legal institution of assignment.

---

[78] See Rengeling 2008:1147–1149; Huber 2001:239.
[79] See Groß 2006, vol. I, § 13 Nos. 89–90; Stadler 2002:17–18.
[80] See, with further references, Rengeling 2008:1147–1150; Menzer 1997:50; Menzer 1998:823–827 who, however, are in favour of assignment.

4.5.3 Institutional framework

The role of the Federal Government in the field of radioactive waste management focuses on policy-making and generic regulation. The Atomic Energy Act contains various empowerments for the Government or the competent Minister for the Environment, Protection of Nature and Nuclear Safety to promulgate regulations for specifying the law (Sections 12 and 54), Moreover, the competent Minister exercises supervisory functions in the relationship to the states (Länder) that are charged with important executive functions.

At federal level, the Federal Agency for Radiation Protection (Bundesamt für Strahlenschutz – BfS), established in 1989, is vested with major executive tasks. The BfS is a superior federal agency. It constitutes a separate organisation but is subject to directions from the Ministry for the Environment (Sections 1(1) and 3 Act on the Establishment of the Federal Agency for Radiation Protection)[81] and therefore exposed to the exertion of political influence. Apart from supporting the Ministry in preparing legal and administrative provisions and conducting safety-relevant research (Section 2 of the Act of 1989) the BfS has various management and regulatory tasks (Section 23 Atomic Energy Act). The agency is charged with the construction and operation of the final repository for high level radioactive waste and the storage of nuclear fuel (Section 23(1) Nos. 2 and 1 Atomic Energy Act). Among the numerous regulatory functions of the agency are the competences for authorising the transport of nuclear fuel and substances that are major sources of radioactivity, authorising the storage of nuclear fuel outside state storage and deciding on an assignment of its management tasks (Section 23(1) Nos. 3, 4, 5 and 2 Atomic Energy Act) as well as administering various registries prescribed by the Act and conducting radiation-relevant investigations (Section 23(1) Nos. 6–8 Atomic Energy Act).

The state authorities are competent for the construction and operation of state interim storage facilities, all other authorisations, in particular planning permits for repositories and supervision of nuclear facilities and other activities except final disposal facilities operated by the BfS (Sections 9a(3), 1st sentence, 19 and 24(1), (2) Atomic Energy Act). That the BfS in its capacity as manager of the final repository is not subject to supervision by the State authorities but supervises itself, is not expressly spelt out in the Nuclear Energy Act. However, this can be derived from Sections 19(1), (5) and 23(1) No. 2 of the Act. The State authorities act on behalf of the Federal Government and have to follow directions given by the latter (Section 23(1) Atomic Energy Act).

---

[81] Section 3 of the Act that subjects the Agency to directions only concerns activities exercised in the area of competence of other ministries but reflects a general principle whereby a superior federal agency is not independent. See generally Britz 1998:116; Rupp 1990:387.

The concentration of management and supervisory functions regarding the operation of high level radioactive waste repositories in one and the same authority, namely the BfS, raises the question whether this is consistent with Article 20(2) of the Joint Convention whereby executive functions that relate to the management of radioactive waste and regulatory functions shall be separated. The Federal Government defends this institutional set-up with the argument that the supervisory function is vested in a separate department within BfS that is independent and not subject to directions regarding the exercise of this function (Bundesrepublik Deutschland 2008:95). Independence is merely based on an organisational order and not legally ensured. However, it would seem that the Convention (as well as the new Draft Directive on radioactive waste management) does not prescribe a particular model of separation of the relevant functions. Therefore the German organisational set-up still appears to be consistent with the mandate of the Convention to ensure an effective independence.

Moreover, the influence the Federal Government can exert on the state authorities raises problems. The agency model on which Section 23(1) Atomic Energy Act is based entails federal powers of giving individual directions to the state authorities; the constitutional limits to the exercise of these powers are negligible.[82] Administrative practice shows that the Federal Government from time to time uses its powers to give directions as regards important nuclear issues on which there are divergent opinions. Although day-to-day activities of the state authorities are unaffected, one can hardly speak of an assurance of effective independence of the state authorities from the Federal Government. In the framework of the binding Article 20(2) of the Joint Convention this is problematic because the Federal Government indirectly, namely through the BfS that is subject to its directions, also is responsible for radioactive waste management. If one considers promotional interests for nuclear energy to be covered by the term "management" there would also be a merger of regulatory and promotional, functions (see also Groth 2002:105–114). In any case, the German institutional set-up is not consistent with the farther-going requirements of the relevant IAEA recommendations (and arguably also the Draft Directive on nuclear waste management) on independence of regulatory functions from management and promotional functions.

For the future, in the light of the Joint Convention, the IAEA safety requirements and the draft EU Directive on the safety of nuclear waste management, reforms of the institutional framework for the regulation and management of high level radioactive waste are advisable. The objective of these reforms should be the creation of an effectively independent regulatory agency that is at least responsible for regulation, siting and supervi-

---

[82] See BVerfGE 81, 310, 332ff; BVerfGE 84, 24, 31ff; BVerfGE 102, 167, 172; Lange 1990; Steinberg 1990.

sion in the field of radioactive waste management. If one opts for this solution, the present commingling of regulatory and management functions within the Federal Agency for Radiation Protection would be abandoned in favour of a clear separation of functions. The management of the repository would have to be entrusted to either a new federal authority or under Section 9a(3), 3rd sentence Atomic Energy Act to a mandatory public law or private law corporation. The latter solution is preferable because even under the present system the Federal Agency for Radiation Protection is highly dependent on the technical expertise and competence of industry for operating the repository. For reasons of accountability and public trust, both the State and the public utilities that directly or indirectly (through subsidiaries or joint ventures) operate the 17 nuclear power plants in Germany could hold shares in this body.[83]

It is another question whether the powers of the Federal Agency for Radiation Protection should be extended to all matters of regulation which would mean that the execution of the Atomic Energy Act by the States as agents of the Federal Government would have to be abolished. While the agency model makes sense in a decentralised system such as it exists in relation to nuclear power plants as well as interim storage facilities, it is less appropriate for a single national radioactive waste disposal facility. However, it would be difficult to renounce to the administrative expertise of the States, in particular the State of Lower Saxony. Moreover, the controversial general discussion on the execution of the Atomic Energy Act by the States as agents of the Federal government has not rendered clear results in one or the other direction.[84]

At federal level, there are a number of independent commissions established by organisational orders that exercise advisory functions and issue guidelines with respect to radioactive waste disposal. These are in the field of nuclear safety including that of waste disposal facilities the Reactor Safety Commission (Reaktor-Sicherheitskommission – RSK) with respect to endangerment by ionising radiation the Radiation Protection Commission (Strahlenschutzkommission – SSK) and in the area of radioactive waste disposal the Waste Disposal Commission (Entsorgungskommission – ESK). These commissions have a pluralistic composition and are obliged to impartiality; in the more recent past, the pluralistic element in the form of full representation of stakeholders including objectors to nuclear energy has been strengthened. Moreover, there is a Nuclear Technology Committee (Kerntechnischer Ausschuss – KTA) associated to the German Federal

---

[83] Concerning the legal problems of a mandatory public law corporation of waste generators for final disposal see Kirchhof 2004:311; Selmer 2005:142–152; Waldhoff 2005:162–167; Hoppenbrock 2009:58–62.

[84] See Burgi 2005:247–253; Leidinger/Zimmer 2004:1005; from a practice-based perspective see Cloosters 2005:183–191; Reneberg 2005:173–181; Niehaus 2004:363–376; Rauscher 2004:377–392.

Ministry for the Environment, Nature Conservation and Nuclear Safety. Its task is to determine technical safety rules as to which the members agree that they can be deemed to be generally recognised. The Committee is composed of representatives of producers, operators, experts and authorities. The relevant organisational order does not contain any language about the degree of independence of this body. At state level, there is a States' Committee for Nuclear Energy (Länderausschuss für Atomkernenergie) that unites the competent State authorities but also includes a representative of the Federal Ministry for the Environment. Its mission is to harmonise the position of the states regarding the exercise of their administrative competences. The Committee may adopt guidelines that are also circulated by the Federal Ministry for the Environment.

4.5.4 Strategies

4.5.4.1 Basic strategic options

The basic strategic options of high level radioactive waste management in Germany are set out in the Atomic Energy Act. The Act prescribes the direct final disposal of radioactive waste, which is also true for low- and medium-level radioactive waste. Reprocessing is prohibited. Further, the Act prescribes interim storage of high level radioactive waste at or near the nuclear power plants where it is generated. Finally export of radioactive waste is in principle prohibited.

However, some elements of the basic strategic options are left to political decision. While the Atomic Energy Act prescribes the direct final disposal of high level radioactive waste (Section 9a(3) of the Act), the kind of final disposal – underground or above ground, the kind of host rock, retrievable or unretrievable – is not explicitly determined in the Act. At political level, Germany has for long favoured geological repositories, and some language in the provisions of the Act on planning permission for final disposal facilities, especially the reservation of a separate permit under the Federal Mining Act (Section 9b(5) No. 3), reflects this stance. As regards the issue of retrievability, which is relevant for spent nuclear fuel, but not for sealed liquid waste from reprocessing, the concept followed at political level until 1998 and also expressed in the planning permit for the disposal of low and medium level radioactive waste in the repository "Schacht Konrad"[85] was that radioactive wastes should be disposed of permanently without considering the possibility of a later use or removal in case of adoption of a new strategy. The new (not final) Safety Requirements of the Ministry for the Environment for the final disposal of high level radioactive wastes of September 2010 (BMU 2010a) require retrievability in the narrow sense that

---

[85] Planfeststellungsbeschluss "Schacht Konrad", 22 May 2002, point C II. 2.1.2.9-2, www.endlager-konrad.de.

the technical possibility of future "salvage" for 500 years after closure of the repository must be ensured, while long-term safety is given priority.[86]

According to the jurisprudence of the German courts, this gap of legalisation regarding the kind of final disposal of radioactive wastes is not inconsistent with the "essential questions" doctrine. It is true that this constitutional legal doctrine requires that essential questions of public policy, even if not mandated by reservations contained in fundamental rights, must be adopted by a parliamentary act. However, in the opinion of the German Federal Administrative Court, the statutory provisions set forth in Sections 9a (3), 1st sentence, and 9b (4) Atomic Energy Act regulate all "essential questions".[87] Further specification can be left to the responsibility of the executive. Although one may sustain that this position amounts to an abdication of parliament and does not sufficiently take account of a basic safety issue of society affecting numerous future generations (Ladeur 1989:243–244; Steinberg 1991:435), one will have to accept it as given for the time being.

As part of the reform of nuclear energy law in 2002, the delivery of spent nuclear fuel for reprocessing has been prohibited as from 1 July 2005 (Section 9a (1), 2nd sentence Atomic Energy Act). Before 2002, the old version of Section 9a (1) of the Act as adopted in 1976 had introduced a priority for reprocessing.[88] Originally, the salt dome in Gorleben in Lower Saxony was envisaged as the site of integrated waste management consisting of reprocessing, waste conditioning, interim storage, and final disposal. This concept was given up in response to wide-spread local opposition. Then in the 1980s, a separate reprocessing facility in Wackersdorf in Bavaria was developed that progressed to the extent that a first partial construction permit was granted in 1985. However, the German operators of nuclear power plants abandoned this project and turned to exporting spent nuclear fuel to France and the United Kingdom for reprocessing there. In legal literature, it has been discussed whether the prohibition of reprocessing is consistent with the freedom of interstate nuclear trade set forth in Article 93 Euratom Treaty.[89] However, since the member states have retained the power to determine their own nuclear energy policy (including entirely abandoning nuclear energy), they cannot be compelled by European law to open a waste management option that is contrary to their domestic nuclear energy pol-

---

[86] BMU 2010a; see as to previous drafts ESK 2009; Thomauske 2008:603.
[87] BVerwG, Neue Zeitschrift für Verwaltungsrecht 2007:837, para. 7 – Schacht Konrad II; in a related context already BVerfGE 49, 89, 129 – Kalkar I; BVerwGE 72, 300, 316 – Wyhl.
[88] See OVG Lüneburg, Energiewirtschaftliche Tagesfragen 1982:147; VGH München, Deutsches Verwaltungsblatt 1985:799, 801.
[89] Kloepfer 2004, § 15 No. 274; Wahl/Hermes 1995; Posser/Schmans/Müller-Dehn 2003, § 9a Nos. 204–206, all with further references. The Commission which in a letter of 12 December 2001 had taken the view that German law was in violation of Article 93 Euratom Treaty did not pursue the matter further.

icy. Therefore, the prohibition of export for reprocessing is based on cogent reasons in the meaning of the case law of the European Court of Justice regarding the market freedoms.

As regards transboundary movements of nuclear materials, Germany in principle adheres to the concept of domestic waste management. The export of spent nuclear fuels or radioactive wastes for final disposal or spent nuclear fuels for reprocessing is not formally prohibited but normally legally impossible. All residual radioactive substances (both spent nuclear fuel and radioactive waste) originating from nuclear power plants must be delivered to the operator of the final disposal facility in Germany (Section 9a(1), (2), 1st sentence Atomic Energy Act). The import of spent nuclear fuels from foreign nuclear power plants for reprocessing in Germany is not permitted since there are no reprocessing facilities in Germany and therefore a use in conformity with the law (Section 3(1) No. 2 Atomic Energy Act) cannot be ensured. The import of radioactive wastes from foreign nuclear power plants for final disposal in Germany is excluded by the very fact that the final disposal facilities that have to be constructed by the Federal State are geared to radioactive waste that is generated in German nuclear power plants. However, the rather complex provisions of Section 9a(1b), (1c) Atomic Energy Act allow the taking back of residual radioactive waste from foreign reprocessing of spent nuclear fuel that was generated in Germany. Further exceptions from the export and import prohibitions obtain with respect to radioactive substances that have not been generated in nuclear power plants (Section 3 Nuclear Energy Act, Section 19 Radiation Protection Regulation).

Dumping of radioactive waste at sea is prohibited by the acts for the ratification of the London and Helsinki Conventions.

4.5.4.2 *Statutory safety requirements*

The choice of technical strategies for ensuring the safety of final disposal of high level radioactive waste in Germany is not subject to a dense network of regulation but, rather, largely lies in the responsibility of the executive. As stated, the Federal Administrative Court[90] is of the opinion that the concept of final disposal of radioactive waste need not be fixed by statute as its essential elements are laid down in Sections 9a(1) and 9b(4) Atomic Energy Act. The planning permission requirements set out in Section 9b(4) Atomic Energy Act as well as, regarding ground water, the requirements of Sections 8, 12 and 48(2) Water Resources Management Act constitute the framework for developing the strategy of final disposal of high level radioactive waste.

Under Section 9b(4) of the Atomic Energy Act the planning permission may only be granted where precautions against harm caused by the construction and operation of the facility, as are necessary according to the

---

[90] BVerwG, NVwZ 2007:833 para. 7 – Schacht Konrad I.

state of science and technology, have been taken and the necessary protection against disturbances of operations and interventions by third persons or natural events is ensured. Moreover, an impairment of the public interest by the facility must be excluded and no other public law provisions, especially as regards the compatibility of the facility with the protection of the environment, may rule out its authorisation. The competent authority must ensure best possible prevention of significant risks and precaution against low risks; harm associated with the activity must be practically excluded. Moreover, already a "potential of concern", that is, a scientifically plausible reason to believe that future harm might occur, triggers the duty to protect. More or less the same is true for the protection of groundwater against radiological and chemical-toxicological risks that have to be considered separately (Section 19 Water Resources Management Act). In view of the serious hazards for groundwater associated with a HLW repository for high level radioactive waste its operation must be considered to be liable to cause a more than insignificant modification of water quality. Therefore it constitutes a constructive use of waters in the meaning of Section 9(2) No. 2 Water Resources Act.[91] According to Section 48(2) of the Act, substances may only be deposited in such a way that there are no grounds for concern that an adverse modification of ground water quality may occur. This provision also requires best possible prevention of significant risks and precaution against low risks.[92] It can be enforced through the permission procedure established by Sections 8 and 12 of the Act.

These requirements in the first place apply to short-term risks that in any case have the highest priority among neighbours and other potentially affected persons. However, as the Federal Administrative Court has held in one of its decisions regarding the final disposal facility for low and medium level radioactive waste in the former mine "Schacht Konrad",[93] these requirements in principle also encompass long-term risks presented by a repository after closure. Although not consistent with the wording of the Act, the "operation" of the facility that is subject to planning permission also includes the "passive operation" during the post-closure phase as long as the facility is designed to provide protection to human health and the environment. Risks caused in the post-closure phase are still causally related to the operational phase.[94] In the general law relating to waste disposal, this question has been explicitly regulated without changing the

---

[91] See BVerwG, Zeitschrift für Wasserrecht 1983:222, 223; BGH, Umwelt- und Planungsrecht 1982:342, 342; OVG Münster, Zeitschrift für Wasserrecht 1996: 72, 473; Czychowski/Reinhardt 2010, § 9 Nos. 86, 87.

[92] Cf. BVerwG, Neue Zeitschrift für Verwaltungsrecht 2007:837, para. 19 – Schacht Konrad II.

[93] BVerwG, NVwZ 2007:833 para. 60 – Schacht Konrad I; to the same extent BVerwGE 105, 6, 20 – Morsleben.

[94] Sellner/Hennenhöfer 2007: part 12 No. 315; Rengeling 1995:90–91; Näser/Oberpottkamp 1995:138–140.

terms "construction" and "operation" (see Sections 31(1), 32(1) No. 2, (3) Life Cycle Economy and Waste Act). In this respect, intrusion after closure of the facility and radioactive pollution of migrating ground water present the most important potential risks. The same problem exists with respect to the scope of application of the Radiation Protection Regulation. Section 2(1) No. 1 lit. c of the Regulation only refers to the construction and operation of repositories, not to the post-closure phase. In parallel to long-term safety in general, the Regulation should at least be applied as a guideline, keeping in mind the difficulties of proof that arise with a growing time frame.[95] The safety criteria for repositories include the standards for radiological protection applicable in the post-closure phase, taking the view that the Radiation Protection Regulation is not applicable (BMU 2010a:11).

Risks that remain when all the requisite measures have been taken ("residual risks") are considered to be tolerable and must in principle be borne by the population as a common burden of civilisation. However, the administrative courts recognise that the residual risk may be further reduced at the discretion of the executive. The limits between mandatory precaution against low, including long-term, risk and discretionary reduction of residual risk are not clear-cut and subject to some divergences of opinion, even among the administrative courts.[96] This controversy is of less practical relevance in the field of long-term safety than of short-term one because in the former case the relevant decisions are subject to judicial review only to a very limited extent. Neighbours or other third parties such as conscientious objectors to nuclear energy would not have standing to challenge the position of the administration that best possible precautions against future harm by the repository at long term have been taken.[97] Nevertheless it makes a difference whether the design of the long-term safety strategy for final disposal of high level radioactive waste has more or less elements that only are discretionary. Moreover, the situation might be fundamentally changed in the future if the European Court of Justice decides that the German transposition of the Public Participation Directive (Directive 2003/35) by the Environmental Remedies Act is not consistent with the mandate of the Directive to introduce a fully-fledged association suit in environmental matters.[98] Then environmental NGOs would be able to challenge the legality of decisions on long-term safety of the reposi-

---

[95] In favour of an analogy: Näser/Oberpottkamp 1995:140; contra Wollenteid 2000:333–344.
[96] See BVerwGE 104, 36, 44ff – Obrigheim on the one hand, and BVerwGE 131, 129 para. 25–33 – Standortzwischenlager Brunsbüttel on the other.
[97] BVerwG, NVwZ 2007: 833 para. 60 – Schacht Konrad I.
[98] Pending case C-115/09 – Bund Umwelt und Naturschutz Deutschland (BUND)/ Bezirksregierung Arnsberg; see the reference to the European Court of Justice for a preliminary ruling by OVG Münster, Neue Zeitschrift für Verwaltungsrecht 2009:987.

tory even if the requirement of long-term safety does not establish rights of individuals.

4.5.4.3 Administrative rules

With respect to the operation of nuclear power plants there are a number of administrative rules (guidelines) adopted by the Reactor Safety Commission that reflect the low degree of legalisation in the field. In particular the Safety Guidelines for Nuclear Power Plants of 1977, the Guidelines for Pressurised Water Reactors of 181/1996 and the Disturbance Guidelines of 1983 (see Roller 2004:119–120) constitute the real substance of the German nuclear policy regarding nuclear facilities. This regulatory package is about to be replaced by new safety criteria (see Schneider 2009:554; BMU 2009:607). A draft set of safety criteria that consists of 12 modules has been developed since 2004. Based on an agreement between the Ministry for the Environment and the relevant State authorities, the draft criteria will be subject to comprehensive consultation and practical testing between July 2009 and October 2010 before the new criteria will be adopted.

These administrative rules are not applicable to the final disposal of high level radioactive waste. However, some of their basic elements are.[99] Thus it is clear that during the normal operation phase of the repository the dose limit values must be complied with and the remaining exposure of people and the environment to ionising radiation must be minimised (Sections 6, 46, 47 Radiation Protection Regulation). With respect to disturbances and accidents, the guidelines distinguish between 4 safety levels: retention and reduction of ionising radiation, control of exposure by the design of the facility, protective measures against rare events, such as accidents, sabotage, terrorist attacks and other external events, and limitation of the effects of rare events to the facility (facility-internal emergency protection).[100] It is not quite clear if these latter two levels have to be attributed to mandatory precaution or to discretionary reduction of residual risk. There is a fairly recent judicial tendency to expand the scope of mandatory precaution.[101]

In September 2010, the Ministry for the Environment published "Safety requirements for the final disposal of high level radioactive waste" (BMU 2010a). The document is not yet final although the requisite concerting with the State authorities has been carried out. It shall be developed further in dialogue with the State authorities.

---

[99] BVerwG, NVwZ 2007:833 para. 54 – Schacht Konrad I; BVerwG, Neue Zeitschrift für Verwaltungsrecht 2007:837, para. 7 – Schacht Konrad II.

[100] This requirement was contained in a previous version of the Nuclear Energy Act (Section 7(2), second sentence, inserted in 1994) as a measure of "further precaution" not subject to judicial review at the request of neighbours. Its removal in 2002 does not have major legal consequences.

[101] To this extent with respect to external interventions such as terrorism and sabotage see BVerwGE 131, 129, para. 14–33 – Standortzwischenlager Brunsbüttel, in contrast to the earlier decision BVerwGE 104, 36, 44ff – Obrigheim.

Largely following international consensus, the draft establishes a systematic hierarchy between paramount safety requirements, safety principles, the safety concept, safety criteria (protective criteria), proof of safety, requirements for the design of the repository and the Safety requirements distinguish between probable scenarios, less probable developments and improbable ones. In the former case an extremely stringent effective dose of 0.01 mSv per year, in the second case of 0.1 mSv per year is set. These standards that apply to human exposure above ground are considered to be met if discharges from the repository at the margin of the host rock, estimated on the basis of an exposure model comprising 10 persons, cannot result in an effective dose of more than 0.1 mSv or 1 mSv per year respectively. As regards improbable developments, only the use of best available technology or other kind of optimisation is required. The time frame is set at 1 million years although it is admitted that beyond 10,000 a robust prognosis is not possible (BMU 2010a:11–14). The quantitative, probabilistic approach regarding the definition of the scenarios and tolerable risk that had been used in the previous drafts was given up. It had the weakness that it required sufficient factual information based on past experience. Given the existing uncertainties and even ignorance about the development of governance structures, political and social stability and technological capacities over a period of potentially several hundred thousand years such factual basis does not exist (see ESK 2009:18–19; Thomauske 2008:603).

4.5.4.4 Programmes

When formally adopted, the safety requirements may have the legal character of an administrative rule or be just a simple recommendation (see Näser/ Oberpottkamp 1995:140). However, the question is whether they should not be integrated in a more comprehensive plan or programme for radioactive waste management. The draft Directive on the safety of nuclear waste management (cf. section B 4.3.1) proposes to introduce a member state obligation to establish waste management programmes. Similar requirements exist in Switzerland. Moreover, with respect to non-radioactive wastes Section 29 Life Cycle Economy and Waste Disposal Act mandates the determination of a waste management plan that sets objectives of waste disposal and recycling as well as suitable disposal sites. Although the draft Directive does not require a comprehensive reference document, one could consider laying down all elements of the radioactive waste disposal strategy not regulated in the Atomic Energy Act, including the safety requirements, as well as interim targets, time-tables to reach the targets, the procedure for siting and ultimately the sites so determined in a radioactive waste disposal programme or plan. The advantage of such a planning instrument especially for high level radioactive is that it provides an incentive for a systematic, coordinated approach for solving the problems presented by final disposal of such. Moreover, it would increase transparency of governmental policy, facilitate public participation and contribute to public acceptance.

The planning instrument could either be a purely political programme or a sectoral plan to be set forth by an amendment of the Nuclear Energy Act and patterned on the waste management plan provided for by Section 29 Life Cycle Economy and Waste Act. The latter model of a "legalised" plan has the advantage that it may accord site selection legal effects in the subsequent authorisation procedure for the repository as well as with respect to financial responsibility for the site selection costs. Since sectoral environmental planning, although space-related, is an aspect of environmental regulation, it can be based on the environmental competences set out in the Federal Constitution.[102] The political practice clearly supports this stance. The exclusive federal legislative competence for nuclear energy also encompasses the disposal of radioactive wastes (Article 73(1) No. 14 Federal Constitution). Therefore, the Federal State has the competence to amend the Atomic Energy Act to the extent that a sectoral planning instrument would be introduced.

However, since a formal sectoral plan is more important as regards the siting of repositories that precedes the planning permission for construction and operation, one could also consider introducing two reference documents, a political programme relating to the strategies and a formal plan that covers the siting process (see section B 4.5.5.8).

4.5.4.5 *Important legal controversies*

Given the uncertainties presented by the extremely long-term nature of final disposal of HLW, it seems clear that the judicial formula whereby the causation of harm must be practically excluded only provides limited guidance even if one postulates that all relevant data must be sufficiently ascertained and that uncertainties of prognoses may not serve as an indication that there is a lower risk The question is how to determine the time frame of precaution and the reliability of protective measures to be taken. In legal literature several formulas have been developed to cope with this problem. Rengeling[103] takes the view that the prognosis must be reasonable and the time frame determined according to the standard of practical reason. This seems to be close to the general standards established by the courts. Barth and Schulze (Barth/Schulze 2008:42) suggest that a conscious causation of harm to future generations or the environment must be excluded. According to these authors the time frame must be such that sufficiently precise assumptions on the safety and security of the repository are possible. The planning permit for the repository for low

---

[102] See Kunig 2003, vol. 3, 5th. ed., Art. 75 No. 34; Schmidt-Bleibtreu, Hofmann & Hopfauf 2011, Art. 74 No. 363; Ramsauer 2008:945; Bothe 2001, Art. 30 No. 34.
[103] Rengeling 1995:65–69; to the same extent Näser/Oberpottkamp 1995:140–141.

and medium level radioactive wastes in "Schacht Konrad"[104] follows the same line of reasoning.

However, there is a certain flaw in these positions. While it appears cogent and in line with the judicial formula of practical reason that reasonability should govern the safety and security analysis and the measures to be taken, the authors do not offer a sufficient answer to the problem that with increasing time over the several thousands of centuries the repository is supposed to function it becomes more and more difficult and even impossible to exclude that harm to human health and the environment can be caused (see sections B 1.3.5, B 1.4, B 2.4 above). Therefore a certain modification of the judicial formula appears apposite. This is justified because the standard of practical exclusion of harm has been developed as an emanation of fundamental rights and apart from that only constitutes a specification of the basic requirement of best possible prevention of risks and precaution against potential risks. Its transfer to the state goal to protect the environment for future generations is not a matter of course. It is amenable to an adjustment to take account of uncertainties over very long periods of time. The same is true for the interpretation of Section 9b (4) in conjunction with Section 7(2) Nos. 3 and 5 Atomic Energy Act. In this sense, the Federal Administrative Court in the first decision on "Schacht Konrad"[105] has pointed to the "imponderability" of ensuring long-term safety and, rather than referring to the formula of practical exclusion of future harm, has only required best possible long-term prevention of significant risk and precaution against low risk. Moreover, while formally inconsistent with the mandate of Article 11(vi) Joint Convention whereby future generations should not have to bear risks that are considered to be intolerable for the existing generations, such an adjustment simply reflects the impossibility to see into the future over several thousands of centuries. Therefore one can state that a practical exclusion of harm must be ensured for a period of time as long as reasonably feasible considering the present possibilities of prognosis, science and technology. For the remaining time, a reduced standard of care appears appropriate whereby the risk of future harm does not need to be practically excluded but must be reduced to a reasonable extent. This constitutes best possible prevention of significant risk and precaution against low risk.

In the framework of determining the final disposal strategy, a further question arises, that is, whether the choice of the radioactive waste disposal strategy is subject to an optimisation requirement. In this context, the minimisation principle of nuclear law has to be considered. Section 6(2) Radiation Protection Regulation requires that, independent

---

[104] Planfeststellungsbeschluss "Schacht Konrad", 22 May 2002, point C II. 2.1.2.9-2, www.endlager-konrad.de.
[105] BVerwG, NVwZ 2007:833 para. 60 – Schacht Konrad I.

from compliance with dose limitations, the exposure or contamination of man and the environment originating from a nuclear activity including a repository for radioactive waste must be kept as low as possible, considering the state of science and technology. This minimisation requirement is also applicable to operational disturbances, as can be concluded from Section 49 (1), (2) Radiation Protection Regulation. Rare events such as accidents (which do not play a particular role with respect to repositories for radioactive waste) and external interventions are not comprised. In this respect, the Regulation confines itself to mitigation of the consequences of such an event (Sections 51–53). However, in these latter cases the statutory standard of best available prevention of harm and significant risk and precaution against low risk (Section 9b (4) in conjunction with Section 7 (2) Nos. 3 and 5 Atomic Energy Act) must be directly applied. It entails a minimisation requirement, although in this respect not subject to judicial review at the request of neighbours. If one rejects the application of the Radiation Protection Regulation to the post-closure phase of a repository, at least this statutory requirement also encompasses this phase.

It follows from these considerations that the development of the final disposal strategy is also subject to the principle of minimisation. Therefore such final disposal strategy must be selected that provides, in combination of all of its elements, the best protection. However, it is to be recalled that according to the established jurisprudence of the German administrative courts, risk assessment and the decision on tolerability of risk is the primary responsibility of the executive. The administrative courts grant the executive a broad margin of discretion in making such choices. It is not the administrative courts but the executive that bears the ultimate responsibility for the safety and security of the repository for high level radioactive waste and therefore also decides on the relevant disposal strategy. This qualifies the relevance of the principle of minimisation.

4.5.5 Site selection

4.5.5.1 Statutory requirements

The selection process for the site of a repository for radioactive wastes in Germany is characterised by a low degree of legalisation. At best, one can name the anticipated effects of the planning permission procedure for the construction and operation of the final disposal facility prescribed in Section 9b (1) Atomic Energy Act. In the normal planning permission procedure ("plan determination procedure"), unless there are mandatory requirements, the competent authority possesses planning discretion and its decision results from a process of weighing all concerns affected by the decision, including the consideration of alternatives. However, the Federal Administrative Court in 2007 in its judgements on the repository "Schacht

Konrad"[106] has held that the planning permission under nuclear energy law does not entail any discretionary element but is a "legally bound" decision. These holdings are controversial, but, of course, will govern political and administrative practice. Therefore, as regards the site of the repository, in the planning permission procedure for the construction and operation of the repository, only the suitability of the site, that is, the question whether the permit requirements (Section 9b(2) of the Act) are complied with at a particular site, has to be determined. In particular, under normal operational conditions, in case of disturbance of operations and external events harm to human health and the environment must be practically excluded. The additional permit requirement of compatibility with environmental protection and the obligation to carry out an environmental impact assessment (Section 9b(2), (4) No. 2 Atomic Energy Act) do not either result in the need to consider alternatives. Section 6(3) No. 5 of the Environmental Impact Assessment Act requires the inclusion of alternatives in the assessment only if and insofar as alternatives have been considered. Finally, the parallel application of water law (Section 19 Water Resources Management Act) and mining law (Section 9b(5) No. 3 Atomic Energy Act) does not either lead to the consideration of alternatives. The relevant permission requirements of the Water Resources Management Act (Sections 8, 12, 48(2) of the Act) do not provide for the mandatory consideration of alternatives, although the competent authority has discretion in this respect and could do so. As regards mining law, a repository is both a waste disposal facility and a mining facility that poses specific mining safety problems. However, the relevant provisions of mining law relating to the authorisation of the operation plan (Sections 55(1), 48(2), 52(2a), 126(3) Federal Mining Act) do not either require the competent authority to consider alternative sites.

The site selection procedure that precedes the planning permission procedure for the construction and operation of the repository for the time being is essentially unregulated and governed by political discretion.

4.5.5.2 *Administrative practice*

The "siting history" regarding the selection of a site for a repository for high level radioactive waste in Germany is well documented[107] but nevertheless the subject of controversial descriptions and evaluations. Between 1974 and 1976 the Federal Government started a process for selecting a site that originally also included a reprocessing, a waste conditioning and an interim storage facility ("integrated waste management centre"). Applying geological, environmental and planning criteria, three salt domes in Lower Saxony

[106] BVerwG, NVwZ 2007:833 para. 27 – Schacht Konrad I; BVerwG, Neue Zeitschrift für Verwaltungsrecht 2007:837, para. 13–14 – Schacht Konrad II.
[107] See Tiggemann 2010:607–615; Tiggemann 2004:394; Bluth 2008; Bluth/Schütte 2008; Möller 2009:279–295, 309–314; Grill 2005.

and one site in Schleswig-Holstein came into the final selection. The latter was then discarded because of proximity to the border with East Germany. However, in response to protests against, and disturbances of, (proposed or already started) investigations by neighbours and nuclear activists, the Federal Government in 1976 suspended all investigations. Independent from the federal process the State Government of Lower Saxony in 1976 conducted its own selection process. At federal level alternative host rock formations and sites in the whole Federal Republic were screened, while Lower Saxony limited the investigation to the State territory and, as a result of the federal investigations that had favoured three salt rock sites in Lower Saxony, concentrated on rock salt. The selection process consisted in the staged use of geological, safety-related, environmental and spatial planning criteria that were not fixed in advance but developed *ad hoc*. In application of these criteria, the original number of 140 sites in Lower Saxony was reduced to 23, then 14 and finally four candidate sites, that is the three sites that had emerged from the federal selection process and in addition the salt-dome Gorleben. A comparison of these sites rendered a preference for Gorleben because all other sites fulfilled at least one exclusion criterion. It is to be noted that Gorleben was very early brought into the selection process by the same expert body that had conducted the federal selection process. In February 1977 the State Government determined Gorleben as preliminary site for the repository. The Federal Government that had originally objected to Gorleben for military reasons and favoured another site ultimately agreed to the selection of Gorleben.

The exact circumstances that led to the concentration on Gorleben in 1977 are controversial. Both safety and regional development opportunities were the most important considerations. Also political considerations such as the expectation of local acceptance or at least an easier handling of protests seem to have played a certain role in favour of Gorleben. Some commentators claim legitimacy for the process because the selection criteria applied do not essentially diverge from the criteria recently proposed by the Ministry for the Environment.[108] Others denote the process as arbitrary and illegitimate; from a safety point of view they also assert that the size of the covering rock is insufficient (Geulen 2008:385–387; Appel 2008). It is clear that the selection process was a technocratic process that did not provide for openness and broad public participation although the affected municipalities could participate and after the decision had been taken an attempt was made to explain its reasons, especially in the Gorleben hearing of 1979 (that led to giving up the idea of an integrated waste management centre and concentration on final disposal at Gorleben; Grill 2005). However, this corresponded to the *Zeitgeist* and was not inconsistent which applicable law.

---

[108] Bluth 2008; Bluth/Schütte 2008; for a positive assessment see also Tiggemann 2010:612–613.

It is only fairly recent that there have been proposals to establish a formal procedure that aims at selecting an optimal site in an open, staged process that is oriented at seeking public involvement and securing consensus or at least public acceptance.

In 1979 and 1980 at Gorleben investigations from above ground including drillings were conducted. Pursuant to three authorisations granted under mining law, since 1983 preparations for underground investigations took place. The authorisation for underground investigations was based on an assessment by the Federal Physical-Technical Agency (PTB). Critics allege that Gorleben did not meet the agency's own criteria and that the positive assessment was due to the exertion of political influence. In order to clarify the relevant circumstances, the German Parliament set up an enquiry commission.[109] Following the construction of the pits that started in 1986, since 1995 the underground investigations were carried on. These works met with increasing resistance by neighbours and objectors to nuclear energy, especially in the region but also throughout Germany. After the electoral victory of the Social Democratic Party and the Green Party in 1998 the phasing out of nuclear energy was set on the political agenda. In the framework of the "Atomic Consensus" between the Federal Government and the generators of nuclear energy of 14 June 2000[110] a moratorium for underground investigation at Gorleben of at least three, at most ten years was agreed upon. Since then the Ministry for the Environment pursued the concept of establishing a new, open process for selecting an optimal site for the repository for high level radioactive waste. The underlying idea apparently was to create the prerequisites for making an abandonment of Gorleben possible.

However Gorleben already serves, besides Ahaus and Morsleben, as a central interim storage facility for HLW from reprocessing. As regards low and medium level radioactive waste, a former iron ore mine, "Schacht Konrad" at Salzgitter in Lower Saxony, was selected. The planning permission for this repository was granted in 2002 and has become legally binding pursuant to three judgements of the Federal Administrative Court rendered in 2007.[111]

In 1999, the Ministry for the Environment established a" Committee on a Selection Procedure for Repository Sites" (Arbeitskreis Auswahlverfahren Endlagerstandorte – AkEnd) that in late 2002 submitted a proposal for a new site selection procedure that was divided into five procedural stages.[112]

---

[109] "Ausschuss zu Gorleben", Frankfurter Allgemeine Zeitung of 26 February 2010:4.

[110] "Atomic Consensus" of 14 June 2000, Chapter IV, No. 4.

[111] BVerwG, NVwZ 2007:833 – Schacht Konrad I; BVerwG, Neue Zeitschrift für Verwaltungsrecht 2007:837 – Schacht Konrad II; BVerwG, Neue Zeitschrift für Verwaltungsrecht 2008:841 – Schacht Konrad III.

[112] AkEnd 2002; for a description of the model see also Nies 2005:93–98; Brenner 2005:100–106.

The Working Group proposed that on the basis of a "white map" of theoretically suitable host rocks, regions and sites all over Germany five sites should be pre-selected and investigated from above ground and out of these two sites from underground. Gorleben would be part of the process. At the end of the process an optimal site should be selected in accordance with safety, environmental and socio-economic criteria (exclusion and balancing criteria), without any advance prejudicial decisions. "Optimal" is meant as a relative term; optimal is a site that is deemed to meet best the criteria developed, in particular provides best possible long-term safety. The public and the affected regions and municipalities should be involved at all stages of the process both with respect to the safety/environmental and regional development aspects. To that end, various institutional arrangements and referenda were proposed. The successful termination of the procedure would depend on the willingness of the region and municipality to host the repository. In order to secure this willingness to participate, major emphasis is laid on promotion of regional development. In case of a possible failure of the process the Government and Parliament should decide on the further procedure.

This concept was followed in a modified form by two draft proposals of the Ministry for the Environment for site selection, the latter of which was circulated in 2006 (BMU 2006). The draft of 2006 called for the selection of sites that, in accordance with safety criteria determined in advance, provides a higher safety potential than Gorleben. The concept was modified in the discussion paper "Site Selection Procedures" circulated in February 2009 (BMU 2009a). The discussion paper proposed four procedural steps, that is, determination of selection criteria, carrying out of simplified and eventually more exhaustive safety analyses for identifying and assessing potential alternatives for Gorleben, eventually underground investigations, and determination of the site that provides the highest safety level by parliamentary act. A legalisation of the selection process as such – apart from the final decision to be taken by a parliamentary act – was not envisaged. Of course, on the road to a legally binding planning permit for the construction and operation of the repository the legally prescribed planning permission procedure must be followed so that the entire procedure would entail more than four steps.

After the Federal elections in late September 2009 that brought a conservative-liberal coalition into power, a new course toward nuclear energy has been taken. The Coalition Agreement of the ruling parties of November 2009[113] provided that the Gorleben moratorium will be terminated and the underground investigations at this site be resumed. At the end of September 2010 the competent state authority of Lower Saxony granted the BfS a prolongation of the existing framework operation

---

[113] Koalitionsvertrag, November 2009:29.

permission under mining law and later on ordered the immediate execution of this decision. The underground investigations at Gorleben were then resumed. The primary aim of the investigation is to determine the suitability of the salt rock and establish a basis for site-specific safety assessments.[114]

The choice of the procedure aroused some controversies, even within the government. According to the jurisprudence of the Federal Administrative Court and the prevailing opinion,[115] the continuation of underground investigations did not require a planning permission under atomic energy law even if parts of the pits can be used later for the repository. These investigations are a preliminary step aimed at identifying a suitable site and do not constitute the first stage of a tiered planning permission procedure. Moreover, a prolongation of the existing framework operation permission under mining law is in principle considered not to be subject to an environmental impact assessment and the accompanying public participation since the project was commenced before the enactment of the EIA Act.[116] However, it is not entirely beyond reasonable doubt that the long interruption of works at Gorleben and changes of the relevant factual conditions such as a decrease of the volumes to be deposited should not have an influence on this question. If an entirely new permission were needed, the amended Federal Mining Act would have required an environmental impact assessment. In any case the Ministry for the Environment is determined to ensure a comprehensive and open public participation accompanying the whole underground investigation process including an international peer review while alternative sites shall not be considered for the time being.[117]

4.5.5.3 Important legal controversies

There is an extensive legal debate on the problems raised by site selection for a high level radioactive waste repository in which both positions regarding existing law and proposals of a legal-political nature have been sustained. This debate relates to a variety of issues, in particular the question of whether the site selection process must or should aim at selection an optimal site, considering all reasonable alternatives, or a suitability analysis is sufficient, participation and the design of the procedure, the issue of voluntarism on the side of the host municipality, the role of parliament and the necessary or desirable degree of legalisation of the procedure.

---

[114] BMU/BfS 2010; see also BMU 2010b.
[115] BVerwGE 85, 54, 61 – Gorleben; BVerwGE 100, 1, 10 – Gorleben; Näser 2009:18; Haedrich 1986, § 9b Nos. 21–31; contra: Breuer 1984:36ff, 53–54.
[116] BVerwGE 100, 1, 6–9 – Gorleben.
[117] BMU 2010a; "Gorleben wird weiter erkundet", BMU 2010b:251.

4.5.5.4 The planning permission procedure: suitability of the site vs. consideration of alternatives

The perhaps most controversial issue concerns the question of optimality of site selection or – less pretentious – the requirement of considering alternative sites. Since there is a close relationship between site selection and the planning permission procedure for the construction and operation of the repository, the first question is whether the position taken by the Federal Administrative Court in the judgements on the "Schacht Konrad" repository[118] to the extent that in this procedure only the suitability of the site is to be scrutinised really is in conformity with the Atomic Energy Act. This is denied by a number of commentators with quite plausible arguments.[119]

(1) The wording of Section 9b(4) of the Act militates against the Court. This provision says that the authorisation may only be granted if certain prerequisites set out in Section 7(2) of the Act are fulfilled and must be denied if the further prerequisites listed in Section 9b(4) Nos. 1 and 2 are not fulfilled. It does not say that the planning permit must be granted under the conditions set out in the Act. This kind of statutory language in a procedural law that is denoted as "planning permission" has always been considered as to confer on the competent authority planning discretion insofar as the relevant mandatory requirements do not limit this discretion. An explicit conferral of planning discretion is not essential. A pertinent example are the closely related Sections 31, 32 of the Life Cycle and Waste Act with respect to waste deposits that contain a similar wording and have always been understood by the Federal Administrative Court[120] as to entail a combination of mandatory requirements and planning discretion. The same is true of Section 68 Water Resources Management Act 2010 that concerns modifications of watercourses.[121]

(2) The precedent of the Federal Mining Act[122] to which the Federal Administrative Court refers for support of its construction of the law is not pertinent because the wording of the relevant provisions is different.

(3) Planning discretion may be deemed appropriate in view of the purpose of the Act. The siting of a repository for radioactive waste, due to its underground extension, its impacts on the neighbourhood and its extremely long-term nature, always implies an important decision on the utilisation of space, which is the very *raison d'être* of the planning permission pro-

---

[118] BVerwG, NVwZ 2007:833 – Schacht Konrad I; BVerwG, Neue Zeitschrift für Verwaltungsrecht 2007:837 – Schacht Konrad II; BVerwG, Neue Zeitschrift für Verwaltungsrecht 2008:841 – Schacht Konrad III; to the same extent: Gaentzsch 2005:117–122; de Witt 2005:127–129; Kühne 2008:365–367.

[119] Ramsauer 2008:946ff; Roßnagel/Hentschel 2004:295; Rengeling 1984:43ff; Rengeling 1995:46ff; Haedrich 1986, § 9b Nos. 1, 3; see also Hennenhöfer 2010:365–366.

[120] BVerwGE 97, 143, 147; furthermore already BVerwGE 81, 128, 132–33; 90, 42, 37–48; 90, 96, 99–100; see Paetow 2003, § 32 Nos. 5–11.

[121] BVerwGE 55, 227.

[122] BVerwGE 127, 259; BVerwGE 127, 272; BVerwGE 126, 205.

cedure. Limiting the procedure to a mere decision on the suitability of the site would deprive the procedure of an essential space-related element. (4) The legislative history of the provision militates in favour of planning discretion.[123] (5) The integration of the planning permission into a staged substantive planning system, which is missing in case of Section 9b Atomic Energy Act, is not an indispensable element of a planning permission procedure that entails planning discretion.

(6) It is true that the nuclear planning permission in its present version by itself does not justify expropriation which is commonly deemed to be an element of planning permission proper. However, the planning permission includes the mining permission (Section 9b (5) No. 3 Atomic Energy Act, Sections 57b (3), 2nd sentence, 126(3) Federal Mining Act) and to this extent does allow expropriation.[124] Moreover, the 12th Amendment of the Nuclear Energy Act (new Sections 9d and 9e), adopted by the Bundestag on 28 October 2010 but not yet law, has re-introduced an empowerment for expropriation as well as an obligation of landowners to tolerate preparatory works, so that this argument will no longer be valid in the future.

(7) The argument that the executive is vested with the ultimate siting responsibility (Gaentzsch 2005; de Witt 2005) does not militate against the recognition of planning discretion – to the contrary because it would increase the rationality of the decision-making process without impairing this responsibility.

However, even if one recognises the existence of planning discretion the competent authority could not search by itself for alternatives and under no means in the whole Federal Republic. The competent authority does not possess an independent planning competence in the sense that it could explore itself alternatives and impose another site on the applicant;[125] rather, it could only scrutinise the proposed site in the light of alternatives that suggest themselves under the concrete circumstances and eventually deny the application because other sites are evidently preferable. In this context it is to be noted that due to the site-dependence of the safety and environmental impacts of the repository and the enormous costs of investigations the requisite depth of considering alternatives is necessarily limited. As recognised by the courts with respect to other planning permission procedures,[126] a screening process based on available information is sufficient; there is no requirement of constructing several underground characterisation facilities in order to explore

---

[123] See Bundestags-Drucksache 7/4794:9 (consideration of all concerns affected).

[124] This is ignored by Gaentzsch 2005:119.

[125] Roßnagel/Hentschel 2004:291–292, 294; Gaentzsch 119–20; to the same extent with respect to disposal of non-radioactive waste: Paetow 2003, § 32 Nos. 5–11, 60–63.

[126] BVerwGE 131, 274, para. 135; BVerwGE 107, 142, 149; BVerwGE 100, 238, 250; with respect to radioactive waste repositories: OVG Lüneburg, Deutsches Verwaltungsblatt 2006:1044, 1050 – Schacht Konrad.

alternatives. It is not far-fetched to assume that the Gorleben selection process already meets these requirements. Moreover, under no means could the selection process extend beyond the territory of the relevant state. This is excluded by the limited territorial competence of the highest state authorities that are competent for making the decision on the project (Section 24(2) Atomic Energy Act).[127] Therefore, one of the major objectives of a site selection procedure that is oriented at optimisation and distributional justice between the states of the Federal Republic would be missed.

All told, even if the recognition of planning discretion in the framework of Section 9b (4) of the Atomic Energy Act may appear reasonable, its repercussions on a site selection procedure that precedes the planning permission procedure would be slight. Although the absence of considering alternatives in the site selection procedure may make the application for the planning permit deficient to the extent that the application must be denied, this is bound to be a rare case. The incentives for the operator of the future repository to devise a nation-wide selection procedure that entails optimisation and the consideration of alternatives would be negligible. The solution to the problem must be sought elsewhere.

4.5.5.5 General spatial and special environmental planning as instruments of considering alternative sites

Section 9b(4) No. 2 Atomic Energy Act provides that planning permission for a repository must be denied when other public law provisions, in particular regarding the compatibility with the environment, militate against the construction and operation of the facility. Such provisions can be contained in legally binding spatial plans that have been determined in the framework of general spatial planning. The consideration of alternatives is a constituent element of spatial planning, the more so since the determination of such plans requires a strategic environmental assessment that engenders alternatives (Section 8(1) Spatial Planning Act in conjunction with Section 14g, Annex 3 No. 1.5 Environmental Impact Assessment Act). Therefore the question is whether optimisation requirements regarding the repository for high level radioactive waste do not ensue from spatial planning law.

In Germany, general spatial planning focuses on the states (Länder) while federal spatial planning is underdeveloped. It is evident that the siting of an important space-relevant facility such as a repository for radioactive waste must be considered in the spatial plans of the states that are concerned. However, the State of Lower Saxony in its recent land use programme[128] fixed Gorleben as site of the repository without considering

---

[127] OVG Lüneburg, Deutsches Verwaltungsblatt 2006 – Schacht Konrad.; Roßnagel/Hentschel 2004.

[128] Verordnung zur Änderung der Verordnung über das Landes-Raumordnungsprogramm Niedersachsen, 2008, Materialienband, 172, http://cdl.niedersachsen.de/blob/images/C44213379_L20.pdf.

any alternatives. This was justified by the consideration that there was an inconsistency between the territorial scope of spatial planning at state level and the location of possible alternative sites; Lower Saxony could not reach beyond its territorial borders. Moreover, it was thought that Lower Saxony had to defer to the paramount special environmental planning by the Federal Government. Some authors (Gaßner/Neusüß 2009:348) even take the view that a special planning procedure for site selection at federal level must be established in order to enable the states to comply with their obligation of considering alternatives in the framework of the prescribed strategic environmental impact assessment. However, Section 14g (1), (2) No. 8 of the Environmental Impact Assessment Act only requires the identification, analysis and assessment of reasonable alternatives. Any option outside the planning area the environmental impacts of which cannot be appropriately analysed and assessed by the planning authority due to territorial restraints of its powers would not not a reasonable alternative. In any case, the inconsistency between the planning area and the location of possible alternatives points to the need to elevate the geographical reference area from the states to the Federal Government. Given the fact that site selection for an onerous national infrastructure facility such as a repository for high level radioactive waste entails sharp distributional conflicts between the potential host states, it appears evident that state planning does not provide for an appropriate answer and only federal spatial planning would be capable of solving the relevant conflicts.

However, the problem is that general spatial planning at federal level is underdeveloped and it is difficult under constitutional law to strengthen it. In accordance with the present Spatial Planning Act, the Federal Government can establish a federal land-use plan that, however, only specifies principles for the spatial development of the whole federal territory and does not contain concrete targets (Section 17(1)). Moreover the Federal Government can determine national siting concepts with respect to maritime and inland ports and airports (Section 17(2)). The idea underlying the latter provision is that the appropriate planning area for infrastructure projects of national importance that cannot be sited all over the country is the federal level. This might suggest an inclusion of repositories for radioactive waste in the list of facilities subject to federal planning. However, it is to be noted that the projects that are presently on the list represent more the type of desirable projects each state wants to have while in the case of repositories for radioactive waste there may be a negative siting conflict because the project probably is undesirable in each state.

In any case, the federal siting concepts are not binding on the states so that the advantage of extending the list of federal siting concept is bound to remain limited. The lack of a binding force of the federal siting concepts reflects limitations of federal competences in the field of execution of federal laws as well as the power of states to deviate from federal spatial planning law. The establishment of a binding federal land-use plan would constitute an act of execution. However, Article 83 of the Federal Constitution establishes the general principle that federal law is executed by the states on their own behalf unless the Constitution provides for or admits federal execution. The exception contained in Article 84(1), 5th sentence whereby in case of a particular need the Federal State can regulate the administrative procedure is not applicable. Other exceptions do not obtain so that only a problematic recourse to an unwritten executive competence by virtue of the "nature of the matter"[129], that is, the cogently federal nature of execution in case of radioactive waste repositories, could be considered. Moreover, under Article 72(3) No. 4 Federal Constitution the states can deviate from the federal Spatial Planning Act. Even if by an amendment of the Spatial Planning Act radioactive waste repositories were included in the list of candidates for a federal siting concept and this concept were declared to be binding, the states could arguably exclude this binding effect by state legislation. This result could only be avoided by again taking recourse to the competence by virtue of the "nature of the matter" that would be exclusive and therefore pre-empt the state powers to deviate from federal legislation. To sum up, however attractive spatial planning as an instrument of considering alternative sites for radioactive waste repositories may appear at first glance, this route should not be pursued further.

This does not mean that the potential of spatial planning law for introducing elements of optimising the site selection is already exhausted. Besides general spatial planning, also sectoral environmental planning must be considered. However this only is an option for the future. In contrast to quite a number of environmental and infrastructure laws that contain sectoral planning instruments, including "advance planning" for projects that are subject to a planning permission, the Atomic Energy Act does not provide for any sectoral planning. As already stated (section B 4.5.4.4), the Federal State has the competence to amend the Atomic Energy Act to the extent that a sectoral planning instrument for nation-wide site selection could be introduced that precedes the planning permission for construction and operation. If one follows the Federal Administrative Court[130], it is within the responsibility of the legislature and the executive to decide whether a site selection procedure that aims at considering alternative sites shall be institutionalised or not.[131] The present limitation to a particular site and review of its suitability in the planning permission procedure is not inconsistent with the Constitution and the Atomic Energy Act. Rather, it is

---

[129] BVerfGE 3, 407, 425ff; Badura 2010, part D No. 79-80, part G No. 30; Kölble 1962:660.

[130] BVerwG, NVwZ 2007: 833 – Schacht Konrad I; BVerwG, Neue Zeitschrift für Verwaltungsrecht 2007: 837 – Schacht Konrad II; BVerwG, Neue Zeitschrift für Verwaltungsrecht 2008: 841 – Schacht Konrad III.

[131] To the same extent Gaetzsch 2005:118; de Witt 2005:133.

a procedural expression of the discretion granted the state organs to decide on tolerability of risk. The responsibility of the legislature and the executive also relates to the choice between legalisation and a merely political site selection procedure organised by BfS or the Ministry for the Environment. In this respect, the "Site Selection Concept" of the Ministry for the Environment of 2006 (BMU 2006, para. 9, 30) had advocated a formal regulation of the procedure. Later on, this idea was abandoned so that the present state of an entirely political site selection process would have lasted. Although this is legally permissible, considering the extremely long-term and complex nature of risk and requisite safety and the need to generate as much knowledge as possible, a legally structured procedure would be clearly preferable.[132] The disadvantage of a political site selection process coupled with a simple suitability analysis in the planning permission procedure also is that full consideration by the planning permission authority and judicial review of the results of the site selection process are not ensured. This would be even the case if risks and hazards originating in the operational phase are at issue and to this extent the decision would be clearly subject to judicial review. All told, the better arguments militate in favour of a legalised site selection procedure.

4.5.5.6 *The constitutional obligation to protect the environment as a mandate for considering alternative sites*

Another possible legal base for establishing an obligation to consider alternatives in a nation-wide site selection process is the "state goal" to protect the environment set out in Article 20a Federal Constitution which also covers all aspects of safety of radioactive waste repositories that go beyond existing generations. It could either establish a duty directly imposed on the BfS as the body responsible for developing the repository or serve as justification for an amendment of the Nuclear Energy Act. The administrative courts and commentators have elaborated several key elements of the duty to protect, in particular the inherent precautionary principle, the polluter pays principle and the principle of sustainable development.[133] Sometimes, also the principle of non-deterioration and optimisation of environmental protection are mentioned (Murswiek 2009, No. 47). Relying on this latter expression of the duty to protect, it has been sustained that in view of the particularly serious risks and hazards associated with radioactive waste disposal Article 20a Federal Constitution obliges the state to devise a site selection process that aims for the selection of an optimal site, that is, a site that

---

[132] Ramsauer 2008:944, 946; strictly in favour of "legalisation"; Ladeur 1989:47–248 who, for the very reasons named in the text, sees a violation of the "essential questions" doctrine.

[133] See BVerwG, Neue Zeitschrift für Verwaltungsrecht 1998, 952; BVerwG, NVwZ 2007: 833 para. 60 – Schacht Konrad I; Murswiek 2009, Art. 20a Nos. 47–51; Kloepfer 2005, No. 36, 55; Epiney 2010, No. 73–74.

is associated with the least remaining long-term risks.[134] However, since the duty to protect does not establish a priority for environmental protection and is subject to balancing with other social concerns including administrative practicality, it appears doubtful whether such conclusion is tenable. In any case, the duty to protect the environment accords the legislature a broad margin of political discretion as to the need for protection and the measures to be taken. It is only when existing regulation is evidently insufficient for an effective protection that the courts consider themselves as entitled to disregard the political prerogative of the legislature. Therefore it would seem that the site selection concept of suitability of the site does not violate Article 20a Federal Constitution.

4.5.5.7 Best possible prevention of risk and precaution against potential risk as mandate for considering alternative sites

Finally, it is being asserted that existing statutory law gives rise to an obligation of BfS, in its capacity as responsible entity for developing the repository, to carry out a site selection process aiming at an optimal site before submitting the application for planning permission. The statutory bases of this concept are the goals provisions of the Atomic Energy Act (Section 1 No. 2) and the planning permit requirements (Section 9b (4) in conjunction with Section 7(2) Nos. 3 and 5 Atomic Energy Act) themselves. As stated, in accordance with these provisions, the competent planning authority must ensure best possible prevention of risks and precaution against potential risks including long-term risks; harm associated with the activity must be practically excluded.

Roßnagel and Hentschel (2004:295) want to derive a direct siting responsibility of the BfS from Section 9b Atomic Energy Act as interpreted in the light of the state duty to protect set out in Article 20a Federal Constitution. In accordance with of these provisions, BfS is said to have a responsibility to search for an optimal site entailing a systematic investigation of appropriate candidates and selection of a site that, as a result of balancing all interests affected, proves to be associated with the least risk. It is only when such a procedure has been concluded that the BfS shall be legally entitled to apply for planning permission. Gaßner and Neusüß (2009:349) assert that the very notion of residual risk as developed by the German courts mandates a site selection process in which the competent authority must aim for an optimal site, considering all reasonable alternatives. Referring to the consideration of the Federal Constitutional Court[135] that such potential risks that remain after all appropriate measures to practically exclude them have been taken (tolerable or "residual" risks) are "inescapable" and must be borne by

---

[134] Gaßner/Neusüß 2009:348. However, it is not entirely clear whether the authors (speaking both of the "state" and of "organs" of the state) suggest a solution on the basis of existing law or make a proposal for the creation of new law.

[135] Supra note 71.

the public as a common burden of civilisation, they conclude that unless the optimal site has been selected the relevant remaining potential risk is not inescapable. In essence, these proposals amount to inventing a site selection procedure preceding the planning permission procedure on the basis of norms that do not provide for such a procedure. This is a task of the legislature, not of the interpreter of the law. Moreover, it would seem that the argument relying on inescapability of risk accords a merely descriptive statement a normative meaning. The key concept of the risk evaluation and management strategy embodied in Sections 9b (4), 7 (2) Atomic Energy Act is not inescapability of risk but rather best possible prevention of risk and precaution against potential risk.

And yet, the recollection of the prerequisites for granting planning permission as the *locus decidendi* may indeed inspire solutions for nuclear law reform which, however, would have to aim at developing a special sectoral planning instrument rather than enriching the planning permission procedure. The objective of all authorisation procedures set out in the Atomic Energy Act is to secure best possible prevention of significant risk and precaution against low risk. This objective is also reflected in the minimisation obligation set out in Section 6(2) Radiation Protection Regulation. Since the Regulation conditions minimisation both on feasibility according to the state of science and technology and the circumstances of the case, it constitutes a variant of the principle of optimisation as recognised in the international discussion. This requirement does not only govern normal operations but also disturbances (Section 49(1) Radiation Protection Regulation) and also applies to external events. Considering only the normal permit prerequisites, it appears odd that the minimisation principle should not also apply to site selection. Best possible prevention and precaution, if extended to all cases where nuclear safety of repositories is at issue, should also mean selecting the relatively best site,[136] i. e. a site that is associated with the lowest short- and long-term risk as achievable under the circumstances.[137]

Conceptually, one can only escape this conclusion if one plays the card of "practical exclusion" of risk according to the state of science and technology against direct recourse to best possible prevention and precaution. Following this line of reasoning, one could argue that the standard of practical exclusion of potential risk is not amenable to a comparative assessment. Either the risk is practically excluded at a given site according to the state of science and technology or it is not. There is – so one could continue – no mandate to compare relative residual risk in the site selection process once the limit of practical exclusion has been reached. A decision between these two possible stances is facilitated by a reflection on the status of the

---

[136] Expression taken from AkEnd 2002:95.
[137] To the same extent Hoppenbrock 2009:52–54, 56, who, however, bases this stance on the proposition that the planning permission procedure already entails the consideration of alternatives.

minimisation principle set out in Section 6(2) Radiation Protection Regulation. Minimisation of exposure and contamination does not belong to the category of requisite precaution that can be claimed by the individual because it aims at reducing collective risk. However, it is a measure commonly attributed to the precautionary principle rather than reduction of residual risk.[138] It would make sense *de lege ferenda* to also apply this principle to the site selection process. This seems to be in accordance with the more recent international development. Both the IAEA and – less clearly – the ICRP relate the optimisation requirement also to siting, geological features and design of a repository, although the emphasis is clearly laid on reduction of exposure to planned and accidental radiation.

However, there are other, "practical" arguments that shed some clouds on the wisdom of searching for the best possible site. The safety of a repository is a product of a combination of geological and engineered barriers at a particular site. Based on the international consensus as to the particular importance of geological barriers, one could indeed assert that one should select the best possible geological barrier and couple it with the best engineered barriers (cf. AkEnd 2002:95ff). However, this combination does not necessarily provide the best possible safety since the nature of the host rock and the local geophysical conditions may place limits on the additional safety that can be provided by the engineered barriers. A strict preference for the geological barrier may not be tenable. What ultimately counts is the safety of the whole system (NEA 2010:17, 20). Suitability of a site is a combination of several properties of a host rock, so that a balancing between favourable and less favourable properties may be necessary (cf. section B 1.3). There always is the theoretical possibility of finding a better site. However, for practical reasons, an infinite continuation of site investigations is not possible. Therefore, optimisation through siting arguably cannot be pursued absolutely but only in the form of a planning directive – as in its classical area of application (radiation protection) the minimisation principle is not absolute. Nevertheless, an open site selection procedure could contribute to a more balanced decision-making process, especially since not only safety-based but also spatial planning considerations are relevant.

4.5.5.8 Elements of new sectoral planning instrument for site selection

As already stated, a new sectoral planning instrument for site selection of a HLW repository could be patterned on the existing model of waste management planning under § 29 Life Cycle Economy and Waste Act and be inserted in the Atomic Energy Act. With the differentiation between the waste management plan that, among others, selects suitable sites for waste deposits, on

---

[138] Supra note 76.

the one hand, and the planning permission procedure for individual sites, on the other, the Life Cycle Economy and Waste Act postulates that a tiered planning process is useful. It is submitted that this is also true for HLW disposal.

An important legal question that needs to be clarified relates to the relationship between the federal sectoral planning decision, on the one hand, and the subsequent construction and operating permission as well as the general spatial planning decisions at state and local levels on the other hand. In the discussion on proposals to develop a new site selection procedure it has been asserted that the interface between the selection procedure and the legally required planning permission procedure is not well defined; the mere participation of the competent permission authority in the selection procedure was considered to be not sufficient.[139] This assertion is not unfounded, the more so since all proposals for a site selection procedure provide for a quite extensive investigation and evaluation of the suitability of the sites compared by the planning authority.

There are a variety of different rules in this field that range from the principle of pre-emption by federal planning to that of temporal priority to that of mere concerting to a duty to consider the sectoral plan. However, the legislature is entitled to make the final site selection decision to be in principle binding on the relevant state authority. A pertinent example in a related field is the site selection for federal highways that precedes the planning permission procedure. The siting decision that is taken by the competent Federal Ministry in concert with the State ministries does not only bind the planning permission authority but is also paramount to state and local spatial plans (Section 16(1), (3), 2nd sentence Federal Highways Act).[140] Another example of a similar tiered planning process in which federal authorities are not involved is provided by the sectoral waste management plan to be established under the Life Cycle Economy and Waste Act (Section 29). This sectoral plan shall determine, among others, suitable sites for waste disposal facilities. Depending on the legal nature of the plan – non-binding or binding – the plan must either be merely considered or is binding in the following planning permission procedure. Section 29(1) No. 5 of the Act requires that the planning permission may not be in conflict with a binding plan. This requirement is interpreted to the extent that siting alternatives can no longer be considered, provided the determinations of the plan are deemed to be definitive.[141] It is well established, although

---

[139] See Brenner 2005:109–13; Ossenbühl 2004:1140–1141; Sendler 2001:192; Gaßner/Neusüß 2009:351–352.

[140] BVerwGE 62, 342, 346; BVerwG, Verkehrsrechtliche Sammlung vol. 37, 154; Steinberg et al. 2000, § 4 Nos. 21–22, § 8 Nos. 83 et seq. However, the relationship of Section 16 Federal Highways Act to the objection procedure under Section 5 Spatial Planning Act is not quite clear.

[141] BVerwGE 81, 128, 133, 135–36; BVerwGE 81, 139, 146–47; BVerwGE 101, 166, 169–170; Paetow 2003, § 29 Nos. 70–73; Dolde 1996:528.

not totally uncontroversial, that an exhaustive waste management plan that has been declared to be binding has this legal effect. However, the sectoral plan does not definitely decide on the suitability of the site, i. e. its conformity with the permission requirements. Rather, it remains the task of the planning permission authority to decide on this question and the authority can reject the application for lack of suitability although the siting plan had in principle affirmed it. Although it also ensues from the jurisprudence of the Federal Administrative Court[142] that the framework for the planning permission procedure becomes the narrower the more specific the determinations of the waste management plan are, the Court has underlined the competence of the planning permission authority to deny the suitability of the site. In this respect it appears that the waste management plan is not equated to a permission-in-principle (Vorbescheid) as provided in Section 7a Atomic Energy Act;[143] the latter binds the permission authority with respect to questions decided and beyond contains a preliminary positive general assessment of the project.[144]

In order to avoid legal uncertainties of this kind, in devising the new sectoral planning instrument for HLW disposal the extent of the binding effect of the site selection must be clearly set out in the amendment of the Atomic Energy Act. This could be done by providing that in the planning permission procedure the determinations and findings of the site selection procedure must merely be considered but that the planning permission procedure is limited to the proposed site – a legal situation that already ensues from the present jurisprudence. One could add that the BfS or other competent agency or body can only initiate the planning permission procedure on the basis of the site selection plan.

A sectoral planning instrument for site selection would need to be executed by a federal agency. Although state execution on behalf of the Federal Government is the rule in the field of nuclear energy (Art. 87c Federal Constitution), Article 87(3) Federal Constitution allows the establishment of separate federal superior agencies and execution by these agencies in fields in which the Federal States has legislative competences. Under the prevailing opinion[145] this competence also covers nuclear energy because it is special to Article 87c Federal Constitution, at least where tasks of national

---

[142] Supra note 123.

[143] But see Gaßner/Neusüß 2009:352.

[144] See with respect to the similar partial permission under Section 7b Atomic Energy Act: BVerwGE 72, 300, 308–309; BVerwG, Neue Zeitschrift für Verwaltungsrecht 1999: 1231, 1232; Sparwasser/Engel/Vosskuhle 2003, § 7 Nos. 211 et seq.; as to the parallel Section 9 Federal Emission Control Act: BVerwGE 121, 182, 189–190; OVG Münster, Nordrhein-Westfälisches Verwaltungsblatt 1990: 93.

[145] Bull 2001, Art. 87c No. 16 et seq.; Britz 1998:1167; Uerpmann 2003, Art. 87c No. 9; Windhorst 2009, Nos. 32–33; contra: Hermes 2008, No. 20; see also BVerfGE 14, 197, 210–215.

importance are concerned. Moreover, Article 87(3) Federal Constitution is not confined to the establishment of a new federal superior agency but also entails the extension of administrative competences of an existing one.[146] Therefore it would be possible to confer the executive powers for site selection in the framework of special sectoral planning on an existing federal agency. In view of the importance of site selection, the competence to steer and supervise the selection process should be entrusted to BfS. Moreover, an independent expert commission that represents all stakeholders and a pool of experts should be associated to the agency. These bodies could accompany the selection process or at least decisive stages of it. This would provide openness of the agency to civil society and access to expertise. It would make an important contribution to better public acceptance.

Although one can consider the waste management plan set forth by the Life Cycle Economy and Waste Act as a model, the new site selection procedure for HLW repositories should be more structured, i.e. formally divided into several stages. Moreover, the statutory requirement of a strategic environmental assessment (SEA) should be formally built into the procedure. Waste management plans are subject to the SEA requirement insofar as they establish a frame for a later planning permission that in turn requires an environmental impact assessment (Section 14(1) No. 2 in conjunction with Annex 3 No. 2 Environmental Impact Assessment Act – EIA Act). Even without an express extension of the Annex 3 No. 2 of the EIA Act a site selection plan for HLW repositories would be subject to an SEA under Section 14(2) EIA Act; however, with the introduction of the new planning instrument the SEA requirement could and should be explicitly established. Moreover, elements of voluntarism and compensation should be introduced. Finally a "Gorleben clause" should ensure that underground investigations at Gorleben can continue.

In determining the stages of the site selection procedure, the proposals especially of the "Committee Selection Procedures for Repositories" (AkEnd, cf. AkEnd 2002) the Öko-Institut[147] and the Institute for Technology Assessment and Systems Analysis (ITAS, cf. Hocke et al. 2010, issue 3) can be used for orientation. One should distinguish between the following stages:

(1) Choice between different procedural options, development of site selection criteria and decision on other questions of national importance such as retrievability;
(2) *Screening phase I*: Based on existing geological data, especially the BGR list (see section B 1.3.6, Fig. B.24), exclusion of areas which are evidently unsuitable;

---

[146] BVerfGE 14, 197, 211; Lerche 1992, No. 175, Jarass/Pieroth 2011, Art. 87 No. 14.
[147] Öko-Institut 2007; short version: Kallenbach-Herbert/Barth/Brohmann 2008:72–78.

(3) *Screening phase II*: Based on existing geological data, selection of sites and areas[148] with particularly favourable geological conditions;

(4) *Screening phase III*: Screening under the EIA Act, i. e. determination of the framework for investigation for purposes of the SEA; if necessary, specification of criteria regarding above ground investigations (in the framework of criteria set by the Government); concerting with stakeholders about structuring the process;

(5) *Site/area selection for surface-based investigation*: Based on geological exclusion criteria and balancing criteria (geological and regional planning criteria), selection of several sites or areas for above ground investigations; determination of the investigation programme;

(6) Surface-based investigations;

(7) *Site/area selection for underground investigation*: Assessment of the results of above ground investigations; input and assessment of the results of underground investigations at Gorleben; if necessary, specification of criteria regarding underground investigations (in the framework of criteria set by the Government); investigation of development potentials of affected regions and municipalities; decision on sites or areas for underground investigation;

(8) Underground investigations, including development of repository concept and safety analysis;

(9) *Site/area selection I*: Assessment of the results of underground investigations, the repository concept and safety analysis; consideration of alternatives; environmental report under the EIA Act; decision on compensatory regional and local development concepts in favour of the affected municipality and region; selection of the site or area for the repository and establishment of the site selection plan.

(10) *Site/area selection II*: Eventually ratification of the site or area selection plan by Parliament; incorporation into a parliamentary act.

Although the stages described constitute discrete phases of the selection process, one also has to take into account that there may be certain overlaps. This is in particular true for the stages 6 and 7 as well as the stages 8 and 9. The concerned parties may not wish to wait for the complete conclusion of the surface-based and underground investigations before they enter into an assessment of their results and the consequences to draw from them. Therefore, the process may also be organised so as to allow a continuous discussion accompanying progress made in the investigations. Moreover, the relevant stages should be flexible and entail return options when the results reached in one phase do not justify entering into a new phase.

The first step of the site (or area) selection procedure should consist in a choice between the basic options for site selection, the development of site

---

[148] In case of clay as host rock, a larger area would have to be selected.

selection criteria and decision on other relevant question of national importance such as retrievability.

There are four basic options for site selection: (1) development of an
international solution, (2) continuation of underground investigations at
the salt dome at Gorleben and consideration of alternatives only if and when
there arise serious doubts as to its suitability; (3) continuation of underground investigations at Gorleben, but in parallel to that immediate search
for alternatives both with respect to host rocks and sites or areas, so that a
fall-back position is in place if and when Gorleben proves to be unsuitable
or there are robust indications that another site or area is clearly superior;
and (4) an entirely open selection procedure on a "white landscape" with or
without Gorleben depending on whether or not Gorleben meets the selection criteria.

In the first place, no international solution is credible if Germany is not
also prepared to host a common repository. This means that the national
site selection process cannot be halted. In making a decision on the three
remaining options, it is legitimate to consider a number of relevant factors
that may militate for or against a simple resumption of underground investigations at Gorleben. On the one hand, the investment of public money
that has already been made at Gorleben, the probable loss of time associated with an entirely new site selection procedure based on a "white map",
the existing knowledge from underground investigations at Gorleben, the
positive assessments by more recent expert reviews and the deliberate inaction on the side of the previous Government are to be considered. On the
other side of the scale, the broader factual base of a new selection procedure
and in particular possible gains in legitimacy and public acceptance are relevant. Whether the quality of the previous selection procedure on Gorleben
is a factor that can be put on the former or rather on the latter side of the
scale remains to be seen.

There also is a compromise solution. It would expand the decision-making horizon both with respect to alternatives (host rocks and sites/areas)
and input into the process from stakeholders and the population at large but
does not amount to further postponing underground investigations at Gorleben. Although fraught with some public acceptance problems, it presents
a pragmatic answer to the dilemma in which the policy-makers are presently caught.

Since the Federal Government has already decided to resume underground investigations at Gorleben without considering alternatives, it
might not appear to be politically opportune to reopen the question at all.
However, the opposition parties are determined to overturn this decision
in the future if they win the next elections and there is wide-spread, partly
highly militant resistance against Gorleben, considered (for right or wrong)
by many as a site that was selected in an illegitimate and unfair technocratic procedure. Therefore, it would be politically unwise to shut the doors

to the discussion on site selection. Besides the question of public acceptance, it is to be pointed out that the objective of all authorisation procedures established by the Atomic Energy Act is to ensure best possible prevention of significant risk and precaution against low risk. Although the Government bears the ultimate responsibility for nuclear risk assessment and management, this principle suggests that a sincere attempt should be made to select a site that is associated with the lowest short- and long-term risk as achievable under the circumstances. Finally, the spatial and regional development considerations that governed the selection of Gorleben at the end of the 1970s, have lost much of their persuasiveness in the meantime so that the question of spatial environmental justice arises anew. Since it will be difficult to achieve substantive distributional justice with respect to a siting decision that will be taken only once, fairness of procedure is of particular importance.[149]

For these reasons, the choice among the three remaining procedural options should be reopened, subject to broad public participation in the form of an organised public debate, and then the question as to which procedural option to follow should be decided anew by the Federal Government.

Since HLW site selection is a process that aims at developing a repository that in turn ultimately is subject to a requisite planning permission, it should be a matter of course that the statutory permit requirements must exert an anticipated effect on the selection process. Even if the legislature is free to set forth additional criteria for the selection process, especially those related to spatial planning, there are strong reasons of consistency that militate for a certain degree of parallelism. To this extent, the siting criteria cannot be merely based on the results of public participation in the site selection process but must be derived from the law. Moreover, since the executive is granted the responsibility for the determination and assessment of risk arising from nuclear facilities, it cannot simply delegate the task of developing site selection criteria to the participatory process. It should be underlined that, only as long as public participation remains in its traditional role as an instrument to improve official decisions and does not usurp decision-making functions, there is no conflict with the principles of parliamentary democracy and the responsibility of the executive. In other words, the expectation of public acceptance as such, from a legal point of view, should not serve as a decision-making criterion.[150] This means that the selection criteria must primarily be developed by the legislature and the executive and the participatory process can only contribute to developing and specifying them.

Although the establishment of the site selection plan should ultimately not depend on voluntary participation of the affected regions and munic-

---

[149] Kloepfer 2009:53–54, 218–219; but see Epp 1998; Bora/Epp 2000:1–35.
[150] To the same extent, although exclusively on grounds of existing statutory law: Hoppenbrock 2009:55, 57.

ipalities, an attempt should be made to achieve such participation. Apart from involvement of the regions and municipalities in the assessment and decision-making process (see section B 4.5.5.9), the procedure should be shaped accordingly. This could be done by providing that in the stages 5, 7 and 9 those regions or municipalities are to be primarily considered that have declared their willingness to participate in the selection process in the respective stage, provided they meet the selection criteria in the relevant stage. In order to avoid that the declaration of willingness to participate may lead to unforeseeable obligations and have unintended deterring effects, the declaration to participate in one stage should, however, in principle not be binding with respect to the following stage. It is to be noted that according to some participation models agreement to participate in an underground investigation practically includes a commitment to host the repository (although the possibility of a non-binding vote for withdrawal is conceded). Since our proposal accords the declaration of willingness to participate only an attenuated role, there is no reason to disallow a withdrawal of the municipality after the underground investigations have been carried out. In such a case the process has to be continued irrespective of voluntarism.

As an element of distributional justice, the municipality and region that host the repository and – either voluntarily or against their will – have to bear the economic and environmental disadvantages associated with it such as risks and nuisances through increased transport movements, loss of land values and image losses should be offered development opportunities to offset these disadvantages. Therefore the amendment of the Atomic Energy Act that would introduce the site selection procedure should establish legal bases for granting such compensation. The investigation of the regional development problems and potentials must be part of the procedure and the relevant development concepts should be set forth in the sectoral siting plan.

If, as proposed here, the government follows the second option, a special "Gorleben clause" should be inserted in the radioactive waste disposal plan in order to ensure that underground investigations at Gorleben can be continued and if they and the associated development of the repository concept and safety analysis render positive results Gorleben can be determined as the site for the final repository for HLW. To this end, it should be established that underground investigations at other sites or areas shall only be carried out if and when there are serious doubts about the suitability of Gorleben or, based on the results of surface-based investigations of alternative host rocks and at alternative sites or areas there is robust evidence that other sites or areas may fulfil the site selection criteria evidently better than Gorleben. The obligation to fully consider alternatives would be limited accordingly. The "Gorleben clause" may diminish the public acceptance of the site selection; it is asserted that a credible or balanced site selection requires at least underground investigations at two locations.[151] This may be so but constitutes only one facet of the problem. The "Gorleben clause" constitutes an element of distributive justice: If no municipality is willing to host the repository, it is not unfair to site it at a location that had previously been selected applying criteria that even in the light of present knowledge cannot be considered as unreasonable. Moreover, the clause counteracts undue strategic behaviour directed against progress of the selection and construction process. The "Gorleben clause" does not amount to a final prejudicial decision in favour of Gorleben but leaves this decision open. However, it should be conceded that the site selection procedure would primarily operate as fall-back position in case of failure of Gorleben.

Finally it remains to be seen whether an involvement of the Parliament in the establishment of the site selection plan would be consistent with the Constitution and advisable as a matter of policy. Article 19(1), 1st Sentence of the Federal Constitution states that where a fundamental right is limited by or on the basis of a law, this law must be general and not only be applicable to a particular case. However, this prohibition of "measures laws" is not absolute. The Federal Constitutional Court[152] has taken the view that such individual laws are permissible where the facts of the case are singular and the regulation is based on reasonable considerations. In view of the national importance of developing a HLW repository there can be no doubt that these requirements are fulfilled. In addition, it should be noted that the siting plan as such is not yet the basis for any intervention into fundamental rights which only is effectuated by or on the basis of the planning permission decision.[153] Viewed under the perspective of the division of powers as set forth in Article 20(2) Federal Constitution, there is no strict delimitation between the legislature and the executive. In principle any question can be the subject of parliamentary legislation. In particular, there is no prohibition of sectoral planning through legislation where such planning constitutes a guiding decision relevant for the whole federal state and therefore appropriate for legislation.[154] This would be clearly the case with respect to the determination of a single site for all HLW generated in the Federal Republic.[155] A co-decision by the Bundesrat, the representation of the states, does not appear to be warranted. The Federal Government only exercises its exclusive competences under Article 73 No. 14, Article 87(3) Federal Constitution and the new sectoral planning act can provide for participation by

---

[151] AkEnd 2002:2, 4, 64, 75; Hoppenbrock 2009:55; Piontek 2004:272–273; see also Kloepfer 2009:146.
[152] BVerfGE 85, 360, 374; BVerfGE 25, 371, 399; BVerfGE 99, 367, 400; more stringent: Krebs 2003, Art. 19 No. 11; Huber 2010, Art. 19 Nos. 61–63.
[153] Contra Gaßner/Neusüß 2009:352.
[154] BVerfGE 95, 1, 16; BVerfG, Neue Zeitschrift für Verwaltungsrecht 1998:1060, 1061.
[155] Contra Gaßner/Neusüß 2009:352.

or, as in the case of federal highway planning, concerting with the states in the site selection process itself.

It is another question whether a site selection by parliamentary act is desirable as a matter of policy. It certainly increases the legitimacy of site selection and is able to speed up procedures. A negative factor on the balance could be seen in the fact that judicial review of a parliamentary act is limited to constitutional violations (Gaßner/Neusüß 2009:352). However, according to our proposal the binding force of the act would be limited to the exclusion of alternatives. The duty to consider the determinations and findings of the site selection process does not compromise the competence of the planning permission authority to comprehensively assess the suitability of the site according to the legal requirements. Therefore there are no strong arguments that militate against a parliamentary vote on site selection.

4.5.5.9 Public participation

Section 9b (5) Atomic Energy Act in conjunction with Sections 72–75, 77 and 78 Administrative Procedure Act and Sections 4–7, 8–13 and 17 Atomic Law Procedure Regulation contains rules on public participation that are limited to the stage of the planning permission procedure in which the competent authority, on the basis of the application and the results of an environmental impact statement, decides on the construction and operation of the repository. This decision includes the suitability of the site which, therefore, must be also subject to public participation.

As regards the site selection procedure proper, there is no provision for public participation. The only statutory rules that could be relevant here are the rules on public participation in the planning procedure for determining a spatial plan that encompasses the siting of a repository for radioactive waste. However, it has already been stated that presently general spatial planning, due to its limitation to the territory of a single state, does not constitute an appropriate planning instrument for site selection. If, as proposed here, a special sectoral planning procedure was introduced, certain participatory requirements would follow from the SEA process set forth in the EIA Act. Public participation is provided at two stages: In the "screening" stage the competent authority only has to consider comments by the public that are known to it (Section 14 f (2), 2nd sentence EIA Act). In the preparation of the decision on the plan, the draft plan and the environmental report are subject to formal participation by the public concerned (publication, opportunity to make comments, duty to consider the comments; Sections 14i, 9(1) to (1b) EIA Act). A public hearing only is required where the relevant sectoral planning act provides so. Moreover, affected authorities including municipalities have to be consulted.

As long as the site selection process was designed as an essentially technocratic process, there arguably was little need for institutionalising or strengthening public participation. This changed when the siting issue became more politicised, in particular when under the previous Government formed by a coalition of the Social-Democratic Party and the Green Party new concepts of site selection as a staged process towards identifying and developing an optimal site were introduced. In this framework, various proposals for a new site selection procedure entailing new forms of public participation and voluntarism have been made. The major elements of the new procedure have already been presented (see section B 4.5.5.3). Both the proposal of the "Committee Selection Procedures for Repositories" (AkEnd 2002:54ff, 205ff) and the draft proposals of the Ministry for the Environment (BMU 2006; BMU 2009a) provide for a high degree of structured, institutionalised dialogue between, and participation of, all stakeholders. The same is true for proposals made by the Öko-Institut in 2007 and by ITAS made in 2010 that are based on an analysis of decision-making processes for major, complex projects (Öko-Institut 2007 and Hocke et al. 2010).

For example in the AkEnd model information platforms, centres of competence, citizen forums, round tables for regional development and a control group that supervises compliance with the procedure are provided. The Öko-Institut model in a way is even more complex because it distinguishes between an organised national and regional/local participation process (national conceptual partnership, regional/local planning and development partnership). The various proposed institutional arrangements relate to both levels. At regional level there shall be a presidency/secretariat with an information office, a representative planning and development council, citizen forums and a regional expert's pool.

The procedure is based on the belief that an open process based on information and rational discourse together with prospects for regional development can generate voluntary acceptance by regions and municipalities to become a host site for the repository. It relies on (in reality quite ambiguous) experience gained in Sweden, Finland, Switzerland and the United Kingdom. Indeed it seems to be undisputed among social scientists and is also asserted by some German nuclear administrators that a more open selection process is needed to create legitimacy of the selection process and move the siting problem out of its present impasse.[156]

However among lawyers scepticism against the proposed participation models prevails.[157] Relying on the constitutional attribution of decision-making powers to the government, a major criticism is voiced against the claimed equivalence of substantive safety criteria, on the one hand, and the

---

[156] See the contributions in Hocke/Grunwald (eds.) 2006; NEA 2007; Jordi 2006; ILK 2005:15; generally Pünder 2005:71.
[157] See Brenner 2005:109–113; Ossenbühl 2004:1140–1141; Sendler 2001:92. However, it is to be noted that in the teams of the Öko-Institut and ITAS lawyers were represented.

results of public participation and willingness to host a repository, on the other. The same argument is made with respect to the participative development of the relevant siting criteria. Moreover, the procedure is considered to be overly complex and cumbersome. Finally, in view of the fundamental dissent in German society about nuclear energy the assumption that the procedure can generate a consensus is considered as illusory ("idyllic procedure", "naïve", "doomed to failure").

This criticism, although overstated, should inspire a participation model for the sectoral site selection plan that is more "down to earth". In particular, recognising that improved participation constitutes an element of political legitimacy and is not inconsistent with the principles of a parliamentary democracy, one cannot take it for granted that it will really increase public acceptance. It also may have the contrary effect of mobilising more resistance and even public unrest. This is at least true as long as the final phase-out of nuclear energy is not ensured (which would only be the case once the last power plant has been definitely shut down) and therefore the site selection process is confronted with a "negative" alliance between fundamental opposition against nuclear power and aversion against a "bad development" in the neighbourhood. In this respect the German situation is clearly distinguishable from that in other nuclear countries.

Some elements of a slimmer procedure have already been introduced, namely the role of the expert commission and the pool of experts associated to the Federal Agency for Radiation Protection, the primacy of selection criteria issued by the Government and a reduced role of voluntarism. The most important reform element is the attempt to grant the public, i. e. all stakeholders as well as the public at large, a greater role in preparing decisions without delegating major decision-making powers and overloading the procedure by creating too many participatory institutions.

In the stage 1 that concerns the basic procedural options, an organised national debate appears as the most appropriate means of participation. In the stages 2 and 3, only geological criteria are relevant. Here, nation-wide information and an opportunity to make comments via the internet would be sufficient. However, in the stages 4, 5, 7 and 9 where the important decisions associated with local and regional effects are taken conventional participation entailing publication of the relevant documentation, an opportunity to comment and eventually a public hearing, all occurring at a rather late stage of the process, are not sufficient to ensure a rational decision and promote public trust, an attitude of common problem-solving and ultimately public acceptance or at least toleration of the relevant decisions. This is even true if conventional participation were extended to all these stages.

Rather, conventional participation should be complemented and partly replaced by new, more "deliberative" elements without making the procedure overly complex. Such elements are an active information of the population concerned about the need for, and basic problems associated with, HLW disposal, an early opportunity for the stakeholders and local population to coop-erate in the shaping of the structure and planning of the process in its various stages (stage 5) and the establishment of a permanent local dialogue forum for each site that remains in the "competition". The local dialogue forum that would be composed of representatives of stakeholders and the local pop-ulation would play a central role in all relevant phases of the process. It is designed as an instrument to facilitate common conflict-solving. Its tasks would also include the investigation and assessment of compensatory devel-opment opportunities. To these ends, the forum could make technical exper-tise available from the national pool of experts and organise public debates and closed discussion rounds ("round tables") according to the discreet issues (e. g., complexity, uncertainty, ambiguity) that arise, while a formal issue-related differentiation is not advisable. The chairperson of the forum should be a qualified, neutral person that could play the role of a mediator.

In stage 9 the publication of the draft siting plan and the environmental report and the granting of an opportunity to comment on them are man-datory requirements already established by the EIA Act. The new proce-dure cannot be substituted for these requirements. Even at the other rele-vant stages, individual comments outside organised participation via the dialogue forum should always be admitted.

4.5.6 Construction and Operation

4.5.6.1 Statutory requirements

As already stated, the German legal regulation of final disposal of radio-active waste focuses on the stage of the construction and operation per-mission for the repository. Section 9b (4) Atomic Energy Act prescribes for the permission of the construction and operation of a radioactive waste repository a planning permission (literally: "plan determination decision"). The prerequisites for authorising the facility are partly derived from Sec-tion 7(2) Atomic Energy Act that applies to nuclear power plants, partly for-mulated independently in Section 9b (4) Atomic Energy Act itself. Under Section 9b (4) in conjunction with Section 7 (2) Nos. 3 and 5 of the Atomic Energy Act the planning permission may only be granted where precau-tions against harm caused by the construction and operation of the facil-ity, as are necessary according to the state of science and technology, have been taken and the necessary protection against disturbances of operations and interventions by third persons or natural events is ensured. Moreover, an impairment of the public interest by the facility must be excluded and no other public law provisions, especially as regards the compatibility of the facility with the protection of the environment, may rule out its authorisa-tion. The competent authority must ensure best possible prevention of risks and precaution against potential risks; harm associated with the activity must be practically excluded.

This standard does not only apply to the design and other technical features of the facility as well as the methods of deposit and other modes of operation. It is also applicable to the site. In the planning permission procedure, the competent authority also decides on the suitability of the site. A preceding authorisation granted under the Federal Mining Act to carry out underground investigations does not yet constitute a binding decision on the suitability of the site.[158] It only provides factual information regarding its suitability that can be used in the planning permission procedure.

As already stated, the planning permission is a "legally bound" decision; unlike in most other planning permission procedures, the competent authority does not possess a planning discretion, with the consequence that in this phase of decision-making an evaluation of siting alternatives does not take place. It is only the suitability of the site in the sense of its conformity with the permit prerequisites that must be scrutinised.[159] However, according to the jurisprudence of the Federal Admininistrative Court,[160] the competent authority has a margin of appreciation of facts and broad statutory terms that must be respected by the courts. This would be confirmed by the proposed new provisions on the sectoral site selection plan. Moreover, the justification for the option of disposal in a geological repository is not reviewable because the Atomic Energy Act has already decided on that.[161]

In particular, in accordance with Section 9b(4) in conjunction with Section 7(2) Nos. 3 and 5 Atomic Energy Act,[162] it must be ensured that during normal operations the dose limit and emission values of the Radiation Protection Regulation (Sections 5, 46, 47) are complied with. The facility must be designed in such a way that in case of improbable, but not practically excluded disturbances the relevant limit values for disturbances (Section 49 Radiation Protection Regulation) are not exceeded. In both cases the minimisation obligation under Section 6(2) Radiation Protection Regulation is applicable in addition. Possible discharges of ionising radiation arising from disturbances of operations of the repository must be prevented or reduced and in any case controlled. Moreover, measures for protection

---

[158] BVerwGE 85, 54, 58 – Gorleben; BVerwGE 100, 1, 9–11 – Gorleben; BVerwG, BVerwG, NVwZ 2007: 833 para. 31 – Schacht Konrad I; BVerwG, Neue Zeitschrift für Verwaltungsrecht 2007: 837, para. 14 – Schacht Konrad II.

[159] BVerwG, NVwZ 2007: 833 para. 31 – Schacht Konrad I; BVerwG. Neue Zeitschrift für Verwaltungsrecht 2007: 837, para. 14 – Schacht Konrad II.; Rengeling 1995:57ff; Sellner/Hennenhöfer 2007, part 15, Nos. 312-313; as to Section 7(2) No. 6 Nuclear Energy Act also Kloepfer 2004:50, § 15 No. 86.

[160] See the decisions cited supra note 74.

[161] BVerwG, NVwZ 2007: 833 para. 35 – Schacht Konrad I; BVerwG, Neue Zeitschrift für Verwaltungsrecht 2007: 837, para. 15 – Schacht Konrad II. By contrast, the need for a further repository arguably could be reviewed; see BVerwG. Neue Zeitschrift für Verwaltungsrecht 2007: 837, para. 16 – Schacht Konrad II.

[162] See BVerwG, NVwZ 2007: 833 para. 54 – Schacht Konrad I; BVerwG. Neue Zeitschrift für Verwaltungsrecht 2007: 837, para. 7 – Schacht Konrad II.

against external interventions such as intrusions and other external events according to the state of science and technology must be taken.

Section 9b (4) Nos. 1 and 2 Atomic Energy Act also mandates the consideration of the public interest and compliance with other public law provisions, especially regarding the protection of the environment. In this context, especially the results of the prescribed environmental impact assessment will play a role. However, "environmental protection" in the meaning of these provisions does not include the protection of groundwater. Although the planning permission has a "concentration effect", Section 19 Federal Water Resources Management Act provides that water management is not included in the planning permission. Rather, a separate water permit is required which, however, is to be granted by the planning permission authority. At the level of substantive law, the prerequisites of the Federal Water Resources Management Act relating to groundwater pollution must be respected. This is especially true for the prohibition of a deterioration of groundwater (Section 47(1) No. 1 of the Act) and of a deposit of substances that gives grounds for concern that an adverse alteration of groundwater quality may arise (Section 48(2) of the Act). The latter principle of "no concern" is interpreted broadly and also covers long-term risk. It is understood in the sense of best possible prevention of significant risk and precaution against low risk.[163]

Transport risks are not to be considered.[164] This position is in conformity with the tradition of German environmental law that considers any transport to and from a facility once it is commingled with general transport to be outside the scope of the facility-related permission procedure. Section 4 Atomic Energy Act requires separate transport permission. This somewhat artificial fission into two separate procedures is not normally considered as acceptable by the population living near a potential repository. Under the perspective of acceptance this is important since transport risks associated with the operation of the repository are of greatest concern for the population.

Most of the requirements discussed are also applicable to long-term safety and security as well as compatibility with the environment. The post-closure phase is covered by the permit prerequisites of the Atomic Energy Act even if the wording of the applicable statutory rules on "operation" might suggest the contrary.[165] However, in applying these requirements certain modifications may be mandated by the very long-term nature of the relevant problems and the ensuing problems of proof. For example, since an

---

[163] BVerwG, Zeitschrift für Wasserrecht 1981: 87, 88/89; Zeitschrift für Wasserrecht 1983: 222, 223; Volkens 1993:105ff; Czychowski/Reinhardt 2010, § 48 No. 26 with further references.

[164] BVerwG, NVwZ 2007, para 40-42 – Schacht Konrad I.

[165] BVerwG, NVwZ 2007, para 60 – Schacht Konrad I; see the discussion in section B 4.5.4.1.

exclusion of risks to groundwater due to geological changes in the host rock over long periods of time will be very difficult to demonstrate, it is sufficient to reduce this risk as far as reasonably achievable. The Radiation Protection Regulation is not directly applicable but can at least serve as a guideline for establishing special post-closure radiation protection requirements.

The planning permission can be divided into two partial permits. However, the applicable Section 74(3) Administrative Procedure Act sets a rather rigid framework for partial permits that does not fully respond to the decision-making situation in the case of radioactive waste repositories. This suggests, as envisaged in the site selection proposals of the Ministry for the Environment (BMU 2006, BMU 2009a), certain changes of Section 9b Atomic Energy Act.

4.5.6.2 Administrative Practice

The specification of the statutory requirements normally is effectuated by administrative rules. As already stated, such administrative rules exist with respect to nuclear power plants in the form of safety guidelines and guidelines for disturbances, but not yet for radioactive waste repositories. For the time being, the existing guidelines can be used for orientation as far as the operational phase of the repository is concerned.[166] As stated, the Safety Requirements for the final disposal of high level radioactive waste in a repository[167] address various issues of the operational and long-term safety and security of the repository including limit values for radiation protection. However, they are not yet final. It remains to be seen how they will be further developed during the ongoing underground investigations at Gorleben.

4.5.6.3 Controversial legal questions

It appears that as yet there is little disagreement about the legal requirements for a repository under normal operating conditions. By contrast, apart from the general postulate of long-term safety and the discussion about the time scale of long-term safety, the post-closure phase has found less attention, one reason being that the geological and technical barriers of the repository are deemed to normally ensure that the dose limit values for ionising radiation can be complied with over a long period of time. Relevant concerns are in particular raised by possible disturbances due to the entering of groundwater into the repository (which could still be considered as an internal event) and external events such as intrusion by people, sabotage, terrorist attacks, explosions, earthquakes and the like.

With respect to nuclear power plants, it is recognised that the operator must also take preventive and precautionary measures against rare events and organise a facility-internal system of emergency protection. The guide-

---

[166] BVerwG, NVwZ 2007: 833, para 55 – Schacht Konrad I.
[167] See section B 4.5.4.3.

lines on disturbances also cover this configuration. What is controversial under the perspective of access to judicial review, though, is the attribution of these obligations to either requisite precaution or discretionary minimisation of residual risk. As already stated, there is a recent tendency by the Federal Administrative Court to expand the scope of mandatory precaution and thereby strengthen the legal position of neighbours.[168]

As regards the post-closure phase of the repository for high level radioactive waste, for the time being the question of judicial review is of little relevance because here long-term safety is paramount. This is deemed to be outside the scope of judicial review because it transcends the life of presently living persons. In the future, it could become the object of association suits if the European Court of Justice rejects the German limitation of association standing to the assertion of individual rights. Thus the problems presently concentrate on the substantive question as to what extent the obligations imposed on operators of nuclear power plants can also be transferred to the operator of a repository. In principle, this question is to be answered in the affirmative. Many of the necessary adjustments to the special problems of long-term safety and security of repositories after closure are already contained in the Safety Requirements of 2010 and will be further developed in the future.

4.5.6.4 *Participation*

In the planning permission procedure, public participation is provided and an environmental impact assessment must be carried out. Section 9b(5) Atomic Energy Act primarily to the rules on public participation set out in Sections 72–75, 77 and 78 Administrative Procedure Act. In modification of these rules, Sections 4–7, 8–13 and 17 Atomic Legal Procedure Regulation are declared to be applicable. The environmental impact assessment requirements are laid down in Section 9b (2) Atomic Energy Act in conjunction with the Environmental Impact Assessment Act. Based on the publication of the permit application and the results of the environmental assessment made by the applicant, everybody can make comments and raise objections. There is an obligatory public hearing which, however, is not public and is not designed to fully discuss the project but, rather, limited to a discussion of the objections raised (Sections 8(2), 12(1) Atomic Legal Procedure Regulation). Moreover, comments made by interested authorities shall be discussed (Section 73(6), 1st sentence Administrative Procedure Act).[169] Apart from these weaknesses that in practice are normally over-

---

[168] See supra B 4.5.4.2.
[169] Since Section 9(1), 3rd sentence of the Environmental Impact Act provides that participation must at least be equivalent to Section 73(3), 1st sentence, (4)-(7) Administrative Procedure Act, it is submitted that the provision on consultation of interested authorities is applicable although a literal reading of the reference in Section 9b(5) Atomic Energy Act might render the contrary result.

come by raising a great number of different objections, it is to be noted that the system of public participation, from a legal point of view, is deficient for two reasons. Regarding the site the planning permit only decides on its suitability and alternative sites are not to be discussed. Moreover, as questions of long-term safety are not subject to subsequent judicial review, the legal position of objectors in the process is not particularly strong.

4.5.7   Financing

The Atomic Energy Act, in keeping with the polluter pays principle, places the financial responsibility for final disposal of radioactive waste on waste generators, especially the operators of nuclear power plants.[170] The operators in their capacity as persons obliged to deliver the radioactive wastes to the repository must pay fees or market prices[171] for the utilisation of the repository (Section 21a Atomic Energy Act). Moreover, in their capacity of having an advantage from state-organised final disposal they are obliged to make financial contributions for covering the necessary expenses for planning, acquisition of land, facility-related research and development, underground investigation, maintenance of premises and facilities as well as the construction, extension and retrofitting of the repository (Section 21b(1) Atomic Energy Act).

The notion of necessary expenses is to be interpreted broadly. It comprises all expenses for the planning, facility-related research and development, investigation as well as the construction of the repository. Also costs for the development of the repository are covered since this development is necessary for creating the prerequisites for the grant of the planning permission.[172] However, with respect to site selection, the prevailing opinion makes important restrictions. Some authors sustain that selection costs are not facility-related, arguing that only the planning and investigation of a concrete facility is covered. Others consider these costs not to be necessary or not to confer an advantage as long as the lack of suitability of Gorleben is not established or another site is evidently superior.[173]

The Act does not couple the planning costs to a concrete facility. The requirement of facility-related investment could only be concluded from the attachment of the financing obligation to the advantage derived from the repository. However, site selection is a process that aims at creating the prerequisites for providing such an advantage. Therefore it is submitted that the expenses are necessary if and insofar as mandated in accordance with

---

[170] For a detailed analysis see Hoppenbrock 2009:45–62.
[171] The latter is relevant where the operation of the repository is assigned to a private entity.
[172] Haedrich1986, § 21b No. 1; Kraß 2004:263ff; Hoppenbrock 2009:49–50.
[173] de Witt 2005:134; Waldhoff 2005:155; Kirchhof 2004:317; Kraß 2004:263–264; Ossenbühl 2004:1132–1133; see also No. IV 7 of the "Atomic Consensus" of 14 June 2000.

the relevant statutory site selection procedure.[174] If the procedure aims at identifying an optimal site, the selection expenses must be covered. This is also true where during the selection process there is a change from a procedure geared to suitability of the site to one that aims at selecting an optimal site. Otherwise the operators of nuclear power plants could exercise a controlling influence on the choice of the procedure, which would not be consistent with the ancillary role of the rules on financing. This argument is reinforced by the fact that Section 21b (4) Atomic Energy Act denies a claim for restitution of payments when a repository is ultimately not constructed or not operated. That Section 21b(3), 3rd sentence requires fixing the contributions so as to cover the expenses that can be charged according to principles of accounting, does not militate against a broad interpretation of the notion of "necessary" expenses. This provision only concerns the meaning of expenses, not that of necessity. However, expenses incurred in the framework of a merely political selection process cannot be considered as necessary expenses. If, as the Federal Administrative Court has held, the planning permission procedure does not entail the consideration of alternatives, the executive cannot dispose of the procedure politically with a view to have alternatives considered at the expense of the waste generators.[175]

The details of the financial responsibility of operators can be determined by regulation for which the Atomic Energy Act gives some specifications (Sections 21a (2), 21b (3) of the Act). Pursuant to the latter empowerment, the Regulation on Advance Payments for the Repository of 1982 has been promulgated. It provides that advance payments for the annual expenses already incurred must be made and the operators must establish reserves in their balance sheets for covering future expenses. The establishment of a fund is not provided nor is a financial security required. A disposition over the assets needed to cover the reserves during the financial year remains possible, which means that the relevant generator has additional liquidity at the expense of the taxpayer. Although part of the waste disposal costs are already covered by the current payments, in view of the long time frame of high level radioactive waste disposal the German financing model of establishing reserves for future payments does not provide sufficient protection against insolvency or dissolution of a waste generator.[176] This is especially true for the – admittedly highly improbable – case of a major accident that goes beyond the amount of 2.5 billion Euros that are secured

---

[174] To the same extent Piontek 2004:270–271; in the result also Hoppenbrock 2009:52–57 (however, based on the author's opinion that the planning permission procedure entails the consideration of alternatives).

[175] In this respect there is agreement with the authors cited in note 170. As to the legal problem of a mandatory public law corporation of waste generators for final disposal see authors cited supra note 84. Even such a model could not overcome the constitutional obstacle that an advantage for radioactive waste generators is required.

[176] See Wüstemann 2004:277–310.

by financial security and the cover obligation of the state under Sections 13–15, 34 Atomic Energy Act.[177]Most nuclear power plants in Germany are not directly operated by one of the four big public utilities but rather by their subsidiaries and joint ventures in the framework of contracts of domination and transfer of profits. In principle, these enterprise contracts establish the liability of the parent companies (§ 302(1) Stock Corporation Act). However, in case of contract termination due to insolvency or dissolution of the subsidiary the liability of the parent companies is limited to obligations that have already arisen (§ 303(1) Stock Corporation Act). It is doubtful whether the abstract statutory obligation to cover future, not yet foreseeable costs of radioactive waste disposal can be deemed to have already "arisen". Whether in such a case the parent company would voluntarily step in even if the operations of the subsidiary are terminated is a matter of business judgement and cannot be predicted. In an international comparison, there are some arguments that militate for fund models. However it cannot be overlooked that fund models absorb liquidity and engender the risk of a loss of, or part of, the capital of the fund on the financial markets. Therefore, a solution that engages the financial responsibility of the respective parent companies appears preferable. Section 4(3), 4th sentence Federal Soil Protection Act already contains a liability concept that aims at holding parent companies liable for soil damage, although it is rather limited. Extending this model, it should at least be provided that the parent companies of the operators of nuclear power plants, notwithstanding the existence of enterprise contracts, shall be liable by way of "disregard of legal entity" for covering the future costs of radioactive waste disposal where payments cannot be secured from the subsidiary or joint venture that operates a nuclear power plant. Of course, the existing statutory liability of the parent company for the debts of the subsidiary in case of a contract to transfer the profits would remain unaffected.

---

[177] For details see Bordin/Paul 2008:271–292; Cloosters 2008:293–306; Müller-Dehn 2008:321–332; Hoppenbrock 2009:135, 207–212.

5 Guidlines for a socially acceptable and fair site selection

5.1 Introduction

The process of selecting a site for the final disposal of high-level radioactive waste is an issue which has fuelled extremely heated debates in Germany ever since the commissioning of the Kahl nuclear power plant, Germany's first experimental plant, in 1960. For forty years now, representatives of a wide variety of interest groups, such as civil movements, ecological groups, as well as representatives from the world of science, politics and industry have tried to agree on a mutually acceptable concept for the selection of the site for the final disposal of high-level radioactive waste. There are manifold reasons why no agreement has been reached to date and why the conflict remains unresolved – and they are not least rooted in the nature of the issue itself (Hocke/Renn 2009). This chapter first takes stock of the perception of the final nuclear waste disposal problem, continues with a conflict diagnosis and concludes with a discussion of the various options for dealing with this conflict.

5.2 Key issues of the debate

The question of final disposal mobilises people, not just in Germany, but around the globe. It is closely connected with the overall debate about the future of nuclear power. Due to the complexity, uncertainty and ambiguity of final disposal risks, three essential challenges must be tackled:

- *Point 1: Disagreement amongst experts.* Amongst natural and technological scientists who have thoroughly explored the issue of final disposal there is little doubt that deep geological depositories provide the best and safest form of final high-level radioactive waste disposal. However, an intense debate continues regarding the practical implementation of such deep geological repositories. It raises such questions as: which type of geological formation (bedrock, salt) is particularly suitable? Which factors commend and which speak against the suitability of a site? Which fundamental concepts are preferable – retrievability or irretrievability? The scientific discourse primarily concentrates on the question of our flexibility to act: how can technological systems compensate human error, organisational learning capacity over long planning and trial periods, as well as the ability to adapt to current scientific and technological knowledge at any given time be ensured? Those questions are fiercely debated; they do not raise any serious doubt, though, about the fundamental question of the technical viability of deep geological repositories. However, in the public at large, this scientific debate is often perceived as proof of the project's lack of maturity and even as an indicator for a lack of scientific reliability and seriousness. If the experts cannot even agree, who else could be prepared to guarantee safety over such a long period of time? The doubt which is clearly stated in all the relevant surveys feeds off this perception that even science cannot offer a conclusive final disposal concept.

– *Point 2: Mobilisation potential.* Most industrialised countries experience a high mobilisation and amplification potential within society. Final disposal moves and mobilises many people, creating intensive involvement and a highly emotional charge. This triggers a worldwide stigmatisation effect around the question of all things nuclear, entailing a polarisation and politicisation of the issue, particularly with regard to current nuclear power policy (Hocke/Renn 2009; Flynn 2003). Introducing or phasing out, re-introducing, phasing out again – these are contentious issues which play an important role, directly or indirectly, in the assessment of the final disposal issue. The question of the final disposal is thus symbolically exaggerated: it is no longer primarily about technological viability, or even about long-term safety, but rather about fundamental perspectives of societal development. It is a question of whether society wants to continue pursuing central, highly efficient and energy-dense but risky power generation or if it prefers decentralized technologies which are low in energy-density, not necessarily low-risk but locally restricted? Going nuclear as Robert Jungk put it once, symbolises more than just a technology, it is fundamentally a way of living (Jungk 1986). And that explains where the discrepancies between the different fronts are located.

– *Point 3: Discrepancies between experts and the layperson's perception.* In 2009, a survey was carried out in the USA amongst experts and the USA general population regarding the weight of different risks (NEA 2010a). It revealed that on average the population believed nuclear waste repositories to pose the absolute highest risk, even higher than driving a car. Amongst experts, however, final nuclear waste repositories lay in the lower third. What we see here is an immense discrepancy between the perception of laypeople and the opinions of experts.

5.3   Factors of risk perception

5.3.1   Risk as an imminent threat

From a psychological point of view, a final nuclear waste repository is perceived as a technological risk to the environment and human health. The risks of nuclear power generation as well as final nuclear waste disposal are associated with the semantic pattern of "the sword of Damocles", even though from a scientific viewpoint, after the termination of the storage, few scenarios are conceivable which could trigger a sudden case of damage.

Semantic patterns are not unlike filing cabinets in how they function. If one is confronted with a new risk or has taken in new information regarding risk, most people try to pigeonhole this information in an existing category. The "sword of Damocles" pattern is one of them. It is about techno-logical risks with a high potential for damage but a very low probability of fully coming to bear. The technical safety philosophy usually aims at low-ering the probability of the occurrence of such a failure, so that the combi-nation of probability and extent of damage decreases drastically. The sto-chastic nature of such an event, however, makes it impossible to predict the time of its occurrence. Subsequently, the event may theoretically occur at any point in time, even though the probability for each of these points is extremely low.

For those concerned with the perception of rare random events, prob-ability plays but a minor role: the randomness of the event is the factor which is perceived as particularly threatening by most people. Exam-ples of risk sources from this category are large technical plants, such as nuclear power stations, final repositories for radioactive waste and other danger potentials devised by humans, which can, in an emergency situ-ation, have catastrophic effects on human beings and the environment (Renn et al. 2007:80ff).

The idea that an event could hit the affected population at any random point in time creates a feeling of being in danger and having no control. Instinctively, most people can mentally (in real life this may be question-able) cope much better with danger if they are prepared for and attuned to it. In the same way that people tend to be more afraid at night than during the day (although the objective risk of being harmed is much higher dur-ing the day, one may be caught off guard by potential dangers more eas-ily during the night), most people feel more threatened by potential dan-gers which strike them unexpectedly and offhand. Dangerous events which occur either regularly or which allow for sufficient time to take measures to defend oneself after the occurrence of the precipitating event seem less threatening. Thus the extent of the risk in the present understanding is a function of three factors: *the randomness of the event, the expected maxi-mum scale of the damage and the time span available for damage control* (Renn 2008:117ff). The rarity of the event, i.e., the statistically derived expected value, is irrelevant in comparison. To the contrary, frequent events tend to signalise a continuous sequence of cases of damage for which one can prepare in a trial and error procedure.

The perception of a risk as an imminent disaster often affects the evalua-tion of technological risks. Particularly where communication is concerned, it is important to know that a decreased probability of damage occurrence has no effect on the acceptance of the risk. The safety engineers of nuclear power plants have had to painfully learn and accept this insight through-out the years. Those who have lectured on the safety of nuclear power plants know from experience that sooner or later someone from the audience will ask: "Could it happen tomorrow?" The honest answer is of course, "Yes, it could". The person will then quite certainly answer: "I don't want it then."

Where the perception of large-scale technological risks is concerned, it is the randomness of it which is the real problem.

5.3.2   Risk as a creeping danger

Another factor is the creeping danger of radioactivity, which defies sensory perception. This is another semantic pattern which often generates anxiety (Renn et. al. 2007:83). In the context of this risk pattern, people rightly assume that scientific studies can detect creeping dangers in good time and discover causal relationships between activities or events and their latent effects. Unlike the technological-medical risk concept, the probability of such an event is not interpreted as a significant deviation from the naturally given variation of such events (i.e., it can no more be explained as a random event), but as a degree of certainty with which a singular event can be ascribed to an external cause.

The knowledge that one can potentially develop cancer from being exposed to ionising radiation at least legitimises people's suspicion that every form of cancer occurring in the vicinity of a nuclear plant has been caused by radioactive radiation. Those who have cancer or must witness a family member or friend suffer from it naturally seek explanations. Metaphysical explanatory models have lost validity in our secularised world. At the same time, the random occurrence of cancer as the best possible explanatory model according to today's knowledge does little to satisfy our mental desire for a "meaningful" explanation (Kraus et al. 1992). How bleak a perspective it is to be the victim of a random disease distribution mechanism. However, if a concrete reason is known, such as radiation exposure, the occurrence of the disease makes sense, at least subjectively. If subjectively no fault can be found with the patient's behaviour (such as smoking or alcohol abuse) and third party negligence may be used as the cause of the disease, the disease may even fulfil a social purpose in alarming potential future victims to fight against the root of the evil (Renn 1997).

The highly emotional debate typically surrounding risks of this type must be examined against this psychological backdrop. The human capacity for compassion enables us to potentially identify with the victim. Risk analyses which prove a certain probability of a creeping risk due to emissions cause us to identify with the victim affected by the risk. While the risk analyst uses stochastic theories to characterise the relative hazards of events which do not allow for causal relationships between singular triggers and their effects (thus creating a distance to one's own field of knowledge), the layperson sees in them proof of the culpable involvement of societal actors in causing life-threatening diseases.

In the case of the semantic pattern of the "creeping danger", the people affected must rely on information provided by third parties. Usually they can neither perceive the hazards sensorially nor check the claims of various experts which tend to be contradictory (Renn 2005). When laypeople evaluate such risks they will be faced with a key question: can I or can I not trust the institutions that provide me with the necessary information? If my assessment tells me I can't, then I will be uncompromising in my call for zero risk. This is because if I must rely on third-party information for my risk assessment yet I do not trust the third party, I will not leave myself open to a cost-benefit balance. Then I want call for zero risk. The desire for zero risk due to a lack of trust can be observed worldwide. If, on the other hand, I am undecided as to whether or not I can trust, peripheral aspects take on particular importance. Suddenly factors which are fundamentally irrelevant for the cause become important, such as the expert's red tie which the viewer doesn't like. In that case the listener has no other option but to trust distribution based on peripheral features because he cannot judge the risk of being harmed by radiation himself. He must either trust one party or none at all.

Evaluation processes triggered by risk semantics are mostly invariant to information or communication offers. At most, targeted information and education will inspire people to critically reflect upon their own judgement. Otherwise, however, these patterns can be expected to influence people's perception of final nuclear waste repositories to a great extent.

5.4   Consequences of the population's risk perception

The starting situation determines the conditions for a future solution in the question of nuclear waste disposal. The final waste disposal issue is highly emotionally charged; it unleashes in people the fear of an imminent threat and a creeping danger as an element of their perception. What empirical evidence exists on this issue?

In all existing surveys, final nuclear waste disposal solutions rank very high in the public perception of what poses a threat. This is true everywhere in the world, even, interestingly enough, in Finland where the problem of final waste disposal has been largely resolved on a political level despite these public concerns (European Commission 2005; 2008).

The complexity of this situation becomes clear when considering the results of a representative survey of the German public from the years 2001 and 2002: during the time of the survey, roughly 65 % of those interviewed assumed that over the next decade a final nuclear waste repository would be established, while 81 % objected to such a repository being created in their vicinity (cf. Stolle 2006:197). This classic *NIMBY syndrome* ("Not in my backyard!") is typical for site selection processes for large-scale technological and risk-related facilities (cf. Fredriksson 2000; Rosa 1998). In principle, the necessity of such a technology is endorsed by people, but with the condition that it be as far removed from their domicile as possible. The NIMBY syndrome is closely connected with potential risks for the population as perceived in the context of such technologies.

Surveys on stakeholder mobilisation show worldwide differences, from which much can be learned. Some countries such as Finland and Sweden have made progress with regard to finding solutions for final nuclear waste disposal. It is not impossible to reach an institutionally satisfactory solution which is tolerable for the majority of the population if the right approach is chosen. However, it is not easy to find an approach which gains acceptance. And success is never guaranteed. If, however, the wrong approach is chosen, failure is certain.

5.5 Conflict diagnosis: what conflicts dominate the problem of final waste disposal?

In the light of a public perception situation characterised by scepticism and anxiety, it is hardly surprising that certain societal groups take up this topic and introduce it actively into the political and societal debate. Here, the conflict[178] over the final disposal of high-level radioactive waste happens on several levels. It entails factual as well as evaluative dimensions and touches upon fundamental convictions regarding technological and social change. On the one hand, a *conflict about preferences and normative evaluations* becomes evident. This conflict manifests itself both in crises of legitimacy and dissatisfied justice claims, whereby different actors pursue diverging goals and are divided when choosing the procedure with which to reach those goals. Classical approaches, such as the top-down approach, i. e., individual legal institutions making decisions for the overall population, lose persuasiveness within those parts of the population which are directly affected by them (i. e., those living in the vicinity of a planned final waste repository). The legitimacy of authorised decision-makers is increasingly questioned by large sections of the population, with their decisions in fact rejected.

In addition to this political crisis of legitimacy there is a severe division amongst the relevant actors on their respective *understanding of justice*. There are very diverse and sometimes controversial answers to the question of how to conduct the process of selecting a site for final high-level nuclear waste disposal in a fair way. Characteristically, there is no single concept of justice but rather a multi-dimensional construct of justice (cf. Deuschle 2007:50ff), which can be interpreted in a duly complex fashion. According to Sabbagh (2002:44ff), the concept of justice can in principle be broken down into the following subcategories: opportunity, performance, need or generational justice. Opportunity justice is characterised by a – both formally and socially – equal distribution of opportunities to participate and find fulfilment in both political and public life.

---

[178] In this contribution we define a conflict as social situation in which the conflicting parties prefer different courses of action which are mutually exclusive or at least obstruct each other. Cf. section B 3.1.

Performance justice, in contrast, emphasises that performance must be adequately rewarded. Justice of needs demands solidarity of the community in order to support those who are unable to lead a dignified life by their own means. Finally, the claim for justice also requires a dimension of generational justice which can be fulfilled if the opportunity to satisfy one's needs is equal for both the current and the future generations. The justice construct is thus extremely complex due to its multi-dimensional nature which allows for a multitude of sometimes contradictory interpretations. Nowhere is this diversity of concepts of justice as vividly and virulently evident as in the question of the final disposal of radioactive waste which must be safely insulated from the biosphere for many millenniums.

This margin of interpretation is additionally widened by contradictory assessments of the factual starting situation. This *conflict of facts* manifests itself in disparate scientific reports, debates and counter-expertises. Dissent regarding facts leads to *conflicting evidence and expertise* (for the term "Expertendilemma" cf. Nennen/Garbe 1996) which, on the one hand, stifles the integrative power of science, while, on the other hand, shatters the public's trust in scientific expertise. Can any of the experts at all be trusted, and if so, which one? Instead of gaining scientific orientation with regard to which waste disposal option carries the lowest risks for people and the environment, the external observer is left feeling confused. To remedy this very deficiency is one of the main motivations behind this report (see also the technological, philosophical and legal chapters).

Especially when the unambiguous nature of the empirical data set is contentious, the interests of individual actors become all the more clearly visible. Since their specific interests are incongruent, even polarising, the *potential for conflict* escalates. The extent of the conflict is dependent upon the degree to which the satisfaction of needs is perceived by each party. If polarising interests collide, such as in the selection of a specific region for the construction of a final nuclear waste repository, it is evident that to satisfy the interests of all the involved parties will hardly be feasible parties. This is particularly the case if expert opinions differ concerning the factual hazard of such a repository and if the necessity of a repository is contentious or at least thematically linked with the future fate of phasing out nuclear energy. Therefore it is important that an agreement be reached amongst experts regarding a fundamental strategy and its implementation.

In order to deal with these different conflicts constructively and arrive at a factually sound and politically acceptable solution, two requirements must be met: a dialogue-based risk communication on the one hand and on the other, a policy based upon a fairness- and discourse-oriented balance of interests.

5.6   A fundamental requirement: effective risk communication

In democratic societies, every citizen is entitled to the public legitimisation of decisions which affect his life and health (cf. Renn et al. 2005:11). This legitimisation of decisions, however, requires communication. The goal of such communication should be to provide the population with the opportunity to become "risk literate". The term *risk literacy* refers to the ability to make an informed risk assessment based on knowledge regarding the factually verifiable consequences of risk-causing events (cf. Renn et al. 2005:11). This does not, however, entail comprehensive information concerning all potential risks, but rather entitles people to a suitable basis of communication. According to the U. S. National Research Council, such a basis of communication enabling citizens to become risk literate can be defined as an interactive process which promotes the exchange of information and opinions of individuals, institutions and groups. The communication includes both information regarding a specific risk along with information addressing concerns, anxieties, attitudes or reactions related to that particular risk (cf. U. S. National Research Council 1989:21). Thus communication is defined as a reciprocal, dialogue-oriented exchange process. A look into the past demonstrates that this definition of communication, particularly in the context of the final disposal debate, has not always existed (Hocke/Renn 2009): Up until the late 1980s, communication with regard to the final disposal debate was largely unilateral. Science experts and the authorities informed the population about potential events based on probability statements. Communication was considered a sort of educational task whose only purpose was to inform the population about certain circumstances. This style of communication did not include a dialogue with far-reaching feedback processes from the side of the population.

In a second communication phase, the communication task was extended by a pedagogical element. Probability statements were linked with warnings and admonitions, in order to actively induce people to change their behaviour. This communication phase was characterised by one-sided communication, with citizens mostly being considered powerless.

Only recently, during the third communication phase, *two-way communication* can be increasingly observed. This form of communication is characterised by the active involvement of all parties potentially affected by a decision as well as all interested actors in the communication process. Furthermore, two-way communication always means dialogue-based communication. According to Bohm, in such cases dialogue must be defined as meaningful communication which allows the involved parties to articulate and mutually explore individual and collective communication requirements, ideas, convictions and feelings (cf. Bohm 1998). This definition of dialogue emphasises the reciprocal open process of two-way communication. Active participation in the communication, construction and evaluation of decisions and measures are essential parts of two-way communication. Thus, the goal of this communication phase is to build trust between the individual actors, by means of a comprehensive understanding of the problem as well as a mutual willingness to learn, thereby establishing a basis for universally acceptable decision-making.

5.7 Approaches to conflict management

Even if the involved parties engage in an intensive risk dialogue, as has occurred to an increasing extent in recent times, it certainly does not mean that the conflictual situation is resolved. Rather, the establishment of a basis of communication is a prerequisite, but by no means the sufficient condition for reaching a universally acceptable solution. In principle, four approaches to the final disposal of high-level radioactive waste can be outlined based on current conditions:

5.7.1 Top-down approach

In this approach, the representatives elected within the democratic system have the sole right of decision-making. By virtue of their office they act in the best interest of the people. An active participation of the population is only provided, if at all, to a very limited extent. However, this solution is also based on transparent risk communication involving the population. Citizens are permitted to voice their opinions during hearings, but without being granted a say in the final decision-making process. The decision-makers must also prove that all objections have been duly dealt with. Then, however, it is in the hands of the decision-makers to make a decision while being obliged to disclose all arguments for and against it.

5.7.2 Top-down and bottom-up mix (muddling through)

This approach relies on minimum consensus (Muddling Through) which emerges from the political opinion process (Lindblom 1959; 1965; Willke 1995). The only options that are considered legitimate are those which bring up the least amount of opposition within society. In this kind of management, societal groups may influence the process of political decision-making to the extent at which they provide proposals which offer connectivity, i. e., which are adapted to the language code and processing style of the political control systems, and mobilise public pressure. In politics then, the proposal which best holds its ground in the competition of proposals will be accepted, i. e., the proposal which entails the least loss of support by interest groups for the political decision-makers. Previous debates on final waste disposal seem to conform most closely to the Muddling Through approach. Depending upon the extent of public pressure, the question is first explored, then postponed and finally decisions which have already been made are revoked. Therefore, muddling through can only lead to a successful conclusion via a polarised debate if the explosive nature of the topic slowly dies down and a general fatigue regarding the issue sets in. Whether or not the factually most sensible and most morally sustainable result prevails is another matter.

5.7.3 Bottom–up approach: discursive site selection

The third option is based on a discursive solution and an attempt at a fair negotiation of the site selection between the different groups involved (Habermas 1991:68ff; 1992:101ff; Renn et al. 2007:188ff). Discursive methods claim to duly account for the rational nature of human beings and to deliver more just and competent solutions to existing problems. No matter which specific claims we associate with discursive processes, they must be structured according to certain rules in order to ensure their effectiveness (for instance to provide constructive solutions to problems in an appropriate and fair manner keeping open more than one possible decision) and to prevent, as far as possible, strategic behaviour amongst the participants. In principle, the legitimisation of collectively binding norms depends upon three conditions: the agreement of all parties involved, a substantial justification of the statements delivered within the discourse as well as a suitable compensation for negatively affected interests and values (Habermas 1981, vol. 1:369ff).

However, one can also choose a strategy of evasion and simply ship the waste abroad. Although most involved parties consider the export of nuclear waste abroad to be morally questionable and are thus sceptical about such a solution, a situation of political paralysis combined with muddling through can render an export solution the only remaining form of conflict resolution upon which consensus may be reached. A joint European solution for waste disposal would be considered morally unobjectionable as long as all the producers of waste accept their fair share of the solution, for instance, that Germany offers potential final waste disposal sites to the same extent as other countries. Deciding which site's turn it is should then be done in a democratic process and according to aspects of technological suitability and social viability. A one-sided relocation of the problem abroad would hardly be acceptable from an ethical point of view.

5.8 A plea for a new beginning with a combined solution

Which of the three options, then, should be pursued? The Muddling Through approach which, as we see it, has dominated all dealings with the subject so far appears the least convincing. Its decision-making process is mostly determined by random constellations, often causes a loss of the legitimacy of the involved institutions, destroys trust in the system and leads, if at all, to solutions which satisfy very few people. Furthermore, a regulatory policy aimed only towards factual acceptance will sooner or later become entangled in contradictions because collective acceptance behaviour is often inconsistent and changeable (Gethmann/Mittelstraß 1992:21).

Simultaneously the affected population becomes increasingly weary of a policy offering neither a clear course nor a claim to leadership. Disenchantment with politics occurs as a consequence.

The obvious alternative of the state legal authority is in principle capable of solving the conflicts. But, as Löfstedt (2005) has shown, the legitimacy of this solution depends upon our trust in the power of judgment and neutrality of the legal decision-maker. Moreover, the perceived fairness of a decision-making process is essential for the acceptance of decisions (Linnerooth-Bayer/Fitzgerald 1996). Both those requirements are questionable in the case of final waste disposal in Germany. The local population in the vicinity of the Gorleben repository has mostly lost their trust in the government decision-makers' power of judgement and neutrality and, since there is currently only one site under discussion, the question of fairness in selecting a site is highly controversial. In short, a decision made by a strong state would come at the high political price of losing legitimacy and the population's trust in the system, a price which few politicians are prepared to pay.

The European solution is only acceptable if all countries which produce a significant amount of radioactive waste enter into a fair competition for finding the most suitable site, and on the basis of this they negotiate an appropriate compensation to be paid by the beneficiaries of the solution to the site communities. However, this very solution is dependent upon a fair prior selection procedure including all of Europe. But we are still miles away from that.

This leaves us with the discursive procedure: in an ideal world this type could be suitable for providing reasonable arguments for the site selection process as well as for the selection itself; however, in the actual case, the requirements for a fair discourse are difficult to fulfil. In a highly politically charged and polarised debate, many strategically-oriented parties have an interest in paralysing practical politics, e. g., by forcing endless marathon sessions through countless procedural motions and peripheral contributions to the discussion (Wiedemann 1994:180; Schönrich 1993). In such cases, "the dictatorship of persistence" (cf. "Die Diktatur des Sitzfleisches" by Weinrich 1972) eventually determines which arguments end up being acceptable. The population is generally disappointed and made insecure by such discourse which starts out with high aspirations and ends in trivial realisations.

What can be done, then, to reach a legitimate solution in this muddled situation? In this context, legitimacy means that a collectively binding site selection can be based on arguments which are comprehensible and acceptable in terms of commitment for those who and are or will be affected by the consequences, but had no opportunity to participate in the decision process.[179] Comprehension and commitment are usually dependent upon

---

[179] "The acquisition, provision and use of political power according to norms based upon discursive arguments and borne consensually by the relevant collective." (Münch 1982:267)

decisions being perceived as fitting the problem while being fair and efficient. First of all, the process of site selection must, directly or indirectly,
ensure fair access for all groups involved. It must be capable of claiming
authority both through the power of the arguments and the inclusion of all
relevant values and interests (cf. Dahl 1989:108ff, Münch 1982:213ff). If we
aim to reach legitimacy in this sense, the following basic demands on regulatory policies in the area of nuclear waste disposal emerge (Hocke/Grun-
wald 2006):

- The selection process must undergo a fundamental reform: additional
  sites which appear technologically and geologically apt must be determined in Germany (or Europe), and a choice from them must be made.
  Alternatively, clear criteria must be determined for a site to meet in order to be accepted as suitable. Only thereafter can we test whether or not
  Gorleben fulfils these criteria. This procedure is meant to ensure that
  the criteria are not consciously or unconsciously chosen in order to fit
  Gorleben. The expert commission must consist of highly-qualified international experts in order to bestow the selection process with credibility and expertise.
- The selection process must be borne to a large extent by fundamental
  consensus amongst the population, i.e., the selection criteria must be
  pre-determined by an expert commission as well as be factually and politically convincing.
- The complete selection procedure must be transparent and comprehensible (effective risk communication criterion).
- The selection procedure must appear fair (all shared value and interest
  groups involved have a say), competent (the problem is treated appropriately and with the necessary expertise) and efficient (the means or costs
  of the decision must be proportionate to the objective) to non-participants.
- The selection itself must be comprehensible and intersubjectively justifiable whilst reflecting the plurality of moral concepts of the affected population in the sense of a fair consensus or compromise.

If we seriously aimed at fulfilling all these demands for legitimising the site
selection process, a single political regulatory instrument would certainly
not suffice. Rather, such wide-reaching decisions call for a sequence of different regulatory instruments, each of which would cover a different aspect
of these criteria.

5.9   Concrete steps towards site selection

If we propose to take into account the above outlined basic requirements
for a legitimate and fair decision, a number of different procedures must
be combined in a specific way: they must achieve both the highest possible effectiveness of the solution (regarding health and environmental risks,

prevention of negative long-term consequences, and cost-efficiency) and, simultaneously, the utmost legitimacy of the decision process. To this end, there are several options which are based in varying degrees upon a mixture of a top-down and discursive approach. We will now outline one option which appears particularly attractive in this respect:

The objective should not be to find the most suitable site for final nuclear waste disposal from among a number of options but rather to determine a site which is suitable, according to objectifiable criteria, for safely storing high-level radioactive waste over the scheduled period of time. Setting the criteria for determining the aptitude of a site for storing high-level radioactive waste should be conducted according to a transparent procedure which would include an international peer review. Technical, geological and spatial planning criteria would be simultaneously developed and transferred to the aptitude test of the Gorleben site (Schenkel/Gallego Carrera 2009), similar to the assessment plan in Swiss procedures for determining sites. Criteria and threshold values would be defined so as to indeed be applicable to Gorleben; however, they would in no way put into question the openness of the aptitude test procedure.

To this end, a group of experts (examination committee) should be convened whose task it would be to determine the criteria for the aptitude test and, at the end of the exploration, to assess whether or not the criteria had been fulfilled. Since this group of experts would carry a high degree of responsibility and would depend upon a high level of legitimacy in this conflict-filled situation, proven scientific, i.e., technological expertise, independence and balance would be important. In order to ensure this, the following conditions would have to be fulfilled:

- The committee should be housed in an independent institution. This could be, for instance, the National Academy of Sciences.
- The members of the committee should be appointed by the German Federal President.
- Members should be proposed by relevant scientific organisations, civil society groups and operators. The selection of the members should be conducted by the institution which will house the committee.
- The sessions of the committee should be open to the public. Representatives of governmental authorities, operators and civil society groups should be allowed to participate in the meetings as guests.
- The committee should be granted the right and the necessary resources to obtain expert advice if deemed necessary and to conduct hearings. Furthermore, individual results of the aptitude test could be subjected to an international peer review.
- The recommendations of the committee should be non-binding. However, the decision-maker would have to produce extremely convincing arguments in order to deviate from the committee's vote.

At the same time as the test procedure, alternative sites should be determined on the basis of existing knowledge and previous explorations; preparations for on-site examinations would need to be made. If the examination of Gorleben indicated that it was not a suitable site, alternative locations could be immediately explored in depth. This simultaneous search would save time and would increase the credibility of the test procedure, too. The staggered intervals of the on-site explorations would certainly be justifiable from a security policy standpoint if the interim storage facilities were upgraded accordingly.

The scientific committee would also be required to work towards criteria according to which alternative sites in Germany could be selected. Such criteria would concern geological host formations, geological aptitude and other technological and scientific topics. The objective would be to find two to four alternative sites which could be explored on site if Gorleben were to fail the aptitude test. The identification of alternative sites would quite definitely bring about major protests within the affected population. But since the on-site exploration would only begin after the results of the Gorleben tests had been presented, this protest could be expected to be less virulent if an intensive exploration were to begin at an early stage.

A more extensive participation of civil society groups would only make sense after a decision had been made as to whether Gorleben was a suitable site or not. If Gorleben were assessed as suitable by the scientific committee and if this assessment were to be accepted by the responsible authorities, immediate citizen participation should be put in motion for the local and regional implementation of the decision. At that point the question would not be "if" but "how". The proposals made by the Working Group for the Selection of Repository Sites (AkEnd) could be taken up. It would be particularly important not to keep uncertainties secret, particularly regarding long-term effects, but to address them openly and to ensure compensation by promoting economic or location development. We are not talking about the sale of indulgences here or about paying people to carry the risk. Rather, those who in future would bear the uncertain consequences and burdens for the general public should receive recognition and support. This procedure would show that accepting uncertainty is respected and honoured. We cannot expect people to silently "swallow" the uncertainty, and therefore unbiased dialogue forums with the groups affected by the consequences would be preferable. We can learn from Sweden in this respect, where such forums are conducted locally. In some cases, experts are involved as sources of knowledge and information. Local does not mean amateur. Professional dialogue facilitation is indispensable. A requirement for the success of such local forums is a "national willingness to pay" since, of course, they do not come free. However, direct payments should be avoided with other forms of compensation, such as establishing an infrastructure, educational opportunities or attractive leisure time facilities being considered instead.

If the recommendation were to say that Gorleben was not suitable, the scientific committee augmented by additional experts would be asked to supervise the current explorations of alternative sites and ensure that the comparative selection was conducted according to scientific and technological aptitude criteria, spatial planning aspects and fairness. If more than one site turns out to be eligible, it is advisable to start a public involvement process for site selections that should not only involve experts, but also representatives of the local communities in questions as well as representatives of civil society.

All activities concerning further decision-making must be accompanied by a proactive, dialogue-based communication programme. Usually, communication cannot resolve a conflict in an amicable way, but it does make a considerable contribution to keeping the tone of the conflict civilised and helping those people who have not yet made up their minds to get enough relevant information so as to reach an informed opinion on their own.

5.10 Conclusions

At the beginning of a new initiative for resolving the question of final nuclear waste disposal we must concentrate on reducing complexity by means of a scientific consensus as to the best selection procedure, coping with uncertainty through fair offers to those who will have to live with the consequences of this uncertainty, and treatment of ambiguity through an open and sincere discourse on objectives regarding the future of our energy supply. Not least, a societal discourse is required concerning a future vision of Germany's path towards sustainability in economic, ecological and social terms.

All activities aimed at participation and communication must be conducted in a timely manner but without undue time pressure and hastiness. In planning law terms, the period available until a repository is built will, even optimistically, take at least a few decades. Thus the time schedule must include sufficient time reserves for participation and communication. The time scale attached in section B 1.5 of this report takes into account additional time periods for this purpose.

It cannot be judged in advance whether or not the considerable time and effort which this combination of procedures necessitates is worthwhile. The procedure of searching for a final nuclear waste disposal is so highly emotionally charged that a rational way of negotiating would be hard to achieve. However, if we assume that the top-down method is not politically viable, there is no real alternative. Considering the growing disenchantment with politics and the increase in legitimacy deficits in modern societies, developing new and complex decision-making forms is unavoidable; not only must they meet the usual criteria of democratic procedures, but their content must also be justifiable and transparent for outsiders. There could be no other topic as ideally suited for this as the final disposal of nuclear waste.

C Annex and apparatus

1 Annex 1: Some fundamental data for the assessment of radiation risk and radiological protection

1.1 Introduction and dosimetric quantities

For the estimation of radiation risk the assessment of the radiation dose and the knowledge of the dose response for certain radiation effects are necessary. The basic quantity for the radiation dose is the absorbed dose, D, given in Gy, which can be measured exactly by physical means. For radiological protection the absorbed dose is averaged over a whole organ or tissue and then the dose conversion factors can be evaluated from measurements of external radiation fields in specific organs and tissues. For internal exposures the incorporated radionuclides have to be determined by either whole body counters or scintillation counters or measurements of radioactivity in the blood, faeces or urine. Dose coefficients for the organs and tissues can then be calculated with the help of biokinetic models. By these procedures the absorbed dose $D_{T,R}$ is determined in an organ or tissue (T) for the radiation (R).

However, the magnitude of the radiation effects is not only dependent on the physically absorbed radiation dose but also on the radiation quality which is determined by the kind of radiation (α-, β-particles, photons, neutrons etc.) and the radiation energy. In exact experimental designs the relative biological effectiveness (RBE) is determined for these reasons. The RBE is obtained by forming the ratio of the dose of a reference radiation divided by the experimental radiation. For such a calculation those radiation doses will be used which cause the same magnitude of biological effects. From such RBE-values the radiation weighting factors have been derived (see section B 2.2) (ICRP 2007) which are used for the calculation of the equivalent dose (H) and the effective dose (E).

1.2 Microdosimetric considerations

1.2.1 Physical considerations

With the transfer of energy of ionising radiation within materials, energy absorption takes place, covalent bonds are broken and ions or radicals are formed. These reactions can directly occur in biologically essential molecules like DNA (direct radiation action) or with water molecules of which the number in living cells is largest. In the latter case radicals of the water (cf. H, OH) are formed which then can react with other molecules like DNA and lead to corresponding damaging events with chemical reactions (indirect radiation action). The transfer of energy occurs in cells and tissues in discrete energy packages. For the development of stochastic radiation effects the changes in the DNA by both direct and indirect radiation actions are considered important.

The principal physical unit in order to describe the energy deposition in organs and tissues is the absorbed radiation dose given in Gy. For radiological protection the energy dose is averaged over the tissue which is considered. This averaged energy dose, however, does not describe the large variability of energy absorption in micro-regions which results from the stochastic nature of energy deposition, from events in individual cells and molecules especially when the energy dose is considered in the low dose range.

In the medium to high dose ranges of low LET radiation (100 mGy and higher) a relatively homogeneous exposure of cells and tissues occurs with low LET radiation (β- and γ-rays). However, this changes in the low dose range when the effects of single ionising particles have to be considered. On the cellular and sub-cellular level microdosimetric considerations have to be introduced under these conditions. As it is assumed that the DNA located in the cell nucleus is the important, sensitive target for the development of the stochastic health effects, the absorbed dose in a single cell nucleus has to be taken into account in order to understand the mechanism of ionising radiation in the low dose range. This dose amounts in average to 1 mGy for cobalt-60 γ-rays (Co-60) radiation when in average one ionising particle of this radiation passes through the spherical cell nucleus with a diameter of 8 μm (ICRU 1993; UNSCEAR 2000). When a tissue with many cells (several hundred millions cells per g tissue) receives an averaged dose of 1 mGy 63.2 % of the cells in the tissue will be hit, 36.8 % of the cells experience the track of one particle. This is equivalent to 58.2 % of the hit cells. Further cells will be hit by several particles (Tab B.25). From these considerations, it follows that with an average number of one track per cell 18.4 % of the cells will receive two tracks. On the other hand, with a radiation dose of 1 mGy 36.8 % of the cells will not be irradiated at all. When the radiation dose is further decreasing the number of cells without a hit will increase. Thus with a dose of 0.1 mGy 90.5 % of the cells will be unhit.

If the energy deposition in a single cell nucleus is sufficient for the induction of radiation damage and an interaction between damaged cell nuclei is not necessary for the cancer development, it is very probable that a dose effect relationship without a threshold dose exists. However, the possibility must also be considered that especially for low LET radiation at least two independent particles have to pass the cell nucleus in order to develop a radiation damage. The number of particle tracks, which pass through the cells, follows a Poisson distribution. The average number of tracks of ionising particles is proportional to the absolute dose.

The microdosimetric arguments for a low dose should be evaluated with respect to linearity of the dose effect relationship for such biological effects for which the radiation effects are induced only in those cells which have been passed by at least one ionising particle. This is apparently the case for cell killing, for the induction of chromosomal aberrations and of mutations in single cells. It is, however, unclear whether this phenomenon is also valid for the transformation of normal to malignant cells. It further has to be considered that unhit cells can show an altered gene expression when a cell was hit in the neighbourhood. Thus, an increased expression of the protein p21 was observed in unhit cells (Little 2000). Such an effect is called "bystander effect". Many studies of this phenomenon have been undertaken during recent years. However, until now it is not clear what will be the impact of bystander effects for the development of health effects after radiation exposures. The mechanism of these phenomena is unexplained up to now.

The situation with respect to dose distribution is somewhat different for the exposure to densely ionising radiation with high LET. α-Rays have a very short range in tissue that is dependent on the energy of the α-particles which are formed through the radioactive decay of the corresponding radioactive isotopes. Thus for α-particles which are formed through the radioactive decay of ^{226}Ra, ^{238}U, ^{239}Pu and others with energies of up to about 7.8 MeV a maximal range of around 80 μm is observed in mammalian tissues. For 5 MeV α-particles of ^{239}Pu the maximal range is around 40 μm in biological tissues. If one considers that the diameter of cell nuclei of human cells is in the range of 5 to 10 μm and the diameter of the cells in the range of 10 to 30 μm this demonstrates that α-radiation can reach in average around 1 to 2 and maximally up to 5 cell layers from their place of origin.

The energy, which is deposited by one single α-particle passing through the cell nucleus is extremely variable. The energy dose can vary from small doses (in the range of mGy) up to more than one Gy even between microregions of the same cell nucleus. These considerations demonstrate clearly that the definition of average tissue doses is an oversimplification for energy deposition especially of high LET radiation. Individual cells in a tissue will experience very different radiation doses. It is therefore very important how the α-emitting radioactive isotopes are distributed within the tissue. Very frequently only the surfaces of an organ within a body will be reached by such radiation qualities especially when the radioactive substance is located in a neighbouring tissue or organ. This is valid for instance for radon and its radioactive decay products in the lung and for ^{239}Pu as well as ^{226}Ra, which are deposited in the skeleton.

Under these circumstances only less than 0.2 % of the cell nuclei are hit by an α-particle if the cells of a tissue receive in average a dose of 1 mGy (equivalent to 20 mSv) of α-radiation, while more than 60 % of the cells are hit by Co-60 γ-radiation at the same average tissue dose (absorbed dose). With such an average radiation dose of α-radiation, around 99.8 % of the cells experience no radiation event (Tab. B.25 of section B 2.7.1). On the other hand, when an α-particle (with energy of around 5 MeV) hits a cell nucleus a high-energy deposition will occur in the corresponding cell nucleus on average in the range of 370 mGy. In individual cell nuclei, the dose of such a radiation quality can reach values up to 1000 mGy (UNSCEAR 2000; Streffer et al. 2004).

Tab. B.25: Proportions of a cell population traversed by tracks for various average doses from γ-rays and α-particles (approximately 1 mGy for γ-rays and 370 mGy for α-particles per track passing through a cell nucleus on average) (UNSCEAR 2000)

Mean tracks per cell	Percentage of cells in population suffering					
	0 track	1 track	2 tracks	3 tracks	4 tracks	>5 tracks
0.1	90.5	9	0.5	0.015	–	–
0.2	81.9	16.4	1.6	0.1	–	–
0.5	60.7	30.3	7.6	1.3	0.2	–
1	36.8	36.8	18.4	6.1	1.5	0.4
2	13.5	27.1	27.1	18	9	5.3
5	0.7	3.4	8.4	14	17.5	56
10	0.005	0.05	0.2	0.8	1.0	97.1

Thus in the low dose ranges (average tissue doses of 1 mGy or smaller) ionisation events will occur only in a small percentage of the cells and the number of hit cells depends significantly on the radiation quality (radiation energy and type of radiation). This means that small doses cannot really be defined based on these microdosimetric considerations and they are very heterogeneously distributed on the cellular and sub-cellular level.

Each track of a radiation with low LET like γ-rays induces only a relatively small number of ionisations at the passage through a cell nucleus of medium size. – The ionising events are caused by secondary electrons, which are released by the interaction of the photons from γ-rays with cellular molecules. – Thus, in average around 70 ionisations occur through these electrons at a nuclear passage of a γ-quantum of Co-60 γ-radiation. This corresponds to an average observed energy dose of 1 mGy in the cell nucleus as it has been described before. The large variability of these processes has already been discussed. In contrast radiation exposures with high LET, cf. 4 MeV α-particle radiation, leads to many thousand ionisations and therefore yields a relatively high dose in an individually hit cell nucleus. In case of such a radiation, around 25.900 ionisations occur when the particle passes through a cell nucleus. This corresponds to an observed dose of around 370 mGy (UNSCEAR 2000).

For depositories of high level radioactive waste a dose limit or constraint of 0.1 mSv per year is foreseen for the late phase after closure. This means for low LET radiation that such a radiation dose will cause a hit in about 10 percent of the cells per year. In the case of 1 nSv, as calculated in Tab. B.25, around 1 cell in a million of cells will be hit.

1.2.2 Biological considerations

Another possibility exists in order to describe the low dose range based on biological effects. After the exposure to low let radiation (γ-rays, β-rays) the extent of radiation effects can be described by dose effect relations with a linear and a quadratic term of the dose cf. For chromosomal aberrations, somatic mutations and cell transformation. Further, it has to be regarded that biological effects, which are observed after radiation exposures, like chromosomal aberrations, mutations or cancer, already occur without any radiation (spontaneous effects). For this reason, a constant term "c" has to be considered additionally in possible equations. A dose effect curve can then be written in the form:

$E(D) = \alpha D + \beta D^2 + C.$

In this formula α and β are constant coefficients for the linear and quadratic term of the dose respectively. These coefficients vary for different endpoints and possibly also for various defined radiation conditions. Such dose effect relationships have been studied after radiation exposures especially for chromosomal aberrations, mutations and cell killing. Frequently α/β-ratios of around 200 mGy have been observed for Co-60 γ-radiation. This corresponds to a medium radiation dose after which the linear and the quadratic terms contribute to the radiation effects to about the same extent. From such a value it results by calculation that the action of radiation increases in a linear way in the low dose range with radiation dose up to around 20 mGy, as the contribution of the quadratic term is low in this dose range (UNSCEAR 2000; Streffer et al. 2004). Then the contribution of the quadratic term amounts to around 9 % of the whole radiation effect. Even after 40 mGy the contribution of the quadratic term is only around 17 % of the total radiation effect.

On this basis and convention a radiation dose in the range of 20 to 40 mGy has been called a low dose (UNSCEAR 2000). In an earlier UNSCEAR report (1993) experimental data were analysed for the carcinogenesis after irradiation (especially of mice) with various dose rates of a low LET radiation. It was concluded and proposed based on these data that a dose rate of around 0.06 mGy per minute can be considered as a low dose rate when the exposure lasted for some days or even weeks. With such a dose rate, the induction of the tumour frequency was reduced in comparison to higher dose rates when equally total doses were compared. With smaller dose rates than 0.06 mGy per minute, no further reduction of the tumour rate per dose unit was obtained. The UNSCEAR committee therefore concluded that a dose rate of 0.05 mGy per minute could be considered as a low dose rate.

The evaluation of epidemiological data for carcinogenesis in humans resulted that a radiation dose of less than 100 mGy (mSv) of low LET radiation was considered as a low dose by UNSCEAR (1993), as no radiation effect can be observed in this dose range for general populations (both sexes, all age groups). The international committee of the United Nations (UNSCEAR) defined doses below 200 mSv as a low dose UNSCEAR (1993). Under the assumption that also for humans a linear-quadratic dose effect relationship may exist for carcinogenesis it can be concluded from the observed epidemiological data that within a dose range of 200 mGy the quadratic dose term is responsible for around 10 % of the effect (UNSCEAR 1993).

With respect to waste depositories radiation doses in the range of 1 mSv and lower are discussed for individual persons of the public during the operational phase and of 0.1 to 0.3 mSv for the later phase. Further we have seen from the estimates of Marivoet et al. (2008) that radiation doses in the range of nSv per year and TeraWatt-hour(e) have been calculated. These doses range in all cases in the very low dose region on biophysical as well as on biological considerations.

1.3   Development of health effects after radiation exposure

It has already been described that two principal dose response curves have been described. For the radiation effects in the low dose range (<100 mSv) which are relevant for all phases of waste depositories only a linear dose response without a threshold (the LNT model) is relevant. The effects which have to be considered under these conditions are hereditary effects and the induction of cancer.

For hereditary effects radiation effects can only be extrapolated from animal experiments. With these studies the so called doubling dose has repeatedly been estimated. This is the radiation dose which induces just as many hereditary damages (mutations) as already manifested with a radiation exposure. For the doubling dose a value of 1 Gy has been estimated for low LET radiation and chronic exposure. All attempts to evaluate the genetic risk for humans have not indicated that the given value for the doubling dose leads to an underestimation of the genetic risk in humans (UNSCEAR 2001). All the experimental data can be best described by a linear dose response curve without a threshold. Such a shape of the dose response is also possible on the basis of mechanistic considerations.

The situation is more critical for the induction of cancer as it has already been pointed out. There are critical discussions whether the LNT model is also valid for the induction of cancer (Beir 2005; Tubiana et al. 2005; ICRP 2007). Experimental and epidemiological evidence has been described for such a dose response but it has also been strongly disputed (Tubiana et al 2005). There is no scientific proof for the dose response of cancer induction in the dose range below 100 mSv. During recent years a number of biological processes have been studied which may modulate the dose response especially in the low dose range. The discussion will be focussed on these questions in the following. Nevertheless for radiological protection the LNT model is used for the extrapolation of radiation risk from high and medium radiation doses to low dose ranges (ICRP 2007; Streffer 2009).

1.3.1 Epidemiological findings and their limits

The most extensive epidemiological studies after exposure to ionising radiation are the investigations of cancer incidence and mortality of the survivors of the atomic bombing in Hiroshima and Nagasaki. With the recent data cohorts of 86,572 survivors with 9,335 cancer deaths and 105,427 survivors with 17,448 primary cancer diseases were analysed which came to more or less the same conclusions (Preston et al. 2003; Preston et al. 2007):

- Up to radiation doses of 2 Sv the data can be described by a linear dose response curve without a threshold.
- A statistically significant increase of cancer (all solid cancers) is observed after radiation doses >120 mSv.
- The excess relative risk per Gy (Sv) is about 0.47 for persons at the age of 70 years and exposure at age of 30 years averaged over both sexes.
- Women are more radiosensitive than men by a factor of about 1.7.
- Children and adolescents are generally more radiosensitive than adults.
- Strong differences exist with respect to the radiosensitivity between the different organs and tissues.

These studies are the basis from which ICRP derived the risk factor of 5×10^{-2} per Sv for stochastic effects after exposure to low LET radiation in

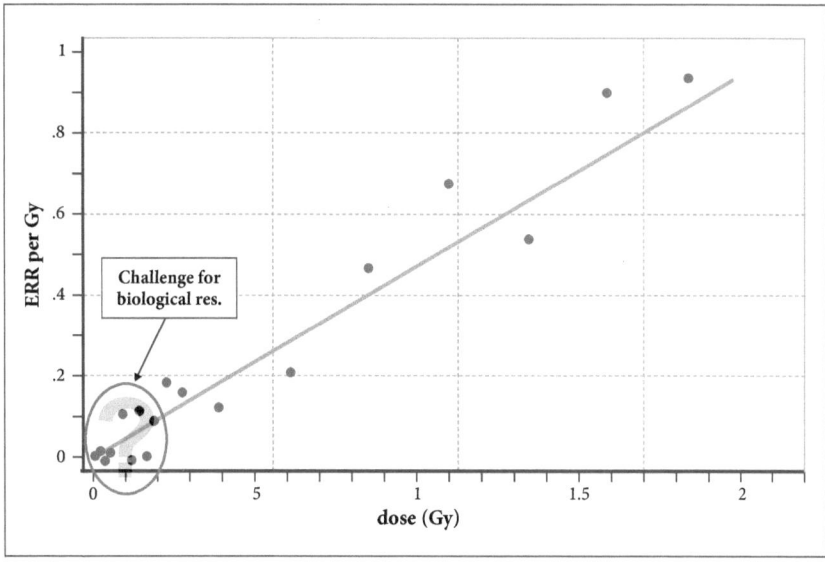

Fig. C.1: Estimated excess relative risk (ERR) in Hiroshima and Nagasaki. In the low dose range possible radiation effects cannot be measured as the scatter of "spontaneous" cancer is larger than the possible radiation effect. The decisive problem is: no specific signature exists for radiation-included cancer. The radiation effect must be evaluated statistically.

the low dose range with low dose rates and of 10^{-1} per Sv for high LET radiation (ICRP 2007).

Quite a number of other epidemiological studies about the induction of stochastic effects and especially of cancer in humans after exposure to ionising radiation are compatible with the data from Hiroshima and Nagasaki. This is the case for investigations on nuclear workers (Cardis et al. 2007; Muirhead et al. 2009), for the population at the Techa River exposed to radioactive releases from the Russian fabrication of atomic weapons (Krestinina et al. 2005) and for populations living in regions with high background radiation (Sugahara et al. 2005). In all studies no significant increase of cancer induction has been found in the low dose range (<100 mSv). The data which have been obtained with the studies on the atomic bomb survivors show fluctuations around the linear dose response below doses of about 100 mSv (FigC.1) (Streffer 2009). This can be explained by two possibilities:

(1) No cancers are induced after exposures to such low radiation doses.
(2) Cancers are induced after these low doses but the effect is so small that it is hidden by the fluctuations of the spontaneous occurrence of cancer (Fig. C.2).

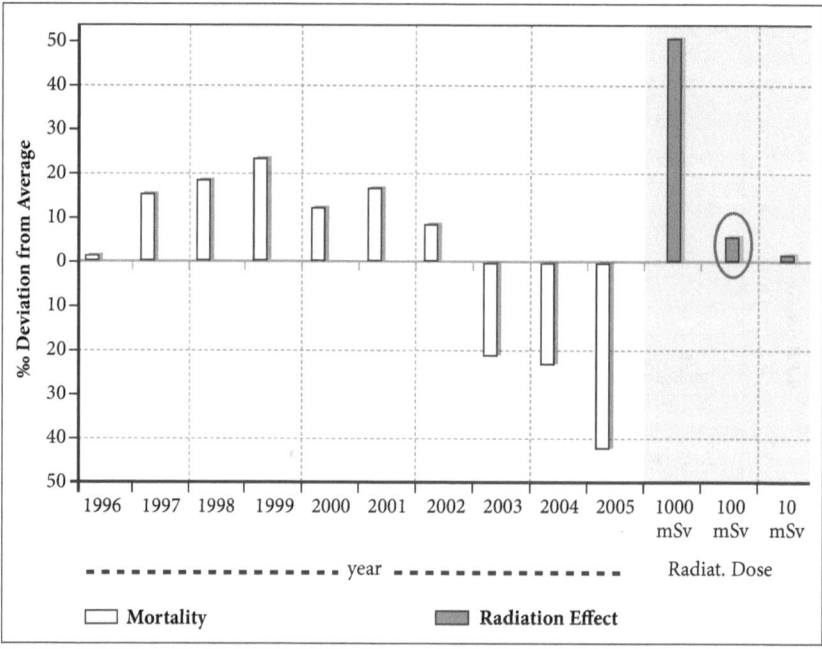

Fig. C.2:   Deviation of cancer mortality from the average (‰) in 1996–2005 (SEER-USA) and radiation effect. (ICRP 2007; Streffer 2009) The radiation effect < 100 mSv is covered within the "noise" of the spontaneous cancer.

In Fig. C.2 the large fluctuations of the annual cancer rate can be seen which has been observed even with the large population of the USA and which is represented as the deviation from the average cancer mortality (SEER) over ten years. In comparison to these values the expected cancer mortality after radiation doses (low LET, low dose rate) of 1,000, 100 and 10 mSv is shown. It is obvious that the possible radiation effect of doses <100 mSv cannot be discovered as the fluctuations of the background cancer rates are larger than the radiation effect in these low dose ranges. An individual cancer which may have been caused by ionising radiation can by no means be distinguished from cancers which originate from endogenous or other unknown causes ("spontaneous" cancer or background). There does not exist a specific signature for radiation which would make such a distinction possible. The clinical appearance and all pathological, cellular as well as molecular features of radiation induced cancers which have been studied so far do not give any indication for a difference. It appears that the evaluation of the mechanism of carcinogenesis can bring clarification whether cancers can be induced by low or very low radiation doses and how the dose response curve looks like in the low dose ranges. It is a great challenge for radiobiological research to contribute to the solution of these questions.

1.3.2 DNA damage and repair

The present view is that the genome of a cell, the DNA, is the primary target for ionising radiation in order to induce stochastic effects including cancers and there is strong experimental evidence for this assumption. Intensive studies have been undertaken to evaluate the DNA damage. The prominent changes after exposure to ionising radiation are:

- Breaks of the polynucleotide strands, there can occur single strand breaks (SSB) or double strand breaks (DSB),
- Base damage, either a DNA base is completely lost or a base is radiochemically altered.

Analyses of the track structure and of the distribution of ionisation events in the DNA helices revealed that clusters of damage occur after exposure to ionising radiation: Very frequently damaging events occur in the direct neighborhood to an SSB or DSB therefore and form a "complex SSB" or a "complex DSB" (Fig. C.3). Since forty to fifty years it is known that these DNA damages can be repaired in living cells by different, very sophisticated enzymatic pathways. The complex regulation and the efficiency of these processes are dependent on the type of the DNA damage. In general the DNA repair of DSB and especially of complex DSB is slower and more difficult than that of other damage types. With DSB also misrepair can occur. Misrepaired DSB may be involved in the initial steps for the development of cancer (Streffer 2009).

These mechanisms are not fully understood until now. Until about fifty years ago it was assumed that DNA is a stable molecule in order to maintain a healthy organism throughout lifetime. It was a firm dogma that any damage in the DNA is an irreversible process which leads either to a mutation or to cell death. Today it is well-known and proven that DNA is quite a labile molecule and the stability of the genome of the organism can be maintained throughout lifetime only by DNA repair. These processes are an essential part of evolution in nature. The occurrence of clustered DNA damage is unique for ionising radiation (Goodhead 1994). Chemical toxic agents generally do not generate such clustered complex DNA damage in the low dose range. The damaging events of such chemical agents are usually isolated events in the low dose range. Further the quantitative distribution of the various damage types is dependent on the radiation quality. Low LET radiation (β-, γ-, X-rays) induces less DSB and especially less complex DSB than high LET radiation (α-rays, neutrons). This is apparently the reason for the general observation that DNA damage of high LET radiation is repaired slower and less efficient than damage of low LET radiation and therefore high LET radiation leads to higher radiation effects than low LET radiation when equal absorbed doses are compared. Fig. C.3 shows DNA damage in human lymphocytes after irradiation with 2 Gy X-rays (100 percent at time zero) at different times of incubation for DNA repair thereafter. It is demonstrated that in healthy persons the DNA damage (DSB) very efficient and within few hours to a large extent.

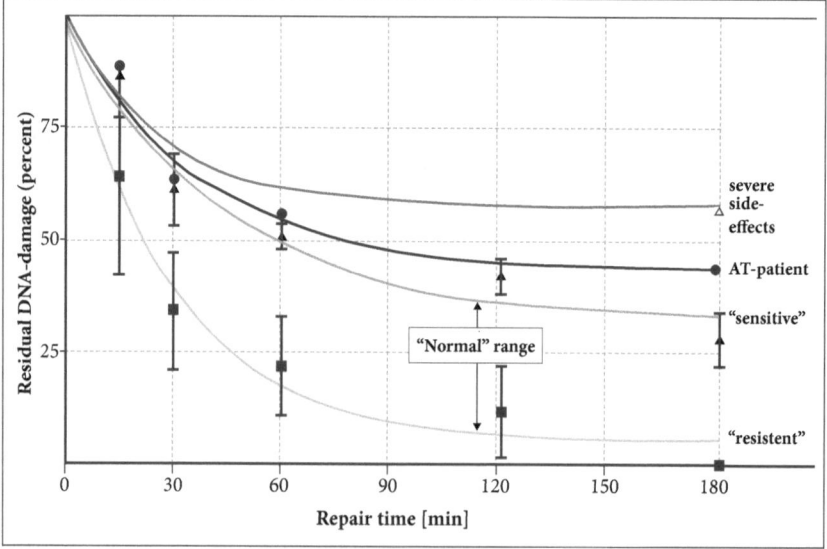

Fig. C.3:   Repair Kinetics of DSBs in lymphocytes of humans (range of healthy persons, patient with the genetic predisposition ataxia telangiectesia, AT and a further patient with severe side-effects an unknown genetic predisposition of increased radiosensitivity) (Streffer 2009)

Further it has been shown that the efficiency of DNA repair is dependent on the genetic disposition. The radiosensitivity of individuals can differ widely due to the genetic disposition (Fig. C.3). Most humans fall with respect to their radiosensitivity into a certain range with a Gaussian distribution. However, some individuals have been observed with a very strong increase of radiosensitivity by cellular studies and clinical observations. These persons show a strong repair deficiency (AT-patient and patient with "severe side effects") (Fig. C.3) (CRP 1998). With these individuals all deleterious radiation effects are enhanced.

1.3.3 Dose modifying phenomena

Extensive biological studies have demonstrated during recent years that several biological phenomena ("New Biology") can modulate the dose response in the low dose range. These phenomena may also modify the dose response curve in various ways in the dose ranges where no significant epidemiological data on cancer induction are available (Fig. C.4). Very important phenomena are DNA-repair processes which have already been discussed. Further adaptive response, apoptosis, bystander effects, genetic disposition, genomic instability, hyperradiosensitivity and immune response have to be mentioned. Some of these phenomena will be discussed in the following.

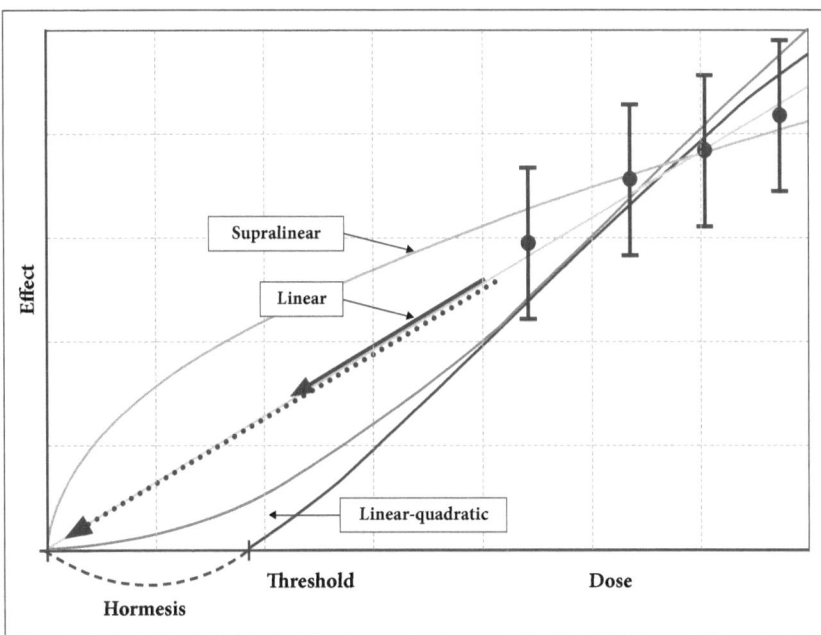

Fig. C.4:   Possibilities of extrapolation into the lower dose range

Adaptive response has been frequently observed during the last 20 years with many organisms starting with bacteria up to mammalian organisms including humans (UNSCEAR 1994). In general biological objects, usually cells like bacteria or human lymphocytes, are irradiated with a low radiation dose (adapting dose in the range of 5 to 200 mGy), about four to 24 hours later a higher dose (challenging dose in the range of 1 to several Gy) is given and then the biological effects (with lymphocytes usually chromosome aberrations) are measured. In parallel the effect of the challenging dose only is measured. Quite often the radiation effect is reduced with the combination of adapting dose plus challenging dose in comparison to the effect of the challenging dose alone. The cells have become more resistant against ionising radiation within the interval, they are adapted. Apparently the DNA repair has become more efficient by adaptation. Such effects have been shown in many cases throughout the whole animated nature with prokaryotic as well as eukaryotic organisms.

However, these effects can be very different between individuals (Streffer 2009). The adaptive response is apparently dependent on the genetic disposition. No adaptive response was observed in cells from individuals with hyperradiosensitive syndromes like Ataxia telangiectesia (AT). Several studies have shown that no or very little adaptive response developed with high LET radiation. During prenatal development no or little adaptive response could be observed and further it has been found that adaptive response decreases apparently with age. It has also to be considered that very distinctive conditions with respect to the seize of the adapting dose and its dose rate, the time interval between adapting and challenging dose and other parameters have to be kept within certain limits in order to observe adaptive response. Thus it can be concluded that adaptive response is a very important biological phenomenon of high scientific interest. However, it has a number of limitations, it is not an universal phenomenon, it does not operate in generality under all conditions.

Apoptosis is a very powerful cellular mechanism to eliminate damaged or no longer needed cells e. g. during prenatal development by triggered intracellular processes. It can be increased after radiation exposure and it is assumed that apoptosis may also eliminate malignant cells so that the cancer risk is reduced. It has further been shown that small radiation doses can induce an adaptation to increased apoptotic activities but again this differs very much between individuals (UNSCEAR 2000; Streffer 2009). Apoptotic cell death is induced by complex intracellular signal transduction mechanisms which is triggered and regulated by a number of molecular factors (e. g. the tumour suppressor p53). These factors are also connected sometimes to the cycle of cell proliferation. At these branching points the cell can decide to undergo apoptotic cell death or proliferation. In many cancers the tumour suppressor p53 or other regulating factors are inactivated by mutation or other translational processes. In these cells apoptosis is reduced and thus also the mechanism of cell elimination by apoptosis does not work. Therefore apoptosis can certainly be a mechanism to reduce the development of cancer after radiation but there are a number of situations were this mechanism does not operate.

For a long time it was accepted that the radiation-induced chromosomal damage is expressed at the first mitosis taking place after a radiation exposure. Nowadays it is well-known, however, that this is not the case but it has been clearly demonstrated that new chromosomal aberrations also appear at later mitotic cell divisions (UNSCEAR 2006). It was even more surprising that an increased number of chromosomal aberrations was found in many cell systems 20 to 30 cell generations after a radiation exposure. The cells had developed an increased "instability of the genome". Such effects have been found in many cell systems and organisms (in vivo and in vitro) during the last 20 years (UNSCEAR 2006). Genomic instability and its increase after radiation exposure seems to be a very important mechanism for the development of cancer, as this process of cancer development is a multistep process in which several mutations have to follow. In cell systems with increased genomic instability the probability for mutations is enhanced. Such effects can be observed in dose ranges of 10 to 100 mSv.

Besides these phenomena extensive experimental studies have been performed during last years on the so-called bystander effects. Thus it has been observed in cell cultures with single cell irradiation that not only the exposed cells show a response but also unexposed neighbour cells (UNSCEAR 2006). These bystander effects have been mainly studied with cells in vitro. They may lead to an enhancement of the radiation effects in vivo. However, also protective effects have been discussed in this connection. Nevertheless all these phenomena can have the ability to modify the dose response in the low dose range. In which way this could happen is unclear until now. It should further be stated that in the development of these radiation effects epigenetic effects are involved although the mechanisms for bystander effects and for the increase of genomic instability are not clear at all. These phenomena are intensively studied in large research projects (e. g. EU-project NOTE) in order to find its impact on the dose response in the low dose range and to formulate a "new paradigm" for radiological protection. The complexity of carcinogenesis is by far not understood until now a new approach, considerations of system biology may be helpful in this situation (Streffer 2009). In order to get some insight into the effects of very small radiation doses the knowledge of the mechanism of cancer development is of utmost importance.

1.3.4 Mechanism of carcinogenesis and association with genomic instability

The present concept about the mechanism of cancer development is roughly the following: The initial events are changes/damage of DNA e. g. by ionising radiation which may be repaired completely or the damaged cell starts to proliferate with either unrepaired or misrepaired DNA. In the latter case the daughter cells will carry a mutation, further proliferation can lead to cell transformation, malignant cells are formed. These cells may stay silent for many years, during which they can be removed by apoptosis or immune defence. However, also further mutations by radiation or facilitated by genomic instability may alter the regulation of cell proliferation which stimulate the whole process to result in pre-cancer stages.

After further cell proliferation and mutations a carcinoma in situ is formed which then can develop to cancer with metastases. Thus, in summary, the development of cancer is mainly accomplished by several successive mutations and extensive cell proliferation (Fig. C.5). It is assumed that a cancer develops from one malignant cell. A cancer diagnosed in the clinic has around one or several billion cells.

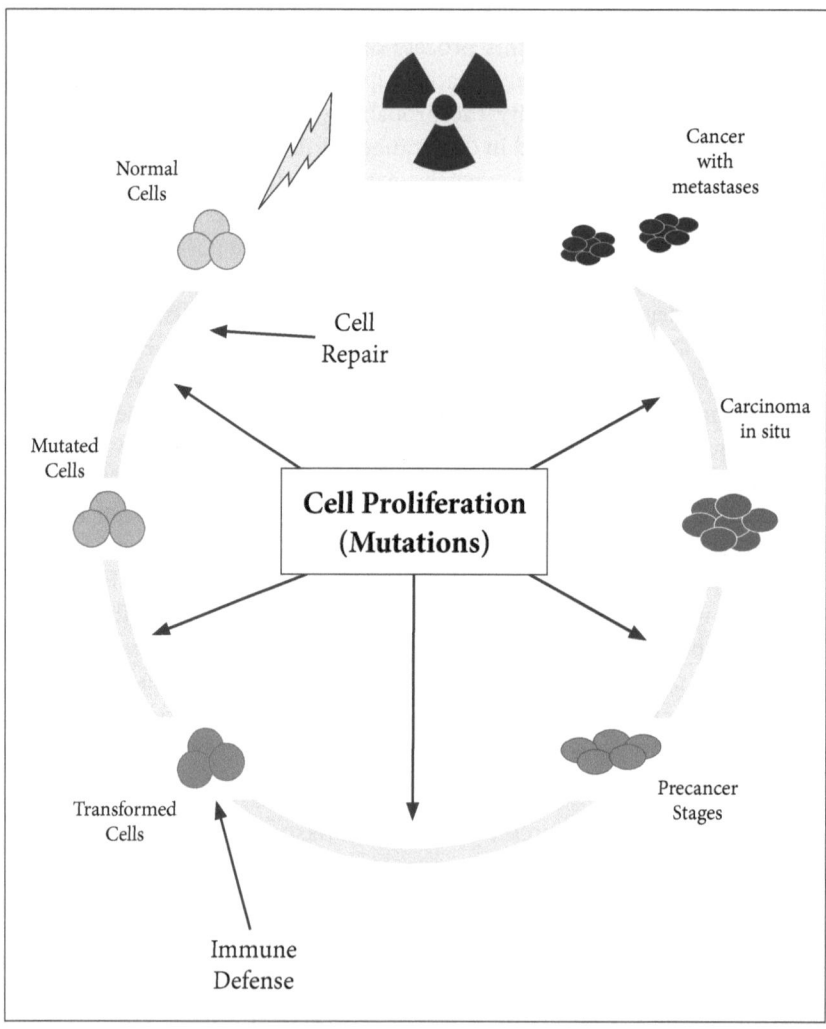

Fig. C.5:   Mechanism of cancer development (modified from Streffer 2009)

Clinical experience and experimental studies have shown that several syndromes with specific genetic predisposition for high radiosensitivity exist which have been described genetically with their molecular features: Ataxia telangiectasia, Bloom's Syndrome, Fanconi Anemia, Li Fraumeni Syndrome, Neurofibromatosis and Retinoblastoma (ICRP 1998). All individuals with these syndromes show proneness for cancer, reduced DNA-repair and/or regulatory changes of the cell cycle as well as increased genomic instability (Streffer 2010). These data demonstrate a strong evidence for a causal association between genomic instability and cancer. It has been estimated that individuals with these predispositions for increased radiosensitivity may develop cancer by a factor of around five in comparison to individuals of the "normal" population.

In summary: Epidemiological studies are important in order to evaluate quantitative risk factors for cancer after radiation exposure, however, they will not solve the open question about the risk in the low dose range (<100 mSv). Biological studies show effects (e. g. DSB; chromosome aberrations) down to dose ranges of several to fifty mSv which is lower than it can be observed with epidemiology. These studies support the view that no threshold exists for certain effects like mutations. How much such effects contribute to the development of health effects has to be solved. Observations of radiation effects <1 mSv appear impossible due to background effects by endogenous processes and radiation effects from natural sources. Genomic instability is associated with the development of cancer. It is increased in all individuals who have a high radiosensitivity. Studies of "new biology"-processes modulate and interact with the development of late radiation health effects. They may lead to a modification of the LNT model but the impact and in which way cannot be foreseen in the moment.

Radiation-induced cancer is dependent on many factors. It differs from organ to organ according to organ-specific, regulatory biological processes. Therefore the dose response is different for various cancer entities etc. For a uniform system in the low dose range (both sexes, all ages, all sensitivities, all radiation qualities) LNT with reference values appears to be the only way to go for prospective radiological protection with the appropriate safety. In a number of cases this model certainly leads to overestimation of the risk but this should be accepted for today. The LNT model principally expresses that radiation exposures induce health effects like cancer even after small or very small doses. However, in these dose ranges the probability of such effects becomes very small and possible effects are completely overshadowed by the cancer development induced by endogenous processes and other exogenous factors. It also has to be realized that life on earth will be exposed by ionizing radiation from natural sources even in the far future. The natural exposures include low LET as well as high LET radiation like α-particles. The ionization processes and its further implications on the development of biological effects cannot be distinguished from the reaction of radioactivity from nuclear installations.

2 Annex 2: Legal questions – comparative experience in selected countries

The following chapter compares in greater detail the regulatory experience important nuclear countries have made in managing high level nuclear waste. It encompasses the United States, France, the United Kingdom, Switzerland, Sweden, Finland, Japan and Spain. The structure of the presentation corresponds to that of the analysis of German law in chapter 4 so as to facilitate the comparison between the respective countries as well as with Germany for the reader.

2.1 United States

2.1.1 Sources of regulation

In the United States, the most important statutory text regarding the management of high level radioactive wastes is the Nuclear Waste Policy Act (NWPA) of 1982 (42 U.S.C. § 10101 et seq.) which was set into force in early 1983 and amended several times (see NEA 2008a:14; NEA 2009f:8–11, 25–31). The Act was extensively amended in 1987 and further amended in 1992 and 2005 by enactment of the Energy Policy Act of 1992 (Publ. L. 102-486) and the Energy Policy Act of 2005 (Publ. L. 109–58) (NEA 2008a:14). The NWPA establishes the U. S. policy for disposing of high level radioactive waste and spent nuclear fuel, it empowers the Secretary of Energy to select, construct and operate the final disposal facility and authorises the Environmental Protection Agency (EPA) to establish regulations and permit requirements for the construction and operation of the facility. Besides, the National Environmental Policy Act (NEPA) of 1969, as amended in 1995 (42 U.S.C. § 4321 et seq.), plays an important role with respect to financing commitments, siting decisions and construction and operation permits since all major Federal action that has a potential adverse effect on the environment is subject to the requirement of carrying out an environmental assessment or environmental impact assessment. By contrast, the Atomic Energy Act of 1954 (42 U.S.C. § 2011 et seq.), although in principle applicable to final disposal facilities, is largely superseded by special regulation. The Act empowers the EPA to set general environmental radiation protection standards and governs certain aspects of the operation of radioactive waste repositories.

The three authorities responsible for nuclear waste have adopted a number of regulations and guidelines that specify the provisions of the NWPA, in particular special standards for the repository at Yucca Mountain (Disposal of High Level Radioactive Wastes in a Geological Repository at Yucca Mountain Nevada, 10 CFR Part 63, and Public Health and Environmental Radiation Standards for Yucca Mountain Nevada, 40 CFR Part 197). The general NRC standards for protection against radiation (10 CFR Part 20) and the EPA radiation protection standards, called "Radiation Protection Programs" (40 CFR Subchapter F [190 Series]) are not applicable (42 U.S.C. § 10141).

As regards high level radioactive wastes originating from military use, there is a special regime which will not be dealt with here.

2.1.2 Responsibilities

In view of the long-term and high risk associated with the disposal of high level radioactive wastes for human health and the environment the NWPA vests the primary responsibility for the disposal of such wastes in the government. The disposal of such waste is declared as a matter of federal policy (42 U.S.C. § 10131(a)(2),(4) NWPA). In particular it is the Federal Government that shall select, construct and operate the final disposal facility. The extent of the state duty to protect is defined in the NWPA at least in general terms. In the view of Congress, radioactive wastes create potential risks and require safe and environmentally acceptable methods of disposal. Adequate protection from nuclear hazards must be provided so as to ensure that such wastes do not adversely affect public health and safety and the environment for this and future generations (42 U.S.C. § 10131(a)(1),(b)(1) NWPA; NEA 2009f:2). Since hazard in the USA legal terminology is equivalent to a mere potential risk, one has to interpret this pronouncement as an expression of the precautionary principle.

The role of industry is limited to considering the waste dimension in operating the facilities that are a source of radioactive wastes and managing such wastes before final disposal, especially by way of interim storage before acceptance for disposal. Beyond, the polluter-pays principle governs the financial responsibility for high level radioactive waste management. Under the NWPA (42 U.S.C. § 10131(a)(4)) the operators of nuclear facilities and other owners of radioactive waste must pay the full cost of implementing the waste disposal programme, especially the selection, construction and operating costs of the relevant interim and final disposal facilities.

2.1.3 Institutional framework

The institutional framework of radioactive waste management is relatively simple (see NEA 2009f:2–7). Within the administration, the Ministry competent for radioactive waste disposal is the Department of Energy (DOE). In particular, the DOE has the responsibility to select, develop, construct and operate a geological repository for high level radioactive wastes. The Nuclear Regulatory Commission (NRC) is an independent agency that has the competence for regulating the technical aspects of construction and operation of high level radioactive waste repositories. This task includes the granting of the requisite permissions. The Environmental Protection Agency (EPA) has responsibility to develop generally applicable standards for the protection of health and the environment against radiation originating from operations and disposal.

There are also institutions of a more technical and scientific character that play a role in regulating radioactive waste disposal. The National Academy of Science (NAS) has the task to give science-based recommendations to EPA regarding the development of standards. The Nuclear Waste Technical Review Board (NWTRB), created by the 1987 Amendments, is charged with evaluating the technical and scientific validity of activities undertaken by the DOE at the Yucca Mountain site.

2.1.4 Strategies

The basic legal requirements for establishing strategies for disposal of high level radioactive waste can be derived from the statutory language of the NWPA, especially its policy pronunciations (goals provisions), provisions on site selection and the permit requirements (See NEA 2009f:13–15). In particular, the Act provides that the siting, construction and operation of final repositories must

> provide a reasonable assurance that the public and the environment will be adequately protected from the hazards posed by high level radioactive wastes as may be disposed in a repository (42 U.S.C. § 10131(b)(1)).

Already in its original version, the NWPA has selected a very simple basic option of managing high level radioactive waste, namely the final deposit of all such waste in a deep geological repository (42 U.S.C. § 10131(b)(1)). This option implies that – in contrast to some European states – no attempt will be made to reduce the quantity and radioactivity of the relevant wastes and increase the energy yield of the relevant nuclear material by reprocessing, partitioning and transmutation. As far as can be seen, this "once through" option was not controversial in the early legislative process in 1982 and not fundamentally challenged later on. However, the NWPA mandates continuing research on alternative options (42 U.S.C. § 10202). In the more recent discussion, especially after the new administration has taken office, an alternative option in the form of a prolonged interim storage for 100 years has emerged. It is too early to assess whether this will lead to a policy shift which would have to be effectuated by amending the NWPA.

Moreover, the NWPA contains a clear pronouncement in favour of retrievability of radioactive waste deposited in the repository (42 U.S.C. § 10142). The repository shall be so designed and constructed to permit the retrieval of the wastes during an appropriate time of operation (not after closure). The motivations underlying this policy are reasons of public health and the purpose of permitting the recovery of the wastes.

Since the United States is a large producer of high level radioactive wastes, the possibility to deposit such wastes abroad is not considered as a realistic and responsible possibility. On the other hand, the American solution is "introvert" also in the reverse sense that the USA is not prepared to import high level radioactive wastes from other countries for deposit in the USA, except waste from reprocessing of American spent nuclear fuel abroad. In particular, the two direct neighbour countries Canada and Mexico have their own final disposal programmes. This position, too, may be revised under the new administration in favour of a regional solution that includes Canada and Mexico. Disposal of radioactive waste at sea was prohibited in 1993 in accordance with the London Dumping Convention (NEA 2008a:16).

The NWPA only contains broad statutory terms that need to be specified by the executive through regulations and administrative guidance. Thus the primary responsibility for regulating the management of high level radioactive waste is vested in the executive. It is to be noted that in the USA, based on the instigation of the Republican administration, the site selection process had relatively early pointed to the choice of Yucca Mountain as the final repository, although there was considerable resistance from the side of the state of Nevada and its citizens. Already the 1987 Nuclear Waste Policy Amendment Act directed the DOE to only study Yucca Mountain. Therefore, important policy documents regarding the radioactive waste management strategy of the USA that have been adopted after this date are not formulated in general terms but limited to this repository. The two relevant regulations are the Yucca Mountain-specific radiation protection standards promulgated by the EPA in 2001 (40 CFR 197) and the corresponding licensing regulations promulgated by the NRC in 2004 (10 CFR 63).[180] The licensing regulations provide that the repository shall be constructed at 1000 feet (= 333 metres) below surface. The disposal strategy is based on a multiple barrier concept composed of geological and engineered barriers and aiming at ensuring the safety of the whole system. In the Yucca Mountain decision, this renouncement of safety redundancy has been met with judicial approval since it was considered to be covered by the discretion the NWPA accords the agency.[181] In particular, the Court of Appeal rejected the argument that under the Act the geological barrier needed to be the primary barrier against water contamination and heat from the containers.

In prescribing a new process for developing the disposal regulations for Yucca Mountain, both in the field of radiation protection and technical requirements, the Energy Policy Act (EnPA) of 1992 (Publ. L. 102–486) set forth important guidelines for agency action. The process heavily relied on scientific input from the National Academy of Science (NAS). The license standards EPA and NRC were to develop were to be "based on and consistent with" the NAS recommendations (§ 801 Energy Policy Act). The NAS was mandated to provide recommendations to EPA as to the following questions: (1) Whether health standards based on doses to individual members of the public from releases to the accessible environment will provide a reasonable standard for the protection of health and safety; (2)

---

[180] The latter incorporate the former; see NEA 2009f:14.
[181] Nuclear Energy Institute v. EPA, 373 F.3d 1251:1289 (D.C. Cir. 2004).

whether it is reasonable to assume that a system of post-closure oversight of the repository can be developed based on active institutional controls that will prevent an unreasonable risk of breaching the repository's engineered or geological barriers or increasing the exposure of individuals to radiation beyond allowable limits; and (3) whether one can make supportable scientific predictions on the probability that the repository's engineered or geological barriers could be breached as a result of human intrusion over a period of 10,000 years (NEA 2009f:27). The NAS published its recommendations "Yucca Mountain Standards" in 1995.[182]

However the administration did not follow the NAS recommendations on the time-scale of high level radioactive waste management. NAS had recommended a compliance period of more than 300,000 years from the time the waste is sent to the repository. This was based on the assumption that high level radioactive waste takes more than 300,000 years to reach its estimated peak risk. According to EPA and NRC, the Yucca Mountain protection concept was only to be designed as to protect people and the environment against radiation up to 150 µSv/yr for 10,000 years; beyond that time, the regulations only required consideration of future risk and the carrying out of an environmental impact assessment without setting any concrete standard (40 CFR § 197.20, 10 CFR § 63.311).

Portions of the environmental and safety standards established by EPA and NRC were challenged by the host State of Nevada, environmental organisations and – limited to ground water standards – industry before the federal courts, among others on the grounds that the period of time for which compliance with the standards must be secured is disproportional to the potential risk presented by high level radioactive wastes. On 4 July 2004, the U. S. Court of Appeals for the District of Columbia[183] ruled in a 100 pages judgement that EPA's 10,000 year safety standard on radiation containment at the site and NRC's licensing standards based on this compliance limit were indeed arbitrary and inconsistent with the recommendations by the NAS. The U. S. Supreme Court refused to review the Court of Appeal decision.[184]

Following the court holding, in 2008 EPA revised the radiation protection regulations and set forth an additional, less stringent standard of 1 mSv for the period of time beginning after 10,000 years, but within the period of geological stability (40 CFR § 197.20(2)).[185] The NRC licensing standards were amended accordingly in 2009 (10 CFR § 63.311). [186]

---

[182] US National Research Commission, Committee on Technical Barriers for Yucca Mountain Standards, Yucca Mountain Standards, 1995; see also Vandenbosch/ Vandenbosch 2007.
[183] Supra note 182, 1266–1273, 1298–1301.
[184] In technical terms: It did not grant "certiorari".
[185] 73 Federal Register 61256, 15 October 2008; see NEA 2009f:14.
[186] 74 Federal Register 10811, 13 March 2009.

In order to gain public acceptance, the NWPA stresses the need for State and public participation in the planning and development of repositories. As regards the preparation of the regulations that determined the waste disposal strategy for Yucca Mountain, there were broad opportunities for public participation that have been used by the host State of Nevada and environmental organisations. Under the Administrative Procedure Act (5 U.S.C. § 701 et seq.), in rule-making procedures the competent agency must first publish a notice in the Federal Register describing the proposed rule. The agency must then allow interested parties an opportunity to make comments for at least thirty days. There are no particular requirements regarding the quality of the "interest". Everybody, including States, non-governmental organisations, industry and scientists, can participate in the procedure. Finally, together with the promulgation of the rule, the agency must publish a concise general statement of the basis and purpose of the rule.

2.1.5 Site selection

The NWPA of 1982 sets forth requirements for the selection of – originally two – sites for a repository for high level radioactive waste and for the siting process that are based on science and safety. The siting must "provide a reasonable assurance that the public and the environment will be adequately protected from the hazards posed by high level radioactive wastes as may be disposed in a repository." DOE must select sites that are suitable and likely to meet the license criteria to be established by the competent agencies (42 U.S.C. § 10133(b)(1)(A)). The legislature originally departed from the proposition that the selection process would be an open process where several alternatives were to be reviewed. Although there is no language in the Act to the extent that an optimal site should be selected, this aim was implicit in the procedure established by the Act. With respect to each of the two repositories envisaged, the DOE had to screen at least five candidate sites and select out of this number three sites that qualified for further investigation, respectively (§ 112(a),(b) NWPA; see NEA 2009a:5). However, already in 1982 there was a clear focus on Yucca Mountain. The Act requires the DOE to consult throughout the siting process with the affected states and Indian tribes and accorded them the right to veto the site selection. Congress could overturn this veto. Moreover, a full environmental impact assessment including alternative sites and subject to public participation was required for all sites recommended to the President (§ 112(b)(D) NWPA 1982).

Pursuant to the statutory mandate, in the years between 1982 and 1987 the DOE screened and characterised a number of candidate sites that were located on federal land. There was quite an extensive public and legal debate on the allocation of the benefits and burdens of the use of nuclear power to the different regions in the USA involved in the site selection process.

The schedule established for the siting of two final repositories for high level radioactive waste was overturned by the 1987 amendments. The legislature narrowed the look to the Yucca Mountain site in Nevada, not the least because quite some budgetary means had already been used to develop Yucca Mountain. The DOE was directed to study the Yucca Mountain site alone and to report to Congress between 2007 and 2012 on the need for a second repository. From a legal point of view the concentration on Yucca Mountain does not mean that other sites were definitely discarded. Under the 1987 Amendments the preference for Yucca Mountain is subject to the proviso that, depending on the outcome of further investigations at Yucca Mountain, one may return to alternative sites. However, the 1987 Amendments limited the scope of the environmental impact assessment prescribed by the Act to the extent that the need for the repository, alternatives to geological disposal and alternative sites were not to be considered (42 U.S.C. § 10134(a)(1)(D), (f) NWPA). By contrast, under general law set forth by NEPA the environmental impact assessment would also scrutinise alternatives and the long-lasting effects on future generations (42 USC § 4332(2) (c)).[187] In the framework of the environmental impact assessment, the NRC also reviews the socio-economic impacts of the relevant decisions. This includes considerations of environmental justice, that is, the equitable distribution of environmental quality among different social categories of the population.

Since 1987, the site selection procedure consisted in the further development of Yucca Mountain. In February 2002 the President submitted his recommendations to Congress recommending Yucca Mountain as the site for the development of a repository. The State of Nevada vetoed this recommendation (NEA 2008a:14). However, the veto was overridden and the recommendation approved by Congress by Joint Resolution 87 and the site selection became law on 23 July 2002 (Publ. L. 107–200). In the law suits described above, the U. S. Court of Appeal for the District of Columbia[188] confirmed the legality of the site selection. In particular, it rejected the argument raised by Nevada in view of the imbalanced burdens imposed on that state that a constitutional principle of state equality existed and was violated. The Court also held that procedural and substantive errors that might have occurred in the selection process and affected the recommendation were remedied by the sovereign decision taken by the U. S. Congress.

The next step taken by the DOE was to prepare and submit a license application to NRC in conformity with the revised radiation protection and the licence regulations. An application to construct the Yucca Mountain repository was made in June 2008. However, the new U. S. administra-

---

[187] NRDC v. Administrator, Energy Research and Development Adm., 451 F. Supp. 1245, 1264 (D.D.C. 1978).
[188] Supra note 182, 1301–1308.

tion had strong objections based on principle against the new repository. Among them are doubts about the suitability of the tufaceous limestone at Yucca Mountain as a geological barrier, concerns about the imbalanced and unfair character of the procedure that led to selecting Yucca Mountain as a repository, a preference for voluntary commitments of host municipalities and ideas about a new management strategy for high level radioactive waste disposal. In March 2009, following an announcement by the President that the budget for Yucca Mountain would be cut down considerably,[189] the DOE (Chicago News 27 February 2009:1) declared that the Yucca Mountain programme will be scaled back to those costs necessary to answer inquiries from the NRC in the pending permit procedure. Moreover, it applied to the Nuclear Regulatory Commission to withdraw the application for the Yucca Mountain repository. The new administration is prepared to devise a new policy for radioactive waste disposal including that of high level waste that in particular addresses the issue of fair distribution of the burdens of radioactive waste disposal on a number of states. To this end, it convened a "Blue-Ribbon Commission on America's Nuclear Future" composed of high-ranking politicians and experts to evaluate alternative approaches. This panel shall provide the opportunity for an open public debate on the available options and make recommendations that may be the basis for an amendment of the statutory framework for high level radioactive waste management (NEA 2009f: 13–14; DoE 2010). Its first report is due on June 2011. It remains to be seen what will be its future after the loss of the democratic majority in the House of Representatives in the elections in November 2010. Already before the elections there have been attempts to anticipate the Commission's work by making legislative proposals for restructuring the allocation of responsibilities for radioactive waste management between the executive and the nuclear industry.

2.1.6 Construction and operation

Pursuant to the NWPA, the construction of a repository for high level radioactive wastes requires an authorisation that will be granted by the NRC (42 U.S.C. § 10134(b)(4)). The NWPA provides that the construction of final repositories must

> provide a reasonable assurance that the public and the environment will be adequately protected from the hazards posed by high level radioactive wastes as may be disposed in a repository (42 U.S.C. § 10131(b)(1); to the same extent: 10 CFR § 63.31(a), 63.304).

As already stated, this pronouncement is an expression of the precautionary principle. According to the Atomic Energy Act, all phases of nuclear facility operation, including that of waste disposal facilities, are subject to pub-

---

[189] Message of the President, A New Era of Responsibility, 26 February 2009:65; see World Nuclear News, "Obama dumps Yucca Mountain", 27 February 2009:1.

lic health, safety and environmental constraints (42 U.S.C. § 2013(d)). Any possession of radioactive waste in a disposal facility requires an authorisation which will be granted if adequate measures for protecting public health and safety have been taken. The competent authorities must ensure that in managing such waste public health, safety and the environment are protected considering cost, and EPA radiation protection standards complied with (42 USC § 2111(b), 2114(a)). As yet, these basic statutory requirements have been specified, geared to Yucca Mountain, in great detail by the license regulations of NRC as well as the radiation regulations of EPA. After the abandonment of the Yucca Mountain site these regulations will have to be amended although it can be expected that they will be maintained in major parts.

In essence, the construction permit prerequisites entail the major elements of the USA disposal strategy for high level radioactive waste regarding design, methods of disposal, technical construction, and quality assurance. Moreover, compliance with the radiation protection standards and adequate emergency planning must be demonstrated (10 CFR § 63.31(a), 63.32). The final decision is taken on the basis of a cost/benefit analysis that also includes technical alternatives (10 CFR § 63.31(c)). As regards the operating permit, the NRC Licensing Regulation requires that the construction has been substantially completed, the operation does not cause an unreasonable risk to human health and the environment and precautions against possible emergencies have been taken (10 CFR 63.41). For the latter purpose the operator must submit a safety report in which emission levels during various hypothetical accident situations are predicted (NEA 2008a:13).

An environmental impact assessment is also necessary in the construction permit procedure (42 U.S.C. § 10134(f)(4), 10 CFR § 63.31) and there are broad opportunities for public participation.

2.1.7   Financing

The NWPA provides that the operators, especially the owners of spent nuclear fuel and other high level radioactive waste, are obliged to pay the full cost of interim storage and final waste disposal (42 U.S.C. § 10131(b) (4), (5); see NEA 2009f:18–19; NEA 2008a:13). To this extent, U.S. law is based on the polluter pays principle. Under the Act, the full management programme for waste disposal must be borne by industry through a fee on the commercial generation and sale of nuclear electricity (42 U.S.C. § 10222(a) and (c)). The charge is set at 1/10 cent per KWh. The proceeds from the charge go into a fund (Nuclear Waste Fund) which covers DOE's disposal activities, including site selection and development activities, through appropriations made by the Government.[190]

---

[190] See NEA 2008a:14; Standard Contract with the operators of nuclear power plants, 10 CFR Part 961.

2.2 France

2.2.1 Sources of regulation

France has relatively early developed a regulatory framework for radioactive waste management (NEA 2009g:1, 3–5, 11–16; NEA 2003a:21). The Act No. 91-1381 relating to Radioactive Waste Management Research ("Loi Bataille") was the first special law for the regulation of high level radioactive waste management in France. It contained the objectives of radioactive waste management, provided for a research and development programme for the management of radioactive waste by specifying three research directions (partitioning and transmutation of long-lived radioactive elements in the waste, retrievable or non-retrievable disposal in deep geological formations, long-term surfacestorage) and set conditions for the construction and operation of an underground laboratory. In major parts, the law has been superseded by more recent legislation, especially by the Act No. 2006-739 of 28 June 2006 on the Planning of Sustainable Management of Radioactive Wastes (Planning Act, codified at Article L.542-1 to L.542-14 Code de l'environnement) and the Act No. 2006-686 of 13 June 2006 on Transparency and Security in the Field of Nuclear Activities (TSN Act, only partly codified in Articles L.121 to L.123).[191] The Planning Act lays down the national policy, the organisation, financing and controls in the field of radioactive waste management. It also mandates the executive to establish a national radioactive waste management plan. The TSN Act contains general regulation on the safety and decommissioning of nuclear facilities and the administrative framework for nuclear activities, especially competencies, supervision and sanctions; moreover, it lays down general rules relating to the transparency of the decision-making process. As regards radiological protection, the Public Health Code (Article L.1333) and the Labour Code (Article L.231) apply.

In addition, some general laws are relevant. This is in particular true for the Act No. 75-633 (Article 541 Code de l'environnement) on Waste Disposal and the Recycling of Materials which in principle also applies to radioactive waste disposal. As regards public participation Articles L.121 to L.123 Code de l'environnement contain supplementary regulation.

In the years 2009 and 2010 France has introduced fundamental reforms of the system of environmental governance. The reform process started with the Act No. 2009-967 of 3 August 2009 concerning the programming for carrying out the Environmental Summit (Loi Grenelle environnement I) and was specified on 12 July 2010 through the new Act No. 2010-788 on implementing the former law (Loi Grenelle environnement II). Both laws are not of direct relevance to high level radioactive waste management as the scope of application of these laws does not include nuclear energy.

---

[191] The reason is that apart from nuclear waste disposal the law relating to nuclear energy has not been codified.

However, the general governance structures established by the new legislation and inserted into Articles L.121 to L.123 Code de l'environnement and the Municipal Code, in particular the strengthening of public participation through comprehensive information of the public, internet participation and round table debates of the executive with economic, social and environmental stakeholders on all issues of sustainable development, are in principle applicable. To this extent the new regime will also influence future developments in the field of high level radioactive waste management.

As regards the regulatory level, Regulation 2007-1557 on Nuclear Installations that regulates the licensing and control of nuclear facilities including final repositories for radioactive waste is limited to short-lived low and intermediary level radioactive waste. High level radioactive waste is not covered. A special regime for this type of waste will only gradually be developed in the ongoing process of developing a final repository. In the field of radiological protection, Article R.1333 Public Health Code and Article R.231 Labour Code set forth more specific rules, especially limit values. Articles R.121 to R.123 Code de l'environnement contain detailed regulation on public participation that specify the broad statutory rules of the legislative part of the Code.

2.2.2 Responsibilities

The major responsibility for high level radioactive waste management in France is vested in the authorities and public enterprises.. The competent authorities such as the Ministry for the Environment, Energy and Sustainable Development and the Nuclear Safety Authority are responsible for policy-making, delegated legislation, regulation, control and information to the public. The duty to protect imposed on the state is set out in the Planning Act. Article 1 of the Planning Act (Article L.542-1 Code de l'environnement) describes the objectives of the Act as sustainable management of radioactive materials and wastes with due regard for the protection of personal health, safety and the environment. In view of the general provisions of the Code de l'environnement (Article L. 110-1(II)), to which also Article 1(I) TSN Act explicitly refers, this is to be construed as to comprise the precautionary principle. Also Article L.542-10-1 Code de l'environnement refers to this principle. A particular concern of the Planning Act is to reduce the burdens of future generations. A public enterprise is responsible for the construction and operation of final disposal facilities. The operators and waste owners are financially responsible. They have to cover the full costs of high level radioactive waste management.

2.2.3 Institutional framework

The institutional framework is largely set forth in the Planning Act of 2006 (NEA 2009g:3–11, 26; NEA 2009b:13–14). The Act vested the highest political and administrative responsibility for radioactive waste management in the Government, especially the Ministry for the Environment, Energy, Sustainable Development and Oceans (Ministère pour l'Écologie, l'Energie, le Développement durable et la Mer – MEEDDM) with its General Directorate for Energy and Radioactive Materials. In particular, the Ministry is competent for policy-making. The adoption of ministerial regulations in the field and the grant of authorisations for the final repository. The Act also established the Nuclear Safety Authority (Autorité pour la sûreté nucléaire – ASN) as an independent agency charged with the preparation of Ministerial decrees, regulation, control and information to the public (Article L.542-10 Code de l'environnement). This agency is competent for developing the national plan for high level radioactive waste management, issuing technical regulations and reviewing draft ministerial decrees as well as applications for the grant of construction and operation permits for the final repository.

The National Radioactive Waste Management Agency (Agence nationale pour la gestion des déchets radioactifs – Andra) has major waste management functions (Article L.542-12 Code de l'environnement). In particular, its task is to select the site and develop, construct and operate the final repository (NEA 2009g:5, 8). Andra which was established in 1991 is a public enterprise for industrial and commercial purposes and due to this its legal status enjoys a major degree of independence. However, the state retains supervisory powers and can exercise quite some influence on Andra's activities. Andra is also charged with tasks in the fields of research and development.

The Planning Act has also established a new advisory body. This is the National Review Board (Comité national d'évaluation – CNE)[192] that has to assess progress of the regulation and implementation process under the national radioactive waste management plan. Within ASN, standing expert groups have been established, among them the Standing Group of Experts on Wastes (Groupe permanent d'experts pour les déchets – GPD). This expert group is composed of scientists who represent all stakeholders. Moreover, the Institute for Research on Radioactive Safety (Institut de recherche sur securité nucléaire – IRSN) that was founded in 2001 is to provide technical support and especially critically review developments and risks in the field of radioactive waste management.

The Atomic Energy Commission (Commission d'énergie atomique – CEA), a rather large technical agency, is charged with research, development and innovation in the field of nuclear energy including waste disposal. It plays a central role in laying the scientific and technical foundations for the development of a repository for high level radioactive waste as well as for partitioning and transmutation.

---

[192] Often called "CNE 2" in order to indicate that it is not identical to its predecessor having the same name; NEA 2009g:5.

2.2.4   Strategies

The basic legal requirements for the development of management strategies for the disposal of high level radioactive waste can be derived from the goals provisions of the Planning Act and the TSN Act that are both governed by the precautionary principle. The objective of these statues is to avoid harmful effects to human health and the environment caused, among others, by nuclear activities. In particular, radioactive wastes shall be managed in a sustainable manner so as to protect human health, ensure nuclear safety and security and protect the environment (Article L.542-1 Code de l'environnement).

Regarding the basic options for high level radioactive waste management, the nuclear energy policy of France is characterised by a more differentiated and process-oriented position than that of most other nuclear states (NEA 2009g:21; NEA 2009b:8–10). By and large, the relevant options are laid down in the Waste Planning Act. This act in principle mandates final disposal of high level radioactive waste in a deep geological repository (Article L.542-10-1 Code de l'environnement). However, in providing so it lays down a priority for reprocessing of spent nuclear fuel; consequently, the definition of wastes that are to be disposed of in a repository excludes those residual substances from nuclear activities that, considering technical and economic conditions, can be reprocessed (Article L.541-1-1 Code de l'environnement). Moreover, partitioning and transmutation are held open and are being developed as future options. Their feasibility will be assessed by 2012 with a view to set a prototype transmutation facility into operation by 2020. Investigations of long-term storage are also envisaged. Special rules apply to historic pollution, especially those from uranium mines.

It is also to be noted that the Planning Act, as a result of a public enquiry made before the adoption of the Act, requires reversibility of final disposal in a repository for at least 100 years (Article L.542-10-1 Code de l'environnement). It is envisaged that the reversibility approach will evolve in the future. The Parliament is to vote on a "Reversibility Act" before the granting of the construction and operation permission for the repository and on a "Closure Act" before the granting of the closure license for the repository. This flexibility mechanism of "reversibility" which was developed by Andra[193] goes beyond mere retrievability (see also section B 1.3.7). It is based on the precautionary principle and designed to ensure that a reversal of the present policy in the light of future preferences and needs as well as technological and design developments remains possible. This concerns both the relevant basic options and the technical strategy of disposal and includes the option of going back one or more steps during construction and operation of the repository.

---

[193] Andra 2005; more specific the contributions in Andra 2010; see also NEA 2009b:10.

Foreign high level radioactive waste is not accepted for final disposal. By contrast, France continues to import spent nuclear fuel for reprocessing provided the residual radioactive waste is taken back by the state of export (Articles L.542-2 and 542-2-1 Code de l'environnement). The dumping of radioactive waste at sea is prohibited in accordance with the London Dumping Convention.

As regards the technical management strategies for high level radioactive waste, the discussion has as yet concentrated on the selection of the fundamental strategic options as well as site selection. There is a guidance document on safety objectives for the post-closure phase that has to be considered in developing the disposal site.[194] The document provides that compliance with the permissible dose limitation of 0.25 mSv per year for long-term exposure associated with certain or very probable internal and external events must be objectively estimated within a period of proof of 10,000 years, based on a demonstration of the stability of the geological environment of the repository. After this period of time, in view of the increasing uncertainties, the quantitative estimations may have to be supplemented by qualitative ones. The dose limitation of 0.25 mSv per year is retained as the relevant reference for verifying safety. As regards uncertain but plausible events, probabilistic approaches shall be used and, where not feasible, a qualitative evaluation of all relevant factors shall be carried out, which however, may not lead to tolerating major exposures. In all cases, the ALARA principle shall be considered.

The details of the management strategy have still to be worked out.[195] According to the ASN, the management strategy must cover all risks presented by high level radioactive waste, considering not only radiological, but also chemical and biological hazards and risks presented by the waste, laying particular emphasis on the sources of uncertainties and using conservative assessments of the performance of the safety system. Waste management comprises the whole waste management route from the operation of a nuclear facility to final disposal including the post-closure phase with all intermediary stages. It also entails the limitation of the volume of waste generated, of its noxiousness and of the quantity of residual radioactive materials contained therein.

2.2.5 Site selection

The Planning Act has introduced a new procedure for selecting the site of a repository for high level radioactive waste. The notion of site is understood as a larger area in which a repository can be located. The Act mandates the establishment of a national radioactive waste disposal plan and sets a sched-

---

[194] ASN 2008:7–8 (replacing the guide document of 10 June 1991, RFS III.2.f).
[195] See also the research agenda set out by Article 11 of the Regulation No. 2008-357 on the application of Article L.542-1-2 Code de l'environnement.

ule for the selection process (Articles L. 542-1-2 and L.542-10-1 Code de l'environnement). The basic prerequisites of site selection follow from the goals provisions of the Planning Act. Specification is left to the political process.[196]

As regards the process of site selection, the Planning Act only sets objectives and milestones. It is to be implemented by the National Plan for the Management of Radioactive Materials and Waste that is to be updated every three years. The first plan was published in May 2007 (ASN 2006; see also ASN 2008, chapter 16). It was updated by the second plan that was published in June 2010 (ASN 2010:62–66, 110). The first plan concentrates on other kinds of radioactive waste but also addresses the future development programme for selecting the site of the repository for high level radioactive waste, describing the steps to be taken in the years between 2007 and 2012 and the principal areas of future research (ASN 2010:108–114). The operational work is based on a special development plan for the project. The second plan likewise focuses on other kinds of radioactive waste but also describes the steps already taken and the results achieved so far regarding the siting of a HLW waste repository.

In France, the process of site selection had already started in 1988 with the first preliminary site investigations which, however, were already terminated one year later for lack of public acceptance. In 1991 the selection process was resumed with the selection of four voluntary candidate sites for investigations and the possible siting of underground laboratories (three in lay formations, one in a critalline formation). In 1996 license applications for the construction of four underground laboratories were filed. After public enquiries and reviews by CNE und GPD, in 1999 one underground laboratory for investigation in a clay formation (argillite) was authorised and later on constructed in the east of France (Meuse/Haute Marne – Bure). The laboratory produced important results relating to geological issues. It led to a basic feasibility (safety) report by Andra, the "Dossier 2005 Argile". By contrast, attempts to find a voluntary candidate for a second underground laboratory in a granite formation failed so that Andra limited itself in this respect to general research (Andra 2005a). In the years 2005 and 2006, various agencies and expert groups, especially ASN, IRSN, GPD and the Office for Technology Assessment of Parliament as well as OECD/NEA evaluated the results of the finding process. In 2005, the Ministry for the Environment, through ASN, published a first draft National Plan for the management of radioactive materials and wastes. Moreover, a public debate was organised and conducted between September and December 2005 which was concluded with a report in early 2006. On the basis of this input and following the assessment mandate by the 1991 Waste Act, the two laws – the Planning Act and the TSN Act – were debated and adopted in June 2006.

---

[196] See for the following text: NEA 2009g:22.

The Planning Act provides for continuation of the studies and research works at Bure in order to choose a site and design for the final repository. Although not formally fixed, it is understood that the final site will in the "transposition zone" of 250 square kilometres in the area where Bure is located. Since underground characterisation of the host rock is required and this is to be done at Bure,[197] the site must be in a rock formation to which the research results gained at Bure can be transferred. Following this rationale, the "transposition zone" has recently been further narrowed by Andra to a "Zone of particular interest for a profound investigation" of 30 square kilometres.[198] According to the present planning, the formal final site selection decision is envisaged for 2013 and the application for the construction permit by 2015.

As a means to improve public acceptance at national and local levels, public participation in the selection process has been considerably extended by the Planning Act and the TSN Act. While previously only the municipalities and districts (départments) were able to participate, especially through a local information and oversight committee (Comité local de l'information – CLI) established under the 1991 Waste Act, the new laws institutionalised a broader and more structured participation with respect to HLW disposal. It consists in particular of reviews by, and inputs from, a standing stakeholder group established around the area of an underground laboratory and an underground repository (Comité local de l'information et de suivi – CLIS; Articles L.542-13, R.542-25 to L.542-30 Code de l'environnement) and the national assocociation of these local groups (Association nationale des commissions locales d'information – ANCLI) as well as public debates organised by the National Debates Commission before important policy decisions are taken. This Commission was established and its tasks described by the Act No. 2002-276 as modified by the Act Grenelle II (codified at Article L.121-1 to L.121-15 Code de l'environnement).

However, in contrast to the selection of a site for an underground laboratory that requires the concerting with the municipalities affected (Article 542-5 Code de l'environnement), voluntarism is not laid down as a mandatory requirement for the selection of a site for a repository (Article 542-10-1 Code de l'environnement). Preparatory investigations that engender the removal of soil or rocks, drillings or seismic investigations are permissible without the agreement of the affected municipalities and landowners on the basis of the Law on compensation for public works of 29 December 1892 (Article 542-7 Code de l'environnement). However, voluntarism is a political principle applied in the practice of Andra and the competent authorities.

---

[197] Decree of 3 August 1999, as amended by Decree of 23 December 2006.
[198] See Andra 2009; ASN supported this decision, see Comment No. 2010-AV-0084 of 5 January 2010.

As a means to promote public acceptance at local level, the law provides for the establishment of a development association (groupement d'intérêt public) composed of the state, the operator, the affected municipalities and departments. The task of the development association, among others, is to conduct spatial planning and development activities and provide public information about scientific issues related to radioactive waste disposal (Article L.542-11 Code de l'environnement).

As regards certain long-lived low level radioactive waste (graphite and radium bearing waste), the new policy of open participation and partnership with the host municipalities has proved to be quite successful. There were about 40 municipalities that showed an interest out of which three candidates shall be selected (see NEA 2009g:21; NEA 2009b:9). However, although one might take this is an indication that the site selection of the repository for high level radioactive waste will also proceed quite smoothly, the recent experience in the field of HLW is not encouraging. Out of the three municipalities whose territory is covered by the 30 square kilometres "zone of particular interest for profound investigations" one municipality whose territory is needed has refused to participate. Consequently, the selection process is presently in an impasse.

2.2.6 Construction and operation

The construction and operation of a final repository for high level radioactive waste constitutes an establishment of a "basic nuclear facility" that requires an authorisation. Moreover, the start of operations requires a permit. The authorisation can only be granted when, considering the present state of science and technology, it can be demonstrated that the site, design, construction, operation, maintenance and surveillance of the facility prevent or sufficiently limit the risks for human health and the environment presented by the facility (Articles L.542-1 and L.110.1 Code de l'environnement, Article 2(I) TSN Act). In particular, it must be ensured that the radiation protection standards will be complied with and adequate precautions for preventing accidents and emergencies based on a safety report must be taken (Article L.542-10-1 Code de l'environnement, Article 29 I in conjunction with Article 28 (II) TSN Act).

The permission procedure is complex. The Planning Act (Article L.542-10-1 Code de l'environnement) provides that in contrast to the general procedure governing basic nuclear facilities the application for the authorisation must be preceded by a public debate under Article L.121-1 Code de l'environnement and will be reviewed by the National Review Board and ASN. It is only after Parliament has adopted a law regarding the conditions of reversibility that the Government can deliver the authorisation. This must be done after a public enquiry has been carried out. In addition, Articles 6, 8, 9 Regulation No. 2007-1557 in conjunction with Articles L.122 and 123 Code de l'environnement are applicable. It follows from these provisions that in the permit process also an environmental impact assessment must be carried out. The authorisation is not granted as an individual decision, but, rather, as a regulatory decision in the form of a governmental decree after consultation of the Conseil d'État (décret en Conseil d'État; Articles L.542-6 and L.542-7).

2.2.7 Financing

In keeping with the polluter-pays principle, under Article 20 Planning Act (not codified) and Article 2(II) No. 2, Article 29(I) TSN Act the financing of decommissioning, storage and management of radioactive waste is the responsibility of the operators of nuclear activities (NEA 2009g:31–33; NEA 2009b:15). The operators are obliged to establish separate reserves to ensure coverage of future expenses for the disposal of their wastes as from 2007. According to Article 20 of the Planning Act, the feeding of the reserves shall be based on periodic cost estimates by the operators and a review by the National Financial Evaluation Commission. The financial responsibility of the operators also covers all costs that are incurred by Andra, which means that also expenses for site selection and investigations at the underground laboratory are encompassed.

Moreover, the Planning Act creates two funds to be administrated by Andra, one destined for the financing of research and development, the other for covering the costs of radioactive waste disposal (Articles L.542-12-1 and 542-12-2 Code de l'environnement). The first fund is financed by part of the special tax imposed on operators of nuclear power plants, the second by voluntary contributions. The details are set out in the Regulation No 2007-343 relating to securities for the financing of nuclear burdens and a Ministerial Order of 21 March 2007. Finally, the operators have also committed themselves to pay a charge as support for economic development in the host municipalities of the radioactive waste repository.

2.3 United Kingdom

2.3.1 Sources of regulation

In the United Kingdom, there is no comprehensive regulatory text that regulates the management of radioactive waste[199] and the Government does for the time being not consider it necessary to enact bespoken legislation (DEFRA 2008: 38). Rather, there are a number of statutes of a general nature that are applicable. The Nuclear Installations Act 1965 (NIA 65) also applies to radioactive waste disposal facilities (Section 1(1) b, iii). The Radioactive Substances Act 1993 (RSA 93) provides that any disposal of radioactive waste, including waste generated at, or handled in, a nuclear site,

---

[199] See DEFRA 2008:39–45; IAEA 2008a:43–58; NEA 2009h:point 2.1; NEA 2003c:13–14; NDA 2010a:44–46.

requires an authorisation (Sections 13, 14, 16).[200] The Health and Safety at Work Act 1974 contains broad powers in the field of protection of health of workers and the public at large. In the field of radiation protection, the Ionising Radiation Regulations 1999 and the Regulations on Justification of Practises Involving Ionising Radiation 2004 are applicable. The Environmental Protection Act 1990 empowers the competent minister to set appropriate provisions of the Act relating to waste disposal (Part II) into force for radioactive waste (Part II Section 78). The goals provisions of the Environment Act 1995 must also be considered in applying the former statutes. The legislation on spatial planning (Town and Country Planning Act 1990, in the future eventually also the Planning Act 2008) and environmental assessment is applicable as regards siting decisions. The Energy Act 2004 is relevant with respect to administrative organisation. Finally, the Freedom of Information Act 2002 and the Environmental Information Regulations 2004 may be relevant as far as public participation is concerned.

2.3.2   Responsibilities

In the UK, the major responsibility for radioactive waste disposal is vested in the Government. British law recognises that Government has a duty to protect the population and the environment from hazards and risks associated with radioactive waste disposal. However, the degree of legalisation of this duty to protect is low. Only the foundations of the governmental duty to protect are set out in the major statutes that govern disposal of high level radioactive waste. In accordance with the Radioactive Substances Act radioactive waste must be disposed of safely in an appropriate way. The competent authorities can attach conditions to any permit they consider fit to ensure that the necessary precautions are taken (Sections 16(8), 18(1), (2) RSA93). The Nuclear Installations Act refers to the necessary safety under normal circumstances and in case of accident or emergency (Section 4(1) NIA65). The requisite authorisations will only be granted if the applicant demonstrates sufficient safety from adverse impacts to the population and the environment. In taking decisions on waste disposal sustainable development, the environment and cost and benefits of any action must be considered (Section 4(1), (3) Environment Act 1995). Specification is the responsibility of the executive (cf NEA 2009h:point 2.1, 2.3; NEA 2009c:3–5). Agency discretion prevails.

It follows from administrative practice that the precautionary principle is well recognised in the field of high level radioactive waste management (see IAEA 2008a:137–138, 185–186, 189–190). A fairly recent guidance document[201] describes the fundamental protection objective as to ensure that

---

[200] The High Activity Sealed Sources and Orphan Sources Regulations 2005.

[201] Managing our Radioactive Waste Safely – CoRWM Recommendations to UK Government, July 2006; see NEA 2009h:point 2.3.2.1.

all disposals of radioactive waste are made in a way that protects the health and interests of people and the integrity of the environment, at the time of disposal and in the future, inspires public confidence and takes account of costs. For specification, it sets out a number of principles such as non-discrimination of future generations, optimisation, protection against non-radiological hazards, avoidance of unreasonable reliance on institutional controls and information and public participation.

The responsibilities for the management of high level radioactive waste (outside nuclear facilities) were originally divided between the authorities and industry. While the authorities were responsible for policy-making, regulation and control, United Kingdom Nirex, a company whose share-holders were the producers of radioactive waste, was charged with the development, construction and operation of a final repository for high level radioactive waste (Mathiesen/Dalton/Russel 2005). However, in 1985 the ownership of Nirex was transferred to two Government departments, in 2006 to the Nuclear Decommissioning Authority and in 2007 the company ceased to exist as a separate entity. Since then, the development, construction and operation of a repository for nuclear waste has also been a governmental responsibility. In keeping with the polluter-pays principle, industry has to cover the costs of radioactive waste management.

2.3.3 Institutional framework

The ministerial responsibility for radioactive waste is vested in various Government departments. These are primarily the UK Department for Energy and Climate Change (DECC), the Department for Business, Enterprises and Regulatory Reform (BERR), the Department for Environment, Food and Rural Affairs (DEFRA) and the Department for Work and Pensions (DWP), together with the respective executive agencies of the two latter ministries, the Health and Safety Executive (HSE) with its Nuclear Installations Inspectorate (NII) and the Environment Agency (EA).[202] The executive authorities have direct responsibility in their areas of competence and are in this sense independent. They are primarily responsible for rule-making and other kind of regulation, in particular giving regulatory guidance and granting authorisations for waste facilities. HSE and NII have the competencies for radiation protection and the grant of nuclear site licences while the Environment Agency focuses on the environment and grants the requisite disposal authorisations under the Radioactive Substances Act. Since 2006 the Nuclear Decommissioning Authority (NDA) which was established by the Energy Act 2004 has been responsible for developing and

---

[202] See NEA 2009h:point 1.2; NEA 2009c:7–8; DEFRA 2008:14, 37; IAEA 2008a:12–14, 59–60, 67, 69 (also for the following text). The different legal and administrative structures of Scotland are not dealt with in this paper unless of particular relevance.

implementing final disposal of high level radioactive wastes including site selection (see NEA 2009h, point 1.3.1, 1.3.2; IAEA 2008a:13, 59–69).

For a long time, the Radioactive Waste Management Advisory Committee (RWMAC) was the major advisory body regarding all matters of radioactive waste management. In 2003 a new independent Committee on Radioactive Waste Management (CoRWM) was established and the work of the RWMAC suspended. CoRWM was reconstituted in 2007. It is composed of scientists and technicians from academia, government and also stakeholder organisations. CoRWM is now the major advisory body of the UK Government for all questions relating to radioactive waste management both of a political and regulatory nature. It is to provide independent scrutiny of Government and NDA's proposals, undertake its work in an open and consultative manner, engage with stakeholders and be open to a dialogue with them and local authorities (DEFRA 2008:45; IAEA 2008a:8). This institutional set-up reflects the more recent turn of the British policy on radioactive waste disposal from a technocratic toward a democratic decision-making model based on public information, participation, acceptance and voluntarism.

2.3.4   Strategies

The choice of strategies for final disposal of high level radioactive waste (including spent nuclear fuel) is subject to the requirement of the Radioactive Substances Act 1993 whereby radioactive wastes must be disposed of safely in an appropriate way and at appropriate times and all necessary precautions be taken (Sections 16(8), 18(1) RSA93). Moreover, Section 4(1) and (3) of the Environment Act 1995 places general duties on the environmental agencies that are responsible for radioactive waste management. These include a duty to consider sustainable development, cost and benefits of any action, a mandate for special consideration of rural communities as well as a general consideration for protection of the environment. Finally, the permit prerequisites for nuclear facilities relating to design, construction, discharge of substances and emergency prevention under the Nuclear Installations Act 1965 (Section 4(1) of the Act) are of some relevance.

A particular basic option for the long-term management of high level radioactive waste was not laid down in the Radioactive Substances Act. The choice of the relevant option was a matter of much debate in the UK, especially in the framework of the failed attempt of Nirex in the 1990s to develop a rock characterisation facility as a precursor to a repository in West Cumbria (Lake District) near Sellafield. In 2002 the Government established a political process for developing a policy for managing higher activity radioactive waste in the long-term (Managing Radioactive Waste Safely – MRWS process). CoRWM was charged with reviewing the available options and making recommendations to the Government. Based on extensive consultations with the public and stakeholder groups, CoRWM made its recommendations in 2006 (CoRWM 2006) favouring geological disposal coupled with safe and secure interim storage. Retrievability of waste was rejected because in the opinion of CoRWM the safety problems were disproportionate to the possible gains. However, CoRWM emphasised that even if a form of direct geological disposal is chosen, it will be 100 years or so before the waste is completely sealed so that during this period reversal or retrieval would still be possible. In October 2006, the UK Government in principle agreed with this position, deferring the issue of retrievability to decision at a later time; the planning, development and construction of the repository shall be carried out in such a way that the option of retrievability is not precluded (DEFRA 2008: 3, 28). From June to December 2007 the government consulted the public on how to start geological disposal. The White Paper by DEFRA "Managing Radioactive Waste Safely: A Framework for Implementing Geological Disposal" of June 2008 confirmed and specified this position.

However, it is to be noted that the Scottish Government policy is one of continued interim storage coupled with further research and that the Welsh Government has reserved its opinion on geological disposal.

As regards the role of reprocessing spent nuclear fuel, the UK has for long practiced reprocessing spent nuclear fuel and will continue to do so. This means that the notion of high level radioactive waste does not include nuclear material that can be reprocessed. The British position toward a possible international disposal option and the acceptance of foreign waste still is open. Dumping of radioactive waste at sea is prohibited pursuant to the London Dumping Convention.

Since the decision for geological disposal is relatively new, the UK does not yet dispose of a developed technical strategy for final disposal. The various regulatory guidance documents on radioactive waste management already published by HSE, the Environment Agency and the Scottish Environment Protection Agency in the last years (HSE 2007) primarily relate to waste management on licensed nuclear facilities, including interim storage and low and intermediate level radioactive waste. This is in particular true for the Guidance on near-surface disposal of high level radioactive waste (HSE 2009). It is envisaged to gradually develop the final disposal strategy in a staged research and decision-making process under the responsibility of the NDA, taking account of the particularities of the geographic conditions and geological formation of the disposal site once this has been selected (DEFRA 2008:10; see IAEA 2008a:18). It is closely linked to site selection. NDA is to follow a staged implementation approach with decision points that allow a review of the design of the facility and the disposal methods to be applied under the perspective of safety, cost, affordability, environment and sustainability before deciding to move to the next stage. The process to develop the disposal strategy will be participatory. NDA established a National Stakeholder Group for receiving input on national issues (NDA 2010b). It will also establish Site Stakeholder Groups at the interface between the host community, the site operator and the NDA that are designed to review, comment on, and influence strategies, plans and achievements regarding a particular site. NDA will also consult the public on the further development of the final disposal strategy. In 2007, it held a public consultation on a framework for implementing geological disposal. Independent consultants were charged with a thorough review of the available strategic options for various possible formations (Baldwin/Chapman/Neall 2008). Finally, NDA must comply with the statutory consultation requirements regarding further development of the disposal strategy during the site selection process.

The White Paper of June 2008 (DEFRA 2008:27–28 and Annex A) contains some core elements of the final disposal strategy. The repository shall be constructed at a depth somewhere between 200 and 1000 metres below surface in a suitable rock formation. It will be based on a multiple barrier concept that consists of geological and engineered barriers which together minimise the possibility of an escape of radioactivity and ensure that the remaining risk is insignificant compared to natural radioactivity. What one can in addition derive from the existing guidelines for low and intermediate level waste is that the repository shall be passively safe, that is, safety shall be independent from institutional controls and ensure continued isolation of waste from the accessible environment. Moreover, the repository shall be designed, constructed and operated in such a way that it is capable of closure so as to avoid adverse effects on the performance of the containment system (IAEA 2008a:13, 107). There are not yet special dose limitations regarding the post-closure period. One can derive from the regulation of land-based disposal of low and intermediate level waste that a risk objective of 1:1 million coupled with best available technology might be chosen. As regards the period of time within which safety must be demonstrated, the White Paper (DEFRA 2008:27) has avoided any clear commitments. However the White Paper departs from the assumption that high level radioactive waste presents risks for several hundred thousands years that must be managed.

2.3.5 Site selection

Like strategies, selection of a site for a geological repository is subject to the broad goals provisions of the Radioactive Substances Act 1993, the Nuclear Installations Act 1965 and the Environment Act 1995. Since site selection constitutes development under the Town and Country Planning Act 1990, additional requirements under this Act are applicable.

In the UK, attempts at developing sites for final disposal of high level radioactive waste were undertaken very early. Already in the 1980s, Nirex made proposals for radioactive waste repositories at various locations that were abandoned due to local opposition. A second attempt to develop a site for a final repository (for intermediate level waste) in West Cumbria (Lake District) near Sellafield was made in the 1990s. However, the application for an authorisation for a rock characterisation plant was ultimately declined by the Government in 1997 in reaction to widespread opposition in the public and the negative results of the public enquiry. With the CoRWM Recommendation of 2006 and the response of the UK Government in the same year a third, more refined attempt at site selection has been initiated by devising and starting a new process for site selection. It is to be noted that the degree of legalisation of the new process is low.

The new process is a criteria-based, staged process based on voluntarism, that is, communities' willingness to participate, and on partnership between NDA, on the one hand, and the host community, local authorities and wider local interests (neighbouring municipalities and districts), on the other (DEFRA 2008:4–7, 47–69, 76–80; NEA 2009h:point 3.2.1, 3.3). It is suggested that the local interests organise themselves in a community siting partnership that ensures information of the population, in particular regarding community concerns but is not designed to assume decision-making functions; these remain with the competent local bodies. The siting process also entails economic compensation for the host community and wider local interests. Its complex elements are described in the White Paper of 2008. The participation in the process does not entail a commitment by the municipalities to proceed. Rather, withdrawal from the process is possible up until underground investigations are due to start. The site assessment process will be conducted in parallel to discussion with the municipality once it has made an expression of interest. It will be a staged process divided into 6 stages entailing an increasing degree of details and allowing all participants to take review the results of the preceding steps before deciding to move to the next step or not (DEFRA 2008:7, 62). These stages are: (1) Expression of interest by the municipality following a staged internal decision-making process; (2) geological subsurface survey and screening of unsuitable areas subject to criteria established by Government (DEFRA 2008, Appendix B); (3) discussion, consultation and decision to participate by suitable municipalities; (4) desk-based suitability studies in particular areas; (5) surface-based investigation on the remaining candidates; (6) underground investigations. The government is also considering adjusting the legal framework in order to allow for staged permits (see also NDA 2010a:45) ("permissions schedule").

In contrast to the previous attempts at site selection, the crucial criteria for assessing and evaluating the sites are more complex and include socio-economic and ethical considerations. Criteria are the geological setting, potential impacts on people (human health and safety), both present and future generations, potential impact on the natural environment and landscape, effect on local socio-economic conditions, transport and infrastructure conditions and cost, timing and ease of implementation. Considerations of optimality will only come into play if more than one community has made a commitment.

Pursuant to the new concept, the UK Government in the White Paper of 2008 invited municipalities to declare their willingness to participate in the process. As yet, three neighbouring local authorities in West Cumbria near Sellafield have expressed their interest and established the West Cumbria Managing Radioactive Waste Safely Partnership.[203] The discussions with the relevant local authorities and stakeholders are continuing (NEA 2009h:point 3.3.1). One problem is that West Cumbria hosts the Lake District and tourism could suffer from the projects. Moreover, the recent decision of the British Government to build, in response to the requirements of climate protection, new nuclear power plants is burdening the site selection process. As a result of this decision not only the volume of wastes to be disposed of and the duration of the active operation of the repository process become uncertain but also an early solution of the waste disposal problem is fraught with the suspicion that it could encourage prolonging nuclear energy into many decades to come (MacKerran 2008). On the other side of the scale, West Cumbria might be interested in also becoming a site of a new nuclear power plant for economic reasons. It is to be noted that voluntarism is not a mandatory requirement laid down by the applicable legislation. In case of need, the government can rely on the existing statutory empowerments for selecting a site for the requisite repository.

The selection of the site for the repository is concluded with a planning permit under the Town and Country Planning Act 1990. Moreover, the nuclear site license addresses the suitability of the site. Under the Town and Country Planning Act the site needs to be determined in the development plan (requirement of plan-led development), and the applicant is required to secure development consent at three stages (surface-based investigations, underground rock characterisation and construction). The competence is vested in the local authorities, subject to the right of the competent Minister to "call in" the planning application. It is a matter of course that such a call in will occur in a case of national importance such as the siting of a repository for high level radioactive waste. The grant of the nuclear site license which encompasses all aspects of design and construction requires that the site of the repository is suitable and acceptable in respect of impact on the population and the environment as set out in the HSE safety criteria. This includes impact of the facility, emergency planning, protection from external hazards, flooding, earthquakes and other geological factors (IAEA 2008a:101–102).

At the planning and the permit stages, a strategic environmental assessment and an environmental impact assessment, respectively,

---

[203] See Statement by the West Cumbria Managing Radioactive Waste Safely Partnership 2010; British Geological Survey 2010:5–7.

are prescribed whose results must be considered in the relevant decision. Furthermore, a sustainability analysis is required. The whole procedure under the Town and Country Planning Act is subject to public participation, normally in the form of a local public enquiry conducted by an independent inspector. This is even true for the siting of an experimental facility such as an underground rock characterisation facility (DEFRA 2008:43). Whether the reformed planning procedure for nationally important infrastructure projects under the Planning Act 2008 that provides for a single consent system and the competence of the Integrated Planning Commission will be extended to repositories for high level radioactive wastes, remains to be seen. The Government seems to favour such a solution but has also to consider the possible inconsistency with the voluntarism and partnership approach regarding the site selection for the repository (see DEFRA 2008:43; see also NDA2010a:44). As regards the nuclear site license, comprehensive participation is at least possible in the administrative appeal procedure.

2.3.6 Construction and operation

The construction and operation of a repository for high level radioactive wastes requires a nuclear site authorisation under the Nuclear Installations Act 1965 (IAEA 2008a:51–53, 55). The procedure is carried on in parallel to the planning permission procedure but is normally only concluded after planning permission has been given. The granting of the construction permit depends on a satisfactory assessment of all safety aspects, apart from the site in particular with respect to the design, disposal methods, construction, measures for radiation protection and technical precautions against accidents and emergencies (Sections 4 (1) and 7 of the Act). Unacceptable risk must be avoided and generally tolerable risk must be reduced to a level as low as reasonably practicable (ALARP principle) which in general requires deterministic but may also require probabilistic analyses (IAEA 2008a:137–138, 185–191). The details are set out in HSE's "Safety Assessment Principles" of 1992. The operation permit requires in addition that the applicant can demonstrate that the routine operations, maintenance, monitoring and inspections comply with the relevant safety and radiation protection regulations and other requirements. The authorisation procedure is a staged process which contains regular hold points and concluded with the final operation consent.

Moreover, the waste disposal activity is subject to a permit requirement under the Radioactive Substances Act 1993 (Sections 13, 14). In the authorisation process the competent authority can impose all safety related precautions it thinks fit (Sections 16(8), 18(1), (2) of the Act) (IAEA 2008a:97). Also, non-radiological effects on the environment must be taken into account. The government is considering to introducing a staged procedure (DEFRA 2008:39–40).

2.3.7   Financing

The costs of reprocessing spent nuclear fuel and disposal of high level radio-active waste in the geological repository are to be borne by the waste produc-ers, in particular the operators of nuclear power plants (NEA 2009h, point 4.5; NEA 2009c:8). The same is true for regulation-based research under-taken by the operators. The general principle is that every actor finances his own respective functions. However, the Radioactive Substances Act 1993 does not contain provisions on the establishment of reserves and reg-ular review of their adequacy. This is different as regards new power plants. In this respect the Energy Act 2008 established obligations of operators to ensure that they can fulfil their long-term financial responsibility for radio-active waste disposal (NEA 2009h, point 4.5; IAEA 2008a:45, 46, 56).

2.4   Switzerland

2.4.1   Sources of regulation

The major sources of regulation of high level radioactive waste disposal in Switzerland are the Nuclear Energy Act of 2003, as amended in 2007 (SR 732.1), and the Nuclear Energy Regulation of 2004 (SR 732.11).[204] The two legislative texts apply to the siting, construction and operation of disposal facilities and to the handling of radioactive waste before it is accepted for disposal. Besides, the Radiation Protection Act of 1991(SR 814.50) and the Radiation Protection Regulation of 1994 (SR 814.501) regulate the protec-tion of workers and members of the public from radiation originating from disposal facilities. However, the relevant permit requirements are not appli-cable (Article 2(3) Radiation Protection Act). Finally, the Spatial Planning Act of 1979 (SR 700) and the rules of the Environmental Protection Act of 1983, as amended (SR 814.01) relating to the environmental impact assess-ment (Article 9) and specified by the Environmental Impact Regulation of 1988 (SR 814.011) play a role.

2.4.2   Responsibilities

The Nuclear Energy Act establishes the responsibility of the state for the regulation and control of radioactive waste disposal. The state duty to pro-tect is spelt out in the Act in a rather specific manner. In using nuclear energy, man and the environment must be protected from radiation. In particular, precautions must be taken against an impermissible discharge of radioactive substances. All measures shall be taken that are necessary according to experience and the state of science and technology and that contribute to a further reduction of an endangerment as far as appropri-ate (Article 4). Moreover, the Act requires staged safety barriers, emergency

---

[204] see NEA 2009i:1–2, 6–12; Resele 2009:226–36; NEA 2009d:1; NEA 2003b:36 (partly outdated); Bühlmann 2008:612.

measures and security measures against intrusion and theft of radioactive materials (Article 5). Specifically regarding radioactive waste disposal, the Nuclear Energy Act prescribes that nuclear waste management facilities must provide lasting safety and protect human health and the environment against risk from radioactivity (Articles 30(3), 13(1)).[205] These broad statutory terms have been specified by a guideline issued by the Swiss Federal Nuclear Safety Inspectorate on "Protection Objectives for the Disposal of Radioactive Waste" (Guideline ENSI-G03; see NEA 2009i:10–11). This guideline is largely based on the internationally agreed IAEA principles and includes, in particular, the principles of low additional exposure of individuals and the environment, transboundary protection, non-discrimination of future generations as to safety and other burdens, preference for passive long-term safety, and optimisation of site selection, construction, operation and closure.

However, the management of radioactive waste and in particular the development, construction and operation of a geological repository until its closure are the responsibility of the waste producers (Article 31 Nuclear Energy Act; see Hoppenbrock 2009:170–174). Since 1972, the National Cooperative for the Disposal of Radioactive Waste (Nagra), a private law association of operators of nuclear power plants and Federal Government (in its capacity as producer of radioactive waste from public health and research institutions), has had the responsibility for managing radioactive waste generated by the producers (see NEA 2009d:8). It has been active in the preparatory work for establishing a concept for radioactive waste disposal. Private special purpose companies may operate the disposal facilities. Moreover, the financial responsibility for waste management is imposed on the operators of nuclear powers plants and other waste generators.

2.4.3   Institutional framework

The Federal Government (Federal Council), often acting through its Department of Environment, Transport, Energy and Communications, possesses the major competencies in the field of radioactive waste disposal.[206] Apart from its policy making role, it has the competence for establishing the underground geological disposal plan of the Federation, the granting of framework authorisations for a final repository as well as for consents to waste management plans of the operators. The framework authorisation also requires the consent of Parliament. The Swiss Federal Nuclear Safety Inspectorate (ENSI) that originally was a division of the Federal Office for Energy but as from beginning of 2009 has been transformed into an independent organisation under public law (see Resele 2009:228, 239) exercises supervisory functions with respect to nuclear facilities including waste

---

[205] See also NEA 2009i:7–8, 17.
[206] See, also for the following text, NEA 2009i:2–6.

disposal facilities. The Federal Office for Energy also exercises important executive functions in the field of radioactive waste management. In particular, it has the competence to process the site selection for the radioactive waste repositories.

In addition, there are some advisory bodies such as the Federal Commission for Nuclear Safety (Kommission für nukleare Sicherheit – KNS) and the Federal Commission for Radioactive Waste Disposal (Kommission für nukleare Entsorgung – KNE). The latter is in particular concerned with the geological aspects of radioactive waste disposal. The role of Nagra in the management of radioactive waste has already been exposed (see 7.4.5.2).

2.4.4   Strategies

The Swiss disposal strategy for high level radioactive wastes is only partly laid down in a legally binding regulatory text. In essence, its development is the responsibility of the Federal Council as well as Nagra within the framework established by the goals provisions and permit requirements of Articles 1, 4, 13, 30(3) and 31 of the Nuclear Energy Act. (see NEA 2009d:5–6) However, certain aspects of the disposal strategy, in particular regarding the basic disposal options, are set out in the Nuclear Energy Act.

The Act provides that high level radioactive waste as well as long-lived intermediate level waste shall be disposed of in a deep geological repository (Article 31 in conjunction with Article 3 lit. b).[207] Interim storage shall take place in facilities at the premises of the operators. Reprocessing of spent nuclear fuel and export for reprocessing is permitted under narrow conditions (Article 9) and was practiced in the past by sending spent nuclear fuel to reprocessing facilities in France and the United Kingdom. However, with the enactment of the Nuclear Energy Act, the Swiss Parliament established a 10 years moratorium on reprocessing which entered into force in July 2009 (Article 106(4) Nuclear Energy Act) (NEA 2009i:1, 14). Thereby, the option of reprocessing in the future shall be left open. The notion of radioactive waste includes spent nuclear fuel if it is not reused (Article 3 lit. i).

Radioactive waste produced in Switzerland shall in principle be disposed of in Switzerland (Articles 30(2), 34(4) of the Act). However, wastes may be exported under certain conditions for conditioning or in the framework of a bi- or multinational disposal project (Article 34(3) and (4) of the Act). Since Switzerland is a small country, the option of an international solution for final disposal of high level radioactive waste has deliberately been held open.

Moreover, certain elements of the technical disposal strategy are set forth in the Nuclear Energy Act and the Nuclear Energy Regulation. Arti-

---

[207] The legislative technique of the Act consists in a "neutral" disposal obligation (Article 31) and a definition of disposal as conditioning, interim storage and deposit in an underground repository.

cle 37(1) lit. b of the Act provides that high level radioactive waste deposited in an underground repository shall be retrievable until eventual closure of the facility without excessive cost. According to the recommendations of a group of experts that prepared the reform of nuclear law in Switzerland, retrievability shall also be ensured after closure of the facility which means that the political aim is reversibility (EKRA 2000; Resele 2009:227). Article 11(2) in conjunction with Article 10(1) Nuclear Energy Regulation prescribes a redundant safety concept, a priority for passive safety, long-term safety through staged barriers, measures for facilitating surveillance and retrieval and the possibility that the repository can be closed within few years.

The specification of these principles is carried out through a "substantive plan". Article 5 of the Nuclear Energy Regulation provides that the Federation, in a substantive plan, shall lay down the objectives and criteria for storing radioactive waste in underground geological repositories. The substantial planning procedure is based on Article 13 of the Federal Spatial Planning Act which requires the determination of the objectives of the Federation regarding all federal measures that have major spatial impacts. According to the Regulation, the plan is binding on the authorities. Moreover, the ENSI Guideline on "Protection Objectives for the Disposal of Radioactive Waste" (Guideline ENSI-G03) contains some specifications (see NEA 2009i:10–11). The guideline provides for an individual dose limit of 0.1 mSv in a year in the post-closure period; in case of very improbable events, the risk of radiological fatality shall not exceed 1:1 million in a year. The timescale of safety assessment is a period of 1 million years (ENSI 2009:235–36).

Over the years the radioactive waste management strategy of Switzerland and the process for its development and implementation have undergone important changes. In 1985, a Nagra study (NAGRA 1985) advocated a concept of early final disposal in a geological repository after a phase of observation without any post-closure monitoring. In 1988, following a detailed examination of the strategy, the Federal Council concluded that a sufficient guarantee of safety had been demonstrated in principle for, among others, high level radioactive waste although it had doubts whether the extent of the host rock formation investigated by Nagra complied with the relevant safety objectives.

However, in response to wide-spread opposition to the Government's endeavours to select a site for a repository for low and intermediate level waste in the 1990s, the Swiss Federal Council fundamentally changed its radioactive waste management strategy in favour of a cautious, step-wise approach which aims for addressing scientific and technical issues as well as societal concerns (IAEA 2009a:49). According to the Underground Geological Disposal Plan for Radioactive Waste Plan established in accordance with Article 5 of the Nuclear Energy Regulation in 2008 (Bundesamt für Energie 2008; see Resele 2009:226–227, 233), but somewhat in contraction to its Article 11(2) lit. d, high level radioactive wastes shall be subject to long interim storage in order to reach a large degree of heat decrease by decay. Then the wastes shall be safely stored in the underground repository subject to continued monitoring in such a way that the radioactive substances remain retrievable. The new strategy is designed to enable the competent authorities to correct their course of action in the future, for example in response to new technological developments or future demands for nuclear materials. The emplacement of some radioactive wastes in a pilot facility is designed to test predictive models and facilitate the early detection of eventual deficiencies of the system.

2.4.5   Site selection

There are no special statutory provisions regarding the site selection process for a radioactive waste repository. The process is to be defined in the framework of existing law. The goals provisions and permit requirements of the Nuclear Energy Act, especially Articles 1, 4, 13, 30(3) and 31, as well as Article 11 of the Nuclear Energy Regulation, contain some substantive guidance. In essence, any endangerment of man and the environment by ionising radiation shall be avoided. Procedurally, the selection process is conducted in accordance with the Federal Spatial Planning Act (Article 13) in a special planning procedure for the establishment of a "substantive plan". As stated, this procedure to which Article 5 Nuclear Energy Regulation refers shall be used for all federal measures that have major spatial impacts. Where the Federal Government has the administrative competence, it includes the determination of the site of the relevant projects. In the Radioactive Waste Disposal Plan of 2008, Conceptual Part, established under these empowerments, the Federal Agency for Energy has specified the broad statutory terms of the Nuclear Energy Act and set out additional requirements regarding the procedure (see Bundesamt für Energie 2008:33ff; Resele 2009:231, 233).

Selection of sites for a geological repository for radioactive waste in Switzerland has gone through a varied history.[208] This history is deeply influenced by a particularity of the political system of Switzerland, namely the possibility under cantonal law to subject permits granted by the cantonal authorities to a cantonal referendum. Since site selection for an underground repository almost invariably requires an underground exploration of the geological formation, the cantonal mining law regime was applicable with the consequence that site selection practically depended on the outcome of a cantonal referendum. This is how the attempt by Nagra to develop the site for a geological repository for low and intermediate level radioactive waste in a crystalline rock formation at Wellenberg in the Canton Nidwal-

---

[208] See NEA 2009i:8–9, 14–17; NEA 2009d:5–8; Hörnschmeier 2008:602; Aebersold 2008:618; Issler 2006; Hoppenbrock 2009:175–177, 179–182.

den failed twice. In the first time Nagra had still proceeded under the old radioactive waste management strategy, the second time it based its plans on a more sophisticated new step-wise strategy which later on was set forth in the Radioactive Waste Management Plan. However, both times, in 1995 und 2002, the exploration permits of the competent cantonal authority of Nidwalden were annulled by a majority of the cantonal citizens that participated in the referendum; by contrast, the majority of the citizens of the host municipality that profited from the project were in favour of it.

As for repositories for high level radioactive waste, the Project "Guarantee 1985" was also based on the crystalline rock option. Although considered as in principle feasible, it was found that there was not sufficient proof for an adequate safety provided by the available extension of the crystalline rock formation. This led the Swiss Federal Government and Nagra to also include sedimentary rocks, in particular opalinus clay, as geological formation for carrying out research and development as from 1988. After the failure of the siting project for low and intermediate level waste disposal crystalline rock was entirely abandoned as a host rock formation. This failure also had strong repercussions on the whole Swiss policy on radioactive waste management in general and the process of siting geological repositories for high level radioactive waste in particular. In essence, the new policy is based on a consensus that Switzerland needs a repository for its high-level radioactive waste.

In the first place, the stumble stone of cantonal exploration permits subject to the risk of being quashed by a cantonal referendum was removed by the new Nuclear Energy Act of 2003. The Act subjects radioactive waste facilities and geological explorations to a federal permit (Articles 12, 15, 19, 35 and 49) and pre-empts cantonal law. In contrast to previous law,[209] cantonal permits and plans are unnecessary and therefore irrelevant. Cantonal law shall be considered insofar as it does not disproportionately restrict the project. If the Department grants the permit against the objections of the affected canton the latter has standing to challenge the permit before the Federal Court (Article 49(3) and (4)). Moreover, framework permits for underground repositories are subject to an optional national referendum. This concept of federalisation aims at facilitating the making of decisions on radioactive waste disposal. In essence it means that conflicts about the distribution of burdens of radioactive waste disposal are transferred from politics to the judiciary and from cantonal to national politics.

Secondly, Article 11(1) of the Nuclear Energy Regulation and Part I of the new radioactive waste disposal plan, adopted by the Federal Agency for Energy in April 2008, contain basic requirements of a substantive and procedural nature for the selection process that are binding on all authorities. In keeping with the statutory requirements, long-term nuclear safety and the lasting protection of human health and the environment are regarded as

---

[209] See Vallender/Morell 1997, § 9 No. 3 with further references.

the primary substantive criteria for the siting process; spatial planning and economic concerns as well as the attitudes of society have also to be considered but are given a lower weight. With regard to the suitability of the site, Article 11(1) of the Nuclear Energy Regulation focuses on criteria such as a sufficient extent of suitable host rock, favourable hydrological circumstances and long-term geological stability. Based on this set of basic criteria, the radioactive waste disposal plan developed more specific safety criteria for the relevant site (see NEA 2009i:9). Sites where, due to geological factors, doses of no more than 10 μSv per year can be ruled out are deemed to be equivalent so that planning and economic concerns can be considered. At higher possible or expected doses optimisation of safety through site selection is to take place. It is to be noted that, since not only high level but also low and intermediate level radioactive waste is to be disposed of in a geological repository, site selection encompasses all categories of radioactive waste. Also a combined repository is considered as an option.

The selection process is divided into three steps.[210] The first step in selecting a site for the geological repository consists in the determination of possible site regions (preliminary selection). The determination of possible host site regions shall be based on geological criteria. In this respect, the goal of the process is to find an optimal site. Moreover, the spatial planning situation shall be investigated, a safety review conducted and preparations made for regional participation. In late 2008, Nagra proposed six site regions of which three were also suitable for high level radioactive waste.[211] These sites are very close to the German border. In June 2010 the provisional list was confirmed by the Swiss Federal Agency for Energy and between beginning of September and end of November 2010 public participation was carried out. Relying exclusively on a safety-based comparison and the relevant results of public participation, the Bundesrat shall decide mid of 2011 which sites will remain in the "competition". In the second phase a spatial planning assessment, socio-economic studies and a quantitative safety analysis shall be conducted. On the basis of these studies, two sites shall be selected for underground exploration. The third step of the procedure consists in an in-depth investigation of the two selected sites and the preparation of the application for the framework license. The ultimate site selection shall primarily be based on safety and environmental, and eventually also on socio-economic including spatial planning considerations. Moreover, compensation to the host municipalities and region may be provided. It is to be noted that a detailed characterisation of a possible site in one of the proposed site regions, the Zurich Weinland, had already been carried out previously for demonstration of disposal feasibility which is required in Switzerland for

---

[210] See for the following text Bundesamt für Energie 2008:33ff; NEA 2009i:9, 16–17; Resele 2009:233–34; Hoppenbrock 2009:175–176.

[211] NEA 2009i:16–17; NEA 2009d:2009:7–8; Resele 2009:238–39.

continuing the operations of nuclear power plants; disposal feasibility had been confirmed by the Swiss Federal Council in 2006 (Nagra 2002; Resele 2009:236–37). In view of these facts only one further underground investigation may have to be undertaken.

The radioactive waste disposal plan provides that the selection process shall be independent, open, transparent and fair. There shall be full participation of all affected municipalities, cantons and foreign states as well as the population at large in all stages of the process. Local partnerships for communication shall be established. Hearings on the selection of sites from the Nagra list for underground investigations were started in September 2010. The Federal Office for Energy is responsible for the procedure. The decisions that conclude each step of the selection process are taken by the Swiss Federal Government (Federal Council).

The site selection process ultimately aims at providing an accepted basis for a framework license to be granted under the Nuclear Energy Act that determines the site, the categories of radioactive wastes to be accepted and the capacity of the repository (Article 14). The grant of the permission is discretionary and contains political elements. It must ensure the protection of man and the environment and consider spatial planning concerns; moreover, it requires that there is a concept for closure and monitoring and the suitability of the site has been confirmed by the results of geological investigations (Article 13 (1) lit. a, b, c and g).

After the submission of the application for the framework license, the host canton, the neighbouring cantons and foreign states are formally consulted and their interest is to be considered insofar as it does not disproportionately restrict the project (Article 44). Moreover, everybody can make comments and raise objections (Article 46). In the permit process, an environmental impact assessment as provided by Articles 10a – 10d Environmental Protection Act in conjunction with the Environmental Impact Assessment Regulation must be carried out (see also Article 43(2) of the Nuclear Energy Act; Rausch/Marti/Griffe 2004:Nos. 733–736). The authorisation is to be granted by the Federal Council with the consent of the Swiss Parliament. A particularity of Switzerland is that the parliamentary resolution is subject to an optional national referendum (Article 48). However, in contrast to the planned construction of new power plants that is very controversial, it may be expected that the chances for the success of such a referendum are small as long as the requirements of openness and fairness of the site selection procedure are met in the future.

2.4.6 Construction and operation

The construction and operation of the underground repository requires construction and operating authorisations under the Nuclear Energy Act (Articles 15-18, and Articles 19-25, 37, respectively, as specified by Articles 24 and 28 Nuclear Energy Regulation). Closure of the facility also requires an authorisation (Article 39 Nuclear Energy Act). The construction permit determines the site, the planned capacity and the essential elements of technical design. The operation permit determines the permissible capacity, limit values for the discharge of radioactive substances, limit values for the radioactive activity of the wastes to be deposited, measures for environmental monitoring and safety, security and emergency measures.

Both the general and special permit prerequisites must be fulfilled. The general permit prerequisites resemble those set out with regard to the framework permit. In addition to that, the Nuclear Energy Act requires compliance with the principles of nuclear safety and security, quality assurance and in the case of the operation permit compliance with the framework permit and the taking of measures for controlling emergencies. In all cases compliance with the various elements of the disposal strategy as discussed under 8.4.5.4 must be demonstrated. Additional operation permit requirements for geological repositories are set out in Article 37 Nuclear Energy Act. In accordance with this provision the information gained during construction must confirm the suitability of the site and it must be demonstrated that the retrieval of the radioactive wastes until closure of the facility is possible without major cost.

The procedure regarding the construction permit and the operation permit is different to the framework permit procedure. Only the construction permit is subject to the requirement of an additional environmental impact assessment (Annex No. 21 of the Environmental Impact Assessment Regulation). Apart from that, in both procedures only a hearing of the cantons affected is provided for (Articles 53, 61 Nuclear Energy Act). These can further pursue their interest by bringing an action against the decision with the competent federal court. Persons who have a particular interest have the right to make objections (Articles 55, 61 Nuclear Energy Act).[212] The same is true for environmental associations in the procedure regarding construction permits (Article 55 Nuclear Energy Act) since association standing is attached to decisions subject to the requirement of an environmental impact assessment.[213]

2.4.7 *Financing*

Generators of high level radioactive waste, in particular operators of nuclear power plants, are obliged to cover the costs of managing their wastes. In the first place this is true for all management costs incurred during operations such as conditioning, reprocessing and also research and develop-

---

[212] These provisions refer to the notion of "party" under Articles 6, 48 of the Administrative Procedure Act.

[213] An association that has standing under Article 55 Environmental Protection Act based on the EIA requirement is considered to be a party in the meaning of Article 6 Administrative Procedure Act. Article 55 Nuclear Energy Act refers to the latter provision.

ment carried out by Nagra. This includes the costs of site selection. However, also the costs of treatment, interim storage, decommissioning and in the future also underground disposal until closure of the repository must be borne by the waste generators (Article 31(1), (2) Nuclear Energy Act).[214] For that purpose, plant operators must establish reserves in their annual accounts. The disposition over these reserves is limited (Article 82(2) and (3) Nuclear Energy Act. As regards waste disposal costs to be incurred after plant closure, since 2000 the operators of nuclear power plants must make annual payments into a special fund, the Waste Disposal Fund (Article 77(2), (3) Nuclear Energy Act). The payments are based on cost estimates that are adjusted periodically. The details are set out in the Federal Regulation on the Decommissioning Fund and on the Waste Management Fund for Nuclear Facilities of 2007 (SR 732.17).

2.5   Sweden

2.5.1   Sources of regulation

In Sweden, there is no special legislation on radioactive waste management. Rather, the general laws on nuclear activities and radiation protection also encompass waste management. Moreover, general environmental laws are applicable. The degree of legalisation of radioactive waste management is very high.

The most important sources of regulation are the Nuclear Activities Act of 1984 (SFS 1984: 3) and the Radiation Protection Act of 1988 (SFS 1988:220). The former statute was amended in particular in 1987, 1995 and 1999, the latter in 1995 and 1999.[215] The Nuclear Activities Act regulates the siting, construction and operation of nuclear facilities including radioactive waste disposal facilities. The Radiation Protection Act establishes permissible radiation exposure doses for workers and the public at large, including those originating from radioactive waste disposal activities. Moreover, the Environmental Code of 1998 (SFS 1998: 808), especially with its rules of consideration, on environmental quality standards, permitting of hazardous activities and environmental impact assessment, is applicable. The Environmental Code also is relevant for the application of the Nuclear Activities Act and the Radiation Protection Act themselves because these two acts refer in certain respects to the Environmental Code (IAEA 2005:43–45). Finally, the Planning and Building Act (SFS 1987:10) is relevant because the siting of a repository constitutes plan-led development and its construction requires a building permit.

---

[214] See NEA 2009i:18–19; NEA 2009d:10–11; Resele 2009:232; Hoppenbrock 2009:182–190.

[215] NEA 2009j:6–8; see also IAEA 2005:35–42 (partly outdated); Michanek/Söderholm 2009:4088–4092 (limited to nuclear power plants).

At the level of delegated legislation the Nuclear Activities Regulation of 1984 (SFS 1984:14), amended in 1992 and 1999, the Radiation Protection Regulation of 1988 (SFS 1988:293), amended in 1999, are in point. At the end of 2008 the Swedish Government decided to start an enquiry into the possibility of having consolidated regulation in the area of nuclear technology and radiation protection and also achieving a better coordination with the Environmental Code (NEA 2009j:7).

Moreover, a number of administrative regulations were adopted by the two formerly competent specialised agencies, the Swedish Nuclear Power Inspectorate (SKI) and the Swedish Radiation Safety Inspectorate (SSI). As a consequence of the merger of these agencies into the new Swedish Radiation Safety Authority (SSM) in early 2008 these regulations have been set into force anew as SSM regulations. Their wording is identical to that of the previous ones with the exception of the names of the relevant authorities. The most important pertinent regulations are the Swedish Radiation Safety Authority Regulations on Safety in connection with the Disposal of Nuclear Waste Materials and Nuclear Waste (SSM FS 2008:21), on the Final Management of Spent Nuclear Fuel and Nuclear Waste (SSM FS 2008:37) and on the Basic Standards for the Protection of Workers and the General Population against Exposure to Ionising Radiation (SSM FS 2008:51). Furthermore, the SSM Regulations on Safety in Nuclear Facilities of 2004 (SSM FS 2008:1), on the Management of Nuclear Waste Materials and Nuclear Waste in Nuclear Facilities (SSM FS 2008:22), on the Physical Protection of Nuclear Facilities (SSM FS 2008:12) and on the Protection of Human Health and the Environment from Discharges of Radioactive Substances from Nuclear Facilities (SSM FS 2008:23) are to be noted.

The financing of radioactive waste management is specified in the Act on Financing of Management of Residual Products from Nuclear Activities of 2006 (SFS 2006:647). The Swedish Act on Phasing-Out Nuclear Power of 1980 was only reluctantly implemented. Only the two reactors of the nuclear power plant Baersebeck were closed down (Parliamentary resolution, SFS 1997:13320). In February 2009, in response to the sensed urgency of global climate change, the Swedish Government decided to ask Parliament for a reversal of the phasing-out decision. This entailed the permissibility of constructing new nuclear power plants for replacing existing ones. In June 2010 the Swedish Parliament, by a very narrow margin, voted in favour of the proposal.

2.5.2 *Responsibilities*

The responsibility of the Swedish Government focuses on policy-making, regulation and control in the field of radioactive waste disposal while Swedish law places the major responsibility for radioactive waste management in the strict sense on the generators of radioactive waste, especially the operators of nuclear power plants.[216] However, the ultimate responsibility for the safety of radioactive waste management is vested in the state. This principle has been deemed so much a matter of course that the relevant laws do not contain a pronouncement to this extent (NEA 2009j:4–5; IAEA 2005:63).

Both central statutes establish a strict duty of the state to protect. The Nuclear Activities Act requires ensuring the safety of all nuclear operations and the protection of the population and the environment against hazards caused by such operations. The Act is supplemented by comprehensive duties of consideration set out in Chapter 2 Sections 2–4 and 7 of the Environmental Code that are formulated as obligations addressed to the operators but serve as standards for regulation and control by the authorities (Chapter 2, Section 1). These duties embody the requirement of sufficient knowledge, the precautionary principle including the use of best available technology, minimisation of environmental burdens and selection of a suitable site for activities that entail the use of land – all mitigated by the principle of reasonableness. The Code's rules on environmental standards provide that such standards must ensure that significant risks to human health and the environment are not caused. The Radiation Protection Act also contains precautionary protective duties, entailing the principles of justification of exposure, optimisation of protection (exposure as low as reasonably achievable) and dose limitations.

The Nuclear Activities Act, as specified by the SSM Regulations on Safety in Nuclear Facilities, provides that the operator of a nuclear power plant must ensure the safe handling and final deposit of nuclear waste (Section 10). In the past this was practically relevant for treatment and interim storage, but it also encompasses final disposal. In discharging their obligations, a pro-active safety management is mandated (NEA 2009j:4–5; IAEA 2005:63). The radioactive waste producers fulfil their waste management obligations through the Swedish Nuclear Fuel and Waste Management Company (Svensk Kärnbränslehantering AB – SKB), a stock corporation which is jointly owned by industry. The existing repository for low and medium level reactor waste at the nuclear power plant in Forsmark and the central interim storage facility for spent nuclear fuel at the nuclear power plant in Oskarshamn are owned by SKB but operated by the respective nuclear power operators on behalf of SKB. In particular, SKB has also been responsible for developing the site for a final repository for high level radioactive wastes in Forsmark/Östhammar which was endorsed by the Swedish Government in June 2009. SKB will construct and operate the facility when the relevant authorisations will have been granted.

A further particularity of the pro-active approach of Swedish law is the obligation of waste generators, imposed by Sections 11 and 12 of the Nuclear Activities Act, to conduct research and development for radioactive waste

---

[216] NEA 2009j:4–5; NEA 2008b:10–11; IAEA 2005:37, 38, 63–64; Zika 2008:233.

management. This programme is also primarily carried on by SKB. It is periodically reviewed by the Swedish Nuclear Power Inspectorate and the Swedish Radiation Protection Authority. The competent Ministry of Sustainable Development can request changes of the programme and in case of failure revise the authorisation to operate the relevant nuclear power plants.

2.5.3 Institutional framework

The Central Government, especially the Ministry for Sustainable Development, primarily has political and regulatory functions. This includes the granting of the central authorisations, namely the framework permission under the Environmental Code and the nuclear activities permission under the Nuclear Activities Act. Below ministerial level, there used to be two major authorities with executive powers in the field of radioactive waste disposal, the Swedish Nuclear Power Inspectorate (SKI) and the Swedish Radiation Safety Inspectorate (SSI). As from 2008, these two agencies have been merged into the Swedish Radiation Safety Authority (SSM) which is now competent for the regulation and control of the safety of nuclear facilities, including repositories for high level radioactive wastes, and radiation protection.[217]

There also is a central advisory body, the Swedish National Council for Nuclear Waste (KASAM) which is consulted before major policy and regulatory decisions are being taken. It also assesses the results of the research and development programme conducted by industry. In order to improve public acceptance of radioactive waste management and give more weight to its recommendations, KASAM has more recently developed a policy of transparency and open discussion with interested individuals and organisations. The responsibilities of SKB have already been described under 6.4.6.2.

2.5.4 Strategies

The basic legal requirements for developing the strategy for managing high level radioactive waste are set forth in the Nuclear Activities Act and, supplementing it, the Environmental Code. The Nuclear Activities Act provides that all nuclear activities must be conducted in a way that they ensure the safe final storage of radioactive waste. This requirement is to be interpreted in the light of the precautionary principle and the principle of optimisation of protection established by the Environmental Code. These statutory pronouncements are supported by the Swedish Parliament which, for example, advocated the construction of a long-term repository with post-closure supervision for a considerable period of time (IAEA 2005:36).

As regards the basic disposal options, the Nuclear Activities Act provides that high level radioactive waste shall be deposited in a geological

---

[217] NEA 2009j:3–5; as to previous law IAEA 2005:38, 47–52, 53–62; NEA 2005a:7; Zika 2008:233–34.

repository. However, this option is not entirely undisputed in Sweden. As an alternative, deposit in deep boreholes has been discussed (see KASAM, Report 2007). Spent nuclear fuel is only classified as waste when it is deposited; otherwise it is nuclear material that in principle can be reprocessed. However, since reprocessing was abandoned for political and economic reasons, in practice spent nuclear fuel also qualifies as radioactive waste and will be deposited, after encapsulation in a central facility, in the geological repository; IAEA 2005:36).

Moreover, a general Swedish policy is that radioactive wastes generated in Sweden must also be disposed of in the country and foreign wastes are not accepted for final disposal in Sweden (Section 5a(2) Nuclear Activities Act). There are certain exceptions to this rule as regards smaller amounts of wastes. The dumping of wastes including radioactive waste at sea is prohibited by the Environmental Code.

The 2008 SSM Regulations on Safety in connection with the Disposal of Nuclear Waste Materials and Nuclear Waste (SSM FS 2008:21) regulate certain technical aspects of the strategy for managing radioactive waste. They require that the design and construction of the repository must ensure and the safety report for it must certify long-term safety. The Regulation also contains provisions on the quality of barriers, scenario requirements and time-scales for the safety assessment. Under the 2008 SSM Regulations on Radiation Protection (SSM FS 2008:51) the applicable general safety standard is best available technology which is equivalent to the ALARA principle.

The strategy for managing high level radioactive waste is already relatively highly developed in Sweden. It is largely based on concepts developed by SKB and then endorsed by the Government. Since the 1970s, SKB conducted research about the best method to deposit radioactive waste and developed later on the "SKB method" which was refined in the 1990s ("KBS-3 method").[218] The strategy consists of the following three elements: a deposit in vertical holes drilled in a 500 metres deep, stable crystalline rock formation as a geological barrier and a long-lived engineered barrier system composed of copper/steel canisters and sealing by protective bentonite clay (for details see section B 1.3.6). As an alternative, horizontal drift disposal is under discussion. The design is deemed to provide an adequate balance between long-term safety, security and the possibility for retrieval,[219] although a political decision in favour of retrieval has not yet been taken. Safety shall be ensured for at least 100,000 years after closure while the Regulation mentioned above only departs from a time-scale of 10,000 years. The tolerable dose for an individual with the highest exposure shall not be higher than 0.1 mSv and the risk to human health not higher

---

[218] NEA 2005a:3, 4; IAEA 2009b:45; SSM 2009. KBS is an acronym for "nuclear fuel safety" (in Swedish), the number 3 means the third version.
[219] NEA 2005d; SKB 2008; IAEA 2009b:47.

than 1:1 Million during a period of proof of 1,000 years and therefore only a fraction of natural radioactivity. In 2001, the Swedish Government in principle endorsed the vertical three barrier strategy but also asked for continuing research on alternatives. The technical development and safety demonstration is carried out in two laboratories, one of them the underground laboratory and the other a laboratory for testing the safety of the canisters used in the repository.

2.5.5 Site Selection

As stated, in Sweden the development of sites for a deep geological repository for high level radioactive waste is the primary responsibility of SKB. However, in accordance with Chapter 17, Section 6 of the Environmental Code the determination of the site requires a framework authorisation, which enables the competent authorities to exercise a large degree of oversight over the process. The basic site selection criteria can be derived from the goals provisions of the Nuclear Substances Act and the duties of consideration set out in the Environmental Code, especially the requirements of selecting a suitable site and minimising possible harm. Within this framework, SKB has always applied optimisation criteria, although mitigated by considerations of public acceptance.

The selection process[220] already started in the 1990s. SKB conducted feasibility studies for eight sites. Most of these municipalities refused to participate in, or withdrew later on from, the process. In 2000 SKB presented a report which formed the basis for the decision of the Ministry for the Environment (later: Ministry for Sustainable Development) to select three candidates for further geological surveys. Since one of the three candidates did not grant permission for field work, the site investigation and selection process was concentrated on Forsmark in the municipality of Östhammar and Laxemar in the municipality of Oskarshamn, both already sites of nuclear power plants and the latter also of an interim storage facility. Field work consisted in surface-based investigations of geological conditions gained from deep boreholes (tunnel drill to the deepest depth of a future repository). Underground investigations in the strict sense were not undertaken. Finally, in June 2009 SKB decided in favour of Forsmark. This decision was endorsed by the Ministry for the Environment. It is envisaged to apply for the construction license under the Nuclear Activities Act and the Environmental Code in 2010 and carry out site characterisation and construction between 2011 and 2017. The application for the grant of the operation permit shall be submitted in 2016. After termination of site investigation work now focuses on analysing data and compiling the license application documents.

The applicable laws do not provide for a formal site selection process. However, since the construction of a repository constitutes plan-led devel-

---

[220] See NEA 2009j:13; NEA 2005a:3, 4, IAEA 2005:14–15.

opment under the Planning and Building Code, the area of the selected site must ultimately be covered by a development plan. In the preparation of the plan, a strategic environmental assessment must be carried out and public participation is provided for. Apart from that, at the stage of the framework permission under the Environmental Code formal public participation has to take place. Consultations of the public under the environmental impact assessment procedure that in accordance with the Environmental Code must precede the application for the framework permit are already in the final stages.

In the site selection process proper, only a sort of "channelled" participation by the municipalities concerned was established by the Swedish Government and SKB. Since public acceptance was considered to be an important restraint of the search for an optimal repository site, SKB conducted a structured series of consultations with all municipalities that participated in the selection process; NGOs could participate as observers; SKB only continued the process with voluntary candidates. The relevant municipalities themselves tried to secure public involvement and acceptance by organising seminars and hearings of local citizens and NGOs (and were reimbursed for the expenses incurred in these activities out of the nuclear waste fund). Moreover, according to earlier decisions by the Ministry for the Environment, SKB had to consult with the two formerly competent agencies, the SKI and the SSI, supported by advisory expert groups about the planning and implementation of its work. In order to improve transparency and information, the concerned municipalities were invited to participate as observers. Finally, in 2006/2007 KASAM developed a transparency programme that was primarily designed to improve transparency and public participation in the later decision-making process.[221]

2.5.6 Construction and operation

The construction and operation of a geological repository for high level radioactive wastes are subject to several authorisation requirements that are based on the Environmental Code, the Nuclear Activities Act and the Planning and Building Act. By contrast, no authorisation is required under the Radiation Protection Act. Its provisions on radiation protection are applied in the permitting process under the Nuclear Activities Act.

First of all, the Environmental Code provides that projects of major importance (to which category repositories for high level radioactive waste belong) are subject to the requirement of a declaration on their permissibility (Chapter 17, Section 6; see Michanek/Söderholm 2009f). Under the Environmental Code the whole project including the repository together with the neighbouring underground laboratory and encapsulation plant is subject to review. The declaration of permissibility is equivalent to a

---

[221] Based on the RISCOM II-model; see Anderson et al. 2004.

framework authorisation. It has a coordinating character, stating that in principle there are no objections against the site, design and impacts of the project. Both safety, radiological and other environmental impacts are considered. Possible damage must be minimised. If significant damage cannot be excluded, a cost-benefit analysis is required. (Chapter 17, Sections 6, 9 and 10). Moreover, other concerns of the public interest such as need for the facility, labour opportunities and regional development, including compatibility with the municipal development plan under the Planning and Building Act are to be considered. Therefore, the decision also contains political elements and is not of a purely legal-administrative nature.

Moreover, Swedish law provides for separate authorisations of the construction and operation of a repository for high level radioactive waste (deemed to be a hazardous activity) as well as its water-related impacts under the Environmental Code and for separate authorisations of the construction and operation of the repository under the Nuclear Activities Act (see Michanek/Söderholm 2009:4090/4091). These authorisations relate to the site, construction, design and modes of operation of the facility. Under the Environmental Code the whole repository system including the central interim storage facility, the encapsulation facility and the repository are scrutinised. It must be ensured that harmful or detrimental impacts on man and the environment that result from the location, scope, design and emissions of the facility are prevented, taking into account the duties of consideration (Chapter 2, Sections 1, 3, 4 and 7). In the nuclear authorisation procedure, the repository is reviewed separately from the other facilities. The authorisation may be granted when the requirements of nuclear safety and of the protection of human health and the environment are met and no significant risk is presented by the facility (Sections 3–5 of the Nuclear Activities Act). The requirements of the Radiation Protection Act form part of the conditions imposed on the operator.

A clear delimitation of the respective scope of the two categories of permits does not exist. Theoretically one could state that the focus of the permits granted under the Environmental Code is on the general impacts of the project on human health and the environment, while the authorisations granted on the basis of the Nuclear Activities Act concentrate on the technical and scientific aspects of operational and radiological safety. However, there is a considerable overlap as regards radiological impacts, the more so since the duties of consideration set out by the Code, especially the principles of precaution, optimisation, minimisation of risk and reasonableness, are also applicable in the framework of the nuclear authorisations.[222]

---

[222] Michanek/Söderholm 2009:4095, discuss this problem only with respect to the framework authorisation for which, in view of its coordinating nature, they claim a comprehensive coverage.

Finally, the construction of the repository requires a building permission under the Planning and Building Act.

As regards the procedure, at all stages of the process the carrying out of environmental impact assessments is required (under Section 5b of the Nuclear Activities Act by reference to Chapter 6 of the Environmental Code). In this framework there is an opportunity for an early participation by the affected population and organisations, at later stages also by the public at large. Moreover, after submission of the relevant application, the Environmental Code (Chapter 22, Sections 3, 10) and the Nuclear Activities Act provide for public participation. Under the Environmental Code, also public hearings are required.

Agency competences vary considerably (NEA 2009j:9–10; IAEA 2005:45, 48). The application for the environmental permissions is processed and reviewed by the Environmental Court. The final decision on permissibility is taken by the Government, that on the construction and operation authorisation by the Environmental Court (Chapter 21, Sections 1, 25 of the Environmental Code). There is a municipal right of veto against the declaration on permissibility which, however, can be overridden by considerations of public interest ("utmost importance with regard to the national interest"), provided there is no other feasible alternative (Chapter 17, Section 6 of the Environmental Code). Likewise, the planning monopoly of the municipality can be overridden although this hardly happens in practice (Michanek/Söderholm 2009:4093f). The applications for the nuclear permits are reviewed by SSM that makes recommendations to the Government on the permit including conditions to be attached to it. However, the present practice tends to an ever increasing extent in the direction of generic regulations. The authorisations are granted by the Government.

2.5.7 Financing

According to the Nuclear Activities Act (Section 10) the generators of radioactive waste are also financially responsible for waste management.[223] This basic obligation is specified by the Act on Financing of Management of Residual Products from Nuclear Activities of 2006 (SFS 2006:647).[224] Under the Financing Act the operators of nuclear power plants must pay annual charges into the Nuclear Waste Fund in order to cover total expenses of radioactive waste management and disposal. Moreover, they must provide financial guarantees for the payment of future charges. The relevant expenses encompass, among others, decommissioning, and the construction and operation of the encapsulation plant and the repository for spent nuclear fuel and other high level radi-

---

[223] See NEA 2009j:16–19; NEA 2005a:8; NEA 2008b:10–11.
[224] The Act was preceded by acts adopted in 1981 and 1992.

oactive waste. Moreover, the costs of research and development undertaken by SKB or by the operators of nuclear power plants themselves are
covered. Finally, the financing of participation by the affected municipalities and non-governmental organisations is encompassed. The charge is
based on annual cost estimates and calculated on the basis of nuclear
electricity generation.

2.6   Finland

2.6.1   Sources of regulation

In Finland, the general laws that regulate nuclear activities also comprise
radioactive waste management. However, there is special delegated legislation on radioactive waste management.[225] The major sources of regulation are the Nuclear Energy Act of 1987 (Act No. 990/1987), as amended
in 2004 and 2008, and the Radiation Protection Act of 1991 (Act No.
592), as amended in 1998 and 2005, and at the level of delegated legislation the Regulation on the Safety of Disposal of Nuclear Waste of 2008
(Regulation 736/2008) and the Radiation Decree of 1991 (Regulation
1512/1991).[226] At the end of 2008, most of these regulations have been
amended. Furthermore, the Act on Environmental Impact Assessment
of 1994 (Act No. 468/1994) is relevant; in particular the first licensing
of a repository, the framework authorisation (decision-in-principle), but
also the construction permit must be preceded by an EIA. Public information on radioactive waste problems is ensured by the Transparency of
Government Act of 1999 (Act No. 621/1999). With respect to environmental aspects not covered by the Nuclear Energy Act, such as chemotoxic effects on groundwater, the Environmental Protection Act (Act No.
86/2000) is applicable.

The Nuclear Energy Act applies to all nuclear activities and therefore
also regulates radioactive waste management; it defines any facility in
which nuclear waste is handled or stored as a nuclear facility (Sections 1
and 11). The Radiation Act regulates all aspects of protection against the
risks presented by exposure to radioactivity, including those originating
from waste management activities. It is only at the lower level of the hierarchy of norms that special rules for radioactive waste management come into
play. In 1999, a General Regulation for the Safety of Spent Fuel Disposal was
promulgated that has been replaced by the new Regulation 736/2008. This
regulation applies to the deposit of high level radioactive waste in a geological repository. There also are administrative rules (guidelines) adopted in

---

[225] See IAEA 2008b:13 and Annex L.5; NEA 2008c:11–12; id, NEA 2009k:1, 5–6;
Posiva 2009:10–12.
[226] The Regulation for the Safety of Nuclear Power Plants of 2008 (Regulation
733/2008) sets radiation exposure limits, among others with respect to interim
storage, but is not relevant for final disposal.

2001 and 2002, respectively, on long-term safety of disposal of spent nuclear fuel (STUK Guide YVL 8.4) and on the operational safety of a disposal facility for such wastes (STUK Guide YVL 8.5). The guidelines will be revised in 2010 to adjust them to the new legislation (Posiva 2009:10).

2.6.2 Responsibilities

In Finland, the responsibilities for managing high level radioactive waste are shared between the state and industry. The authorities, especially the competent ministries and executive agencies, are competent for policy-making, regulation and control, including the licensing of high level radioactive waste repositories. The Nuclear Energy Act describes the state duty to protect in a rather broad way. In accordance with Sections 1, 5, 6 and 19 of the Act, the management of radioactive waste must be in conformity with the public interest. The authorities have to ensure the safe use of radioactive materials as far as reasonably achievable. They are obliged to enhance protection of human life and health and of the environment according to the state of science and technology. The new Chapter 2A of the Act (Act 242/2008, Sections 7a-7r) lays down a number of very specific nuclear safety and security principles that previously had been established merely politically or in regulations and guidelines of the Government. These include the principles of optimisation of safety, in-depth safety (redundant levels of safety), dose limitations for exposure, accident prevention and preparedness, long-term safety of the final disposal of radioactive waste, crossgenerational equity regarding exposure to radiation from waste disposal, external security and prevention of unlawful interferences.

However, it is to be noted that the role of the state is limited by that of radioactive waste generators. According to Section 9 of the Nuclear Energy Act, the radioactive waste generators are primarily responsible for managing their radioactive waste.[227] They may also be required by the competent Ministry for Trade and Industry and the Radiation and Nuclear Safety Authority to establish a waste management plan (Section 28) and engage in joint waste management measures (Section 29). For that latter purpose the two public utilities in Finland as the major radioactive waste generators have set up the private law company Posiva Oy. This company has assumed the responsibility for preparing site selection and constructing and operating the final disposal facility. The responsibility of the generators ends when the wastes are permanently disposed of in an appropriate manner and the repository has been closed (Sections 31–34; IAEA 2008b:37, 40). Moreover, the operators are obliged to carry out research and development for the safe management of their radioactive wastes. Posiva is charged with conducting that research and development. Finally, the financial responsibility for radioactive waste management is imposed on the operators (Section 28).

---

[227] IAEA 2008b:37–39; NEA 2008c:11; NEA 2009k:1.

2.6.3 Institutional framework

The Ministerial competences for radioactive waste disposal are primarily vested in the Ministry of Trade and Industry (KTM), in the second place in the Ministry of the Environment which often has to be consulted. Apart from policy-making in the field of nuclear energy, the former is competent for promulgating safety regulations, reviewing the waste management activities as to their consistency with the national policy and preparing the grant of the requisite authorisations by the Government (Art. 81 of the Nuclear Energy Act). It has been criticised that this organisational set-up does not fully ensure the independence of the regulatory function from functions entailing the promotion of nuclear energy (EU Peer Review 2009:6–7). The Radiation and Nuclear Safety Authority (STUK) is responsible for setting forth detailed safety regulations, reviewing license applications (safety assessment) and for all control and inspection activities in the field (see NEA 2009k:3–4). There also is an Advisory Committee on Nuclear Safety that also deals with safety issues related to radioactive waste management.

2.6.4 Strategies

The basic requirements of developing a strategy for high level radioactive waste disposal can be derived from the goals provisions of the Nuclear Energy Act. According to Sections 1, 5, 6 and 19 of the Act, nuclear, including waste disposal facilities must be consistent with the public interest and ensure a safe use of radioactive material and wastes as far as reasonably achievable; all measures to enhance safety and the protection of the environment according to the state of science and technology shall be taken. These provisions are further specified by the new Chapter 2A of the Act. According to Section 7b of the Act the safety of a nuclear facility including a waste disposal facility shall be ensured by means of successive levels of protection independent of each other. The disposal of radioactive waste shall be planned in such a manner that the exposure of future generations does not exceed levels considered acceptable at the time of final disposal; it shall be permanent without any need for institutional controls after closure (Section 7h of the Act). As regards non-radiological effects of radioactive waste disposal, for example on groundwater, the Environmental Protection Act with the precautionary and minimisation principles set forth there applies (Sections 4–6, 8 of the Act).

Already in 1983 the Finnish Government adopted the policy that the disposal of spent nuclear fuel and other high level radioactive waste should take place in a geological repository (IAEA 2008b:33–36; NEA 2009e:1). This option is a political one and has been not been laid down in a legal form in the Nuclear Energy Act of 1987. However, the Act departs from the proposition that this option is the one to be followed in the future. Moreover, the new Section 7h of the Act requires "permanent" disposal of radioactive waste from which one could derive a statutory option for geological disposal. Reprocessing of spent nuclear fuel is not provided for ("once-through" option) (IAEA 2008b:14). Therefore the definition of radioactive waste includes spent nuclear fuel.

The Nuclear Energy Act in its amended version establishes the principle that nuclear waste generated in Finland must be handled, stored and disposed of in Finland and that such waste generated abroad shall not be handled, stored and disposed of in Finland (Sections 6a and 6b). Certain exceptions can be made, especially regarding the export and import of small amounts for research purposes and the export of wastes from domestic research reactors.

Dumping of radioactive waste at sea is prohibited by legislation that implements the relevant international conventions (Helsinki and London Dumping Conventions).

On the occasion of deciding on its basic radioactive waste option, the Government decided that the location of the repository shall be 400-500 metres underground in cristaline hardrock (IAEA 2008b:62; NEA 2009k:9). Finland follows a strategy of multiple barriers that consist of a geological barrier and engineered barriers, namely deposit in copper canisters and bentonite clay buffers. This strategy is similar to the Swedish KBS-3 concept, with the variation that in Finland horizontal bore drills are preferred. As now also laid down in Section 7h of the Nuclear Energy Act the barriers must be such that institutional controls after closure are not needed (IAEA 2008b:60).

The time-frame for the safety assessment is several hundred thousand years. In the first several hundred thousand years after closure the annual effective dose for the most exposed individuals shall be below 0.1 mSv and the average annual dose for the whole population significantly lower. Later on, natural radiation provides the standard for the safety that shall be aimed at. Special requirements apply to very improbable events such as human intrusion where an exposure to an individual dose of 0.5 mSv in a year must be less probable than 1:1 million per year. [228]

The framework decision for the repository at Olkiluoto contains retrievability as on option. However, since the reform of radioactive waste management policy in 2008 retrievability of the deposited wastes is no longer set forth in the Nuclear Energy Act or in the Regulation 736/2008 (see EU Peer Review 2009:9).

Regulation 736/2008 and the STUK Guidelines YVL 8.4 and 8.5 specify the relevant requirements. Although Finland has made some progress in

---

[228] See Regulation 736/2008 (that replaces Decision of the Government on the General Regulation for the Safety of Spent Nuclear Fuel Disposal of March 1999: Decision 478/1999; STUK Guideline YVL 8.4 point 2.; IAEA 2008b:59.

developing its waste management strategy for high level radioactive waste, extensive research is still needed to further develop the necessary disposal techniques and safety assessment methods. This is in keeping with the mandate by the Finnish Government which fixed a long-term schedule for the implementation of radioactive waste management.

2.6.5   Site selection

Under Finnish law, the siting of a nuclear facility and hence also of a repository for high level radioactive waste requires a framework authorisation (decision-in-principle) under the Nuclear Energy Act (Section 11). This authorisation confirms that there are no objections based on principle against the site, the basic technical concept of the repository and its possible impacts (IAEA 2008b:61–63). The facility must be consistent with the public interest and ensure a safe use of radioactive material and wastes as far as reasonably achievable. The Act also requires a risk/benefit assessment that includes the need for the facility (Section 14(2) of the Act). In particular, the new guiding principles on safety and security have to be observed, as applicable. Moreover, an EIA is necessary (IAEA 2008b:32). The authorisation can only be granted with the agreement of the affected municipality and needs to be ratified by Parliament (Sections 14(1), 15 of the Act). In the view of the Finnish legislature, this system of checks and balances guarantees a fair and open selection process.

After extensive screening and field research which included several possible candidates, in 1999 Posiva applied for a framework authorisation for Olkiluoto in the municipality of Eurajoki which is already the site of a nuclear power plant and of one of the two Finnish repositories for low and intermediate level radioactive waste at intermediate depth, as the future site for the repository. The choice was primarily based on socioeconomic considerations. The concerned municipality agreed in early 2000, the Government made a positive site decision in December 2000 and Parliament ratified it in 2001. The next steps of the rather smooth procedure were the construction of an underground rock characterisation facility on the site (Onkalo) that started in 2004 and will be its operation. The application for a construction authorisation is expected by end of 2012.[229]

The Nuclear Energy Act provides for a fair degree of public participation that is also furthered by transparency of government as provided by the Transparency Act of 1999. The residents as well as the municipalities in the vicinity of the facility have the right to make comments on the application for the framework decision and on the environmental impact assessment (Section 13 of the Act). The competent agency shall

---

[229] See IAEA 2008b:14–15, 20–21, 61–63; NEA 2009e:3; NEA 2009k:7.

also organise a hearing that is open to the public at large. There clearly is a strong focus on the local population as the latter can veto the project.[230] However, as regards the agreement of the municipality of Eurajoki, it is to be noted that the decision of the Finnish Government to go forward with the construction of a new nuclear power plant was taken after the municipality had agreed to become the site of the repository and the Finnish Parliament ratified the site selection. It is not far-fetched to assume that the outcome of the site selection process might have been different if this decision had been taken earlier.

2.6.6 Construction and operation

The construction and operation of the geological repository requires a construction and an operation authorisation that are to be granted, on the basis of the framework authorisation, by the Government (Section 16 Nuclear Energy Act). An updated environmental impact assessment is required. Furthermore, as regards non-radiological effects, an authorisation under the Environmental Protection Act (Sections 8, 28) is needed. Finally, it is to be noted that the construction constitutes plan-led development under the Building Act of 1958 (Act No. 370/1958) so that a local plan must allow for such type of facility.

Apart from the binding force of the framework authorisation, the permit requirements ensue from the Nuclear Energy Act (Sections 18, 19 and 20, respectively as well as the new guiding principles on safety). In particular, the construction and design of the facility must be such that the site is appropriate from the point of view of safety, environmental protection has been taken into account appropriately, operational safety and physical security have been demonstrated, compliance with the radiation protection standards is ensured and the methods of waste disposal are considered to be appropriate. In the operation permit process scrutiny focuses on operational safety, protection of workers and the population and environmental protection. In both cases, compatibility with the public interest also is a permit prerequisite. The environmental permit is limited to non-radiological effects, especially chemo-toxic effects on groundwater; it is governed by the principles of precaution and minimisation (Sections 4–6, 8 Environmental Protection Act).

STUK is responsible for the safety review of the applications, while the Ministry of Trade and Industry prepares the decision that is taken by the full cabinet. As regards participation, the Nuclear Energy Activities Act refers to the procedure governing the framework permit (Section 23). Moreover, insofar as an environmental impact assessment is required, including the requisite updating of an existing one, there is an opportunity to participate in the decision-making process.

---

[230] IAEA 2008b:63; EU Peer Review 2009:13; Javanainen 2006:19.

2.6.7   Financing

Apart from waste management and research obligations, industry also bears the financial responsibility for radioactive waste management (NEA 2008c:12; NEA 2009k:8–9). Under the Nuclear Energy Act (Section 38), a charge is levied on all radioactive waste producers. The charge amounts to 10 percent of the production costs. It is assumed that it will be included in the price of nuclear energy. According to the 2004 amendment of the Act (Section 38), the payments feed two funds. The specifics are set out in the Regulation on the Nuclear Waste Management Fund of 1988. The State Nuclear Waste Management Fund, administered by KTM, is to cover the costs of decommissioning, interim storage, disposal in a repository and other management measures costs as well the costs for research on, and development of, the repository. The other fund which is administered by the former fund is designed to finance general nuclear research in order to maintain or create sufficient scientific and technical expertise in the field. The calculation of the charge is based on estimates as to the (undiscounted) expenses to be incurred in the future.

2.7   Japan

2.7.1   Sources of regulation

In Japan there are three major laws, partly of a general character, partly specific, that regulate radioactive waste disposal.[231] The Act on Prevention from Radiation Hazards of 1957 (Act No. 167/1957 – Prevention Law), as amended in 1980, 1995 and 2004, applies to any handling of radioactive materials. In essence, it is designed to ensure radiation protection. It is supplemented by the Regulation of the Prime Minister on Prevention from Radiation Hazards (Ordinance No. 56/1960, as amended) and Regulations on the Prevention of Ionising Radiation Hazards at the Work Place (Regulation No. 41/1972, as amended by Regulation 172/2001). The Act on Regulation of Nuclear Source Material, Nuclear Fuel Material and Reactors of 1957 (Regulation Law – Act No. 166/1957), as amended in 2002 (Act No. 179/2002) regulates nuclear facilities; it also applies to radioactive waste disposal facilities. It is specified by the Regulation of the Prime Minister on the Enforcement of the Law on Nuclear Source Material, Nuclear Fuel Material and Reactors of 1957 (Ordinance No. 324/1957), as amended by Ordinance No. 57/2004 and later on in 2008. The Specified Radioactive Waste Final Disposal Act of 2000 (SRWDA or Waste Law – Act No. 117/2000), as amended in 2007 (Act No. 84/2007), contains special regulation for the geological disposal of high level radioactive waste. Pursuant to this Act, new safety regulations for radioactive waste disposal have been promulgated in the same year. Finally, the Environmental Impact Assessment Act of 1997

---

[231] NEA 2008d:8; NEA 2009l:1, 6; NEA 2005c:5.

(Act No. 81/1997) is applicable regarding projects to construct a final disposal facility for radioactive wastes.

2.7.2 Responsibilities

In Japan, the responsibility for high level radioactive waste is divided between the state and industry, the latter playing an important role.[232] The competent authorities are responsible for policy-making, regulation and control of disposal activities including the granting of the requisite authorisations. The state duty to protect only is formulated in the Prevention law in broad terms, incorporating the precautionary principle. The state has to prevent ionising radiation hazards and secure public safety (Section 1 Prevention Law). A similar wording exists in the amended Waste Law. According to Article 51-1 of the Act the state shall ensure public safety by preventing hazards associated with the disposal of high level radioactive waste.

Under Section 40 of the Waste Law, the Nuclear Waste Management Organisation (NUMO),[233] a private law company owned by the radioactive waste producers, is charged with the task of implementing the final disposal of high level radioactive waste. This includes all steps involved in managing such wastes, from the selection of the site and preliminary investigations to construction and operation of the repository to post-closure management. Japan Nuclear Fuel Limited may assume the operational tasks on behalf of NUMO. Moreover, industry is financially responsible for the management of their radioactive wastes.

2.7.3 Institutional framework

The institutional framework in the field of radioactive waste management in Japan is rather complex.[234] A Japanese particularity is that the borderline between regulation and advisory functions in the institutional set-up is not easy to draw. The Ministry for Energy, Trade and Industry (METI, formerly MITI) is competent for the development of the basic policy and regulation in the field of radioactive waste management. It also establishes the high level radioactive waste disposal plan prescribed by the Waste Law. The Ministry has various agencies through which it performs part of its tasks, in particular the Agency for Natural Resources and Energy, the Nuclear and Industrial Safety Agency (NISA) and the incorporated Nuclear Energy Safety Organisation (Radioactive Waste Management and Transport Division). The Ministry competent for Science is responsible for radioactive waste from public research activities. Within METI, the Nuclear and Industrial Safety Agency (NISA) possesses a particular role. It has both regulatory and control functions. Although a part

---

[232] NEA 2008d:9; NEA 2009l:1, 11; NEA 2005c:5, 8.
[233] See NUMO 2011.
[234] See NEA 2009l:2–5; NEA 2005c:8.

of the Ministry, it is supervised by the Nuclear Safety Commission (NSC) which itself is a department of the Prime Minister's Office. Despite this organisational set-up, in view of OECD the independence of NSC from governmental agencies that promote nuclear power is not provided for with irrefutable clarity (OECD 2008:179).

There are a number of advisory bodies that partly also exercise regulatory functions in the field of radioactive waste management. The Atomic Energy Commission is responsible for the development of basic policy in the field of the civilian use of nuclear energy. The Nuclear Safety Commission, established in the Office of the Prime Minister, has regulatory and advisory functions in the fields of radiation protection and radioactive waste management. In particular, it reviews applications for the grant of licenses for radioactive waste facilities. Within the Commission, there also is an Advisory Board on High Level Radioactive Waste Repositories.

2.7.4 Strategies

The Japanese strategy for managing high level radioactive waste is based on the goals provisions of the Waste Law (Section 2) as well as on the prerequisites for granting a construction and operation permit for the repository (Sections 6, 7 and 7-2 Prevention Law). The Waste Law contains a clear statement in favour of geological disposal of high level radioactive waste in a deep underground repository (Section 2).[235] This is remarkable because in Japan particular concerns arise from the complex geology and tectonically active setting of the Japanese territory. The Government and the legislature were of the opinion that these problems could be overcome through careful site selection and engineering solutions. However, presently there are deliberations about a modification of the option chosen by the Waste Law in the direction of prolonged interim storage for about 100 years in the hope of new technological solutions.

High level radioactive waste in particular includes spent nuclear fuel. Reprocessing is neither prohibited nor prescribed. In the practice, spent nuclear fuel is mostly exported for reprocessing, partly reprocessed in Japan. In accordance with the 2007 amendments of the Waste Law, the radioactive waste originating from reprocessing (transuranic waste) will also be disposed of in the geological repository.

Dumping of high level radioactive waste at sea is prohibited since 1993 pursuant to the London Convention of 1972 to which Japan is a party.

For implementing the statutory requirements, the Japanese Government adopted the "Basic Policy on Specified Radioactive Waste Final Disposal" in September 2000 and established the "Specified Radioactive Waste Final Disposal Plan" in 2005. Both policy papers were revised in March 2008 to reflect the inclusion of transuranic waste from reprocessing in the pro-

---

[235] See NEA 2008d:9; NEA 2009l:8.

gramme.[236]. The plan contains, among others, assessments and estimates of the volume of high level and transuranic waste to be disposed of, a time frame for the site selection process and a determination of the size of the repository. Moreover, based on a recommendation by the Advisory Board for High Level Radioactive Waste Disposal, on 1 April 2008 NISA published rules for category 1 waste disposal for nuclear fuel that are to serve as the basis for future regulation.[237] Work on the licensing procedure for high level radioactive waste disposal under the perspective of safety is in progress (NSC 2009).

As many other countries, Japan follows a multiple barriers concept that entails natural and engineered barriers to provide long-term isolation of wastes from the environment (NSC 2009:7–9). Special safeguards are to be taken with respect to hazards presented by earthquakes. The repository shall be located at more than 300 metres underground in crystalline hardrock. A requirement of retrievability is under consideration.[238] In the operation of the repository and with respect to the post-closure period measures must be taken to prevent hazards from radiation in accordance with the dose limits for workers and the public at large set forth in the Prevention Law (Sections 19–21) and the relevant implementing regulations. The details of the waste disposal strategy will be worked out in the future in keeping with the progress obtained in developing a site for high level radioactive waste disposal. The post-closure assessment period (time frame for ensuring safety) has not yet been determined in Japan. However, NSC seems to have a preference for setting as assessment period a period of time for which highly reliable assessment can be made according to best available scientific knowledge (NSC 2004°).

2.7.5 Site selection

The basic procedure for selecting a site for the underground repository is determined in the Waste Law (Sections 6–8).[239] It is a three-step procedure to be implemented by NUMO. The first step consists in a survey of geological disturbances such as earthquakes and other natural phenomena (which play a particular role in Japan) and the selection of preliminary investigation areas (PIAs). In a second step, tests such as borehole surveys and geophysical prospecting shall be conducted in these areas to determine the stability of the geological stratum, resulting in the choice of detailed investigation areas (DIAs). As the third step detailed underground investigations shall be carried out. Based on the results of these investigations, NUMO shall select a site where the final disposal facilities are to be constructed and

---

[236] Press release 14 March 2008. Transuranic waste is radioactive waste with a low activity, but a long half-life.
[237] NEA 2009l:17. Category 1 waste means high level radioactive waste.
[238] See World Nuclear Organisation 2004.
[239] See NEA 2008d:9; NEA 2005c:5–6; NUMO 2004; NUMO 2007.

establish a final disposal plan. All steps are subject to approval by METI based in interim reports to be supplied by NUMO. The time frame for the selection process was determined by METI in the Specified Radioactive Final Disposal Plan. The screening phase is to last from 2002 to 2008, the detailed investigation phase from 2009 to 2013 and the underground investigation phase from 2014 to 2022. Between 2023 and 2027, final site selection shall take place.

The ultimate decision is to be taken by METI, considering the opinion of local authorities that have to be consulted as well as the population at large. In practice, the selection process is based on voluntary participation by the host municipalities. In exchange for the agreement of the concerned municipalities to become the site for a repository, compensation for promoting economic development is envisaged although the Waste Law does not formally provide for compensation. The municipalities participating in, or affected by, the programme as well as the regions concerned are regularly consulted by NUMO and ultimately by METI. The affected population is afforded an opportunity to comment on the project in its various stages. In taking its decisions on the project, METI is to consider the views of the affected municipalities, regions and population. However, a formal obligation to consult and consider the results of participation does not exist. Nor is there any requirement to carry out a formal environmental impact assessment at this stage of the process.

In 2002, NUMO asked municipalities to become candidates for preliminary investigation, indicating the geological exclusion criteria, and afforded those who volunteered an opportunity to comment on its evaluation report. The selection of the areas for preliminary investigation (PIAs) was then made by NUMO and endorsed by METI. The selection process is still in the stage of preliminary investigation. A number of municipalities signalled interest but none of them has as yet made a firm commitment to host the repository. In one municipality participation in the selection process was rejected by a referendum.[240] In 2009, NUMO started another campaign for persuading municipalities to volunteer for become hosts for a repository. It should be noted that Japan already operates two underground laboratories, one in a crystalline rock formation and another in a sedimentary rock formation.

2.7.6 Construction and operation

The construction and operation of the final repository by NUMO requires an authorisation under the Regulation Law (Sections 6, 7, 7–2) (NEA 2008d:8). The authorisation encompasses the site, design and equipment

---

[240] See Tanaka, Strategic Planning for Designated Radioactive Waste Disposal in Japan, University of Tokyo, October 15, 2008; see also Japan Atomic Energy Commission, Framework for Nuclear Energy Policy, October 2005:22.

of the facility. It must be demonstrated that the facility complies with the applicable standards and that radiation hazards are dealt with satisfactorily. Further specifications ensue from Articles 51-1, 51-3 and 51-7 of the Waste Law (as amended in 2007). The grant of the authorisation generally requires that public safety by preventing hazards associated with waste management is ensured. The site, construction, design and methods of disposal must be such that radioactive disasters are prevented.

In the procedure, also an environmental impact assessment is required. According to the Act on Environmental Impact Assessment of 1997 (Act No. 81/1997) the construction of all final waste disposal facilities is subject to the requirement of carrying out an environmental impact assessment (Article 2). The Act also provides for staged participation both of the general public and the affected municipalities and regions. The authorisation is granted by METI, acting upon the advice of NSC.

2.7.7 Financing

The Waste Law provides that the operators of nuclear power plants have to finance reprocessing, decommissioning, interim storage as well as the expenses of NUMO for the development (including the selection) and in the future also the operation of the final repository.[241] The details are set out in the Act on Deposit and Management of the Reserve Fund for Spent Fuel Reprocessing and so forth in the Nuclear Power Generation of 2005 (Act No. 48/2005). As regards waste disposal, the operators have to pay a specific charge, determined by METI every year, into the Radioactive Waste Management Fund. This fund is administered independently from the waste generators by the Radioactive Waste Management Fund Financial Centre.

2.8 Spain

2.8.1 Sources of regulation

In Spain, there is no special regulation on radioactive waste management. Rather, a number of general laws relating to nuclear energy also apply to radioactive waste management.[242] The Nuclear Energy Act of 1964 (Act 25/1964), as amended by Article 6, Additional Provisions, of the Act on the National Electricity System of 1997 (Act 54/1997), contains basic requirements for the use of nuclear energy and radioactive substances, the authorisation of nuclear facilities and safety and radiation protection. It is specified by the Regulation on Nuclear and Radioactive Activities of 1999 (Royal Decree 1836/1999) and the Regulation on Radiation Protection (Royal Decree 783/2001). The establishment of the major executive authority and the financing of radioactive waste disposal are regulated by special laws. Apart from that, general environmental laws are applicable or at least rele-

---

[241] NEA 2008d:9; NEA 2009l:12–13; NEA 2005c:9.
[242] See Ruiz de Apodaca Espinosa 2009:89–111; NEA 2004a:123, 124–129.

vant. This is in particular true for the legislation on environmental impact assessment (Royal Legislative Decree 1/2008), on strategic environmental assessment (Act 9/2006) and on environmental information, participation and access to justice (Act 27/2006, implementing EC Directive 2003/35).

2.8.2 Responsibilities

In Spain, the major responsibility for radioactive waste management is vested in the Government. The responsibility of industry is limited to interim storage of high level radioactive waste and the financing of waste management. In particular, the selection of the site of a repository for high level radioactive waste, its construction and operation are considered to be a state duty and entrusted to a public enterprise, ENRESA.

The state duty to protect is set forth, in broad statutory terms, both in the Nuclear Energy Act of 1964 (Act 25/1964) and the Regulations on Nuclear and Radioactive Activities of 1999 and on Radiation Protection of 2001 (Royal Decrees 1836/1999 und 783/2001). Article 1 of the Act 25/1964 mandates the protection of life, health and property against dangers of nuclear energy and adverse effects of ionising radiation. The authorisation prerequisites and the provisions on radiation protection (Articles 29, 37 et seq.) refer to the notion of risk and safety. Likewise, the two regulations aim at ensuring nuclear safety and protection against risk caused by ionising radiation.

2.8.3 Institutional arrangements

The organisation of the executive regarding radioactive waste management in Spain is relatively simple. The Government, especially acting through the Ministry for Industry, Tourism and Commerce with its Secretariat of Energy, has the ultimate responsibility for nuclear waste policy, regulation and supervision.[243] In particular, it is competent for granting the requisite authorisations for nuclear facilities, including waste disposal facilities. In the authorisation process, the Ministry acts on a mandatory, binding assessment by the Nuclear Safety Council (Consejo de Seguridad Nuclear – CSN). The Ministry for the Environment participates in the authorisation process by providing an environmental impact statement on the relevant projects.

The Nuclear Safety Council, established by the Act 15/1980 (CSN Act), is a large and well equipped authority independent from the administration that reports directly to Parliament. It has broad competences in the fields of nuclear safety and radiation protection, in particular the regulation and supervision of nuclear facilities. Its competence also relates to studying and reviewing plans, programmes and projects during all phases of radioactive waste management (Act 14/1999) and it performs or supports research and development.

---

[243] See for the following text NEA 2004a:124–126; NEA 2005b:7–8; Ruiz de Apodaca Espinosa 2009:98, 108.

The basic management functions are exercised by the National Enterprise for Radioactive Waste (Empresa Nacional de Residuos Radioactivos S. A. – ENRESA). Established in 1980, it is a state-owned company that has the task to develop radioactive waste management programmes according to policy and strategy approved by the Spanish Government. In particular, ENRESA establishes the radioactive waste management plan, is responsible for site selection, construction and operation of repositories and central interim storage facilities for radioactive waste and performs a number of other tasks in its area of competence. ENRESA closely cooperates with CSN without compromising the independence of the latter (NEA 2004a:132). According to the Act 24/2005 ENRESA shall be converted into a public enterprise (Entidad pública empresarial Enresa de gestion de resiuos radiactivos – EPE) once its statutes have been adopted by the Government. However this has not occurred yet.

2.8.4 Strategies

The basic strategic options and technical strategies of high level radioactive waste management in Spain have not been set out in the applicable laws and regulations. The goals provisions and permit prerequisites of the Nuclear Energy Act of 1964 (Articles 1, 29, 37 et seq.) and the Regulations on Nuclear and Radioactive Activities (Articles 14, 17, 20) and on Radiation Protection (Articles 1, 4) only contain a broad "normative programme". Rather, the strategy and the relevant technical solutions will be developed by the "General Radioactive Waste Plan" that in accordance with Article 6bis of the Act on the National Electricity System of 1997 must be determined by ENRESA every four years subject to approval by the Spanish Government.[244] This plan is mandatory and binding on the executive.[245] In the future, it will be subject to the requirement of a strategic environmental assessment under the Act 9/2006 (Article 3). The Sixth Plan was determined in 2006.

The basic strategic option in Spain is final disposal in a geological repository. CSN has also already established general safety criteria for geological disposal. However, the envisaged time horizon of waste disposal planning is very long. The target for having a repository available is the years 2050-60. In the meantime, long-term interim storage of high level radioactive waste is the preferred option. This is based on various reasons. These are, for example, the expected greater flexibility, the approaching need to take back radioactive waste from foreign reprocessing and interim storage, the safety advantages of central interim storage over decentralised one, and the hope that in the next decades new technical solutions might be devel-

---

[244] See Ruiz de Apodaca Espinosa 2009:98–99.
[245] Ruiz de Apodaca Espinosa 2009:99. This – controversial – position is supported by the judgement of the Supreme Court of 17 December 2008, STS 7138/2008, on emergency plans.

oped that enable the choice of an entirely new strategy.[246] In view of foresee-able exhaustion of interim storage capacity for high level radioactive waste on the premises of several nuclear power plants and two individual separate facilities the Spanish Government intends to construct a central interim storage facility very soon.[247] Accordingly, as regards final disposal of high level radioactive waste, the general radioactive waste plan still focuses on research and development, laying the emphasis on the consolidation and actualisation of existing knowledge. In accordance with the option of geo-logical disposal, the plan provides for further (non-site specific) studies on possible host rock formations (granite, clay and salt rock), assessment methodology, management options and the possibilities of retrieval.

In the past, spent nuclear fuel used to be exported for reprocessing. How-ever, this option has been given up and now the generally accepted strategy is direct final disposal of spent nuclear fuel. Separation and transmutation of long-lived radionucleides is also on the research agenda with a view to combine this strategy with radioactive waste disposal (NEA 2005b:4).

Spain has not yet developed a position as to an international solution to high level radioactive waste disposal. Article 31 of the Nuclear Energy Act of 1964 requires an authorisation for any transfer of radioactive mate-rials. An export of high level radioactive waste for disposal abroad appar-ently is not formally prohibited. In the past, spent nuclear fuel was not only exported to France for reprocessing but partly also for interim storage. Dumping of radioactive waste at sea is prohibited in accordance with the London Dumping Convention.

2.8.5 Site selection

In accordance with Articles 12 of the Regulation on Nuclear and Radio-active Activities (Royal Decree 1836/1999, as amended by Royal Decree 35/2008) the siting of a repository for high level radioactive waste requires an authorisation.[248] The authorisation confirms the objective of the facil-ity and the suitability of the site. The more specific authorisation prereq-uisites can be derived from the documentation that must be joined to the application. In particular, the impact of the facility on its environment in terms of nuclear safety and radiation protection must be included (Arti-cle 14(a) and (d) of the Regulation). There is no indication that optimality of the site must be demonstrated. The authorisation is granted by the Min-ister for the Economy and Energy who is bound to a negative safety assess-ment or restrictive conditions provided by CSN. In the proceeding, an envi-ronmental impact assessment must be carried out under the supervision of the Ministry for the Environment; this Ministry is competent for the con-

---

[246] ENRESA 2006:42–44; Ruiz de Apodaca Espinosa 2009:99; NEA 2004a:130.
[247] ENRESA 2006; Ruiz de Apodaca Espinosa 2009:99–102.
[248] Ruiz de Apodaca Espinosa 2009:102–03; NEA 2004a:125.

cluding environmental impact statement (Article 12 EIA Regulation, Royal Decree 1/2008). However, the environmental impact assessment does not cover all aspects relevant in the siting process. Moreover, public participation is required (Articles 9, 10 EIA Regulation). All told, the applicable legislative texts on the site authorisation are far from establishing a site selection procedure in the proper sense.

By contrast, the general waste plan emphasises the need for a staged procedure, public information, participation and dialogue with stakeholders as well as compensation of the host region and municipalities for promoting regional and local development as a means to secure political and societal acceptance of site selection. The CSN Act (Article 14) gives CSN a clear mandate for public information and dialogue with stakeholders. One could consider these pronouncements as antecedents of a more refined political site selection procedure. However, this is only partly confirmed by the site selection process devised for the new central interim storage facility for high level radioactive waste.[249] Although there are indications that the Government is conscious of the acceptance problems associated with site selection and the process is based on voluntarism, it still has strong technocratic elements. It consists of four phases, namely public information, a public convocation of municipalities that are interested in hosting the facility, presentation of candidates, selection and decision by Government. The selection criteria that relate to technical/safety, socio-economic and environmental concerns were developed by an Interministerial Committee (Royal Decree 775/2006). In July 2006 this committee started an information process on the new interim storage facility that lasted until mid-February 2007. Then, due to divergences of opinion within the Government, the process was practically suspended until end of 2009. On 23 December 2009 the Spanish Government decided in principle to go forward with the construction of the interim storage facility. In January 2010 the convocation was published and one expects to be able to decide on the site already within half a year. However, environmentalists and regional governments complain of the lack of scientific, geological and environmental criteria and a neglect of the opinion of the public and the autonomous provinces. Also the experience with the siting of interim storage facilities at the premises of nuclear power plants shows that siting decisions of this kind have a great potential of conflicts between central government, on the one hand, and the autonomous provinces, affected municipalities and the local population, on the other. In this field, a number of court decisions have been rendered about the delimitation of central and provincial competences that, however, have decided in favour of central government.[250]

---

[249] See Ruiz de Apodaca Espinosa 2009:101; Camacho 2010; "Atommüll, ja bitte!", Frankfurter Allgemeine Sonntagszeitung, 24 January 2010.

[250] See Ruiz de Apodaca Espinosa 2009:96, 100–03 and the decisions cited at 102 notes 53 and 54.

2.8.6   Construction and operation

Apart from the authorisation of the site, the repository needs an authorisation for construction and another for operation (Article 12 Regulation on Nuclear and Radioactive Activities of 1999). As can be derived from the required documentation for the application (Articles 17 and 20 of the Regulation), the prerequisites for granting the permissionrelate to nuclear safety and radiation protection; they are similar to that provided for the siting authorisation although geared to design and construction on the one hand, operational safety on the other. At both stages an environmental impact assessment is required. It seems clear on the basis of existing legislation that during the operational phase of the repository in particular the radiation protection standards (dose limitations) and the optimisation principle set forth in the Regulation on Radiation Protection of 2001 (Article 4) must be complied with and adequate safety against possible accidents and external interventions must be demonstrated. The question as to what extent these standards and principles also apply in the post-closure phase has not yet been addressed.

2.8.7   Financing

Until 2005, the financing of radioactive waste management was ensured through a charge added to the tariffs for electricity and therefore directly borne by the consumer. Since then, in accordance with the polluter-pays principle, the financial responsibility for high level radioactive waste management is placed on the operators of nuclear power plants in their capacity as waste generators. They do not only have to bear the costs of their own waste management activities such as interim storage on the premises of the nuclear facilities but are also responsible for interim storage and final disposal undertaken by the state (Article 7, Additional Provisions, Act 40/1994, Article 6, Additional provisions, Act 54/1997 and Royal Decree 5/2005 Ruiz de Apodaca Espinosa 2009: 107-108). They must pay a certain percentage of the proceeds from the sale of electricity into a public fund. The financial responsibility covers all measures provided in the general radioactive waste plan, including costs that will only arise in the future. It must be concluded from the wording of the regulation that also the costs of site selection are covered if a particular process is prescribed in the plan.

3   Abbreviations

ADS	accelerator-driven systems
AKEnd	working group for the selection of repository sites "Arbeitskreis Auswahlverfahren Endlagerstandorte", Germany
ALARA	principle in exposure to radiation and other occupational health risks: as low as reasonably achievable
ALARP	principle in exposure to radiation and other occupational health risks: as low as reasonably practicable
ANCLI	national association of local information Commissions "Association nationale des commissions locales d'information", France
ANDRA	national radioactive waste management agency "Agence nationale pour la gestion des déchets radioactifs", France
ASN	nuclear safety Authority "Autorité pour la sûreté nucléaire", France
AT	ataxia telangiectesi, a neurgenerative, inherited disease
BERR	Department for Business, Enterprises and Regulatory Reform, UK
BfS	Radiation Protection Agency "Bundesamt für Strahlenschutz", Germany
BGR	Federal Institute for Geosciences and Natural Resources "Bundesanstalt für Geowissenschaften und Rohstoffe", Germany
BMU	Federal Ministry for the Environment, Nature Conservation and Nuclear Safety "Bundesministerium für Umwelt, Naturschutz und Reaktorsicherheit", Germany
BNFL	British Nuclear Fuels Limited
Bq	unit of radioactivity, measure of the activity of radioactive material in which one nucleus decays per second
BSK-3	special fuel canister "Brennstab-Kokille Typ 3" used in Germany
BWR	boiling water reactor
CASTOR	cask for storage and transport of radioactive material
CCS	carbon dioxide capture and storage
CEA	government-funded technological research organisation "Commission d'énergie atomique" (Atomic Energy Commission), France
CED	committed effective dose coefficient

CLAB	Swedish central interim storage facility "Centralt Lager för Använt Bränsle"
CLI	local information committee "Comité local de l'information", France
CLIS	local information and oversight committee "Comité local de l'information et de suivi", France
CNE	national review board "Comité national d'évaluation", France
CoRWM	British Committee on Radioactive Waste Management
CSN	nuclear safety council "Consejo de Seguridad Nuclear", Spain
DECC	Department for Energy and Climate Change, UK
DEFRA	Department for Environment, Food and Rural Affairs, UK
DNA	deoxyribonucleic acid
DOE	Department of Energy, USA
DSB	double strand breaks (of the DNA molecule)
DT, R	mean absorbed dose in an organ or tissue (T) by a radiation (R)
DWP	Department for Work and Pensions, UK
EA	Environment Agency, UK
ECC La Hague	compact hull storage "Entreposage de Coques Compactées", La Hague, France
EIA Act	Environmental Impact Assessment Act "Gesetz über die Umweltverträglichkeitsprüfung – UVPG", Germany
EnPA	Energy Policy Act, USA
ENSI	Federal Nuclear Safety Inspectorate "Eidgenössisches Nuklearsicherheitsinspektorat", Switzerland
ENSREG	European Nuclear Safety Regulators Group
EPA	Environmental Protection Agency, USA
ESK	Waste Disposal Commission "Entsorgungskommission", Germany
FR	fast reactors
GIHP, fast	scenario "Gorleben implementation at highest possible pace"
GLFLA, slow	scenario "Gorleben late failure, late consideration of alternatives"
GNEP	Global Nuclear Energy Partnership

GPD	standing group of experts on wastes "Groupe permanent d'experts pour les déchets", France
GRS	technical research and expert organization for nuclear safety "Gesellschaft für Anlagen- und Reaktorsicherheit mbH", Germany
GSF	research center for environmental health "Deutsches Forschungszentrum für Umwelt und Gesundheit", Germany
GWd/t	Gigawatt-days per ton
Gy	Gray, unit for the absorbed energy dose. 1 Gy corresponds to the absorbed energy of 1 Joule per kg mass (tissue)
HABOG	highly radioactive wastetreatment and storage building "Hoogradioactief AfvalBehandelings- en Opslag Gebouw" near Borssele, The Netherlands
HLW	high-level radioactive waste
HSE	Health and Safety Executive, UK
IAEA	International Atomic Energy Agency, Vienna
ICRP	International Commission for Radiological Protection, Ottawa
ILW	intermediate level waste
ILW-LL	long lived intermediate level waste
IRSN	institute for research on radioactive safety "Institut de recherche sur securité nucléaire", France
ITAS	institute for technology assessment and systems analysis "Institut für Technikfolgenabschätzung und Systemanalyse", Germany
JNFL	Japan Nuclear Fuel Limited
KASAM	national council for nuclear waste "Kärnavfallsrådet" Sweden
KBS-3	technology for the disposal of high-level radioactive waste ("kärnbränslesäkerhet", i. e. "nuclear fuel safety"), developed in Sweden
KBS-3H	KBS-3 horizontal
KBS-3V	KBS-3 vertical
km	kilometer
KNE	federal commission for radioactive waste disposal "Kommission für nukleare Entsorgung", Switzerland
KNS	federal commission for nuclear safety "Kommission für nukleare Sicherheit", Switzerland

KTA	nuclear technology committee "Kerntechnischer Ausschuss", Germany
KTM	ministry of trade and industry "Kauppa- ja Teollisuusministeriö", Finland
KW	kilowatt
KWh	kilowatt hour
LET	linear energy transfer, a measure of the energy an ionizing particle transfers along its track through material
LILW	low and intermediate level waste
LILW-LL	longlived low and intermediate level radioactive waste
LILW-SL	short lived low and intermediate level radioactive waste
LLW	low-level radioactive waste
LLW-LL	longlived low-level radioactive waste
LNT	linear no-threshold(-model), a model for predicting the long term, biological damage caused by ionising radiation
LWR	light-water reactors
m	meter
MA	minor actinides
man Sv	unit of collective dose quantities
MEEDDM	ministry for the environment, energy, sustainable development and oceans "Ministère pour l'Écologie, l'Energie, le Développement durable et la Mer", France
METI	Ministry for Energy, Trade and Industry, formerly MITI, Japan
MeV	megaelectron volt, a Nuclear energy unit
mGy	megagray
MOX	mixed oxide
mSv	millisievert (1 mSv = 0.001 Sv)
TiHM	metric tons of initial heavy metal
MWd	megawatt-day
MWe	megawatt electrical
NAGRA	national cooperative for the disposal of radioactive waste "Nationale Genossenschaft für die Lagerung radioaktiver Abfälle", Switzerland
NAS	National Academy of Science, USA
NDA	Nuclear Decommissioning Authority, UK
NEA	Nuclear Energy Agency, agency within the Organisation for Economic Co-operation and Development (OECD)

NEPA	National Environmental Policy Act, USA
NGO	non-governmental organisation
NII	Nuclear Installations Inspectorate, UK
NIREX	United Kingdom Nirex Limited, formerly Nuclear Industry Radioactive Waste Executive, UK
NISA	Nuclear and Industrial Safety Agency, Japan
NPP	Nuclear Power Plant
NRC	Nuclear Regulatory Commission, USA
NSC	Nuclear Safety Commission, Japan
nSv	nanosievert (1 nSv = 0.000000001 Sv)
NUMO	Nuclear Waste Management Organisation, Japan
NWMO	Nuclear Waste Management Organisation, Canada
NWPA	Nuclear Waste Policy Act, USA
NWTRB	Nuclear Waste Technical Review Board, USA
OECD	Organisation for Economic Cooperation and Development, Paris
ONDRAF.NIRAS	
	Agency for Radioactive Waste and Enriched Fissile Materials "Organisme National des Déchets Radioactifs et des matières Fissiles enrichies" (ONDRAF) and "Nationale Instelling voor Radioactief Afval en verrijkte Splijtstoffen" (NIRAS), Belgium
OSPAR	Convention for the Protection of the Marine Environment of the North-East Atlantic
P&T	partitioning and transmutation
PET	positron emission tomography
PIA	area for preliminary investigation
PTB	federal physical-technical agency "Physikalisch-Technische Bundesanstalt", Germany
PUREX	plutonium-uranium extraction, reprocessing method for the recovery of uranium and plutonium from spent nuclear fuel
PWR	pressurised water reactor
RBE	relative biological effectiveness
RSA 93	Radioactive Substances Act 1993, UK
RSK	reactor safety commission "Reaktor-Sicherheitskommission", Germany
RWM	radioactive waste management

RWMAC	Radioactive Waste Management Advisory Committee, UK
SEA	strategic environmental assessment
SEER	Surveillance Epidemiology and End Results Program of the National Cancer Institute (NCI), USA
SF	spent fuel
SFS	swedish act as published in the Swedish statue book ("Svensk författningssamling")
SI	small intestine
SKB	nuclear fuel and waste management company "Svensk Kärnbränslehantering AB", Sweden
SKI	nuclear power inspectorate "Statens Kärnkraftinspektion", Sweden
SL	short lived
SNF	spent nuclear fuel
SRWDA	Specified Radioactive Waste Final Disposal Act (or Waste Law) of 2000, Japan
SSB	single strand breaks (of the DNA molecule)
SSI	radiation safety inspectorate "Statens Strålskyddsinstitut", Sweden
SSK	radiation protection pommission "Strahlenschutzkommission", Germany
SSM	radiation safety authority "Strålsäkerhetsmyndigheten" , Sweden
STUK	radiation and nuclear safety authority "Säteilyturvakeskus", Finland
Sv	sievert – unit for the equivalent dose which is obtained from the absorbed dose multiplied with the radiation weighting factor (wR)
T	tissue
TA	technology assessment
TBL-A	transport cask storage facility "Transportbehälterlager" Aarhaus, Germany
TBL-G	transport cask storage facility "Transportbehälterlager" Gorleben, Germany
tHM	ton of heavy metal
TSN Act	transparency and security in the field of nuclear activities
TWhe	terawatt-hours of electricity

UNSCEAR	United Nations Scientific Committee on the Effects of Atomic Radiation
UOX	uranium-oxide
UO2	uranium dioxide
US NRC	United States Nuclear Regulatory Commission
μSv	micro-sievert (1μSv = 0.000001 Sv)
VGB	european technical association for power and heat generation "Verband der Großkessel-Besitzer e.V.", Essen
VLLW	very low level waste
VSL	very short lived
VSLW	very short lived waste
W	watt
WENRA	Western European Regulator's Association, a network of chief regulators of EU countries with nuclear power plants and Switzerland
wR	radiation weighting factor
w. r. t.	with respect to
ZLN	intermediate storage facility for nuclear waste at Lubmin "Zwischenlager Nord", Germany
ZWILAG	intermediate storage facility for nuclear waste at Würenlingen "Zwischenlager Würenlingen AG", Switzerland

Chemical elements sorted by abbreviations in alphabetical order

Ac	Actinium		Fe	Iron
Ag	Silver		Fm	Fermium
Al	Aluminum		Fr	Francium
Am	Americium		Ga	Gallium
Ar	Argon		Gd	Gadolinium
As	Arsenic		Ge	Germanium
At	Astatine		H	Hydrogen
Au	Gold		He	Helium
B	Boron		Hf	Hafnium
Ba	Barium		Hg	Mercury
Be	Beryllium		Ho	Holmium
Bh	Bohrium		Hs	Hassium
Bi	Bismuth		I	Iodine
Bk	Berkelium		In	Indium
Br	Bromine		Ir	Iridium
C	Carbon		K	Potassium
Ca	Calcium		Kr	Krypton
Cd	Cadmium		La	Lanthanum
Ce	Cerium		Li	Lithium
Cf	Californium		Lr	Lawrencium
Cl	Chlorine		Lu	Lutetium
Cm	Curium		Md	Mendelevium
Co	Cobalt		Mg	Magnesium
Cr	Chromium		Mn	Manganese
Cs	Cesium		Mo	Molybdenum
Cu	Copper		Mt	Meitnerium
Db	Dubnium		N	Nitrogen
Ds	Darmstadtium		Na	Sodium
Dy	Dysprosium		Nb	Niobium
Er	Erbium		Nd	Neodymium
Es	Einsteinium		Ne	Neon
Eu	Europium		Ni	Nickel
F	Fluorine		No	Nobelium

Np	Neptunium		Sm	Samarium
O	Oxygen		Sn	Tin
Os	Osmium		Sr	Strontium
P	Phosphorus		Ta	Tantalum
Pa	Protactinium		Tb	Terbium
Pb	Lead		Tc	Technetium
Pd	Palladium		Te	Tellurium
Pm	Promethium		Th	Thorium
Po	Polonium		Ti	Titanium
Pr	Praseodymium		Tl	Thallium
Pt	Platinum		Tm	Thulium
Pu	Plutonium		U	Uranium
Ra	Radium		Uub	Ununbium
Rb	Rubidium		Uuh	Ununhexium
Re	Rhenium		Uuo	Ununoctium
Rf	Rutherfordium		Uup	Ununpentium
Rg	Roentgenium		Uuq	Ununquadium
Rh	Rhodium		Uus	Ununseptium
Rn	Radon		Uut	Ununtrium
Ru	Ruthenium		V	Vanadium
S	Sulfur		W	Tungsten
Sb	Antimony		Xe	Xenon
Sc	Scandium		Y	Yttrium
Se	Selenium		Yb	Ytterbium
Sg	Seaborgium		Zn	Zinc
Si	Silicon		Zr	Zirconium

4 Bibliography

Adorno TW (1980) Minima Moralia. Reflexionen aus dem beschädigten Leben [=Gesammelte Schriften Vol 4]. Frankfurt/Main

Aebersold A (2008) Auswahlverfahren für geologische Tiefenlager im Spannungsfeld von Gesellschaft, Wissenschaft und Politik, Regeln für die Standortsuche. Atomwirtschaft 53:618–622

Agrenius L (2002) Criticality safety calculations of storage canisters [=SKB Technical Report TR-02-17]. Stockholm, www.skb.se/upload/publications/pdf/TR-02-17.pdf

Åhäll KI (2006) Final Deposition of High-level Nuclear Waste in Very Deep Boreholes. An evaluation based on recent research of bedrock conditions at great depths [=MKG Report 2]. Göteborg, www.nuclearliaison.com/pdfs/MKG_Final_Deposition_of_High_Level_Nuclear_Waste_in_Very_Deep_Boreholes_December_2006.pdf

AkEnd (Arbeitskreis Auswahlverfahren Endlagerstandorte) (2002) Auswahlverfahren für Endlagerstandorte, Empfehlung des AkEnd – Arbeitskreis Auswahlverfahren Endlagerstandorte, December 2002

Anderson K, Grundfelt B, Wene CO (2004) Transparency and Public Participation in Radioactive Waste Management, RISCOM II Final Report, December 30, 2003, SKI Report 2004:08. SKI, Stockholm

Andra (Agence nationale pour la gestion des déchets radioactifs) (2005) Dossier 2005 Argile. Safety evaluation of a geological repository. Châtenay-Malabry Cedex, www.andra.fr/download/andra-international-en/document/editions/270va.pdf

Andra (Agence nationale pour la gestion des déchets radioactifs) (2005) Dossier 2005 Granite. Andra, Paris, www.andra.fr/index.php?id=edition_1_5_2&recherche …4

Andra (Agence nationale pour la gestion des déchets radioactifs) (2009) Projet de stockage géologique profonde réversible, CT8ADF 09 0039. Andra, Paris, www.andra.fr/download/site-principal/document/actualites/ctradp090039-a.pdf

Andra (Agence nationale pour la gestion des déchets radioactifs) (2010) Rendre gouvernables les déchets radioactifs – Le stockage profond à l'épreuve de la réversibilité. Andra, Paris, www.andra.fr/download/site-principal/document/editions/381.pdf

Appel D (2008) Das Gorleben-Moratorium und Argumente für ein Standortauswahlverfahren. In: BMU (2008) Endlagersymposium. BMU, Berlin

ASN (Agence de sûreté nucléaire) (2006) Plan national pour la gestion des matières et déchets radioactifs. ASN, Paris

ASN (Agence de sûreté nucléaire) (2008) Guide de sûreté relatif au stockage définitif de déchets radioactifs en formation géologique profonde afin d'assurer la sûreté après la période d'exploitation du stockage, 12 February 2008. ASN, Paris, www.asn.fr/index.php/S-informer/Publications/Guides-pour-les-professionnels/Gestion-des-dechets-radioactifs/Guide-de-surete-relatif-au-stockage-definitif-des-dechets-radioactifs-en-formation-geologique-profonde

ASN (Agence de sûreté nucléaire) (2008) Rapport annuel 2008. ASN, Paris

ASN (Agence de sûreté nucléaire) (2010) Plan national de gestion des matières et déchets radioactifs 2010–2012. ASN, Paris

Badura P (2010) Staatsrecht: Systematische Erläuterung des Grundgesetzes für die Bundesrepublik Deutschland. 4th ed. C.H. Beck, München

Baldwin T, Chapman N, Neall F (2008) Geological Disposal Options for High-Level Waste and Spent Fuel. Report for the UK Nuclear Decommissioning Authority, www.nda.gov.uk/loader.cfm?csModule=security/getfile&pageid=20941

Baltes B (2009) Sicherheitsanforderungen an die Endlagerung hochradioaktiver Abfälle in tiefen geologischen Formationen. In: Schmidt-Preuß M (ed.) (2009) Deutscher Atomrechtstag 2008. Nomos, Baden-Baden, pp 211–224

Baltes B, Röhlig KJ, Kindt A (2007) Sicherheitsanforderungen an die Endlagerung hochradioaktiver Abfälle in tiefen geologischen Formationen, Entwurf der GRS. Gesellschaft für Anlagen- und Reaktorsicherheit, GRS-A 3358

Baron W (1997) Grundfragen und Herausforderungen an eine partizipative TA, in: Westphalen R (1997) "Technikfolgenabschätzung als politische Aufgabe".München, pp 137–158

Barth R, Schulze F (2008) Rechtliche Rahmenbedingungen bei der Endlagerung. In: Öko-Institut/Gesellschaft für Anlagen- und Reaktorsicherheit, Endlagerung wärmeentwickelnder radioaktiver Abfälle in Deutschland, Anhang: Rechtsgrundlagen. Öko-Institut Darmstadt, endlagerung.oeko.info

BEIR (National Academy of Sciences Advisory Committee on the Biological Effects of Ionizing Radiation) (1990) Health Effects of Exposure to Low Levels of Ionizing Radiation. National Academy Press, Washington D.C.

BEIR (National Academy of Sciences Advisory Committee on the Biological Effects of Ionizing Radiation) (2005) Health Effects from Exposure to Low Levels of Ionizing. BEIR VII, Radiation. National Academy Press, Washington D. C.

Bericht der Bundesrepublik Deutschland für die 3. Überprüfungskonferenz im Rahmen der Gemeinsamen Konvention (2008). BR-Drs. 771/08. Berlin

BfS (Bundesamt für Strahlenschutz) (2008) Unterrichtung durch die Bundesregierung. Umweltradioaktivität und Strahlenbelastung im Jahr 2007. BfS

BGR (Bundesanstalt für Geowissenschaften und Rohstoffe) (2007) Nuclear Waste Disposal in Germany. Investigation and evaluation of regions with potentially suitable host rock formations for a geologic nuclear repository, www.bgr.bund.de/cln_109/nn_335086/EN/Themen/Geotechnik/Downloads/WasteDisposal__HostRockFormations__en,templateId=raw,property=publicationFile.pdf

Blue Ribbon Commission on America's Nuclear Future (2011) What We've Heard. A Staff Summary of Major Themes in Testimony and Comments Received by the Blue Ribbon Commission on America's Nuclear Future to Date. Washington D. C., www.brc.gov/library/reports/BlueRibbon_Report-pages.pdf

Bluth J (2008) Die Auswahl des Standorts Gorleben aus der Sicht des Niedersächsischen Ministeriums für Umwelt. In: BMU (2008) Endlagersymposium. Berlin

Bluth J, Schütte H (2008) Die Auswahl des Standortes Gorleben. Ein Beitrag aus der Sicht des Niedersächsischen Ministeriums für Umwelt und Klimaschutz. Hannover

BMU (Bundesministerium für Umwelt, Naturschutz und Reaktorsicherheit) (2006) Verantwortung übernehmen: Den Endlagerkonsens realisieren, Berlin, 18 September 2006, www.bmu.de/files/pdfs/allgemein/application/pdf/konzept_endlagerkonsens.pdf

BMU (Bundesministerium für Umwelt, Naturschutz und Reaktorsicherheit) (2008) Endlagersymposium. BMU, Berlin

BMU (Bundesministerium für Umwelt, Naturschutz und Reaktorsicherheit) (2009a) Standortauswahlverfahren, Stand: 25.02.2009, www.bfs.de/de/endlager/endlagerung_ueberblick/endlagerausstellung_textfassung/standorte.html/printversion

BMU (Bundesministerium für Umwelt, Naturschutz und Reaktorsicherheit) (2009b) Bund und Länder einigen sich auf Verfahren zur Einführung neuer Sicherheitsregeln für AKW. Umwelt-BMU 7–8/2009:607

BMU (Bundesministerium für Umwelt, Naturschutz und Reaktorsicherheit) (2010a) Sicherheitsanforderungen an die Endlagerung wärmeentwickelnder radioaktiver Abfälle, 30 September 2010, http://www.bmu.de/atomenergie_ver_und_entsorgung/downloads/doc/42047.php

BMU (Bundesministerium für Umwelt, Naturschutz und Reaktorsicherheit) (2010b) "Gorleben wird weiter erkundet". Umwelt-BMU 4/2010:251

BMU/BfS (Bundesministerium für Umwelt, Naturschutz und Reaktorsicherheit/ Bundesamt für Strahlenschutz) (2010) Weitererkundung des Salzstocks Gorleben, November 2010. BMU, Berlin

Bodenez P, Röhlig KJ, Besnus F, Vigfusson J, Smith R, Nys V, Bruno G, Metcalf P, Ruiz-Lopez C, Ruokola E, Jensen M (2008) European Pilot Study on the Regulatory Review of the Safety Case for Geological Disposal of Radioactive Waste. Symposium "Safety cases for the deep disposal of radioactive waste: Where do we stand?", 23-25 January 2007 [=NEA No. 06319]. Paris, www.oecdnea.org/ html/rwm/reports/2008/ne6319-safety.pdf

Boetsch W (2003) Ethische Aspekte bei der Endlagerung radioaktiver Stoffe (Abschlussbericht) [= Schriftenreihe Reaktorsicherheit und Strahlenschutz BMU – 2003 – 619]. Bonn www.bmu.de/files/pdfs/allgemein/application/pdf/schriftenreihe_rs619.pdf

Bohm D (1998) Der Dialog. Das offene Gespräch am Ende der Diskussion. Stuttgart

Bora A, Epp A (2000) Die imaginäre Einheit der Diskurse. Zur Funktion von Verfahrensgerechtigkeit. Kölner Zeitschrift für Soziologie und Sozialpsychologie 52:1–35

Bordin G, Paul M (2008) Die langfristige Sicherheit der Finanzierung von Stilllegung und Entsorgung im Nuklearbereich zur Wahrung der Interessen der öffentlichen Hand. In: Koch HJ, Roßnagel A (eds.) (2008) 13. Deutsches Atomrechtssymposium 2007. Nomos, Baden-Baden, pp 271–292

Bothe M (2001) Artikel 30. In: Denninger E, Hoffmann-Riem W, Schneider JP, Stein D (eds.) (2001) Alternativkommentar zum Grundgesetz, 3rd. ed. Luchterhand, Neuwied

Boutellier C, McCombie C (2006) Zu den politischen und rechtlichen Bedingungen der Errichtung eines internationalen Lagers für hochradioaktive Abfälle. Europäisches Umwelt- und Planungsrecht 4: 136–143

Bredimas A, Nuttall WJ (2008) An international comparison of regulatory organisations and licensing procedures for new nuclear power plants. Energy Policy 36:1344–1354

Brenner M (2005) Das Verfahren der alternativen Standortsuche im AkEnd-Bericht. In: Ossenbühl F (ed.) (2005) Deutscher Atomrechtstag 2004. Nomos, Baden-Baden, pp 99–114

Breuer R (1984) Die Planfeststellung für Anlagen zur Endlagerung radioaktiver Abfälle. Erich Schmidt Verlag, Berlin, Bielefeld, München

British Geological Survey (2010) Managing Radioactive Waste Safely: Initial Geological Unsuitability Screening of West Cumbria, mrws.decc.gov.uk/en/mrws/ cms/disposal/site_selection/initial_screen/west_cumbria/west_cumbria.aspx

Britz G (1998) Bundeseigenverwaltung oder selbständige Bundesoberbehörde nach Art. 87 III 1 GG. Deutsches Verwaltungsblatt 113:1167–1174

Bühlmann W (2008) Swiss Legislation on Radioactive Waste Management. Obligations Under the State Treaty, National Legal Regulations; Role of Players. Atomwirtschaft 53:612–617

Bull HH (2001) Artikel 87c. In: Denninger E, Hoffmann-Riem W, Schneider JP, Stein D (eds.) (2001) Alternativkommentar zum Grundgesetz, 3rd. ed. Luchterhand, Neuwied

Bundesamt für Energie (2008) Sachplan geologische Tiefenlager, Konzeptteil, 2 April 2008. Bundesamt für Energie, Bern, www.bfe.admin.ch/radioaktiveabfaelle/01277/01306/index.html?lang=de&dossier_id=02151,

Burgi M (2005) Die Überführung der Atomaufsicht in die Bundeseigenverwaltung aus verfassungsrechtlicher Sicht. Neue Zeitschrift für Verwaltungsrecht 24:247–253

Camacho E (2010) A dónde van los residuos radiactivos en España?, www.finanzas.com/noticias/empresas/2010-01-23/239379_donde-residuos-radiactivos-espana.html

Cardis E, Vrijheid M, Blettner M, Gilbert E, Hakama M, Hill C, Howe G, Kaldor J, Muirhead CR, Schubauer-Berigan M, Yoshimura T, Bermann F, Cowper G, Fix J, Hacker C, Heinmiller B, Marshall M, Thierry-Chef I, Utterback D, Ahn YO, Amoros E, Ashmore P, Auvinen A, Bae JM, Bernar J, Biau A, Combalot E, Deboodt P, Diez Sacristan A, Eklöf M, Engels H, Engholm G, Gulis G, Habib RR, Holan K, Hyvonen H, Kerekes A, Kurtinaitis J, Malker H, Martuzzi M, Mastauskas A, Monnet A, Moser M, Pearce MS, Richardson DB, Rodriguez-Artalejo F, Rogel A, Tardy H, Telle-Lamberton M, Turai I, Usel M, Veress K (2007) The 15-country collaborative study of cancer risk among workers in the nuclear industry. Radiat. Res. 167:396–416

CERRIE (2004) Report of the Committee Examining Radiation Risks of Internal Emitters (CERRIE), www.cerrie.org

Carrier M (2006) Wissenschaftstheorie: Zur Einführung. Hamburg ²2008

Cloosters W (2005) Atomaufsicht – Bundesauftragsverwaltung oder Bundeseigenverwaltung? Aus der Sicht optimaler Aufgabenerfüllung. In: Ossenbühl F (ed.) (2005) Deutscher Atomrechtstag 2004. Nomos, Baden-Baden, pp 183–191

Cloosters W (2008) Rückstellungsverpflichtungen für Kernkraftwerke – aus der Sicht einer atomrechtlichen Genehmigungs- und Aufsichtsbehörde. In: Koch HJ, Roßnagel A (eds.) (2008) 13. Deutsches Atomrechtssymposium 2007. Nomos, Baden-Baden, pp 293–306

CoRWM (Commission on Radioactive Waste Management) (2006) Managing our Radioactive Waste Safely – CoRWM Recommendations to UK Government, July 2006, corwm.decc.gov.uk/assets/corwm/pre-nov 2007 doc archive/doc archive/introduction/top level key docs/700 – managing our radioactive waste safely, recommendations to government.pdf

Cotton M (2009) Ethical assessment in radioactive waste management: a proposed reflective equilibrium-based deliberative approach, Journal of Risk Research 12, pp 603–618

Czychowski M, Reinhardt M (2010) Wasserhaushaltsgesetz, 10th ed. C.H. Beck, München

Dahl RA (1989) Democracy and its Critics. New Haven

Damveld H, van den Berg RJ (2000) Nuclear Waste and Nuclear Ethics. Social and ethical aspects oft the retrievable storage on nuclear waste.. www.nuclearfiles.org/menu/key-issues/ethics/issues/scientific/nuclear_waste_ethics.pdf

Darby S, Hill D, Auvinen A, Barros-Dios JM, Baysson H, Bochicchio F, Deo H, Falk R, Forastiere F, Hakama M, Heid I, Kreienbrock L, Kreuzer M, Lagarde F, Mäkeläinen I, Muirhead C, Oberaigner W, Pershagen G, Ruano-Ravina A, Ruosteenoja E, Schaffrath Rosario A, Tirmarche M, Tomášek L, Whitley E, Wichmann HE, Doll R (2005) Radon in homes and risk of lung cancer: collaborative analysis of individual data from 13 European case-control studies. BMJ 330:223–229

de Witt S (2005) Struktur und Probleme des atomrechtlichen Planfeststellungsverfahrens. In: Ossenbühl F (ed.) (2005) Deutscher Atomrechtstag 2004. Nomos, Baden-Baden, pp 125–134

Debes M (2006) Evolutions des combustibles et impact sur le cycle. Revue générale nucléaire 2:61–66

DEFRA (Department on Environment, Food, Rural Development and Agriculture) (2008) White Paper "Managing Radioactive Waste: A Framework for Implementing Geological Disposal", June 2008, Cm 7386

Deuschle J (2007) Ein normativ-funktionales Konzept für Nachhaltige Entwicklung: Gerechtigkeit. In: Renn O, Deuschle J, Jäger A, Weimer-Jehle W (eds.) Leitbild Nachhaltigkeit: eine normativ-funktionale Konzeption und ihre Umsetzung. Wiesbaden, pp 49–60

DoE (US Department of Energy) (2010) Secretary Chu Announces Blue Ribbon Commission on America's Nuclear Future, Press Release, 29 January 2010, www.energy.gov/news/8584.htm

Dolde KP (1996) Standortfestlegung kooperierender Entsorgungsträger und abfallrechtliche Planfeststellung. Neue Zeitschrift für Verwaltungsrecht 15:526–532

Drábová D (2009) Development and contents of WENRA reference levels. In: Schmidt-Preuß M (ed.) (2009) Deutscher Atomrechtstag 2008. Nomos, Baden-Baden, pp 183–186

Dutton M, Hillis K, Stansby J, Kennett L, Seppälä T, Macias RM, Röhlig KJ, Haverkate B, O'Sullivan PJ, Mrskova A, Prítrský J, Díaz Terán JA, Valdivieso Ramos JM, Morén L, Hugi M, Zuidema P, King S, Breen B (2004) The Comparison of Alternative Waste Management Strategies for Long-Lived Radioactive Wastes (COMPAS Project). Luxembourg

EKRA (Expertengruppe "Entsorgungskonzepte für radioaktive Abfälle") (2000) Recommendation of the Group of Experts "Disposal Concepts for Radioactive Waste"

ENRESA (Entidad Nacional de Residuos Radiactivos S.A.) (2006) Sexto Plan General de Residuos Radiactivos. ENRESA, Madrid, www.enrcsa.es/files/multimedios/6PGRR_Espa_ol_Libro_versi_n_indexada.pdf

ENSI (Swiss Federal Nuclear Safety Inspectorate) (2009) Spezifische Auslegungsgrundsätze für geologische Tiefenlager und Anforderungen an den Sicherheitsnachweis, Richtlinie ENSI G03, April 2009. ENSI, Brugg, www.ensi.ch/fileadmin/deutsch/files/G-003_D.pdf

ENSREG (European Nuclear Safety Regulators Group) (2009) Report of the European Nuclear Safety Regulators Group, 13 July, Brussels, www.ensreg.org/documents

Epiney A (2010) Artikel 20a. In: von Mangold H, Klein F, Starck C (eds.) (2010) Kommentar zum Grundgesetz, 5th ed. Carl Heymanns Verlag, Köln

Epp A (1998) Divergierende Konzepte der Verfahrensgerechtigkeit. Eine Kritik der Procedural Justice-Forschung, Arbeitspapiere 98-30. Wissenschaftszentrum Berlin

ESK (Entsorgungskommission) (2009) Stellungnahme zum Entwurf des BMU "Sicherheitsanforderungen an die Endlagerung wärmeentwickelnder radioaktiver Abfälle, 29 January 2009, http://www.entsorgungskommission.de/downloads/snsianf290109.pdf

EU Peer Review (2009) EU-27 Regulators Peer Review, Effectiveness of STUK'S Regulatory Infrastructure for Waste Safety. STUK, Helsinki

European Commission (2002) Eurobarometer 56.2: Europeans and Radioactive Waste

European Commission (2002) Nuclear Safety in the European Union, COM (2002) 605

European Commission (2006) Nuclear Illustrative Programme, COM (2006) 565

European Commission (2007) An Energy Policy for Europe, COM (2007) 1

European Commission (2008) Nuclear Illustrative Programme, COM (2008) 726

European Commission (2008) Special Eurobarometer on Radioactive Waste Management, europa.eu/rapid/pressReleasesAction.do?reference=IP/08/1100&type=HTML

Flowers B (1976) Nuclear Power and the Environment [=Royal Commission on Environmental Pollution (ed.), Sixth Report]. London

Flüeler T (2006) Decision Making for Complex Socio-technical Systems. Robustness from Lessons Learned in Long-Term Radioactive Waste Governance. Environment & Policy, vol 42. Springer, Heidelberg, New York

Flynn J (2003) Stigma. In: Pidgeon N, Kasperson RE, Slovic P (eds.) The Social Amplification of Risk. Cambridge (USA), pp 326–352

Foucault M (1971) L'ordre du discours, Paris,

Frankfurter Allgemeine Sonntagszeitung 24 January 2010 "Atommüll, ja bitte!"

Frankena W (1963) Ethics, Englewood Cliffs N. J, ²1973

Fredriksson PG (2000) The Siting of Hazardours Waste Facilities in Federal Systems. The Political Economy of NIMBY. Environmental and Resource Economics 15:75–87

Gaentzsch G (2005) Struktur und Probleme des atomrechtlichen Planfeststellungsverfahrens. In: Ossenbühl F (ed.) (2005) Deutscher Atomrechtstag 2004. Nomos, Baden-Baden, pp 115–124

Gaßner H, Neusüß P (2009) Standortauswahlverfahren und Sicherheitsanforderungen für ein Endlager. Zeitschrift für Umweltrecht 19:347–351

Gen-IV FCCG (Generation IV Fuel Cycle Crosscut Group) (2001) Summary Report FR02-00 Nov. 2001

Gethmann CF (1982) Proto-Ethik. Zur formalen Pragmatik von Rechtfertigungsdiskursen, in: Ellwein T/Stachowiak H (eds): Bedürfnisse, Werte und Normen im Wandel. Band l. München/Paderborn, pp ll3-143 [engl.: Protoethics. Towards a formal pragmatics of justificatory discourse. In: Butts RE, Brown JR (eds.), Constructicism and Science. Essays in Recent German Philosophy. Dordrecht 1989, pp 191–220

Gethmann CF (1991) Universelle praktische Geltungsansprüche. Zur philosophischen Bedeutung der kulturellen Genese moralischer Überzeugungen, in: P. Janich (ed.): Entwicklungen der methodischen Philosophie. Frankfurt/Main, pp 148-175

Gethmann CF (1993) Langzeitverantwortung als ethisches Problem im Umweltstaat, in: Gethmann CF, Kloepfer M, Nutzinger H-G (eds.), Langzeitverantwortung im Umweltstaat. Bonn, pp l-21

Gethmann CF, Mittelstraß J (1992) Maße für die Umwelt. GAIA – Ökologische Perspektiven in Natur-, Geistes- und Wirtschaftswissenschaften 1:16–25

Gethmann CF, Sander T (1999) Rechtfertigungsdiskurse. In: Grunwald A, Saupe S (eds.), Ethik in der Technikgestaltung. Berlin/Heidelberg, pp 117-151

Geulen R (2008) Rechtsprobleme der Endlagerung aus der Perspektive Drittbetroffener. In: Koch H-J, Roßnagel A (eds.) (2008) 13. Deutsches Atomrechtssymposium 2007. Nomos, Baden-Baden, pp 377–388

GIF (Generation IV International Forum Meetings) (2002)

Gonzales EM (2008) Impact of Partitioning and Transmutation on Nuclear Waste Management and the Associated Geological Repositories. In: Euradwaste '08. Seventh European Commission Conference on the Management and Disposal of Radioactive Waste. Conference Presentations, Papers and Panel Reports, cordis.europa.eu/fp7/euratom-fission/euradwaste2008-presentations_en.html

Goodhead DT (1994) Initial events in the cellular effects of ionizing radiations: clustered damage in DNA. Int. J. Radiat. Biol. 65:7–17

Gosseries A, Radiological protection and intergenerational justice, in: Eggermont G, Feltz B (eds.), Ethics and radiological protection. Louvain-la-Neuve, pp 167-195.

Grill KD (2005) Safety, justice, and transparency in nuclear waste management. Atomwirtschaft 50:452–453

Groß T (2006) Die Verwaltungsorganisation als Teil organisierter Staatlichkeit. In: Hoffmann-Riem W, Schmidt-Aßmann E, Vosskuhle A (eds.) (2006) Grundprobleme des Verwaltungsrechts vol. I. C. H. Beck, München, § 13

Groth M (2002) Die Unabhängigkeit der Atomaufsicht. In: Koch HJ, Roßnagel A (eds.) (2002) 11. Deutsches Atomrechtssymposium 2001. Nomos, Baden-Baden, pp 105–114

GRS (Gesellschaft für Reaktorsicherheit) (1998) Handbuch zur Kritikalität. Vol 2, 12/1998

Grunwald A (2010) Ethische Anforderungen an nukleare Endlager. Der ethische Diskurs und seine Voraussetzungen. In: Hocke P, Arens G (eds.): Die Endlagerung hochradioaktiver Abfälle. Gesellschaftliche Erwartungen und Anforderungen an die Langzeitsicherheit. Karlsruhe/Berlin/Bonn, pp 73–84

Grunwald A, Hocke P (2006) Die Endlagerung nuklearer Abfälle als ungelöstes Problem, In: Hocke P, Grunwald A (eds.), Wohin mit dem radioaktiven Abfall? Perspektiven für eine sozialwissenschaftliche Endlagerforschung. Berlin, pp 11–34

Habermas J (1981) Theorie des kommunikativen Handelns. Band 1. Frankfurt/Main

Habermas J (1991) Moralbewußtsein und kommunikatives Handeln. 4. Auflage. Frankfurt/Main

Habermas J (1992) Faktizität und Geltung. Beiträge zur Diskurstheorie des Rechts und des modernen Rechtsstaates. Frankfurt/Main

Haedrich H (1986) Atomgesetz mit Pariser Atomhaftungsübereinkommen. Nomos, Baden-Baden

Haedrich H (1993) Zur Zulässigkeit der Wiederaufarbeitung abgebrannter Brennelemente aus deutschen Kernkraftwerken in anderen EG-Mitgliedstaaten. Neue Zeitschrift für Verwaltungsrecht 12:1036–1044

Hall EJ (1994) Radiobiology for the Radiologist. 4th Edition. J.B. Lippincott Company, Philadelphia

Harrison JD, Muirhead CR (2003) Quantitative comparisons of cancer induction in humans by internally deposited radionuclides and external radiation. Int. J. Radiat. Biol. 79:1–13

Harrison T (2000) Very deep borehole. Deutag's opinion on boring, canister emplacement and retrievability. [=Svensk Kärnbränslehantering AB R-00-35] Stockholm, www.skb.se/upload/publications/pdf/R-00-35webb.pdf

Hartmann N (1925) Ethik. Berlin/Leipzig

Hasinger G (2010) Der Weg zu einem Fusionskraftwerk. EnergieMix 2050 – Konferenz der GeoUnion 19./20. April 2010. Berlin

Hennen L (1999) Partizipation und Technikfolgenabschätzung. In: Bröchler S, Simonis G, Sundermann K (eds.), Handbuch Technikfolgenabschätzung. Berlin, pp 565-571

Hennenhöfer G, Schneider H (2010) 50 Jahre Atomgesetz – Eine Zwischenbilanz. In: Dolde KP, Hansmann K, Paetow S, Schmidt-Aßmann E (eds.) (2010) Verfassung – Umwelt – Wirtschaft, Festschrift für D Sellner, 2010. C.H. Beck, München, pp 347–372

Hermes G (2008) Artikel 87c. In: Dreier H (ed.) (2008) Grundgesetz, vol. 3, 2nd ed. Mohr-Siebeck, Tübingen

Hocke P, Renn O (2009) Concerned Public and the Paralysis of Decision-Making: Nuclear Waste Management Policy in Germany. Journal of Risk Research12:921–940

Hocke P, Barth R, Grunwald A, Kallenbach-Herbert B, Kuppler S, Renn O, Reuß M (2010) Schlüsselentscheidungen und -maßnahmen für einen gehaltvollen Dialogprozess, Technikfolgenabschätzung 19:91–97

Hocke P, Grunwald A (eds.) (2006) Wohin mit dem radioaktiven Abfall? Perspektiven für eine sozialwissenschaftliche Endlagerforschung. Gesellschaft, Technik, Umwelt, Neue Folge 8. Edition Sigma, Berlin

Hofmann H (1981) Rechtsfragen der atomaren Entsorgung. Klett-Cotta, Stuttgart

Hoppenbrock V (2009) Finanzierung der nuklearen Entsorgung und der Stilllegung von Kernkraftwerken. Nomos, Baden-Baden

Horkheimer M, Adorno TW (1988) Dialektik der Aufklärung. Philosophische Fragmente, Frankfurt/Main

Hörnschemeier FG (2008) Endlagerung radioaktiver Abfälle in Deutschland – im internationalen Vergleich. Atomwirtschaft 53:600–602

HSE (Health and Safety Executive) (2007) Joint Regulatory Guidance on Radioactive Waste Management – Part 1, The Management of Higher Activity Level Radioactive Waste on Nuclear Licensed Sites, www.hse.gov.uk/nuclear/wastemanage.htm (visited on 8 October 2009)

HSE (Health and Safety Executive) (2009) Near-Surface Disposal on Land for Higher Activity Radioactive Waste, Guidance on Requirements for Authorisation, www.environment-agency.gov.uk/business/sectors/99322.aspx

Huber PM (2001) Endlagerung als Staatsaufgabe und Betreiberpflicht. Deutsches Verwaltungsblatt 116:239–248

Huber PM (2010) Artikel 19. In: von Mangoldt H, Klein F, Starck C. Kommentar zum Grundgesetz, 6th ed. Vahlen, München

IAEA (International Atomic Energy Agency) (1994) Classification of Radioactive Waste. A Safety Guide. Vienna

IAEA (International Atomic Energy Agency) (1995) The Principles of Radioactive Waste Management, Safety Series No. 111-F. IAEA, Vienna

IAEA (International Atomic Energy Agency) (1997) Joint Convention on the Safety of Spent Fuel Management and on the Safety of Radioactive Waste Management. Vienna, www.iaea.org/Publications/Documents/Infcircs/1997/infcirc546.pdf

IAEA (International Atomic Energy Agency) (2000) Legal and Governmental Infrastructure for Nuclear Radiation, Radioactive Waste and Transport Safety, Safety Requirement GS-R-1. IAEA, Vienna

IAEA (International Atomic Energy Agency) (2002) Organisation and Staffing of the Regulatory Body for Radioactive Waste Management, Safety Guide GS-G-1.1. IAEA, Vienna

IAEA (International Atomic Energy Agency) (2005) 2ndNational Report of Sweden. IAEA, Vienna

IAEA (International Atomic Energy Agency) (2006a) Fundamental Safety Principles, Safety Standards Series No. SF-1. IAEA, Vienna, www-pub.iaea.org/MTCD/publications/PDF/Pub1273_web.pdf

IAEA (International Atomic Energy Agency) (2006b) Geological Disposal of Radioactive Waste, Safety Standards Series No. WS-R-4, jointly sponsored by the IAEA and the OECD Nuclear Energy Agency. IAEA, Vienna, www-pub.iaea.org/MTCD/publications/PDF/Pub1231_web.pdf

IAEA (International Atomic Energy Agency) (2007) IAEA Safety Glossary. Terminology Used in Nuclear Safety and Radiation Protection. Vienna, www-pub.iaea.org/MTCD/publications/PDF/Pub1290_web.pdf

IAEA (International Atomic Energy Agency) (2008a) 3rd National Report of the United Kingdom. IAEA, Vienna

IAEA (International Atomic Energy Agency) (2008b) 3rd National Report of Finland. IAEA, Vienna

IAEA (International Atomic Energy Agency) (2009a) Geological Repositories. IAEA, Vienna

IAEA (International Atomic Energy Agency) (2009b) Geological Disposal of Radioactive Waste, Technological Implications for Retrievability. IAEA, Vienna

IAEA (International Atomic Energy Agency) (2009c) Regulations for the Safe Transport of Radioactive Material [= IAEA Safety Requirements TS-R-1], www-pub.iaea.org/MTCD/publications/PDF/Pub1384_web.pdf

ICRP (International Commission on Radiological Protection) (1977) Recommendations of the International Commission on Radiological Protection. ICRP Publication 26. Pergamon Press, Oxford

ICRP (International Commission on Radiological Protection) (1991) 1990 Recommendations of the International Commission on Radiological Protection. ICRP Publication 60. Pergamon Press, Oxford

ICRP (International Commission on Radiological Protection) (1994) Dose coefficients for intake of radionuclides by workers. ICRP Publication 68. Pergamon Press, Oxford

ICRP (International Commission on Radiological Protection) (1995) Age-dependent doses to members of the public from intakes of radionuclides: part 4. Inhalation dose coefficients. ICRP Publication 71. Pergamon Press, Oxford

ICRP (International Commission on Radiological Protection) (1997) Radiological Protection Policy for the Disposal of Radioactive Waste, Publication 77. Annals of the ICRP 27. Pergamon Press, Oxford

ICRP (International Commission on Radiological Protection) (1998) Genetic Susceptibility. ICRP Publication 79. Pergamon Press, Oxford

ICRP (International Commission on Radiological Protection) (1999) Radiation protection recommendations as applied to the disposal of long-lived solid radioactive waste. ICRP Publication 81. Pergamon Press, Oxford

ICRP (International Commission on Radiological Protection) (2002) Basic anatomical and physiological data for use in radiological protection. ICRP Publication 89. Pergamon Press, Oxford

ICRP (International Commission on Radiological Protection) (2003) Relative biological effectiveness (RBE), quality factor (Q), and radiation weighting factor (wR). ICRP Publication 92. Pergamon Press, Oxford

ICRP (International Commission on Radiological Protection) (2006) Human alimentary tract model for radiological protection. ICRP Publication 100. Elsevier

ICRP (International Commission on Radiological Protection) (2006) The Optimisation of radiological protection: Broadening the process, Publication 101, part 2. Annals of the ICRP 36. Elsevier

ICRP (International Commission on Radiological Protection) (2007) Scope of Radiological Protection Control Measures. ICRP Publication 104. Pergamon Press, Elsevier

ICRP (International Commission on Radiological Protection) (2007) The 2007 Recommendations of the International Commission on Radiological Protection. ICRP Publication 103. Elsevier

ICRP (International Commission on Radiological Protection) (2009) Nuclear Decay Data for Dosimetric Calculations. ICRP Publication 107. Elsevier

ICRU (International Commission on Radiation Units & Measurements) (1993) Quantities and Units in Radiation Protection Dosimetry. ICRU Report 51. ICRU Publications, Bethesda, MD

ILK (Internationale Länderkommission Kerntechnik) (2005) ILK-Empfehlung zur Revitalisierung der Endlagerprojekte Gorleben und Konrad, November 2005. ILK, Augsburg

International Maritime Organisation (1991) The London Dumping Convention – The First Decade and Beyond, IM-532. London

Issler H (2006) Radioaktive Abfälle Schweiz – Strategie und Schritte zu einem langfristig sicheren Umgang. European Energy Conference, 22-23 November 2006, www.entretiens-europeens.org/2006/Isslerspeech.pdf

Janich P, Kambartel F, Mittelstraß J (1974) Wissenschaft als Wissenschaftskritik. Frankfurt/Main

Jans J, Vedder H (2008) European Environmental Law, 3rd ed. Europa Publishing, Groningen

Japan Atomic Energy Commission (2005) Framework for Nuclear Energy Policy, October 2005. Japan Atomic Energy Commission, Tokyo

Jarass HD, Pieroth B (2011) Grundgesetz, 11th ed. C.H. Beck, München

Javanainen J (2006) Nuclear installations licensing and democratic decision making in Finland – A case study regarding the Oikiluoto 3 nuclear power plant unit and the final disposal repository for spent nuclear fuel. International Journal of Nuclear Law 1:19–27

Johnson L (ed.) (2006) Proceedings of a workshop on criticality in the context of the safety case – requirements and approaches. Nagra Arbeitsbericht. Nagra, Wettingen

Jonas H (1984) The Imperative of Responsibility. In search of an ethics for the Technological Age. Chicago/London

Jordi S (2006) Die Anwendung partizipativer Verfahren in der Entsorgung radioaktiver Abfälle. Bundesamt für Energie, Bern

Jungk R (1986) Der Atomstaat. Rowohlt Taschenbuch, Reinbek bei Hamburg

Kallenbach-Herbert B, Barth R, Brohmann B (2008) Anforderungen an die Gestaltung von Öffentlichkeitsbeteiligung. Zur Planung von Endlagerstandorten für hochradioaktive Abfälle. Technikfolgenabschätzung 17:72–78

Kamlah W (1973) Philosophische Anthropologie. Sprachkritische Grundlegung und Ethik. Mannheim

Kamp G (2010) Art. „Konsequentialismus". In: Mittelstraß J (ed.), Enzyklopädie Philosophie und Wissenschaftstheorie. Stuttgart/Weimar, pp 302–305

Kant I [AA IV GMS] Grundlegung zur Metaphysik der Sitten. In: Gesammelte Schriften, Bd. IV. Berlin 1903/1911

Kant I [AA III KrV] Kritik der reinen Vernunft. In: Gesammelte Schriften, Bd. III. Berlin 1904

Kant I [AA VI MS] Die Metaphysik der Sitten. In: Gesammelte Schriften, Bd. VI. Berlin 1907/1914

Kant I [AA VIII Aufklärung] Beantwortung der Frage: Was ist Aufklärung? In: Gesammelte Schriften, Bd. VIII, Berlin 1912, pp 33-42

Kant I [AA VIII Frieden] Zum ewigen Frieden. In: Gesammelte Schriften, Bd. VIII, Berlin 1912, pp 341-386

Kant I [AA VIII Idee] Idee zu einer allgemeinen Geschichte in weltbürgerlicher Absicht. In: Gesammelte Schriften, Bd. VIII, Berlin 1912, pp 15–35

KASAM (Swedish Council for Nuclear Waste) (2007) Report 2007. KASAM, Stockholm

Kass G (2000) Recent Developments in Public Participation in the United Kingdom. In: TA-Datenbank-Nachrichten Nr.3, pp 20-28

Kaul A, Aurand K, Bonka H, Gumprecht D, Harder D, Hardt HJ, Jacobi W, Kellerer AM, Landfermann HH, Oberhausen E, Streffer C (1987) Possibilities and limits for applying the concept of collective dose. Health Phys. 53:9–10

Kern L, Nida-Rümelin J (1994) Logik kollektiver Entscheidungen. München

Kirchhof F (2005) Finanzmethoden zur Refinanzierung der Standortsuche der Endlagerung. In: Koch H-J, Roßnagel A, Schneider JP, Wieland J (eds.) (2004) Atomrechtssymposium 2003. Nomos, Baden-Baden, pp 311–327

Kloepfer M (2004) Umweltrecht, 3rd ed. C. H. Beck, München

Kloepfer M (2005) Artikel 20a. In: Dolzer R, Graßhof K, Kahl W, Waldhoff C (eds.) (Looseleaf ed) Bonner Kommentar zum Grundgesetz, 4th ed. C. F. Müller, Heidelberg

Kloepfer M (2009) Umweltgerechtigkeit. Duncker & Humblot, Berlin

Koalitionsvertrag (2009) Wachstum, Bildung, Zusammenhalt. Koalitionsvertrag zwischen CDU, CSU und FDP, 17. Wahlperiode. November 2009, Berlin

Kölble J (1962) Die Errichtung von Bundesoberbehörden nach Art. 87 Abs. 3 Satz 1 GG. Deutsches Verwaltungsblatt 76:658–662

Kraß G (2004) Verantwortung der Betreiber für die Endlagerung. In: Koch HJ, Roßnagel A, Schneider JP, Wieland J (eds.) (2004) 12. Deutsches Atomrechtssymposium 2003. Nomos, Baden-Baden, pp 257–266

Kraus N, Malmfors T, Slovic P (1992) Intuitive Toxicology: Expert and Lay Judgments of Chemical Risks. Risk Analysis 12:215–232

Krebs W (2003) Artikel 19. In: von Münch I, Kunig P (2003) Grundgesetz, vol. 1, 5th ed. C. H. Beck, München

Krestinina LY, Preston DL, Ostroumova EV, Degteva MO, Ron E, Vyushkova OV, Startsev NV, Kossenko MM, Akleyev AV (2005) Protracted radiation exposure and cancer mortality in the Techa River cohort. Radiat. Res. 164:602–611

Krüger H (1996) Allgemeine Staatlehre, 2. Aufl. Stuttgart

Kühne G (2008) Aktuelle Rechtsprobleme zur Endlagerung. In: Koch HJ, Roßnagel A (eds.) 13. Deutsches Atomrechtssymposium 2007. Nomos, Baden-Baden, pp 361–388

Kunig P (2003) Artikel 75. In: von Münch I, Kunig P, Grundgesetz, vol. 3, 5th. ed. C. H. Beck, München

Ladeur KH (1989) Die Entsorgung der Kernenergie als Regelungsproblem. Umwelt- und Planungsrecht 9:241–248

Lange K (1990) Das Weisungsrecht des Bundes in der atomrechtlichen Auftragsverwaltung. Nomos, Baden-Baden

Lawrence C (1989) Grundrechtsschutz, technischer Wandel und Generationsverantwortung. Berlin

Leggett RW, Bouville A, Eckerman, KF (1998) Reliability of the ICRP's systemic biokinetic models. Radiat. Protect. Dosim. 79:335–342

Lerche P (1992) Artikel 87. In: Maunz T, Dürig G (eds.) (Looseleaf ed)Grundgesetz. C. H. Beck, München

Likhtarev I, Minenko V, Khrouch V, Bouville A (2003) Uncertainties in thyroid dose reconstruction after Chernobyl. Radiat. Prot. Dosim. 105:601–608

Lindbloom C (1959) The Science of Muddling Through. Public Administration Review 9:79–99

Lindbloom C (1965) The Intelligence of Democracy. Decision Making through Mutual Adjustment. New York

Lindell B, Dunster HJ, Valentin J (1998) International Commission on Radiological Protection: History, Policies, Procedures. Elsevier Science, Oxford, www.icrp.org/DownloadDoc.asp?document=docs/Histpol.pdf

Linnerooth-Bayer J, Fitzgerald KB (1996) Conflicting Views on Fair Siting Processes Evidence from Austria and the US. Risk: Health, Safety and Environment 7:119–134

Little JB (2000) Radiation carcinogenesis. Carcinogenesis 21:397–404

Löfquist L (2008) Ethics beyond finitude. Responsibility towards future generations and nuclear waste. Uppsala

Löftsedt R (2005) Risk Management in Post Trust Societies. London

Luhmann N (1969) Legitimation durch Verfahren. Frankfurt/Main

MacKerran G (2008) Das Beispiel Großbritannien: Zur Standortsuche und Endlagerproblematik. In: BMU (ed.) (2008) Endlagersymposium. BMU, Berlin

Marivoet J, Cuñado M, Norris S, Weetjens E (2008) Impact of Advanced Fuel Cycle Scenarios on Geological Disposal. Euradwaste '08 Conference, Luxembourg 20-22 October 2008, ftp://ftp.cordis.europa.eu/pub/fp7/fission/docs/euradwaste08/papers/paper-10-impact-of-advanced-pt-j-marivoet_en.pdf

Marx K, Engels F [MEW 3 Ideologie] Die deutsche Ideologie. Kritik der neuesten deutschen Philosophie in ihren Repräsentanten Feuerbach, B. Bauer und Stirner und des deutschen Sozialismus in seinen verschiedenen Propheten. In: Karl Marx – Friedrich Engels – Werke, Bd. 3, 9–530. Berlin (east)

Mathiesen J, Dalton J, Russel B (2005) Long-Term Radioactive Waste Management in the United Kingdom – Nirex Perspective. Practice Periodical of Hazardous, Toxic and Radioactive Waste Management 9:20–32

Meadows DH et al. (1972) The Limits of Growth. New York

Menzer JK (1997) Privatisierung der atomaren Endlagerung. Pro Universitate, Berlin

Menzer JK (1998) Privatisierung der atomaren Endlagerung. Deutsches Verwaltungsblatt 113:820–827

METI (Ministry of Economy, Trade and Industry) (2008) Revision of the Basic Policy on Specified Radioactive Waste Final Disposal and the Specified Radioactive Waste Final Disposal Plan. Press release 14 March 2008. METI, Tokyo, www.meti.go.jp/english/newtopics/data/nBackIssue20080314_01.html

Michanek G, Söderholm P (2009) Licensing of nuclear power plants: the case of Sweden in an international comparison. Energy Policy 37:4086–4097

Mittelstraß J (2008) Vorwort. In: Gethmann CF, Mittelstraß J (eds.): Langzeitverantwortung. Ethik, Technik, Ökologie. Darmstadt, pp 7 f.

Möller D (2009) Endlagerung radioaktiver Abfälle in der Bundesrepublik Deutschland – Administrative-politische Entscheidungsprozesse zwischen Wirtschaftlichkeit und Sicherheit, zwischen nationaler und internationaler Lösung. Peter Lang Verlag, Frankfurt/Main

Müller-Dehn C (2008) Finanzielle Entsorgungsvorsorge aus Sicht der Betreiber. In: Koch HJ, Roßnagel A (eds.) (2008) 13. Deutsches Atomrechtssymposium 2007. Nomos, Baden-Baden, pp 321–332

Münch R (1982) Basale Soziologie: Soziologie der Politik. Opladen

Muirhead CR, Goodill AA, Haylock RGE, Vokes J, Little MP, Jackson DA, O'Hagan JA, Thomas JM, Kendall GM, TJ Silk, Bingham D, Berridge GLC (1999) Occupational radiation exposure and mortality: second analysis of the National Registry for Radiation Workers. J. Radiol. Prot. 19:3–26

Murswiek D (2009) Artikel 20a. In: Sachs M (ed.) (2009) Grundgesetz, 5th ed. C. H. Beck, München

Nagra (National Association for the Storage of Radioactive Waste) (1985) Projekt Garantie (Project Guarantee). Nagra, Wettlingen, www.nagra.ch – "Publikationen/Technische Berichte"

Nagra (National Association for the Storage of Radioactive Waste) (2002) Project Opalinus Clay: Safety Report, Technischer Bericht NTB 02-05. Nagra, Wettlingen, www.ensi.ch/fileadmin/deutsch/files/Nagra_entsorgungsnachweis.pdf

Näser HW (2009) Atomrechtliche Planfeststellung für Erkundungsmaßnahmen?. Atomwirtschaft 54:18–21

Näser HW, Oberpottkamp U (1995) Zur Endlagerung radioaktiver Abfälle. Deutsches Verwaltungsblatt 110:136–142

National Research Commission (1995) Yucca Mountain Standards. Committee on Technical Barriers for Yucca Mountain Standards. NRC, Washington D. C.

NDA (Nuclear Decommissioning Authority) (2010a) Geological Disposal – Steps towards implementation. Report no. NDA/RWMD/013. NDA, Moor Row/West Cumbria, www.nda.gov.uk/documents/loader.cfm?url=/commonspot/security/getfile.cfm&pageid=40044

NDA (Nuclear Decommissioning Authority) (2010b) Establishment of a National Stakeholder Group, www.nda.gov.uk/Stakeholders/index.cfm – "Stakeholders& Community", "National Stakeholder Group (NSG)"

NEA (Nuclear Energy Agency) Nuclear Legislation in OECD Countries. NEA/OECD, Paris, www.oecd-nea.org/law/legislation/…

NEA (2003a) Legislation – France

NEA (2003b) Legislation – Switzerland

NEA (2003c) Legislation – UK

NEA (2008a) Legislation – US

NEA (2008b) Legislation – Sweden

NEA (2008c) Legislation – Finland

NEA (2008d) Legislation – Japan

NEA (Nuclear Energy Agency) (2009a) Radioactive Waste Management Programmes in OECD/NEA Countries. NEA/OECD, Paris, www.nea.fr/html/rwm/profiles/welcome.html

NEA (2005a) Profile – Sweden

NEA (2005b) Profile – Spain

NEA (2005c) Profile – Japan

NEA (2009a) Profile – US

NEA (2009b) Profile – France

NEA (2009c) Profile – UK.

NEA (2009d) Profile – Switzerland

NEA (2009e) Profile – Finland

NEA (2009f) Report – US

NEA (2009g) Report – France

NEA (2009h) Report – UK

NEA (2009i) Report – Switzerland

NEA (2009j) Report – Sweden

NEA (2009k) Report – Finland

NEA (2009l) Report – Japan

NEA (Nuclear Energy Agency) (2001) Reversibility and Retrievability in Geologic Disposal of Radioactive Waste. Reflections at the International Level [=NEA No. 3140]. Paris, www.oecd-nea.org/rwm/reports/2001/nea3140.pdf

NEA (Nuclear Energy Agency) (2004) The Handling of Timescales in Assessing Post-closure Safety. Lessons Learnt from the April 2002 Workshop in Paris, France [=NEA No. 4435], www.oecd-nea.org/rwm/reports/2004/nea4435-timescales.pdf

NEA (Nuclear Energy Agency) (2004a) Radioactive Waste Management. NEA/OECD, Paris

NEA (Nuclear Energy Agency) (2005) Clay Club Catalogue of Characteristics of Argillaceous Rocks [=NEA No. 4436]. Paris, www.oecd-nea.org/rwm/reports/2005/nea4436-argillaceous-catalogue.pdf

NEA (Nuclear Energy Agency) (2005d) The Regulatory Control of Radioactive Waste Management – Overview of 15 NEA Member Countries – Sweden, www.oecd-nea.org/rwm/rf/

NEA (Nuclear Energy Agency) (2006) The Roles of Storage in the Management of Long-lived Radioactive Waste. Paris, www.oecd-nea.org/rwm/reports/2006/nea6043-storage.pdf

NEA (Nuclear Energy Agency) (2007) Fostering a Durable Relationship Between a Waste Management Facility and its Host Community. NEA/OECD, Paris, www.nea.fr/html/rwm/reports/2007/nea6176-fostering.pdf

NEA (Nuclear Energy Agency) (2007a) Advanced Nuclear Fuel Cycles and Radioactive Waste Management. Paris

NEA (Nuclear Energy Agency) (2007b) Management of Recyclable Fissile and Fertile Materials.Paris

NEA (Nuclear Energy Agency) (2007c) Regulating the Long-term Safety of Geological Disposal. Towards a common Understanding of the Main Objectives and Bases of Safety Criteria [=NEA No. 6182]. Paris, www.oecd-nea.org/rwm/reports/2007/nea6182-regulating.pdf

NEA (Nuclear Energy Agency) (2008) Nuclear Energy Outlook 2008 [=NEA No. 6348]. Paris

NEA (Nuclear Energy Agency) (2009) Considering Timescales in the Post-Closure Safety of Geological Disposal of Radioactive Wastes. NEA/OECD, Paris

NEA (Nuclear Energy Agency) (2010) Optimisation of Geological Disposal of Radioactive Waste. NEA/OECD, Paris

NEA (Nuclear Energy Authority) (2010a) Public Attitudes to Nuclear Power. Report 06859. OECD, Paris

NEA (Nuclear Energy Authority) (2010b) Radioactive Waste in Perspective [=NEA No. 6350]. OECD, Paris

Nennen HU, Garbe D (eds.) (1996) Das Expertendilemma. Springer, Berlin

Netherland's Ministry of Housing, Spatial Planning and the Environment (2009) National report on the joint convention on the safety of spent fuel management and on the safety of radioactive waste management. The Hague, international. vrom.nl/Docs/internationaal/9190.pdf

Nida-Rümelin J (1993) Kritik des Konsequentialismus. München, ²1995

Niehaus G (2004) Die Bundesauftragsverwaltung in der Praxis. In: Koch HJ, Roßnagel A (2004) 12. Deutsches Atomrechtssymposium 2003. Nomos, Baden-Baden, pp 363–376

Nies A (2005) Das Verfahren der alternativen Standortsuche im AkEnd-Bericht. In: Ossenbühl F (ed.) Atomrechtstag 2004. Nomos, Baden-Baden, pp 93–98

NIREX (Nuclear Industry Radioactive Waste Executive) (2002) Description of Long-term Management Options for Radioactive Waste Investigated Information [NIREX Report No. N/050]. Harwell, Didcot

NIREX (Nuclear Industry Radioactive Waste Executive) (2002) What is the Nirex Phased Disposal Concept? Harwell, Didcot, www.world-nuclear.org/uploaded-Files/org/info/Appendices/NirexPhasedDisposalConcept2002.pdf

NIREX (Nuclear Industry Radioactive Waste Executive) (2005) The viability of a phased geological repository concept for the long-term management of the UK's radioactive waste [NIREX Report no. N/122]. Harwell, Didcot, www.nda.gov.uk/documents/upload/The-viability-of-a-phased-geological-repository-concept-for-the-long-term-management-of-the-UK-s-radioactive-waste-Nirex-Report-N-122-November-2005.pdf

NSC (Nuclear Safety Commission) (2004) Technical Topics, Radioactive Waste, www.nsc.go.jp/NSCenglish/topics/radwaste.htm

NSC (Nuclear Safety Commission) (2004a) Commonly Important Issues for the Safety Regulations of Radioactive Waste Disposal, June 2004, www.nsc.go.jp/haiki/page3/050728.pdf (visited on 5 Febuary 2010)

NUMO (Nuclear Waste Management Organisation of Japan) (2004) Development of Repository Concept for Volunteer Siting Environments, NUMO August 2004, www.numo.or.jp/en/index.html

NUMO (Nuclear Waste Management Organisation of Japan) (2007) The New Structured Approach for HLW Disposal in Japan. Structured Project Implementation at Volunteer Sites, NUMO, July 2007, www.numo.or.jp/en/index.html

NUMO (Nuclear Waste Management Organisation of Japan) (2011) Outline of Operations, www.numo.or.jp/en/index.html

NWMO (Canada's Nuclear Waste Management Organisation) (2003) Range of Potential Options for the Long-term Management of Used Nuclear Fuel [=NWMO Background Papers 6–5], www.nwmo.ca/uploads_managed/MediaFiles/658_6-5RangeofPotentialManagementOptionsforUsedNuclearFuel.pdf

NWMO (Canada's Nuclear Waste Management Organisation) (2005) Choosing a Way Forward. The Future Management of Canada's Used Nuclear Fuel. Ottawa, Ontario, www.nwmo.ca/uploads_managed/MediaFiles/341_NWMO_Final_Study_Nov_2005_E.pdf

NWMO Roundtable on Ethics (Canada's Nuclear Waste Management Organization) (2005) Ethical and Social Framework, www.nwmo.ca/uploads_managed/MediaFiles/577_EthicalandSocialFrameworkVersionMarch52005.pdf

OECD (Organisation for Economic Co-operation and Development) (2008) International Energy Agency, Report – Japan. OECD, Paris

Öko-Institut (2007) Anforderungen an die Gestaltung der Öffentlichkeitsbeteiligung im Endlagerauswahlverfahren. Konzept zur Ausgestaltung der Öffentlichkeitsbeteiligung. Abschlussbericht Teil A und Teil B, August 2007, www.oeko.de/oekodoc/273/2007-017-de.zip

ONDRAF/NIRAS (Belgian agency for radioactive waste and enriched fissile materials) (2001) Technical overview of the SAFIR 2 report. Safety Assessment and Feasibility Interim Report 2. Brussels, www.nirond.be/engels/PDF/Safir2_apercutech_eng.pdf

ONDRAF/NIRAS (Belgian agency for radioactive waste and enriched fissile materials) (2009) The Long-term Safety Assessment Methodology for the Geological Disposal of Radioactive Waste. SFC1 level 4 report: second full draft. Brussels, www.niras-afvalplan.be/nieuw/downloads/SAmetho_NIROND%20TR%20 2009_14.pdf

OSPAR (1992) Convention for the Protection of the Marine Environment of the Northeast Atlantic, September 1992

Ossenbühl F (1983) Rechtsanspruch auf Erteilung atomrechtlicher Genehmigungen und Versagungsermessen. Energiewirtschaftliche Tagesfragen 33:665–676

Ossenbühl F (2004) Zur Verbandslösung als Finanzierungsinstrument der atomrechtlichen Endlagerung. Deutsches Verwaltungsblatt 119:1132–1143

Paetow S (2003) § 32. In: Kunig P, Paetow S, Versteyl LA (2003) Kreislaufwirtschafts- und Abfallgesetz, 2nd ed. C.H Beck, München

Piontek N (2004) Endlagersuche und die Finanzierung. In: Koch HJ, Roßnagel A, Schneider JP, Wieland J (eds.) (2004) 12. Deutsches Atomrechtssymposium 2003. Nomos, Baden-Baden, pp 267–275

Posiva (Posiva Oy) (2009) Nuclear Waste Management at Olkiluoto and Loviisa Power Plants – Review of Current Status and Future Plans, TKS-2009, www.posiva.fi/files/1078/TKS2009_Eng_web_rev1_low.pdf

Posser H, Schmans M, Müller-Dehn C (2003) Atomgesetz, Kommentar zur Novelle 2002. Carl Heymanns Verlag, Köln

Preston DL, Ron E, Tokuoka S, Funamoto S, Nishi N, Soda M, Mabuchi K, Kodama K (2007) Solid cancer incidence in atomic bomb survivors: 1958–1998. Radiat. Res. 168:1–64

Preston DL, Shimizu Y, Pierce DA, Suyama A, Mabuchi K (2003) Studies of mortality of atomic bomb survivors. Report 13: Solid cancer and non-cancer disease mortality 1950–1997. Radiat. Res. 160:381–407

Pünder H (2005) "Open Government leads to Better Government" – Überlegungen zur angemessenen Gestaltung von Verwaltungsverfahren. Natur und Recht 27:71–79

Ramsauer U (2008) Planfeststellung ohne Abwägung? Neue Zeitschrift für Verwaltungsrecht 27:944–950

Rausch H, Marti A, Griffel A (2004) Umweltrecht. Schulthess, Zürich, Basel, Genf

Rauscher D (2004) Die Bundesauftragsverwaltung in der Praxis. In: Koch HJ, Roßnagel A, Schneider JP, Wieland J (2004) 12. Deutsches Atomrechtssymposium 2003. Nomos, Baden-Baden, pp 377–392

Rawls, J (1971) A Theory of Justice. Cambridge (USA) [cited according to revised edition 1999)

Rengeling HW (1984) Die Planfeststellung für Anlagen der nuklearen Entsorgung. Carl Heymanns Verlag, Köln

Rengeling HW (1995) Rechtsfragen zur Langzeitsicherung von Endlagern für radioaktive Stoffe. Carl Heymanns Verlag, Köln

Rengeling HW (2008) Organisation der nuklearen Endlagerung unter besonderer Berücksichtigung der Beleihung. Deutsches Verwaltungsblatt 123:1141–1150

Renn O (1997) Mental Health, Stress and Risk Perception: Insights from Psychological Research. In: Lake JV, Bock GR, Cardew G (eds.) Health Impacts of Large Releases of Radionuclides. Ciba Foundation Symposium 203. London, pp 205–231

Renn O (2005) Risk Perception and Communication: Lessons for the Food and Food Packaging Industry. Food Additives and Contaminants 22: 1061–1071

Renn O (2008) Risk Governance. Coping with Uncertainty in a Complex World. London

Renn O, Schrimpf M, Büttner T, Carius R, Köberle S, Oppermann B, Schneider E, Zöller K (1999) Abfallwirtschaft 2005. Bürger planen ein regionales Abfallkonzept. Baden-Baden

Renn O, Schweizer PJ, Dreyer M, Klinke A (2007) Risiko. Über den gesellschaftlichen Umgang mit Unsicherheit. München

Renneberg W (2005) Atomaufsicht – Bundesauftragsverwaltung oder Bundeseigenverwaltung? Aus der Sicht optimaler Aufgabenerfüllung. In: Ossenbühl F (ed.) (2005) Deutscher Atomrechtstag 2004. Nomos, Baden-Baden, pp 173–181

République française (1991) Loi n°91-1381 du 30 décembre 1991 relative aux recherches sur la gestion des déchets radioactifs ("Loi Bataille"), Journal official du 1er janvier 1992, www.legifrance.gouv.fr/affichTexte.do?cidTexte=JORFTEXT0000 00356548&fastPos=1&fastReqId=1600876769&categorieLien=cid&oldAction=r echTexte

République française (2006) Loi n°2006-739 du 28 juin 2006 de programme relative à la gestion durable des matières et déchets radioactifs, Journal officiel du 29 juin 2006, www.legifrance.gouv.fr/affichTexte.do?cidTexte=JORFTEXT000 000240700&fastPos=1&fastReqId=750078383&categorieLien=cid&oldAction= rechTexte

Resele G (2009) Die Sach- und Rechtslage der Entsorgung radioaktiver Abfälle in der Schweiz. In: Schmidt-Preuß M (ed.) (2009) Deutscher Atomrechtstag 2008. Nomos, Baden-Baden, pp 225–239

Roller G (2004) Schadensvorsorge gegenüber auslegungsübergreifenden Störfällen. In: Koch HJ, Roßnagel A (eds.) (2004) 12. Deutsches Atomrechtssymposium 2003. Nomos, Baden-Baden, pp 115–123

Rosa EA (1988) NAMBY PAMBY and NIMBY PIMBY: Public Issues in the Siting of Hazardous Waste Facilities. Forum for Applied Research and Public Policy 3:114–123

Roßnagel A, Hentschel A (2004) Alternativenprüfung für atomare Endlager? – zur Prüfpflicht des öffentlichen Vorhabenträgers im Fachplanungsrecht –. Umwelt- und Planungsrecht 25:291–296

Ruiz de Apodaca Espinosa A (2009) Régimen jurídico de la gestión de los residuos nucleares. Revista de derecho ambiental 16-2:89–111

Rupp HH (1990) Bemerkungen zur Bundeseigenverwaltung nach Art. 87 III 1 GG. In: Maurer H (ed.) Das akzeptierte Grundgesetz. Festschrift Dürig. C.H. Beck, München, pp 387–399

Schärf WG (2008) Europäisches Nuklearrecht. De Gruyter, Berlin

Scheler M (1925) Der Formalismus in der Ethik und die materiale Wertethik. Halle

Schell, T von, Mohr, H. (eds) (1994), Biotechnologie – Gentechnik. Eine Chance für neue Industrien. Berlin, Heidelberg, New York

Schenkel W, Gallego Carrera D (2009) Sachplan Geologisches Tiefenlager. Kommunikation mit der Gesellschaft. Wissenschaftlicher Schlussbericht. BFE, Zürich

Scherer E, Streffer C, Trott KR (1991) Radiopathology of Tissues and Organs. Springer, Berlin-Heidelberg-New York

Schmidt-Bleibtreu B, Hofmann H, Hopfauf A (2011) Grundgesetz, 12th ed. Carl Heymanns Verlag, Köln

Schneider H (2009) Das neue kerntechnische Regelwerk. Atomwirtschaft 54:554

Schönrich G (1993) Bei Gelegenheit Diskurs. Von den Grenzen der Diskursethik und dem Preis der Letztbegründung. Frankfurt/Main

Sellner D, Hennenhöfer G (2007) Atomrecht. In: Hansmann K, Sellner D (eds.) (2007) Grundzüge des Umweltrechts, 3rd ed., part 1. Erich Schmidt Verlag, Berlin

Selmer P (2005) Finanzierung der Standortsuche durch eine öffentlich-rechtliche Körperschaft (Verbandsmodell). In: Ossenbühl F (ed.) (2005) Deutscher Atomrechtstag 2004. Nomos, Baden-Baden, pp 139–152

Sendler H (2001) Überlegungen zur geplanten Atomgesetznovelle. In: Ossenbühl F (ed.) (2001) Deutscher Atomrechtstag 2000. Nomos, Baden-Baden, pp 185–198

Serre J-L (2004) Ya-t-il une éthique de la gestion des déchets nucléaires? Le Point de vue d'un geneticien. In: ANDRA (ed.): Ya-t-il une éthique de la gestion des déchets nucléaires? Châteney-Malabry Cedex, pp 83-105

Shrader-Frechette (2000) Duties to Future Generations, Proxy Consent, Intra- and Intergenerational Equity: The Case of Nuclear Waste. Risk Analysis 20, pp 771– 778

SKB (Swedish Nuclear Fuel and Waste Management Company) (2008) Äspö Hard Rock Laboratory, Annual Report 2007, Report TR-08-10. SKB, Stockholm

Sparwasser R, Engel R, Vosskuhle A (2003) Umweltrecht, 5th ed. C. F. Müller, Heidelberg

Sparwasser R, Engel R, Vosskuhle A (2003) Umweltrecht: Grundzüge des öffentlichen Umweltschutzrechts. 5th ed. C.F. Müller, Heidelberg

SSM (Swedish Radiation Protection Authority) (2009) Review Process. SSM, Stockholm, www.stralsakerhetsmyndigheten.se/In-English/About-the-Swedish-Radiation-Safety-Authority1/

Stadler G (2000) Die Beleihung in der neueren Bundesgesetzgebung. Kovacz, Hamburg

Steinberg R, Berg T, Wickel M (2000) Fachplanung. 3rd ed. Nomos, Baden-Baden, § 4 Nos. 21-22

Steinberg R (1990) Bundesaufsicht, Länderhoheit und Atomgesetz. C.F. Müller, Heidelberg

Steinberg R (1991) Rechtliche Perspektiven der energiewirtschaftlichen Nutzung der Kernenergie. Juristen-Zeitung 41:431–437

Stober R (1983) Atomare Entsorgung und Verfassung. Energiewirtschaftliche Tagesfragen 33:585–597

Stolle M (2006) Die Einstellung zur Endlagerung und die politische Partizipation der Bevölkerung. In: Hocke P, Grunwald A (eds.) Wohin mit dem radioaktiven Abfall? Perspektiven für eine sozialwissenschaftliche Endlagerforschung. Edition Sigma, Gesellschaft-Technik-Umwelt, H. 8, pp 193–216

Streffer C (2009) Radiological protection: challenges and fascination of biological research. Strahlenschutzpraxis 2009(2):35–45

Streffer C, Bolt H, Follesdal D, Hall P, Hengstler JG, Jacob D, Oughton K, Prieß K, Rehbinder E, Swaton E (2004) Low Dose Exposures in the Environment: Dose-Effect Relations and Risk Evaluation. Wissenschaftsethik und Technikfolgenbeurteilung, Band 23. Springer, Berlin

STUK (Finish Radiation and Nuclear Safety Authority) (2002) Operational Safety of a Disposal Facility for Spent Nuclear Fuel. Guideline YVL 8.4. STUK, Helsinki

STUK (Finnish Radiation and Nuclear Safety Authority) (2002) Operational safety of a disposal facility for spent nuclear fuel [STUK- Guide YVL 8.5], www.finlex.fi/data/normit/15315-YVL8-5e.pdf

Sugahara T, Sasaki Y, Morishima H, Hayata, I, Sohrabi, M, Akiba, S (eds.) (2005) High levels of natural radiation and radon areas: Radiation dose and health effects. Elsevier, Amsterdam

Svensk Kärnbränslehantering AB (2000) Förvarsalternativet djupa borrhål Innehåll och omfattning av FUD-program som krävs för jämförelse med KBS-3-metoden. Stockholm, www.skb.se/upload/publications/pdf/R_00_28ny.pdf

Svensk Kärnbränslehantering AB (2006) Long-term safety for KBS-3 repositories at Forsmark and Laxemar – a first evaluation Main Report of the SR-Can project [=Technical Report TR-06-09]. Stockholm, www.skb.se/upload/publications/pdf/TR-06-09webb.pdf

Tanaka S (2008) Strategic Planning for Designated Radioactive Waste Disposal in Japan, October 15, 2008. University of Tokyo, Tokyo

Thomauske B (2004) Wege zur Endlagerung radioaktiver Abfälle in der Bundesrepublik Deutschland. Wird die Verantwortung auf zukünftige Generationen verschoben? atw 49:235–249, www.endlagerung.de/binary.ashx/3753

Thomauske B (2008) Safety Requirements to be Met in Final Storage of Heat-producing Waste – An Evaluation of the BMU Draft. Atomwirtschaft 53:603–611

Tiggemann A (2004) Die "Achillesferse" der Kernenergie in der Bundesrepublik: Zur Kernenergiekontroverse und Geschichte der nuklearen Entsorgung von den Anfängen bis Gorleben 1955 bis 1985. Europaforum Verlag, Lauf an der Pegnitz

Tiggemann A (2010) Gorleben – Entsorgungsstandort auf der Grundlage eines sachgerechten Auswahlverfahrens. Atomwirtschaft 55:607–615

Tremmel J (2009) A Theory of Intergenerational Justice. Oxford

Tonhauser W (2006) The Peer Review Process under the Joint Convention on the Safety of Spent Nuclear Fuel Management and on the Safety of Radioactive Waste Management. Europäisches Umwelt- und Planungsrecht 4:131–136

Tubiana M, Aurengo A, Averbeck D, Bonnin A, Le Guen B, Masse R, Monier R, Valleron A J, de Vathaire F (2005) Dose-effect relationships and the estimation of the carcinogenic effects of low doses of ionizing radiation. English Translation, Académie Nationale de Médecine, Institut de France – Académie des Sciences (March 30, 2005), Joint Report no 2. Edition Nucleon, Paris, http://www.radscihealth.org/rsh/papers/frenchacadsfinal07_04_05.pdf

Uerpmann HP (2003) Artikel 87c. In: von Münch I, Kunig P (eds.) (2003) Kommentar zum Grundgesetz, vol. 3, 5th ed. C. H. Beck, München

UNSCEAR (United Nations Scientific Committee on the Effects of Atomic Radiation) (1988) Sources, Effects and Risks of Ionizing Radiation. UNSCEAR Report to the General Assembly with Scientific Annexes. United Nations, New York, NY

UNSCEAR (United Nations Scientific Committee on the Effects of Atomic Radiation) (1993) Sources and Effects of Ionizing Radiation. UNSCEAR Report to the General Assembly with Scientific Annexes. United Nations, New York, NY

UNSCEAR (United Nations Scientific Committee on the Effects of Atomic Radiation) (1994) Sources and Effects of Ionizing Radiation. UNSCEAR Report to the General Assembly with Scientific Annexes. United Nations, New York, NY

UNSCEAR (United Nations Scientific Committee on the Effects of Atomic Radiation) (2000) Sources and Effects of Ionizing Radiation. UNSCEAR Report to the General Assembly with Scientific Annexes. Vol. II: Effects. United Nations, New York, NY

UNSCEAR (United Nations Scientific Committee on the Effects of Atomic Radiation) (2001) Hereditary Effects of Radiation. UNSCEAR Report to the General Assembly with Scientific Annex. United Nations, New York, NY

UNSCEAR (United Nations Scientific Committee on the Effects of Atomic Radiation) (2006) Effects of Ionizing Radiation. UNSCEAR Report to the General Assembly with Scientific Annexes. United Nations, New York, NY

UNSCEAR (United Nations Scientific Committee on the Effects of Atomic Radiation) (2008) Sources and Effects of Ionizing Radiation. UNSCEAR Report to the General Assembly with Scientific Annexes. United Nations, New York, NY

US National Research Council (1989) Improving Risk Communication. Washington, DC

US NRC (United States Nuclear Regulatory Commission) (1998) Regulatory Guide 3.58, www.nrc.gov/reading-rm/doc-collections/reg-guides/fuels-materials/rg/division-3/division-3-41.html

Vallender K, Morell R (1997) Umweltrecht. Stämpfli, Bern

Vandenbosch R, Vandenbosch F (2007) Nuclear Waste Stalemate. University of Utah Press, Salt Lake City

Vigfusson J, Maudoux J, Raimbault P, Röhlig KJ, Smith RE (2007) European Pilot Study on The Regulatory Review of the Safety Case for Geological Disposal of Radioactive Waste Case Study: Uncertainties and their Management, www.grs.de/sites/default/files/pdf/Pilotstudie_Radioactive%20Waste.pdf

Volkens B (1993) Vorsorge im Wasserrecht. Erich Schmidt Verlag, Berlin, Bielefeld

Wahl R, Hermes G (1995) Nationale Kernenergiepolitik und Gemeinschaftsrecht. Werner Verlag, Düsseldorf

Waldhoff C (2005) Finanzierung der Standortsuche durch eine öffentlich-rechtliche Körperschaft (Verbandsmodell). In: Ossenbühl F (ed.) Deutscher Atomrechtstag 2004. Nomos, Baden-Baden, pp 153–168

Warnecke W, Bonne A (1995) Radioactive waste management: International peer reviews. IAEA Bulletin 37/4:26–29

Weber M (1919): Politik als Beruf; cited according Weber M: Gesamtausgabe Bd.17, ed. Mommsen WJ. Tübingen 1992

Weinrich H (1972) System, Diskurs, Didaktik und die Diktatur des Sitzfleisches. Merkur 8:801–812

WENRA (Western European Nuclear Regulators Association) (2006) Waste and Spent Fuel Storage Safety Reference Levels, Report, www.wenra.org/dynamaster/file_archive/070718/327631f8acb266a075aa84a9a7846167/V1_0_storage_report_final.pdf

West Cumbria Managing Radioactive Waste Safely Partnership (2010) Statement, www.cumbria.gov.uk/news/2010/July/07_07_2010-114024.asp

Wiedemann PM (1994) Mediation bei umweltrelevanten Vorhaben: Entwicklungen, Aufgaben und Handlungsfelder. In: Claus F, Wiedemann PM (eds.) Umweltkonflikte: Vermittlungsverfahren zu ihrer Lösung. Taunusstein, pp 177–194

Wieland W (1999) Verantwortung – Prinzip der Ethik?. Heidelberg

Willke H (1995) Systemtheorie III: Steuerungstheorie. Stuttgart

Windhorst K (2009) Artikel 87c. In: Sachs M (ed.) (2009) Grundgesetz, 5th ed. C.H. Beck, München

Wollenteid U (2000) Zur Langzeitsicherheit von Endlagern. In: Koch HJ, Roßnagel A (eds.) (2000) 10. Deutsches Atomrechtssymposium 1999. Nomos, Baden-Baden, pp 333–344

Working Group on "Scenario Development" (2008) Position of the Working Group on "Scenario Development": Handling of human intrusion into a repository for radioactive waste in deep geological formations. Internationale Zeitschrift für Kernenergie 53:2–4, www.grs.de/sites/default/files/fue/20090107_en_ak_szenarien_sonderdruck_atw.pdf

World Nuclear Organisation (2004) Radioactive Waste Management – Disposal of used fuel and other HLW, www.world-nuclear.org/info/inf04.html (visited on 21 February 2010)

Wüstemann J (2004) Die betriebswirtschaftliche Bedeutung von Rückstellungen für die nukleare Entsorgung. In: Koch HJ, Roßnagel A, Schneider JP, Wieland J (eds.) (2004) 12. Deutsches Atomrechtssymposium 2003. Nomos, Baden-Baden, pp 277–310

Leidinger T, Zimmer T (2004) Die Überführung der Bundesauftragsverwaltung im Atomrecht in Bundeseigenverwaltung. Deutsches Verwaltungsblatt 119:1005–1013

Zaiss W (2009) Die WENRA Reference Levels und die IAEA Safety Standards und ihre Auswirkungen auf das kerntechnische Regelwerk in Deutschland. In: Schmidt-Preuß M (ed.) (2009) Atomrechtstag 2008. Nomos, Baden-Baden, pp 187–198

Zika H (2008) Regulatory Control of Radioactive Waste in Sweden. In: Sneve MK, Kiselev MF (eds.) (2008) Challenges in Radiation Protection and Nuclear Safety Regulation of the Nuclear Legacy, pp 233–236

List of Authors

Gethmann, Professor Dr. phil. habil. Carl Friedrich, studies of philosophy at Bonn, Innsbruck and Bochum; 1968 lic. phil. (Institutum Philosophicum Oenipontanum); 1971 doctorate Dr. phil. at the Ruhr-Universität Bochum; 1978 Habilitation for philosophy at the University of Konstanz. 2003 honorary degree of doctor of philosophy (Dr. phil. h.c.) of the Humboldt-Universität Berlin. 2009 Honorary professor at the Universität zu Köln. 1968 scientific assistant; 1972 Professor of Philosophy at the University of Essen; 1978 private lecturer at the University of Konstanz; since 1979 Professor for philosophy at the University of Essen; lectures at the universities of Essen and Göttingen. Called to the Board of Directors at the Akademie für Technikfolgenabschätzung Baden-Württemberg combined with a full professorship of Philosophy (1991, refused) and to full professorship at the universities of Oldenburg (1990, refused), Essen (1991, accepted), Konstanz (1993, refused) and Bonn (1995, refused). Since 1996 Director of the Europäische Akademie zur Erforschung von Folgen wissenschaftlichtechnischer Entwicklungen Bad Neuenahr-Ahrweiler GmbH (European academy for the study of the consequences of scientific and technological advance). Member of the Academia Europaea (London); member of the Berlin-Brandenburgische Akademie der Wissenschaften; member of the Deutsche Akademie der Naturforscher Leopoldina (Halle); member of the Bio-Ethikkommission des Landes Rheinland-Pfalz. 2006–2008 President of the "Deutsche Gesellschaft für Philosophie e.V." Since 2008: Member of the Deutsche Akademie der Technikwissenschaften "acatech".

Main fields of research: linguistic philosophy/philosophy of logic; phenomenology and practical philosophy (ethics of medicine/ethics of environment/technology assessment).

Kamp, Dr. phil. Georg, 1979–1982 merchandising apprenticeship; 1982–1984 retail salesman; 1987–1993 studies in philosophy und German literature and linguistics in Bochum, Duisburg and Essen; 1993–1998 scientific assistant at the Institut für Philosophie at the Universität Duisburg-Essen; 1993–1998 Ph.D. studies, graduation with a thesis on "Praktische Sprachen. Zur Möglichkeit und Gestaltung des Argumentierens in regulativen Kontexten" at der Universität Essen. 1999–2002 member of scientific staff of the Europäische Akademie GmbH. 2002–2005 freelance consultant, lecturer and editor (during parental leave). 2005–2006 Cooperative Education

"Master of Mediation" at the FernUniversität Hagen. 2005–2006 Co-ordinator of the project "continuo – Diskontinuierliche Erwerbsbiographien und Beschäftigungsfähigkeit in kleinen und mittleren Unternehmen".

Since 2006 member of the scientific staff of the Europäische Akademie GmbH.

Kröger, Professor Dr.-Ing. habil. Wolfgang, studied mechanical engineering, specialized on nuclear technology, at the RWTH Aachen. He received his diploma degree in 1972, his doctoral degree in 1974 and the venia legendi in 1986. From 1974 to 1989 he was employed at the German Research Centre Jülich, finally as director of the Institute for Nuclear Safety Research. His research aimed to develop advanced reactor concepts and related methods for comprehensive safety assessment.

In 1990 Kröger became professor for safety technology (analysis) at the ETH Zurich and director of the Research Department on Nuclear Energy and Safety at the national Paul Scherrer Institut (PSI), Switzerland. At ETH, as director of the Lab of Safety Analysis he has mainly contributed to the development of risk and vulnerability analysis of large scale, complex engineered systems including energy supply infrastructure and to approach risk assessment issues in a multidisciplinary, trans-sectorial way. On his initiative, the International Risk Governance Council was established in 2003 at Geneva and he became its founding rector. At the beginning of 2011 he became managing director of the ETH Risk Center.

Kröger has worked in distinguished international committees and advisory boards, authored numerous publications, contributed to various books and is co-editor of various books. Inter alia he is member of the Scientific Directorate of the German Council on Foreign Relations, individual member of the Swiss Academy of Engineering Sciences and honorary member of the Swiss Nuclear Society.

Rehbinder, Professor Dr. iur. Eckard, Professor emeritus of Economic Law, Environmental Law and Comparative Law at the Goethe-University Frankfurt/Main, Member of the Research Centre for Environmental Law at the same university. He studied law at the University Frankfurt/Main and the Free University Berlin; LLD in 1965 and habilitation in 1968 at the former University. From 1969 to 1972 he was Professor of Law at the University Bielefeld. Since 1972 he has been Professor of Law at the Goethe-University Frankfurt. In the academic year 1981/82 he was dean of the law faculty. Between 1987 and 2000 he was a member, between 1996 and 2000 also the chairman of the German Council of Environmental Policy. Since 1996, he has been a member of the Academy and participated in three further of its research projects. He has been a member of several national and international bodies in the field of environmental law. He was visiting professor in the United States (Ann Arbor und Berkeley) and several times at the European University Institute (Florence). Apart from the doctoral and the habilitation theses, he published or co-authored a number of books, commentaries and numerous essays in the fields of economic law and since 1972 in particular environmental law. He was awarded the Prix Elizabeth Haub in 1978 and the Bruno H. Schubert Prize in 2004, both for pioneering contributions to environmental law.

Renn, Professor Dr. rer. pol. Dr. sc. tech. h. c. Ortwin, serves as full professor and Chair of Environmental Sociology and Technology Assessment at Stuttgart University (Germany). He directs the Interdisciplinary Research Unit for Risk Governance and Sustainable Technology Development (ZIRN) at Stuttgart University and the non-profit company DIALOGIK, a research institute for the investigation of communication and participation processes in environmental policy making. Since 2006 Renn has been elected Deputy Dean of the Economics- and Social Science Department. He also serves as Adjunct Professor for "Integrated Risk Analysis" at Stavanger University (Norway) and as Contract Professor at the Harbin Institute of Technology and Beijing Normal University.

Ortwin Renn has a doctoral degree in sociology and social psychology from the University of Cologne. His career included teaching and research positions at the Juelich Nuclear Research Center, Clark University (Worcester, USA), the Swiss Institute of Technology (Zürich) and the Center of Technology Assessment (Stuttgart). His honours include an honorary doctorate from the Swiss Institute of Technology (ETH Zürich) and the "Distinguished Achievement Award" of the Society for Risk Analysis (SRA). Among his many political advisory activities the chairmanship of the State Commission for Sustainable Development (German State of Baden-Württemberg) is most prominent. Renn is primarily interested in risk governance, political participation and technology assessment. His has published more than 30 books and 250 articles, most prominently the monograph "Risk Governance" (Earthscan: London 2008).

Röhlig, Professor Dr. rer. nat. Klaus-Jürgen, diploma (1985) and Ph.D. (1989) in mathematics at TU Bergakademie Freiberg. 1989–1991 Institut für Energetik Leipzig. Development and application of computer codes for simulating fluid flow and contaminant migration. 1991–2007 Gesellschaft für Anlagen- und Reaktorsicherheit (GRS) mbH. Research and technical advice to the German Federal Ministry for Environment, Nature Conservation and Nuclear Safety (BMU) in safety assessment and safety criteria for radioactive waste repositories, policy, regulatory and licensing questions. 2003–2007 deputy head of GRS' final disposal department. Since 2007, Professor for Repository Systems at the Institute of Disposal Research, Clausthal University of Technology. Research on safety case methodology and analytical assessment of repository systems. Lecturing in the frame of the Master Course "Radioactive and Hazardous Waste Management". Chair of the Integration Group for the Safety Case (IGSC) at OECD/NEA. Member of the Scientific Advisory Board of the French Institut de radioprotection et de sûreté nucléaire (IRSN). 2008–2010 deputy chair of the Radioactive Waste Management Commission (ESK, an advisory body of BMU), and chair of its Committee on Final Disposal. International peer reviews of safety reports produced in France, Sweden, and in the UK.

Streffer, Professor em. Dr. rer. nat. Dr. med. h. c. Christian, studied Chemistry and Biochemistry at the universities of Bonn, Tübingen, Munich, Hamburg and Freiburg; Ph.D. in Biochemistry January 1963. Postdoctoral fellowship at the Department of Biochemistry, University of Oxford; 1971 Professor (C3) for Radiobiology at the University of Freiburg i. Br. From 1974 until 1999 he was full Professor for Medical Radiobiology at the University of Essen, 1988–1992 vice chancellor of the University, 1999 Emeritus. Guest Professorships: 1985 University of Rochester, N.Y., USA; 2000 University of Kyoto, Japan. Honorary Member of several scientific societies, 1995 Honorary Doctor of the University of Kyoto. Streffer is member of the Institute for Science and Ethics of the university of Bonn, Emeritus member of the International Commission on Radiological Protection (ICRP). Several scientific awards, e.g. 2008 Sievert Award of the International Radiation Protection Association (IRPA); 2009 Distinguished Service Award of the Radiation Research Society, U.S.A.

His main research interests are: radiation risk especially during the prenatal development of mammals; combined effects of radiation and chemical substances; experimental radiotherapy of tumours, especially individualization of cancer therapy by radiation.